Advanced and Innovative Approaches of Environmental Biotechnology in Industrial Wastewater Treatment

Maulin P. Shah
Editor

Advanced and Innovative Approaches of Environmental Biotechnology in Industrial Wastewater Treatment

Springer

Editor
Maulin P. Shah
Division of Applied and Environmental
Microbiology Lab
Industrial Wastewater Research Lab
Gujarat, India

ISBN 978-981-99-2597-1 ISBN 978-981-99-2598-8 (eBook)
https://doi.org/10.1007/978-981-99-2598-8

This Springer imprint is published by the registered company Springer Nature Singapore Pte Ltd.
The registered company address is: 152 Beach Road, #21-01/04 Gateway East, Singapore 189721, Singapore

Contents

Microbial Biotechnology for Circular Economy in Wastewater Treatment: Potentials, Technologies, and Challenges

Shreya Sharma and Shilpa Sharma

1 Introduction

The science and environmental engineering sector is at the kernel of a revolutionary shift to succor responsible innovation for realization of circular economies in healthy and ecologically balanced communities. The water sector holds the chief position, by managing water flows, nutrients, and prominent pollutants to safeguard public health and environment, and to ascribe value to employed resources within cities and watersheds (Weissbrodt et al. 2020). The trailblazing models for circular economy steer resource efficiency by closing resource loops, redefining waste as a value, flourishing on avant-garde concepts of sustainable development, exergy, energy, material flow analysis, systems resiliency, life cycle assessments, industrial symbiosis, city metabolisms, and by urban and industrial ecology. The novel terms and conceptions of circular economy are revolutionary since concomitantly espousing the broader discernment of industrial ecology and positioning it into economical perspective to exhort intrigues across societies, communities, municipalities, activity sectors, and citizens (Nielsen 2017). The concept of circular economy has promoted sustainable management of materials and energy by restricting the amount of generated waste and their reutilization as secondary material. The entailment to close the production cycles and permit resource sustainability is driving current production systems toward auto-regeneration, substituting the present 3'R model—reduce, reuse, and recycle, to a more efficient paradigm (Puyol et al. 2017). Interest has lately transpired in engineering microbial communities—multiple interacting microbial populations—to effectuate intricate functions and to be more sturdy to environmental variations (Fig. 1).

S. Sharma · S. Sharma (✉)
Department of Biological Sciences and Engineering, Netaji Subhas University of Technology, Sector 3, Dwarka, New Delhi 11078, India
e-mail: shilpa.sharma@nsut.ac.in

M. P. Shah (ed.), *Advanced and Innovative Approaches of Environmental Biotechnology in Industrial Wastewater Treatment*,
https://doi.org/10.1007/978-981-99-2598-8_1

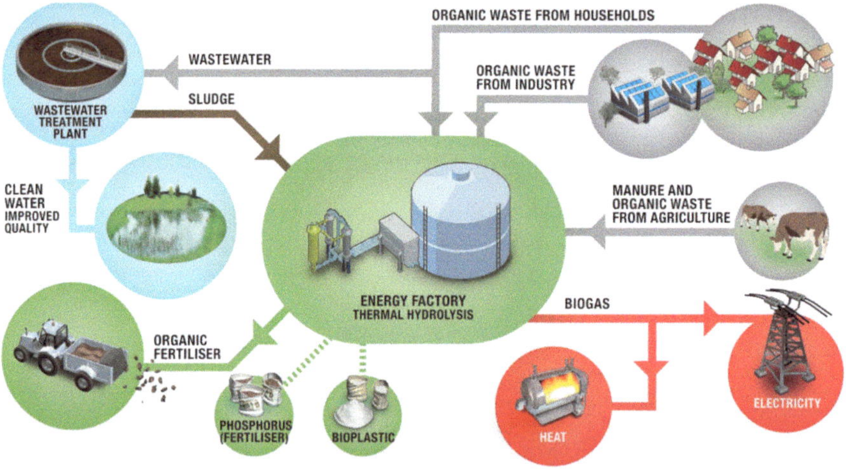

Fig. 1 Microbial biotechnology for circular economy in wastewater treatment (Nielsen 2017)

Microorganisms, the most diverse group of organisms, inhabitant of planet Earth are accountable for sustaining the biogeochemical cycling of nitrogen, carbon, phosphorus, sulfur, and various metals that fuels the Earth (Wu and Yin 2020). Essentially all ecosystem processes are driven by microbes, and numerous processes are bolstered solely by microbes. Deciphering the mechanisms that are cardinal and engender microbial biodiversity is indispensable to prognosticating ecosystem responses to environmental changes and ameliorating bioprocesses, for instance, wastewater treatment. The presiding reactor archetype of choice presently is the activated sludge, however, other types including membrane bioreactors, granules, biofilms, etc., may in some instances be prevailing to these conventional systems. Nevertheless, the process critical microbes are customarily similar, untrammeled of reactor type, and therefore, complete and broad comprehension of the microbial community structure and links to altering environmental conditions will be inclusive and indispensable. Intriguingly, albeit the knowledge about the identity and function of many of these microorganisms, majority are still abysmally characterized and their functions/roles unknown that need to be uncovered for a sagacious cognizance of the systems. Moreover, numerous challenges facing the wastewater treatment sector today are attributable to the rigorous discharging standards and emergence of new pollutants, for example, micropollutants like personal care products, pharmaceuticals, endocrine disrupting chemicals, steroid hormones, pesticides, nanoparticles, etc., and removal of excess of inorganic phosphorus and nitrogen accountable for eutrophication of receptor water bodies have made it difficult for the conventional wastewater treatment technologies to function. Howbeit, we reckon that particular category of microbes may be acculturated to the emanation of new pollutants, proselytizing the removal of these toxic compounds (Mo and Zhang 2013; Kuypers et al. 2018).

About 80% of municipal wastewater is treated utilizing activated sludge process in the developed countries. Majority of the energy requirement (>60%) for the treatment can be attributed to aeration which in further dependent on treatment plant capacity, scheme, and wastewater characteristics (Ghimire et al. 2021). The integration of resource recovery and energy production into clean water production has become part of circular sustainability movement in wastewater treatment. The achievements to date include both lab-scale/pilot-scale or large-scale (industrial-scale) implementations (Kaszycki et al. 2021). Following 100 years of burgeoning of activated sludge process, novel concepts accrediting 'wastewater' as the 'used water' emphasizing on water purification and resources recycling have been put forward. The final effluents can be sustainably and safely reutilized for a variety of purposes, for instance, in agriculture, the sector with the largest consumption of freshwater (Nunes 2021). Changing the consumption model toward a climate neutral, circular economy, and recovery of waste have not only allowed us to think about the value of water as a resource, but also about the substances contained in these effluents, transforming sanitation from a costly service to one that is self-sustaining and adds value to the economy (Cerruti et al. 2021).

It becomes imperative to recognize how microbes have become the focus of operation monitoring scheme in wastewater treatment plants and how elucidation of microbial community structure can facilitate better design and operations of wastewater treatment plants in response to several environmental changes. Apprehension of inherent microbiological aspects of wastewater treatment frameworks design and operation can help access microbiomes and characterize microbes, metabolisms, and interactions at increased resolution and throughput for environmental protection and resource recovery aiding bioinformaticians, molecular biologists, microbiologists, environmental engineers, practitioners, industrialists, etc., to develop common investigation lines.

2 Diverse Microbial Communities and Their Contribution to Pollutants Erasure

The apprehension of history of removal of phosphorous, nitrogen, carbon from wastewater is pivotal to proffer disparate microbial niches to augment functional microbes including polyphosphate accumulating organisms, denitrifiers, nitrifiers, denitrifiers, heterotrophs, and so on. Table 1 expands on elemental cycles, the associated microbes, and cooperative interactions of pollutant removal found in nature. Table 2 expands on microbial diversity in wastewater treatment systems.

Table 1 Elemental cycles, the associated microbes, and cooperative interactions of pollutant removal found in nature

Element cycles	Redox potential	Anaerobic	Aerobic	Some cooperative interactions
Nitrogen cycle 0.34–0.97 V		Ammonia oxidizing bacteria: Conversion of ammonia to nitrite Nitrite oxidizing bacteria: Conversion of nitrite to nitrate Denitrifying bacteria: Conversion of nitrate to nitrogen		Denitrification: Removal of C and N simultaneously Sulphur based denitrification: Removal of S and N simultaneously Denitrifying polyphosphate accumulating organisms: Removal of organic C, N, and P
		Anaerobic ammonia oxidizing bacteria: Conversion of ammonia to nitrogen		
Sulphate reduction	−0.22 to −0.14 V	Sulphate reducing bacteria: Regulation of sulfidogenic bioprocess Reduction of sulfate to sulfite followed by conversion to sulfide		Coupling of methane production and sulphate removal: Addition of conductive materials results in re-enrichment of syntrophic community which in turn can produce methane in sulphate containing environment
Carbon cycle Methanogenesis	−0.37 to −0.12 V −0.43 to −0.25 V	1. Fermenting bacteria: Complex organic substrates including sugar and protein are converted to monomers 2. Acidogenic bacteria: Monomers utilized for production of hydrogen and acetate 3. Methanogens: Consumption of acetate and hydrogen/carbon dioxide for generation of methane	Heterotrophs: Degradation of organic carbon for production of CO_2 and synthesis of biomass	

Table 2 Microbial diversity in wastewater treatment systems

Process	Microorganism	Activity
Nitrogen removal	Aerobic (proteobacterial ammonia oxidizers): Betaproteobacteria ammonia oxidizers including *Nitrosomonas* and *Nitrosospira*, and the Gammaproteobacteria, *Nitrosococcus* Anaerobic ammonia oxidation (anammox) bacteria: *Candidatus 'Brocadia', Candidatus 'Kuenenia', Candidatus 'Scalindua', Candidatus 'Anammoxoglobus'* and *Candidatus 'Jettenia'*	Ammonia oxidation to nitrite, nitric oxide, dinitrogen
	Aerobic nitrite bacteria: *Nitrospira, Nitrococus,* and *Nitrobacter*	Oxidation of nitrite to nitrate
	Alcaligenes, Pseudomonas, Methylobacterium, Bacillus, Paracoccus, and *Hyphomicrobium* Non-convention denitrification: Coupled formation of dinitrogen with anaerobic oxidation of methane by *Candidatus Methylomirabilis oxyfera*, and reduction of nitrous oxide to dinitrogen catalyzed by means of an atypical nitrous oxide reductase by non-denitrifying species, like *Anaeromixobacter dehalogenans*	Denitrification (nitrate to nitrite)
Phosphorus removal	Polyphosphate accumulating organisms: *Acinetobacter,* Rhodocylus related betaproteobacterium *Candidatus Accumulibacter phosphatis, Accumulibacter* (carbon, nitrogen, and phosphorus utilization gene revealed), filamentous actinobacterium *Candidatus Microthrix parvicella*	
Sulphate removal	Sulphate reducing bacteria: *Desulfobacter, Desulfobulbus, Desulfococcus, Desulfocarcina, Desulfomaculum, Desulfonema, Desulfotomaculum, Desulfacinum,* and *Desulfovibro*	Sulphur completely reduced to degradable organic compounds or complete degradation of organics to carbon dioxide
Energy production by microbial fuel cells	*Proteobacteria*, particularly *Beta- and Deltaproteobacteria*, and phyla *Bacteroidetes, Firmicutes,* or *Actinobacteria* (*Clostridium, Desulfobulbus, Pseudomonas, Rhodoferax, Rhodopseudomonas Arcobacter,* and *Geothrix*)	Pollutants removal while generating electricity

3 Microbial Characteristics and How They Can Influence the Wastewater Systems

The benefaction of the microbiome to wastewater ecosystem processes is deployed through miscellany community attributes that are molded by both environmental factors as well as microbial membership. Three perceptible categories of microbial characteristics can be proposed to link the microbes with the processes that they may influence: (a) microbial processes (b) microbial community properties, and (c) microbial membership (Hall et al. 2018).

3.1 Microbial Processes

Microbial processes are the contribution to alterations in pools and fluxes of elements or compounds by the collective metabolisms of the microbiome. Organic carbon mineralization, carbon fixation, phosphorus uptake and immobilization, denitrification, nitrogen fixation, nitrification, etc., are some of the customarily measured microbial processes in wastewater ecosystem. This type of microbial data can be promptly subsumed into system models as they are the representation of crucial sub-processes contributing to a specific ecosystem process. All microbial processes are rates and need bioassay for estimation. The assays employed for estimations are logistically exigent, demand manipulations that inescapably deviate from in situ conditions, and are traditionally dependent on the environment from which the microbiome was sampled.

3.2 Microbial Community Properties

Wide range of microbial characteristics, like the community biomass, or biomass elemental ratios (biomass C:N or C:P ratios), the majority of phylogenetically undifferentiated aggregate sequence-based measurements (for instance, gene abundance, metagenomes, transcriptomes) represent the microbial community properties. Microbial community properties proffer the possibilities of prediction and curtailment of estimates of microbial processes. Microbial community properties can be further divided into community aggregated and emergent properties. As the name suggests, the community aggregated properties are gauged based on traits of their integrant while the emergent properties are conventionally qualities of the whole that is unique and discernible from the additive traits of its constituents. Community aggregated attributes may encompass commonly estimated properties of a microbial community, for example, quantitative PCR assessed functional gene abundance (for model, pmoA, which encodes a subunit of the chemical associated with methane oxidation, can be utilized to appraise potential for methanotrophy and as a phylogenetic marker for methanotrophs). An example of an emergent property is the distribution of traits that influence key microbial processes. Trait-based approaches have a rich history in environment and have been progressively applied to resolve inquiries in different areas of microbial ecology. For instance, take-up of a natural substrate can frequently include the outflow of different qualities—varying among individual living beings—all fit for performing take-up of the natural substrate, but with contrasts in the hidden productivity. The dissemination of the statement of these utilitarian quality variations produces an attribute conveyance that is an emanant property of the microbiome. This new attribute conveyance decides the general execution of the microbiome for that capacity, however, it can't be anticipated from the presence of the creatures presenting that quality utilizing present methods. While portrayal of developing properties might work on the comprehension of microbial cycles, right now, they most frequently can't be assessed or anticipated based on the constituent taxa (that is, participation) alone, and hence should stay a middle person between ecological drivers and microbial cycles. Apprehension of which community properties can be anticipated by participation is a basic exploration question and a significant stage in anticipating how the microbiome adds to framework level processes. Whether a community property is probably going to be a community aggregated characteristic or an emanant property is an astonishing area of research and gives a significant system to propel research at the microbial-environment nexus (Grilli et al. 2017).

3.3 Microbial Community Membership

Albeit the ordinary examination of community membership by sequencing phylogenetic markers or set-ups of phylogenetically rationed protein groupings can distinguish constituent microbial taxa, the immediate coupling of microbial phylogeny to physiology and ecology stays evasive. In general, the scarcity of related physiological information or data on populace aggregates that go with phylogenetic investigations restricts the framework level deduction that is conceivable from examinations of community memberships. In any event, when the physiology of an organic entity is known, apparently numerous metabolic pathways are phylogenetically wide, and that any given microbiome will contain a various arrangement of microorganisms with the qualities that encode a large number of similar normal microbial metabolic pathways, frequently alluded to as utilitarian redundancy. There also appears to be no steady phylogenetic goal at which explicit microbial metabolic pathways are constrained. This restricts our capacity to ascribe microbial cycles to community membership of even somewhat straightforward environmental consortia. While obviously microbial populaces are not arbitrarily dispersed in space and time, and that some microbial attributes are rationed at coarse ordered scales, the physiological instruments of basic non-irregular dispersions of microbial taxa across natural slopes are regularly obscure. This restricted comprehension of the metabolism of most microbial phyla is one component that at present forestalls our capacity to interface a microorganism's overflow in a climate to its part in a related microbial interaction.

4 Understanding Key Microbial Players

The dawn of molecular methods like polymerase chain reaction techniques transformed the study of microbial diversity formerly restricted by our incapacity to grow majority of microbes in culture. Polymerase chain reaction amplification and sequencing have been employed now for decades to scrutinize the existence, diversity, and expression of protein coding, and ribosomal genes in constructed environments, enabling the discernment of key microbes involved (Ferrera and Olga 2016). The polymerase chain reaction technique casts light on the diversity of key microbial players in wastewater treatment plants and how genetic diversity assembles in the context of environmental parameters. Besides, there are innate polymerase chain reaction amplifications and primer-related prejudices that restrict quantitative data that can be extricated from these processes. Conceivably, the most critical step from these is primer design, which may be extraordinarily arduous for highly diverse functional genes. Technique like fluorescent in situ hybridization using group specific rRNA-targeted oligonucleotide probes evades the polymerase chain reaction bias and has been broadly pertained to wastewater microbiology proffering quantitative data of the dominant groups involved in waste removal.

Furthermore, the ascent of the 'omics' era has ensued climacteric point in studies allied to functional and phylogenetic diversity of wastewater treatment systems. The first step for elucidation of functional potential of wastewater pertinent microbes was the whole genome sequencing of isolated microbes. Metatranscriptomics and metagenomics have been applied to divulge the gene transcripts and genomic potentiality of whole community of microorganisms out-maneuvering isolation and polymerase chain reactions. Nonetheless, these methodologies have likewise the impediment that the vast majority of the data is procured from the most plentiful individuals of the community. Notwithstanding treatment frameworks present in general less diversity than regular environments, critical additions in the examination of quality articulation designs require a more noteworthy sequencing speculation and the mix of both metatranscriptomics as well as metagenomics (Crovadore et al. 2017). More exertion is expected to comprehend the useful capacities of microbial networks in wastewater treatments and what qualities are communicated under which conditions to better comprehend the connection between construction and capacity in these microbes conciliated processes.

Notwithstanding nucleic acid based techniques, the study of metaproteomics can usher the distinguishing proof of proteins engaged with known processes and furthermore to the revelation of unknown metabolic cycles and functional roles, since there are still numerous protein families whose capacity is obscure. Metabolomics (the investigation of all cell metabolites) is maybe right now the most un-utilized field of the -omics approaches essentially because of logical intricacy however obviously its development will be basic to comprehend organism microorganism and microorganism atom associations (Mcdaniel et al. 2021).

Regardless of the capability of -omics for concentrating microbial community's functional capabilities, the culturing power of individual species can't be dismissed.

The premise to interpret the microbial world is proffered through culturing and is essential for physiological investigations in a rearranged and controlled climate. These examinations are essential to decipher -omics information and to determine the pertinence of the newfound elements of proteins through -omics. Up to this point, culturing was additionally important to produce reference genomes, yet admittance to the genome of crude living beings is these days conceded on account of the improvement of single-cell arranging and manipulation and entire genome amplification. Notwithstanding, regardless of the colossal advances acquired on information on the character and functionalities of wastewater microbes, we are still a long way from arriving at a full comprehension of these generally intricate microbial networks. The blend of all these methodologies with orderly estimations and exploratory approval (biological systems science) will be the way to accomplish a comprehensive portrayal of microbial consortia liable for squander treatment.

The application of genome resolved metagenomics in amalgamation with one or more additional –omics based techniques including metabolomics, metaproteomics, or metatranscriptomics has been employed as integrated multi-omics techniques in water systems. Fully integrated studies combining the methods for sequential extraction of metabolities, proteins, RNA, and DNA from a microbial community have been reported to enhance the reproducibility of integrated -omics measurements proffering the potential to allow powerful ecological perceptions into engineered water microbiomes and concoct the foundation for more predictable and controllable systems (El Sheekh et al. 2022).

5 Microbial Metabolic Mechanisms

Clarification of biological metabolic mechanisms can enhance the wastewater treatment processes. For instance, interspecies transfer of formate and hydrogen have been contemplated as the normal routes for syntrophic methanogenesis. As of late, some syntrophic microbes and methanogens have been accounted for trading electrons straight by conductive pili or external layer cytochromes for syntrophic methane production (Moritaa et al. 2011; Summers et al. 2010). Since electron transporters are not needed at the time of direct interspecies electron transfer, it was contemplated as a quicker what's more possibly more energy-preserving route for methane production. In this manner, direct interspecies electron transfer might be an essential way for enhancement of energy conversion from wastewater.

By revelation of this new microbial mechanism, a few procedures have been put forward that can be possibly utilized to accomplish the pique of direct interspecies electron transfer to enhance methanogenesis (Lovley 2017). The first is the microbial science based guideline. A high wealth of direct interspecies electron transferable microbes regularly infers the great execution of direct interspecies electron transfer. Advancing direct interspecies electron transfer players by improving their specialties can bring about the predominance of direct interspecies electron transfer pathway in methanogenic frameworks. For instance, the notable microbes with the direct interspecies electron transfer capacity, *Geobacter*, which syntrophically utilize ethanol as the organic substrate for development, could be improved in an up-stream anaerobic slime cover reactor treating bottling works wastewater. Much of the time, high methane production efficiency could be credited to the high wealth of *Geobacter*. Leading the pretreatment of ethanol-type aging might be a helpful approach for developing *Geobacter* species (Zhao et al. 2017). Second, advancing the discharge of extracellular mixtures and changing the syntrophic collaboration could be additionally applied for better direct interspecies electron transfer performance. At long last, application of conductive materials as electron channels in methanogenic frameworks can give a decent outside conductive climate for syntrophic accomplices with the direct interspecies electron transfer capacity. For this situation, electrons let out of syntrophic microscopic organisms can be straightforwardly moved to methanogens by means of conductive materials without reaching intently, upgrading the proficiency of methane production. In the wastewater treatment framework, the expansion of conductive materials could upgrade the conductivity of anaerobic slime, and invigorate the action of respiratory chain also the extracellular electron move pace of syntrophic accomplices, accordingly advancing the methanogenic efficiency.

6 Wastewater Treatment Through Microbial Niche Tuning

Proffering competent niches for explicit microbes can likewise upgrade the evacuation of arising compounds and ease their poisonousness. Traditional ammonia oxidizing bacteria can eliminate endocrine disrupting chemicals employing ammonia monooxygenase, which can corrupt particular sorts of miniature toxins, furthermore heterotrophs could be additionally liable for the degradation of manufactured estrogen. In expansion, the as of late found total smelling salts oxidizing microscopic organisms which could oxidize alkali to nitrate through nitrite were likewise observed to have the option to debase miniature contaminants. Subsequently, by tuning this large number of useful microorganisms, not just customary contaminations will be eliminated proficiently (Fig. 2), yet in addition apparent mixtures will be all around controlled. Then again, appropriate microbial specialties could be applied to ease the organic harmfulness actuated by arising compounds. For instance, the substitute activity of oxygen consuming furthermore expanded anaerobic treatment brought about the improved expulsion of endocrine exercises furthermore better control of organic poisonousness. Different redox circumstances of wastewater under oxygen consuming and anaerobic circumstances may be one reason for advancing endocrine corruption. Moreover, the difference in natural stacking rate could prompt a specialty assortment of organisms too, in this way influencing the expulsion efficiencies (Vo et al. 2016). As of late, it was affirmed that cysteine delivered during the sulfate decrease could mitigate the nano-metal molecule poisonousness. This shows that the microbial connections during natural cycles could be working assorted for accomplishing various purposes.

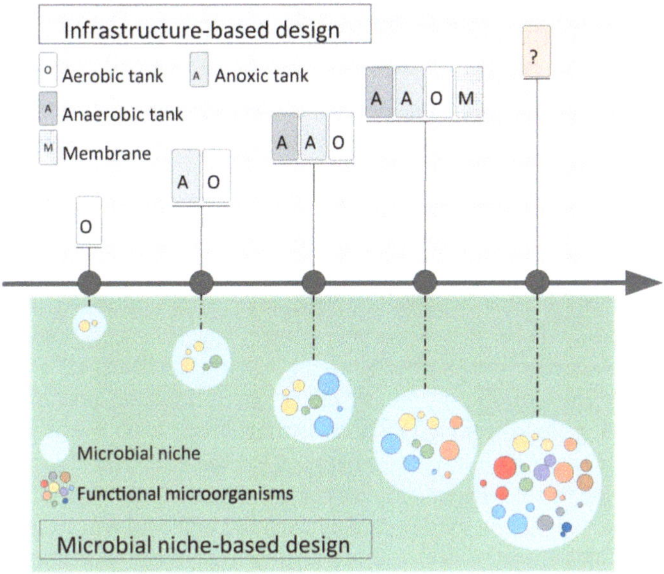

Fig. 2 Developing wastewater treatment plants based on infrastructure-based design or microbial niche-based design (Wu and Yin 2020)

For wastewater treatment, for elimination of organic carbon, one oxygen consuming reactor is satisfactory. While for organic C and N expulsion, anoxic joined with oxygen consuming reactors would be employed. Moreover, for organic phosphous, carbon, nitrogen evacuation, anaerobic, anoxic, as well as oxygen consuming reactors would be embraced. With more kinds of pollutant evacuation, the quantities of natural reactors would be stretched out for wastewater treatment framework plan. Other than organic reactors, the microbial specialty nexus idea ought to be fused during wastewater treatment framework plan. For the upgradation of traditional wastewater treatment plants, novel microbial networks could be investigated and used for addressing new difficulties, including arising intensifies expulsion. Every one of these could be accomplished through microbial specialty improvement to advance assorted microbes in the present wastewater treatment plants other than to assemble new foundation. To accomplish this, it is fundamental to additionally investigate the concealed organic cycles or then again works. For instance, interesting species in natural treatment cycles should be focused to, which might go about as the seed and would be prevailing with shifted environmental circumstances. Now and again, species with a low overflow may likewise contribute a ton to the vital capacity of a microbial framework.

7 Circular Economy for Wastewater Treatment: Application Niche of Microbes

The pertinence of any microbiological treatment is strongly reliant on the stability of microbial ecosystems. Among other distinct research requirements, better apprehension of nitrogen removal apropos emission of greenhouse gas nitrous oxide is a need. There is a need to understand the groups of microbes producing this gas in different systems and how can this be controlled (Ghimire et al. 2021). The probable dissipation of antibiotic resistance genes (ARG) into the environment is also a major concern wherein wastewater treatment plants are considered as the hotspots for transfer of these genes. Furthermore, it is important to carry out enhanced biological phosphorus removal for recovery of phosphors through sludge distribution to farmland or struvite formation. Similarly, another challenge in the daily operations include membrane fouling, foam formation, poor settling because of overgrowth of filamentous microbes. Novel discovered processes, for instance, denitrifying anaerobic methane oxidation process for anaerobic oxidization of methane or novel amalgamations of processes including autotrophic nitrification and denitrification, integrated sulphate reduction for wastewater treatment also seem promising for the development of more sustainable wastewater treatment processes.

Unearthing alternatives for production of biogas for yielding higher levels in value change is also a precedence. Currently, biogas production is the methodology of choice for various wastewater treatment plants, but escalated requirement for several biochemical or high-value products such as production of biohydrogen, or value-added chemicals including lipids, organic acids, alcohols for fabrication of biopolymers/bioplastics should be considered. Table 3 highlights few examples from the literature for resource recovery and associated processes for circular approach in wastewater treatment.

Admittance to healthy drinking water isn't just a significant aspiration yet additionally a fundamental common liberty that since relic has called researchers, specialists, and lawmakers for activity. The acknowledgment that human feces effect the nature of the wellsprings of drinking water set off the improvement of sewage waste frameworks around 3,500–2,500 BC, in urban areas like Ur and Babylon (Richardson and Ternes 2022). In spite of the surprising relapse seen during the middle ages, the ascending of metropolitan and modern agglomerations, and increase in wastewater, has been setting off the advancement of wastewater treatment innovations since the modern upheaval. In contrast to other modern exercises, whose high added esteem items empower high-overall revenues, wastewater treatment might be not focused on, essentially in world districts with restricted pay and ability to put resources into both foundation and activity frameworks. Subsequently, the vast majority of the metropolitan wastewater treatment plants working overall depend upon organic based minimal expense advancements. With a long improvement history itself, this oxygen consuming biologic interaction (activated sludge-based treatment), in full-scale activity starting around 1914, is viewed as the regular standard for wastewater treatment. A century prior the significant test of natural engineers was to foster

Table 3 Resources recovered and the associated processes for recovery for circular economy approach in wastewater treatment

Resource recovered	Raw material	Process for recovery	References
Phosphorus (as struvite/iron phosphate/calcium phosphate)	Thickened/dewatered/digested sludge	Wet and chemical precipitation	Amann et al. (2018), Egle et al. (2016)
Phosphorus (as aluminium phosphide/iron phosphide/calcium phosphate)	Thickened/dewatered/digested sludge	Supercritical water oxidation, acidic/alkaline leaching, and precipitation	Sartorius et al. (2012)
Phosphorus (as struvite)	Thickened/dewatered/digested sludge	Wet oxidation and precipitation	Amann et al. (2018), Egle et al. (2016)
Phosphorus (P-rich slag)	Dewatered sludge briquettes	Metallurgic melt-gassing	Remy et al. (2016)
Heavy metals (Zn, Ni, Cu, Cr)	Leaching effluents from sewage sludge	Supported liquid membranes	Yesil and Tugtas (2019)
Heavy metals (Mn, Ni, Pb, Cr)	Sewage sludge	Electrokinetic extraction with chelant	Tang et al. (2017)
Heavy metal (Cu)	Sewage sludge	EDDS and EDTA	Li et al. (2018)
Heavy metal (Zn, Cu)	Sewage sludge	Extraction with citric acid	Veeken and Hamelers (1999)
Protein	Secondary sludge	Alkaline/acidic hydrolysis	Gao et al. (2020)
Protein	Activated sludge before gravity thickening	Ultrasonication with/without additives like EDTA	Monique et al. (2008)
Protein	Sludge from anaerobic ammonium oxidation	Physico-chemical treatment	Feng et al. (2019)
Protein	Secondary sludge	Enzymatic (thermal/ultrasonic)	Gao et al. (2020)
Lipase (enzyme)	Activated sludge after gravity thickening	Stirring, ultrasonication with additives	Nabarlatz et al. (2010)
Protease (enzyme)	Activated sludge after gravity thickening	Extraction with additives	Karn and Kumar (2019)
Amylase (enzyme)	Activated sludge before gravity thickening	Extraction with additives	Karn and Kumar (2019)
Volatile fatty acids (acetate, butyrate)	Intermediate compounds after anaerobic digestion	Membrane based process (Electrodialysis)	Liu et al. (2020)

(continued)

Table 3 (continued)

Resource recovered	Raw material	Process for recovery	References
Olive polyphenols (hydroxytyrosol)	Olive oil wastewaters	Activated carbon coated with milk proteins	Yangui and Abderrabba (2018)

a wastewater treating framework ready to decrease the heap of promptly degradable natural matter and microbes from sewage. Activated sludge-based treatment frameworks completely accomplish these objectives. However, over one century of modern development and advancement changed drastically our way of life, and therefore, the kind of poisons released in wastewater. These days, wastewater treatment plants are additionally expected to eliminate abundance of inorganic N and P supplements, liable for the eutrophication of the receptor water bodies, and a heap of (possibly) dangerous compound micropollutants, which might present gamble to the oceanic environments and human wellbeing given their intense what's more constant poisonousness. These compound micropollutants of arising perturb, present in extremely low focus, incorporate both normal and xenobiotic mixtures like drugs, individual consideration items, steroid chemicals, medications of misuse, and pesticides, among others. Notwithstanding the substance micropollutants, wastewater treatment plants are presently likewise tested to hinder the arrival of high heaps of natural foreign substances of arising concern, like some pathogenic infection, protozoa, or microbes specifically antibiotic resistant holding onto antimicrobial obstruction genes, into the receptor water bodies. Successful wastewater treatment frameworks are to be sure the essential and significant obstruction between human exercises and the climate, with a critical job on the anticipation of tainting of surface- and groundwater. Definitely, water bodies, for example, streams, lakes, and springs span areas of movement and geologies, for example, when utilized as wellsprings of farming water system water, drinking water creation or living space and wellspring for natural life or food-creating creatures (Gomes et al. 2020; Shah Maulin 2020, 2021a). Tensions to carry out innovations ready to effectively eliminate both compound and natural foreign substances inside the metropolitan water cycle are exacerbated under the environmental change situation. Monstrous withdrawal and utilization combined with erratic atmospheric conditions, like dry season and flood occasions, has been driving not exclusively to freshwater shortage yet additionally to the disintegration of water quality. Freshwater shortage brought the new idea of wastewater treatment plants as reusing units, equipped for creating last effluents that can be securely and reasonably reused for various purposes, to be specific in farming, the area with the biggest utilization of freshwater. Be that as it may, to be reused, treated wastewater should be protected. This implies that the grouping of possible compound and/or then again natural toxins in treated wastewater should not put in danger the ecological and human wellbeing. Consequently, the level of pollution of the treated wastewater decides its end use or site of releases. Updating advances fit for expulsion of N and P supplements from wastewater have been effectively evolved and carried out. These days, full-scale wastewater treatment plants with trains inclining toward the distribution of the blended alcohol among oxygen consuming and anoxic tanks, where ammonification of natural N, nitrification, and denitrification happen as indicated by the oxygen accessibility in every compartment are usually found; and an expanding number of wastewater treatment plant where, notwithstanding the trains alluded to above, distribution incorporates anaerobic reactors leaning toward P granules aggregation are additionally working around the world. All the more as of late, the synchronous C, N, and P expulsion is guaranteed through the oxygen consuming granular slime innovation, given the spatial dispersion of the microbes of the different metabolic gatherings in the different miniature conditions of the granules. Interestingly, with the C, N, and P evacuation, the organic evacuation of compound micropollutants is by all accounts less proficient. Notwithstanding the capacity of an immense number of microbes to debase a wide variety of micropollutants, the low grouping of these mixtures in wastewater may contribute for their low bioavailability in the organic reactors. Subsequently, the optional last effluents of activated sludge-based wastewater treatment plants actually contain various micropollutants at natural troubling focuses (Weissbrodt et al. 2020). Advanced Oxidation Technologies have been suggested among the best answers for the evacuation of substance micropollutants from the optional effluents of Activated sludge-based wastewater treatment plants (Shah Maulin 2020, 2021b; Xia et al. 2018). Countless logical concentrates on have been led around here, to create and enhance tertiary cycles equipped for the proficient expulsion of these impurities from the effluents before release into the receptor water bodies. Among these advancements, ozonation has high perceivability, being carried out in full-scale wastewater treatment plants, for example in Switzerland, a nation that as of late carried out regulation suggesting progressed treatment of wastewater targeting safeguarding the climate. One of the benefits of Advanced Oxidation Technologies is their ability to clean water. Henceforth, other than debasing unwanted substance micropollutants, various logical seat concentrates on exhibited that the components for microbial inactivation utilized by Advanced Oxidation Technologies, for example, the oxidative pressure as it is produced by ozonation, are moreover equipped for diminishing the microbial heap of wastewater. Such promising outcomes opened the chance of updating activated sludge-based wastewater treatment with a last Advanced Oxidation Technologies cleaning step what's more involving the offices as reusing units of metropolitan wastewater. Extra treatment might be expected in a reuse situation, and in that cases, the last treated wastewater might have to go through an adsorption post Advanced Oxidation Technologies treatment step to ultimately eliminate harmful corruption items and to be put away for periods that might fluctuate between couple of hours to certain days, contingent upon the requirements. Henceforth, some studies have been directed to evaluate the microbiological nature of the wastewater after the last Advanced Oxidation Technologies treatment. Reliably, concentrates on zeroed in on the impact of Advanced Oxidation Technologies presume that the microbiota, including antibiotic resistant bacteria (ARB) and genes (Fig. 3), getting through Advanced Oxidation Technologies treatment is equipped for re-regrowth during the capacity period, now and again to values coming to or outperforming those deliberate in the untreated optional gushing. Besides, re-regrowth is joined by the aggravation of the microbial local area, with potential ramifications on the decline of variety, and the abundance of Proteobacteria (Nunes 2021). Among these, bacterial gatherings portrayed as possible vectors of anti-infection obstruction, for example, *Pseudomonas*, have been identified at high family member overflow. Similar peculiarities happen when different advances are applied in the wastewater treatment. Nearly milder cycles like UV254 nm light or even coagulation lead to comparative aggravations. While contrasting various advances, a positive relationship between sanitization adequacy and the transcendence of omnipresent, possibly dangerous, microscopic organisms in the treated put away wastewater appears to happen. Curiously, clean fabricated conditions, where asepsis and successive sterilization are the standard, are described by the transcendence of Proteobacteria. Also, cleaning with forceful specialists appears to lean toward microbiomes encoding capacities related with destructiveness, multi-drug efflux, oxidative pressure, as well as film transport and discharge, which engage cells to secure supplements in exceptionally cutthroat supplement unfortunate conditions.

Such outcomes are not unforeseen. Any cycle decreasing the variety and wealth of microorganisms in a given environment, through actual evacuation of the cells or on the other hand physical or potentially compound inactivation of macromolecules or cell processes, is relied upon to create an environment where intercellular contest for space and supplements is diminished, offering the chance for those that arbitrarily endure the interaction and that are most adaptable and quick to develop, to multiply. Subsequently, among the survivors, those with high ability to develop under the

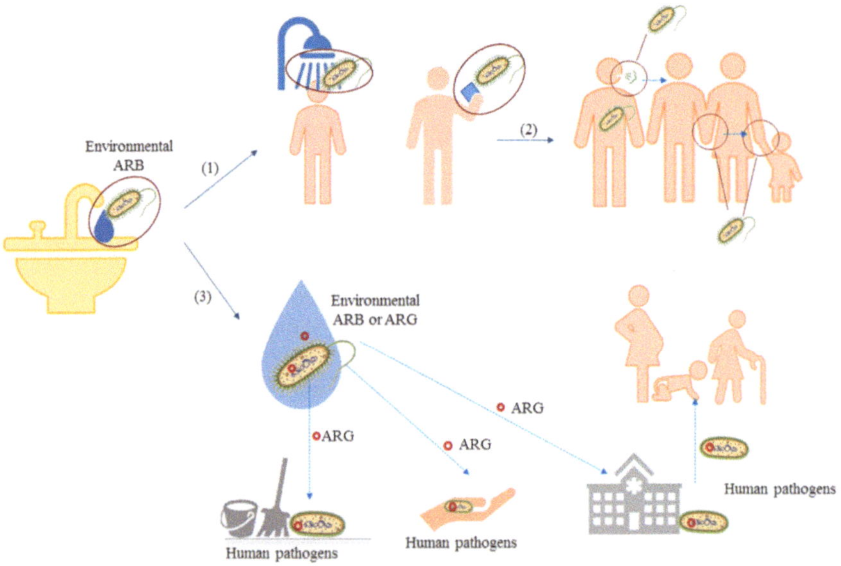

Fig. 3 How antibiotic resistant bacteria and genes may hamper human health. Obtained from Shah Maulin (2020a)

circumstances winning in the sanitized or cleaned framework will flourish. Alternately, the microorganisms with explicit necessities (for example wholesome) or with more slow development rates will be outcompeted. Proteobacteria are notable for their genomic versatility. Some proteobacterial species, like *Pseudomonas aeruginosa*, colonize a wide variety of natural compartments, including mineral water, chlorinated drinking water, surface water and soils, and, surprisingly, human bodies. Part of the achievement of this pervasive pioneering pathogenic species depends upon its ability to trade hereditary data through even quality exchange. Subsequently, *P. aeruginosa* harbors hereditary data which permits cell advancement in a wide variety of ecological circumstances, remembering for the presence of an immense range of antimicrobial mixtures. Hence, other than conveying natural antimicrobial obstruction, *P. aeruginosa* strains are incredible vectors of antibiotic resistant gene spread. The prevalence of microorganisms with these kind of highlights in treated wastewater is consequently not attractive, basically assuming that its further use in agribusiness water system is imagined, given the chance of tainting of the well established order of things. In this unique situation, it very well might be contended that the overhauling wastewater treatment plants with a last sterilization step isn't sufficient to change these offices into wastewater reusing units, furthermore more examinations ought to be completed to plan and execute capacity frameworks equipped for lessening the lopsidedness of the bacterial local area before reuse of the put away treated wastewater (Shah Maulin 2021a). Measures to reestablish the microbial wealth and variety of the sanitized wastewater would forestall the abundance of risky microscopic organisms fitted to couple with very clean oligotrophic conditions, like *P. aeruginosa*, through contest. Such measures could incorporate the vaccination of the sanitized wastewater with adjusted normal microbial networks, with a rich and various phylogenetic and useful gathering of microorganisms. In these networks, organic entities having a place with a wide assortment of animal categories interface through complex connections (mutualism, commensalism, contest, predation, parasitism) guaranteeing metabolic overt repetitiveness and the honesty of supplement cycles and energy streams. Such people group are steady and tough, that is to say, show little unsettling influence and reestablish quickly upon adjustment of the ecological circumstances or attack. Subsequently, techniques, for example, weakening sanitized wastewater with non-contaminated surface water, blending in with flawless silt or soils or release in wetlands would present a sound microbiome in the treated wastewater. Under the present situation, the exogenous microbiome would go about as a security safeguard for the multiplication of the dangerous microorganism enduring the sanitization process, along these lines of the normal human microbiota, our first line of safeguard against the attack of microorganisms. Certainly, microorganisms should have a say on eliminating waste from wastewater. The following exploration steps ought to be arranged toward a superior comprehension of the biotic connections happening in the treated wastewater and mechanical execution of frameworks that can sustain these significant craftsman networks (Shah Maulin 2021b; Yu et al. 2021; Guo et al. 2020).

8 Conclusion and Future Perspective

The sorts of toxins in wastewater decide the improved different useful microbes. To propel the possibility of microbial specialty nexus in explicit ways, the essential is the comprehension of microbial nature, including the capacity, guideline, and connection of known what's more obscure microorganisms. Investigating and once again building assorted microbial specialties especially founded on existence arrangements in wastewater treatment frameworks is the critical method for tackling the expanding difficulties and fulfill future needs. Further explores on microbial specialty based natural biotechnology ought to be done by zeroing in on the accompanying perspectives:

1. Long-term activity frameworks with assorted microbial specialties may assist with finding more obscure utilitarian central members as well as foster the advancement of methodologies. Plus, a few regular habitats, for example, intertidal zones, which can give elective anaerobic and oxygen consuming circumstances, are additionally important microbial hotspots for finding new microbial digestion systems and capacities.
2. It becomes imperative to combine DNA, RNA, and proteomics-based strategies to investigate microbial natural capacities and guideline procedures. Aside from the C, N, P, S cycles, the explanation of other metabolic pathways and the guideline of micronutrients like amino acids and nutrients ought to be additionally thought of.
3. New innovation advancement and application. The disclosure of new advances what's more new standards (e.g., new utilitarian materials and new natural digestion standards) will prosper the improvement of natural wastewater treatment innovations. Interdisciplinary participation is the critical driver to accomplish this reason.

References

Amann A, Zoboli O, Krampe J, Rechberger H, Zessner M, Egle L (2018) Resources, conservation & recycling environmental impacts of phosphorus recovery from municipal wastewater. Resour Conserv Recycl 130:127–139. https://doi.org/10.1016/j.resconrec.2017.11.002

Cerruti M, Guo B, Delatolla R, De Jonge N, Hommes-De Vos Van Steenwijk A, Kadota P, et al (2021) Plant-wide systems microbiology for the wastewater industry. Environ Sci Water Res Technol 7:1687–1706. https://doi.org/10.1039/d1ew00231g

Crovadore J, Soljan V, Calmin G, Chablais R, Cochard B, Lefort F. (2017) Metatranscriptomic and metagenomic description of the bacterial nitrogen metabolism in waste water wet oxidation effluents. Heliyon: e00427. https://doi.org/10.1016/j.heliyon.2017.e00427

Egle L, Rechberger H, Krampe J, Zessner M (2016) Science of the total environment phosphorus recovery from municipal wastewater: an integrated comparative technological, environmental and economic assessment of P recovery technologies. Sci Total Environ 571:522–542. https://doi.org/10.1016/j.scitotenv.2016.07.019

El Sheekh M, El Dalatony MM, Zheng NTY, Sayed E. (2022) Role of microalgae and cyanobacteria in wastewater treatment: genetic engineering and omics approaches. Int J Environ Sci Technol 19: 2173–2194.https://doi.org/10.1007/s13762-021-03270-w

Feng C, Lotti T, Lin Y, Malpei F (2019) Extracellular polymeric substances extraction and recovery from anammox granules: evaluation of methods and protocol development. Chem Eng J 374:112–122. https://doi.org/10.1016/j.cej.2019.05.127

Ferrera I, Olga S (2016) Insights into microbial diversity in wastewater treatment systems: how far have we come? Biotechnol Adv. https://doi.org/10.1016/j.biotechadv.2016.04.003

Gao J, Weng W, Yan Y, Wang Y, Wang Q (2020) Chemosphere comparison of protein extraction methods from excess activated sludge. Chemosphere 249:126107. https://doi.org/10.1016/j.chemosphere.2020.126107

Ghimire U, Sarpong G, Gude VG (2021) Transitioning wastewater treatment plants toward circular economy and energy sustainability. ACS Omega 6:11794–11803. https://doi.org/10.1021/acsomega.0c05827

Gomes IB, Maillard JY, Simões LC, Simões M (2020) Emerging contaminants affect the microbiome of water systems—strategies for their mitigation. NPJ Clean Water 3(1):39. https://doi.org/10.1038/s41545-020-00086-y

Grilli J, Barabás G, Michalska-smith MJ, Allesina S (2017) Higher-order interactions stabilize dynamics in competitive network models. Nat Publ Gr 548:210–213. https://doi.org/10.1038/nature23273

Guo J, Fang F, Yan P, Chen Y (2020) Bioresource technology sludge reduction based on microbial metabolism for sustainable wastewater treatment. Bioresour Technol 297:122506. https://doi.org/10.1016/j.biortech.2019.122506

Hall EK, Bernhardt ES, Bier RL, Bradford MA, Boot CM, Cotner JB et al (2018) Understanding how microbiomes influence the systems they inhabit. Nat Microbiol 3:977–982. https://doi.org/10.1038/s41564-018-0201-z

Karn SK, Kumar A (2019) Sludge: next paradigm for enzyme extraction and energy generation. Prep Biochem Biotechnol 49:105–116. https://doi.org/10.1080/10826068.2019.1566146

Kaszycki P, Głodniok M, Petryszak P (2021) Towards a bio-based circular economy in organic waste management and wastewater treatment—The Polish perspective. N Biotechnol 61:80–89. https://doi.org/10.1016/j.nbt.2020.11.005

Kuypers MMM, Marchant HK, Kartal B (2018) The microbial nitrogen-cycling network. Nat Publ Gr. https://doi.org/10.1038/nrmicro.2018.9

Li R, Zhai Z, Li Y, Yang T, Chen Y (2018) Kinetic study of heavy metals Cu and Zn removal during sewage sludge ash calcination in air and N 2 atmospheres. J Hazard Mater 347:227–232. https://doi.org/10.1016/j.jhazmat.2017.12.061

Liu F, Moustafa H, Salah M, Hassouna E, He Z (2020) Resource recovery from wastewater can be an application niche of microbial desalination cells. Environ Int 142:105855. https://doi.org/10.1016/j.envint.2020.105855

Lovley DR (2017) Syntrophy goes electric: direct interspecies electron transfer. Annu Rev Microbiol 71:643–664

Mcdaniel EA, Aljoscha S, Pinto A, Ziels R, Halkjær P, Mcmahon KD et al (2021) Prospects for multi-omics in the microbial ecology of water engineering. Water Res 205:117608. https://doi.org/10.1016/j.watres.2021.117608

Mo W, Zhang Q (2013) Energy-nutrients-water nexus: integrated resource recovery in municipal wastewater treatment plants. J Environ Manage 127:255–267. https://doi.org/10.1016/j.jenvman.2013.05.007

Monique R, Elisabeth G, Etienne P, Dominique L (2008) A high yield multi-method extraction protocol for protein quantification in activated sludge. Biores Technol 99:7464–7471. https://doi.org/10.1016/j.biortech.2008.02.025

Moritaa M, Malvankara NS, Franksa AE, Summersa ZM, Giloteauxa L, Rotarua AE, Rotarud C, Lovley DR (2011) Potential for direct interspecies electron transfer in methanogenic wastewater digester aggregates. https://doi.org/10.1128/mBio.00159-11
Nabarlatz D, Vondrysova J, Jenicek P, Stüber F, Font J, Fortuny A et al (2010) Ultrasonics Sonochemistry Hydrolytic enzymes in activated sludge: extraction of protease and lipase by stirring and ultrasonication. Ultrason Sonochem 17:923–931. https://doi.org/10.1016/j.ultsonch.2010.02.006
Nielsen PH (2017) Microbial biotechnology and circular economy in wastewater treatment. Microb Biotechnol 10:1102–1105. https://doi.org/10.1111/1751-7915.12821
Nunes OC (2021) The challenge of removing waste from wastewater: let technology use nature! Microb Biotechnol 14:63–67. https://doi.org/10.1111/1751-7915.13711
Puyol D, Batstone DJ, Hülsen T, Astals S, Peces M, Krömer JO (2017) Resource recovery from wastewater by biological technologies: opportunities, challenges, and prospects. Front Microbiol 7:1–23. https://doi.org/10.3389/fmicb.2016.02106
Remy C, Boulestreau M, Warneke J, Jossa P, Kabbe C, Lesjean B (2016) Evaluating new processes and concepts for energy and resource recovery from municipal wastewater with life cycle assessment. Water Sci Technol 73(5):1074–1080. https://doi.org/10.2166/wst.2015.569
Richardson SD, Ternes TA (2022) Water analysis: emerging contaminants and current issues. Anal Chem 94(1):382–416. https://doi.org/10.1021/acs.analchem.1c04640
Sartorius C, Von Horn J, Tettenborn F (2012) Phosphorus recovery from wastewater—expert survey on present use and future potential. https://doi.org/10.2175/106143012X13347678384440
Shah Maulin P (2020) Microbial bioremediation & biodegradation. Springer
Shah Maulin P (2021a) Removal of refractory pollutants from wastewater treatment plants. CRC Press
Shah Maulin P (2021b) Removal of emerging contaminants through microbial processes. Springer
Summers ZM, Fogarty HE, Leang C, Franks AE, Malvankar NS, Lovley DR (2010) Direct exchange of electrons within aggregates of an evolved syntrophic coculture of anaerobic bacteria. Science 330:1413–1416
Tang J, He J, Liu T, Xin X, Hu H (2017) Chemosphere Removal of heavy metal from sludge by the combined application of a biodegradable biosurfactant and complexing agent in enhanced electrokinetic treatment. Chemosphere 189:599–608. https://doi.org/10.1016/j.chemosphere.2017.09.104
Veeken AHM, Hamelers HVM (1999) Removal of heavy metals from sewage sludge by extraction with organic acids. Water Sci Technol 40(1):129–136. https://doi.org/10.2166/wst.1999.0029
Vo J, Castronovo S, Wick A, Ternes TA, Joss A, Oehlmann J et al (2016) Advancing biological wastewater treatment: extended anaerobic conditions enhance the removal of endocrine and dioxin-like activities.https://doi.org/10.1021/acs.est.5b05732
Weissbrodt DG, Winkler MKH, Wells GF (2020) Responsible science, engineering and education for water resource recovery and circularity. Environ Sci Water Res Technol 6:1952–1966. https://doi.org/10.1039/d0ew00402b
Wu G, Yin Q (2020) Microbial niche nexus sustaining biological wastewater treatment. NPJ Clean Water 3:33. https://doi.org/10.1038/s41545-020-00080-4
Xia Y, Wen X, Zhang B, Yang Y (2018) Diversity and assembly patterns of activated sludge microbial communities: a review. Biotechnol Adv 36:1038–1047. https://doi.org/10.1016/j.biotechadv.2018.03.005
Yangui A, Abderrabba M (2018) Towards a high yield recovery of polyphenols from olive mill wastewater on activated carbon coated with milk proteins: experimental design and antioxidant activity. Food Chem 262:102–109. https://doi.org/10.1016/j.foodchem.2018.04.091
Yesil H, Tugtas AE (2019) Science of the total environment removal of heavy metals from leaching effluents of sewage sludge via supported liquid membranes. Sci Total Environ 693:133608. https://doi.org/10.1016/j.scitotenv.2019.133608

Yu J, Tang SN, Lee PK (2021) Microbial communities in full-scale wastewater treatment systems exhibit deterministic assembly processes and functional dependency over time. Environ Sci Technol 55(8):5312–5323. https://doi.org/10.1021/acs.est.0c06732

Zhao Z, Zhang Y, Li Y, Dang Y, Zhu T, Quan X (2017) Potentially shifting from interspecies hydrogen transfer to direct interspecies electron transfer for syntrophic metabolism to resist acidic impact with conductive carbon cloth. Chem Eng J 313:10–18. https://doi.org/10.1016/j.cej.2016.11.149

Activated Sludge Process for Wastewater Treatment

Farzana Yeasmin, Md. Rasheduzzaman, Mohammed Manik, and M. Mehedi Hasan

1 Introduction

One of the most intricate microbial systems ever designed for a specific purpose is the activated sludge process. Now, several diverse heterotrophic and autotrophic groups can coexist and carry out a variety of tasks, including the removal of organic carbon, nitrification and denitrification, increased biological phosphorus removal, etc. To assure the best process performance today, their metabolic activities can be predicted by modeling and individually managed by complex system parameter management (Orhon 2015).

Wastewater discharge into rivers and other major bodies of water became customary during the nineteenth century. 42.3% of the 99 million people living in the United States in 1915 had access to sewage systems, with 26.7% of those systems releasing untreated wastes into lakes and streams and 8.6% into the sea. 7% of these systems released sewage into the waters after some form of treatment. The putrefaction of organic compounds caused by sewage discharge into land or into rivers prompted challenges that sparked suggestions for alternative treatment methods. Granular activated sludge has become more popular as a result of its potential to treat wastewater compactly and effectively. It is well-known that specific environments, such as batch-wise operation with feast-famine feeding, high hydrodynamic shear pressures, and quick settling time, favor dense microbial aggregates allowing activated sludge to form granules (Wilén et al. 2018).

F. Yeasmin · M. M. Hasan (✉)
Department of Applied Chemistry and Chemical Engineering, Bangabandhu Sheikh Mujibur Rahman Science and Technology University, Gopalganj-8100, Dakha, Bangladesh
e-mail: mhasan.acde@gmail.com; mehedi.acce@bsmrstu.edu.bd

Md. Rasheduzzaman · M. Manik
Department of Environmental Science and Disaster Management, Bangabandhu Sheikh Mujibur Rahman Science and Technology University, Gopalganj-8100, Dakha, Bangladesh

M. P. Shah (ed.), *Advanced and Innovative Approaches of Environmental Biotechnology in Industrial Wastewater Treatment*,
https://doi.org/10.1007/978-981-99-2598-8_2

Aerobic granules with stable structure and functionality have been produced in a variety of wastewaters seeded with various sludge sources under various operating circumstances, but the microbial communities that resulted from these processes varied greatly. Granule instability still happens in spite of this.

The majority of the expenses incurred by treatment plants are related to the aeration process of processed wastewater. Due to the vital function that biological breakdown of organic compounds plays in the process, microorganisms, mostly bacteria that convert organic compounds into gaseous end products and water, must receive the adequate amount of oxygen (Skouteris et al. 2020).

In this chapter, various aeration systems, a thorough explanation of the CAS mechanism, progressive development over time, parameters that regulate the process' effectiveness, and several plant design types based on the activated sludge process have all been covered. Additionally, potential alternatives have also been considered along with the process's current limitations.

2 Mechanism of Activated Sludge Process (ASP)

According to Porter (1921), activated sludge treatment can be defined as a biochemical process by which the purification of sewage is accomplished by passing through tanks in which sewage sludge is artificially agitated and intimately mixed with sewage and is supplied with the requisite oxygen for the optimum development of a countless number of nitrifying organisms incorporated in and adhering to the sludge, the final settlement of which causes a distinct clarification of the oxidized sewage, where activated sludge as: "A flocculent sludge of medium brown color enveloped by masses of aerobic organisms possessing the power of rapidly oxidizing and nitrifying sewage, and which, though low specific gravity, settles rapidly."

The term "floc" refers to a "sponge-like mass" that is generated by encasing colloidal particles and bacteria. When injected into incoming sewage, the floc feeds on this material and regains its adsorption capacity.

Initial research revealed three stages of activated sludge activity: clarification of sewage input, reactivation of the sludge, and nitrification of the "sewage and sludge." Sewage clarification only occurred during the initial one to two hours of aeration, and the majority of the improvement in sewage treatment was thought to have been physical in character. Additional aeration in the same tank or in separate "re-activation or re-aeration" tanks may be used to restore the clarifying potential of sludge and nitrification (Orhon 2015). More than a century has been passed since the activated sludge technique had first been started operation and the International Water Association (IWA) celebrated it at the global conference with the theme "Activated sludge - 100 years and counting" in 2014 (Schneider 2014). Its theory has grown steadily in smaller steps over these periods. There are different varieties and modifications of the activated sludge process. The most significant advancement in industrial wastewater treatment, nutrient removal (N and P), and bulking control technology have been achieved over the past tens of years or so.

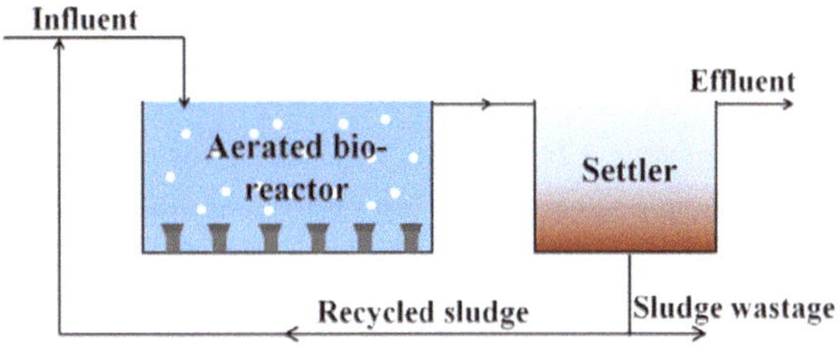

Fig. 1 Schematic diagram of basic ASP (Hreiz et al. 2015)

The infrastructure of a conventional ASP process includes following parts (as shown in Fig. 1) (Hreiz et al. 2015):

A. *Bioreactor,*
B. *Settling tank,*
C. *Sludge recycling line, and*
D. *Sludge wastage line.*

2.1 Bioreactor

The colloidal and dissolved organic matter is consumed by suspended microorganisms in a single bioreactor that runs constantly. Reactor is aerated to allow aerobic biodegradation with dissolved oxygen (DO). A portion of the colloidal and dissolved carbonaceous chemicals are consumed by bacteria to meet their energy needs (catabolism), while a smaller fraction is synthesized into new cellular tissues together with ammonium and phosphorus (anabolism). High height-to-diameter (H/D) ratios are impractical in practice because it is difficult to achieve plug-flow and there is a possibility of inhomogeneous carbon deposition in the granule bed. Recently, activated sludge from a full-scale biological nutrient removal facility was used to inoculate lab-scale reactors running continuous systems with AGS. Selective waste of floccular sludge was carried out in a settler with mixing capabilities between three step-fed mixed anaerobic tanks in sequence, followed by two aerobic tanks. This is a very appealing option since constantly operating AGS systems make it easier to retrofit existing activated sludge facilities, but obtaining stable granules for long-term operation is a challenge.

2.2 Settling Tank

In a settling tank (or a secondary settler or clarifier), treated wastewater is separated from activated sludge (flocculated biomass) gravitational force. Although the effluent overflows into the receiving waterbody, some WWTPs may treat it further (for example, by filtering and disinfecting it) before discharging it. For bacteria to aggregate into flocs (bio flocculation) and allow for gravity separation, a minimum solids retention time (SRT, also known as "sludge age") of roughly 3 days is necessary.

2.3 Sludge Recycling Line

A sludge recycle line allows the bioreactor to retain back a high microbial concentration, hence accelerating the biological nutrient removal by returning the majority of the settled sludge there. The decoupling of the hydraulic retention time (HRT) and SRT, which allows for an effective treatment with only an HRT of the order of ten hours, is the key advantage of sludge recycling.

2.4 Sludge Wastage Line

A little proportion of sludge is removed from the sludge waste line at the bottom of the clarifier in order to stabilize the biomass content in the bioreactor and fix an appropriate SRT. The extra sludge that was removed is then dealt separately.

Many advanced approaches have been suggested as alternatives to the conventional ASP. These contemporary ASPs incorporate combination of aerated and non-aerated reactors, recirculation lines between these reactors, and sludge recycle lines between the settler and the bioreactor. These advanced treatment methods provide improved biological nutrient removal, specifically with respect to phosphorus and nitrogen removal, increased process flexibility and plant layout, and lowered expense, when compared to the basic ASP (Hreiz et al. 2015).

Different crucial processes of the activated sludge population dynamics like removal of organic carbon, reactions and significance of inorganic carbon, reactions of various kinds of nitrogen, biological processes for removing phosphorus, and metabolism of sulfur compounds make the modern activated sludge system a complex biochemical system (Wanner 2020).

The following characteristics of the activated sludge process are mostly common:

I. Variable substrate in terms of chemical profile and particle size,
II. Multispecies biological culture, preferably growing in flocs,
III. Dramatic fluctuations in flows, temperatures, and the content and concentration of influent wastewater,
IV. Capacity to metabolize a large variety of organic molecules, including those containing nitrogen, phosphorus, sulphur, and others.
V. Different reactor topologies have been utilized, such as fully stirred tanks, plug-flow, batch sequencing, oxic, anoxic, anaerobic selections, etc.

Some IAWQ-recommended notation used to describe parameters in biological wastewater treatment process are as follows (Grau et al. 1983):

2.4.1 Substrate—The Energy Source for Living Cells

Three major substrates in aquatic systems:

- Light—energy source of phototrophic microorganisms,
- Organic compounds—biochemical oxidation of organic carbon into carbon dioxide produces organic compounds which are energy source of chemoorganotrophs,
- Inorganic compounds—oxidation of reduced elements like N, S, Fe, and Mn produces inorganic compounds which are fed by chemolithotrophs (chemoautotrophs).

These organic and inorganic compounds act as electron donors in biochemical oxidation–reduction reactions of ASP.

2.4.2 Carbon Source

The synthesis of new biomass involves a source of carbon for the living cells. Carbon can be metabolized in form of:

- Organic carbon: majority of aquatic microbes need organic carbon for cell formation. As a result, the organic carbon serves as both the substrate and the supply of carbon for the chemoorganotrophs' production. In the studies, these bacteria are frequently referred to as (chemo)heterotrophs.
- Inorganic carbon: autotrophs convert dissolved CO_2, carbonate, bicarbonate into organic cell materials.

The organotrophs will typically grow faster and easier than the lithotrophs in complex environments, such as wastewaters, where both organic and inorganic substrates are present. The lithotrophs have a relatively sophisticated method for acquiring energy and generating new biomass. This is a crucial deciding factor that affects the make-up of open mixed cultures like activated sludge.

2.4.3 Nutrients

Only two elements, nitrogen and phosphorus are referred to as nutrients in the context of wastewater treatment. The rationale is that those two substances are thought to be the only nutrients that algae in eutrophic surface waters can feed on. Due to their high concentration in microbial biomass (in activated sludge, 6–8% of the N and 2% of the P are found in relation to dry matter), elements like N, P, and S are referred to as macronutrients from the perspective of bacterial growth. Micronutrients include such elements as Fe, Ca, Mg, K, Mo, Zn, and Co. These elements make up a small mass percentage of biomass, yet they are essential to the biochemistry of cells as components of several enzymes (Eckenfelder 1980).

2.4.4 Biochemical Oxidation and Cultivation Conditions

Biochemical oxidation is involved to produce energy from substrates for cellular maintenance. In the process of bio-oxidation, electrons are transferred from oxidized substrates (donors) to the reduced substances referred to as acceptors. Modern parlance for cultivation conditions is based on several electron acceptors that take part in various metabolic processes.

- **Oxic conditions**—molecular oxygen bonded in a molecule of water receives electrons from both organic and inorganic substrates. In oxic settings, the crucial biochemical reactions are:
  - Organotrophic metabolism involving oxic oxidation of organic compounds,
  - The production of intracellular polyphosphate polymers (this process uses energy released from the organotrophic metabolism of intracellular organic molecules but is independent to an electron transit between donors and acceptors),
  - Chemolithotrophic oxidation of ammonia, of reduced sulphur compounds to sulphate, and nitrification.
- **Anoxic conditions**—Nitrogen with oxidation state +5 and +3 (less commonly) replaces oxygen as the electron acceptor (nitrite) and is reduced to oxidation stage 0 (molecular nitrogen, N_2) by accepting 5 or 3 electrons. Under anoxic conditions, there are primarily two probable reactions:
  - Anoxic chemolithotrophic oxidation of sulfide and elemental sulfur to sulfate,
  - Anoxic organotrophic oxidation of organic molecules (denitrification occurs when nitrate-nitrogen used as an electron acceptor is reduced to nitrogen gas).
- **Anaerobic conditions**—The absence of both molecular oxygen and nitrate/nitrite nitrogen is a marker of anaerobic conditions. In biochemical redox reactions, the electrons can be released from organic substrates to:
  - Organic compounds (by fermentation, such as acido- and acetogenesis; the activated sludge process does not take methanogenesis into account)
  - Sulfate of sulfur by dissimilatory sulphate reduction, where chemoorganotrophic microorganisms produce elementary sulfur and sulfur dioxide.

    Under anaerobic conditions, the stored energy in intracellular polyphosphates can be accessed to create organic intracellular storage materials from the byproducts of anaerobic fermentation. Since the oxidation stage of organic carbon is essentially unchanged in this reaction, electron acceptors are not required.

2.4.5 Cultivation Medium

The exact microenvironment in which cells develop is represented by a culture medium. Substrate material, a carbon source, nutrients, electron acceptors, and other elements are present in the medium. In addition to its chemical make-up, the medium can be determined by factors like temperature, pH, osmotic pressure, ORP, etc. The cultivation medium in biological wastewater treatment is a mixture of treated wastewater and an activated sludge mixed culture. This substance is frequently known as mixed liquor (Eckenfelder 1980).

2.4.6 Microorganisms in Activated Sludge

The key thing to understand of activated sludge as an artificial, dynamic ecosystem that is constantly being influenced by both abiotic and biotic elements. The activated sludge is grown under restricted conditions due to the need to achieve relatively low effluent concentrations of organic molecules (carbon and energy source) and inorganic nutrients. Due to this, there is intense rivalry among different groups of bacteria, and only those that have mutated to the best conditions succeed. The beneficiaries of the competition, controlling the population of activated sludge, can fluctuate since the influencing elements in wastewater treatment plants are not consistent. Since the activated sludge system is exposed to various impacts, the microbial composition of the activated sludges is not consistent. Individual microbial cells do not separate in the cultivation medium but rather develop in aggregates, which is another distinguishing trait of the mixed culture known as activated sludge. The most important characteristic of activated sludges that allows us to use this mixed culture in large-scale installations is its ability to flocculate. In large-scale wastewater treatment plants, the only commercially viable method of separating the biomass from treated wastewaters is gravity sedimentation. The flocculated aggregates (floes of activated sludge) exhibit technologically acceptable sedimentation rates (Randall et al. 1992). In the mixed culture of activated sludge, the capacity to flocculate can be viewed as a key selective pressure. The ability to form floes, or at the very least, to be fixed into them, gives microorganisms the selective advantages over freely growing cells—(a) microbes in flocs are trapped in the activated sludge system, whereas the dispersed, freely growing cells are flushed out and (b) the growth in flocs protects the majority of microbial cells from predators.

Thus, there are two main categories of activated sludge microorganisms. One is named as decomposers, which primarily degrade the contaminants in wastewater by biochemical reactions. The chief members of this group are bacteria, fungus, and colorless cyanophyta. However, due to the low amounts of chemicals in wastewater, osmotrophic protozoa are unable to effectively compete with bacteria. Decomposers, primarily bacteria, make up around 95% of the activated sludge's microbial population. Another type of microorganism is consumers, which consume bacteria and other microorganisms as substrates. This collection of phagotrophic *protozoa* and microscopic *metazoa* is part of the activated sludge microfauna. Microfauna (only 5% of the activated sludge's microbial population) have a minor role in removing organic contaminants and nutrients.

- **Oxic Organotrophic Microorganisms**
  It has been observed that exo- and endoenzymes from the genera *Bacillus*, like—*Pseudomonas, Alcaligenes, Moraxella*, and *Flavobacterium* may break down complex organic substrates. After the mixed culture has been properly adapted to a particular wastewater, bacteria adapted to a particular substrate can be concentrated in activated sludges. Examples of such dedicated bacteria are *Achromobacter spp.* (lipids, acids, and alcohols) and *Proteus spp.* (proteinaceous materials). Along with colorless *cyanophyta* (*cyanobacteria*) and microscopic fungi (*micromycetes*), organotrophic bacteria are active in the oxic oxidation of organic substrates, particularly saccharidic and polysaccharidic substances.
- **Anoxic Organotrophic Microorganisms (Denitrifiers)**
  The activated sludge bacteria are capable of using nitrate-nitrogen as the final electron acceptor in biochemical reactions. At least 40 different types of aquatic microbes have the ability of denitrification. The usual genera of activated sludge organotrophic denitrifiers include *Achromobacter, Arthrobacter, Alcaligenes, Bacillus, Moraxella, Flavobacterium, Hypomicrobium*, and *Pseudomonas*. According to Grabinska-Loniewska, 82–97% of the microbes in activated sludge of systems with an anoxic zone are capable of denitrification. Nitrate respiration, or the initial stage of denitrification, is the only function carried out by activated sludge fungi (Grabínska-Łoniewska 1991; Toerien et al. 1990).
- **Fermentative Bacteria**
  The EBPR (Enhanced Biological Phosphate Removal) mechanism depends critically on the fermentative conversion of organic molecules to volatile fatty acids. In this regard, the literature emphasizes the existence of the genera *Pasteurella* and *Alcaligenes* as well as *Aeromonas punctate* (Grabínska-Łoniewska 1991).
- **Nitrifier Bacteria**
  The nitrifying bacteria are primarily the soil microorganisms. The literature includes the following genera of nitrifiers for aquatic environments:

  - For the oxidation of ammonia—*Nitrosomonas, Nitrosococcus, Nitrosospira*, and *Nitrocystis* and
  - For nitrite to nitrate conversion in the final step—*Nitrobacter, Nitrospina*, and *Nitrococcus*.
    The chemolithotrophic bacteria, *Nitrosomonas* and *Nitrobacter* are regarded as the primary nitrifiers in the activated sludge process (Eckenfelder 1980; Randall et al. 1992).

- **Polyphosphate Accumulating Microorganisms**
  The genus Acinetobacter, which fluorescent antibody techniques used to identify in the isolates from EBPR activated sludges, is responsible for the EBPR mechanism's ability to extract phosphate from wastewater. Other microbes that

could contribute to the EBPR and possess polyphosphate granules come from the genera *Aeromonas, Moraxella, Arthrobacter, Pseudomonas,* and *Klebsiella.* The bacteria that accumulate polyphosphate are commonly referred to as poly-P bacteria in the literature on the activated sludge process. Some poly-P bacteria appear to have the potential to denitrify, or to take up phosphate both in oxic and anoxic environments. Anoxic phosphate uptake is the term used to describe the concomitant denitrification and phosphorus removal process (Cloete and Steyn 1988).

- **Sulphur Bacteria**
  The colorless filamentous bacteria, *Beggiatoa* and especially *Thiothrix*, which are among the many sulphur bacteria, are crucial to the process of creating activated sludge since they can result in bulking issues. It appears that *Thiothrix* is a mixotrophic species.
- **Microfauna of activated sludges**
  The following groups of *protozoa* and *metazoa* are the microfauna of activated sludges:

  i. *Protozoa*: *Flagellates, rhizopods*, and *ciliates* (free swimming, crawling or grazing, attached or stalked) and
  ii. *Metazoa*: *Nematodes, rotifers,* and higher microfauna.
  In biological wastewater treatment, this microfauna is responsible for—enhanced growth of bacterial floc, thorough uptake of substrates, and elimination of dispersed bacteria by adsorption and predation onto the protozoan metabolites.

2.4.7 Granulation Process of the Activated Sludge

Granulation of activated sludge is the outcome of interactions between microbes and sludge particulates that are both biotic and abiotic. This process produces tiny, spherical aggregates with a diameter of 1–3 mm in which the microbial cells are anchored in an EPS matrix. Though many granulation mechanisms have been put forward, no agreement has been reached. Moreover, it is likely that these mechanisms participate in the granulation concurrently and affect the processes separately rather than acting solely. According to literature, filamentous fungus and stalked protozoa are crucial for the conformation of the granular structure, increasing the surface area accessible for bacterial attachment. Granulation proceeds through the following steps (Liu and Tay 2002):

1. Contact between cell-to-cell,
2. Microbial aggregation due to attractive forces between cells,
3. Maturation of the aggregates by forming a matrix of EPS, and

4. Formation of a 3-D structure shaped by the hydrodynamic forces and the microorganisms involved.

Alternatively, granulation is the outcome of dynamic floc/particle aggregation and breakup or micro-colony expansion. Each stage in the production of granules is intricate and impacted by several physical, chemical, and biological mechanisms (Zhou et al. 2014). Different forces and characteristics of the biomass, such as hydrodynamic forces, diffusion, cell mobility, and cell surface features, influence the early phases of the granulation of sludge (Wan et al. 2015). There are several steps involved in a successful granulation process which are described below.

2.5 Cell-To-Cell Contact and Micro-Aggregation

The potential of microorganisms for developing aggregates, which are governed by cellular mechanisms and their physical and chemical properties, is crucial for cell-to-cell interaction and aggregate formation. Hydrophobicity of the cell surface plays a crucial role in the beginning of granule formation. The biomass tends to become more hydrophobic as it granulates. The negative surface charge of granules would be lowered if the protein/polysaccharide ratio was increased, which should lessen electrostatic repulsion between bacterial cells and improve agglomeration (Lin et al. 2010). The cell modifications make it easier for bacteria to anchor together. It is proposed that exopolysaccharides produced by negatively charged bacteria adhering to an inorganic core of calcium and phosphate precipitates can help to promote microbial aggregation. Granules can withstand greater shear forces, but they appear to follow the same colloidal interactions and mechanisms that govern cell aggregation in activated sludge. These factors encompass, but are not confined to, DLVO-type interactions, cation-mediated EPS bridging, hydrophobic interactions, cell surface charge, and water phase surface tension (Tan et al. 2014).

2.6 Extracellular Polymeric Substance (EPS) Production

Methods for evaluating the exopolysaccharides must be improved to distinguish between the various polymers and their interactions within the aggregate in order to comprehend the dynamics of granulation. In a study on the effects of various EPS components on the structural integrity of phenol-fed granules, it was reported that hydrolysis of ß-polysaccharides led to granule disintegration while hydrolysis of extracellular proteins, lipids, and ß-polysaccharides seemed to have no effect. Exopolysaccharides or glycosides have been found to be significantly more adhesive gel-forming agents in flocs than EPS in activated sludge. Exopolysaccharides that resembled alginate were recovered by Lin et al. from aerobic granules grown at a pilot-scale (Wu et al. 2017). With calcium chloride, the exopolysaccharides exhibited gel-forming capabilities, and it was hypothesized that they greatly influenced the granules' hydrophobicity and elastic structure.

2.7 *Cell-To-Cell Communication*

Quorum sensing is a vital factor involved in granulation (QS). QS controls a wide range of bacterial processes. The significance of QS for granulation and granular stability has been demonstrated by recent investigations. For example, it has been demonstrated that granular sludge contains more acyl-homoserine lactones (AHL), a common autoinducer in Gram-negative bacteria, than floccular sludge. AHL concentrations increase with granulation. An increase in QS function during granulation appears to be associated to an increase in the generation of gel-forming EPS, which has a higher hydrophobicity and is implicated in granule aggregation and stability. A strong positive correlation between EPS production, community composition changes, and AHL concentrations was observed when a sequencing batch reactor (SBR) was inoculated with floccular sludge and fed with synthetic wastewater, whereas a correlation between granular disintegration and reduction of AHL content was noticed. Exogenous AHLs that increased the extracellular proteins and autotrophic biomass might be added to nitrifying sludge dominated by Nitrosomonas to improve granulation (Tan et al. 2014; Wu et al. 2017). Quorum quenching (QQ), which suppresses QS, has also a significant effect on granulation. When granulation took place in an SBR inoculated with activated sludge treating synthetic wastewater, QQ was a key controller of QS. In particular, during granulation, a change in the population was seen, with a larger percentage of QQ active bacteria in the floccular sludge and a higher percentage of QS active bacteria in the granules. When Proteinase K was added, the breakdown of extracellular proteins and the consequent dispersion of granules occurred. It likely resulted from the EPS matrix collapsing at the same time when AHL and autoinducer-2 (AI-2) concentrations dropped (Wilén et al. 2018).

2.8 *Granule Development*

After the microbial aggregates have grown, they expand in dimension and the substrate gradients in the granule provide various ecological niches, enabling the coexistence of a diversified population with a variety of uses in wastewater treatment. Operational factors include the kind of substrate, organic loading rate (OLR), COD/N ratio, food-to-microorganism (F/M) ratio, settling duration, solids retention time, and redox conditions that influence the microbial composition of granules. AGS is said to have a higher microbial diversity than floccular sludge because the substrate gradients within the aggregate create more biological niches. Similar functional subgroups of microbes are found in sludge that is both granular and floccular, but in different amounts between a phylum or class between phylogenetic groups level. Typically, AGS systems are used for COD, COD, and N, or the sequential elimination of COD, N, and P. Liu et al. discovered that it took 400 days to generate granules, still with some floccular sludge present, after inoculating a pilot plant fed with genuine wastewater and operated with COD and N removal. There was no significant microbial selection for specific groups of microorganisms that formed granules (Winkler et al. 2013; Fan et al. 2018). Using identical reactors to process alternatively domestic wastewater and synthetic wastewater with the same COD and N load, Fan et al. compared the granulation process. In contrast to the seed sludge, Illumina sequencing revealed a rapid decline in bacterial diversity during granulation and a distinct bacterial community structure in the granules. Strikingly, the microbial community was similar for the two reactors. *Arcobacter*, *Aeromonas*, *Flavobacterium*, and *Acinetobacter*, minor genera in the seed sludge, became dominant in the granules and ammonium-oxidizing archaea (AOA) were gradually washed out, whereas ammonium-oxidizing bacteria (AOBs) and nitrite-oxidizing bacteria (NOBs) were retained (Fan et al. 2018). When turning activated sludge into AGS, the microbial community structure in a full-scale facility that removes COD, nitrogen, and phosphorus demonstrated that the richness of *ß-proteobacteria*, *ß-Proteobacteria*, *Flavobacteria*, and *Cytophagia* increased (Wilén et al. 2018).

2.9 Granule Size Increase and Microbial Stratification Within the Granules

Granule size development is largely unpredictable and depends on a complex interaction of various environmental factors. As a result of the process parameters, granules appear to obtain a particular, more or less stable granule size that is balanced between granule growth, attrition, and breakage. Microorganisms develop both on the granule's exterior and within, which causes variations in shape and size. Bacteria are not evenly distributed throughout the granules since some are more prevalent at the granule's outer layers and others are on its inside. The granular structure is frequently simplified theoretically and mathematically by envisioning granules as a multilayer sphere with diminishing oxygen and substrate gradients from the surface to the core (Fan et al. 2018). This means that, in accordance with these ideas, nitrifiers are found in the oxygen-penetrated outer layers, whereas denitrifiers and phosphate-accumulating organisms (PAOs) are found in the interior layers, as shown in Fig. 2 of an aerobic granule, having several conversion processes for organic matters, nitrogen and phosphorous within separate redox zones.

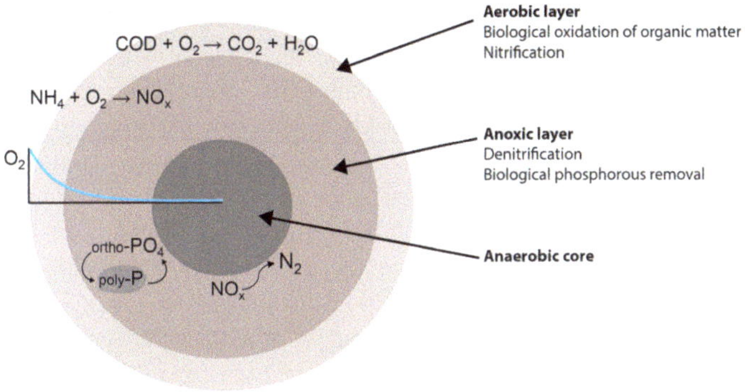

Fig. 2 Schematic representation of an aerobic granule in ASP (Wilén et al. 2018)

The mature granules that removed COD and N contained *Rhodocyclaceae* in their cores, while the outside shell of these granules also contained both aerobic and anaerobic strains. Beginning with *Flavobacteriaceae, Microbacteriaceae, Rhodobacteraceae*, and *Xanthomonadaceae* anaerobic strains became more prevalent as the granulation progressed. Furthermore, nitrifiers formed inside the granule in channels and voids as well as at the oxygen-rich surface utilizing fluorescence in situ hybridization (FISH) and confocal laser scanning microscopy. The presence of AOBs in the interior regions of the granules suggests that the channels are used to transfer both oxygen and ammonia across the granule. Three distinct rates of organic loading produced these findings. It is no surprise that aerobic granules have considerable structural heterogeneity given that biofilms are known to be heterogeneous structures and typically comprise holes, channels, mushroom-like structures, and water-filled spaces. Complex three-dimensional structures of granules were demonstrated using microscopy in conjunction with fluorescence staining methods and image analysis, and it was postulated that granules grow from the inside via outgrowth to create granules rather than aggregation of small microbial colonies (Gonzalez-Gil and Holliger 2014). Shear stresses make it easier for bacteria growing near the granules' surface to become separated than those growing inside.

2.10 Granular Stability

The concentration of suspended solids in the effluent is a crucial factor in the operation of AGS processes because granules have a lower ability to flocculate (to clarify the effluent during sedimentation) than activated sludge, which may result in higher concentrations of suspended solids in the effluent. As a result, additional polishing stages like filtration are required to achieve the effluent limits, especially if tight phosphorus requirements are in place. To maintain a low concentration of suspended solids in the effluent, granule stability is essential. Within the granule, microorganisms compete not just for substrates but also for room to thrive. Different zones are formed by the diffusion of substrates and dissolved oxygen; aerobic zones are created near the granule's outer surface, where fast-growing heterotrophs and filaments thrive, and anoxic and anaerobic zones are created farther within, toward the center. Particles and colloids in the wastewater and EPS production can clog channels that could help with improved diffusion. In the center of granules, starvation, and anaerobic circumstances trigger endogenous cell respiration and cell lysis, which ultimately result in hollow cavities and granule disintegration. Varied OLRs (organic loading rates) produce different rates of biomass growth in the reactor, which may have an impact on the concentration of suspended particles in the effluent. Li et al. noted that the effluent-suspended solids concentration was higher at higher OLR during the startup of three AGS reactors operated at different OLRs, but as the granules matured, the difference in effluent-suspended solids concentration decreased, indicating that the food-to-microorganisms (F/M) ratio plays a crucial role (Winkler et al. 2013).

3 Performance of Activated Sludge Process

3.1 Biological Nutrient Removal

The charge imbalance between the activated sludge floc and the sewage colloids would cause an electrostatic attraction, which could then lead to subsequent purification. According to the "bi-zeolite theory," micro organics were removed through a series of exchange reactions with activated sludge. Iron compounds are crucial for the oxidation process. Rudophs et al. also made reference to Lumb's experiment, in which he investigated the balance of solids removal by activated sludge under laboratory conditions. He came to the conclusion that removal was primarily a physical process because all of the suspended and colloidal matter was recovered as activated sludge. According to a number of studies, enzymatic processes acted as a conduit for the reduction of organic materials (Kocaturk and Erguder 2016). By including an anaerobic phase, simultaneous biological nutrient removal (SBNR) can be achieved. Because phosphate-accumulating organisms (PAOs) are enriched in the presence of volatile fatty acids during the anaerobic phase, elimination of Phosphorous can be improved even more by adding a prolonged anaerobic filling phase. By lowering the DO content and extending the anaerobic feeding phase, PAOs and GAOs were chosen in the study. This process allowed for the conversion of all acetate into storage polymers that degrade more slowly. While glycogen-accumulating organisms (GAOs) predominated in the presence of phosphorus and PAOs in the absence of it, there was no change in how the granules looked. According to the findings, slow-growing microorganisms develop at the surface and require less shear to create dense aggregates at low DO than they do at high ORL (Winkler et al. 2013).

3.2 Nitrogen Removal

The abrupt wash-out of biomass at AGS reactor startup may result in decreased nitrogen removal, lower SRT, and carbon leakage from the anaerobic feeding phase into the aerobic phase. As a consequence, zoogloeal populations and filament populations overgrow. To get effective nutrient elimination, it frequently takes 2–4 months. AOB and NOB are examples of slow-growing bacteria that are susceptible to decreased SRT. The microbial community has reportedly changed rapidly during startup. After 10 days of operation with practically complete nitrification, a progressive reduction in the settling time allowed for quick granulation and a quick start to the nitrification process. The biomass was dominated by granules. Winkler et al. found a difference between the microbial populations of an AGS pilot plant and a full-scale activated sludge plant operating at the same temperature and HRT. The nitrifying granules contained both *Nitrosomonas* and *Nitrosospira*, whereas the predominating AOB in the seeding sludge was *Nitrosomonas*. The microbial communities varied, but they shared a comparable diversity and evenness, indicating that the various communities served the same functional needs. It is crucial to maintain control over DO in order to achieve instantaneous nitrification and denitrification (SND). DO must be sufficient to support nitrification at the outer layer and small enough to protect oxygen from permeating into the deeper anoxic parts of the granule where denitrification can occur (Winkler et al. 2013; Fan et al. 2018). For effective SND, the COD/N ratio is crucial because it exerts substantial selection pressure on either the enrichment of heterotrophs or nitrifying bacteria. Kocaturk and Erguder demonstrated that high COD/N values (7.5–30) resulted in large, fluffy granules that favored heterotrophic growth, while low COD/N values (2–5) produced small, dense granules that contained more slowly growing nitrifiers. They came to the conclusion that a ratio of 7.5 is ideal for sustainable granules (Wilén et al. 2018).

3.3 Phosphorous Removal

Boosting the formation of PAOs is crucial in AGS systems intended for phosphorus removal. The granules will be more stable even at low DO concentrations by choosing slow-growing microorganisms like PAOs. PAOs are more prevalent in activated sludge at lower temperatures (10 °C), whereas GAOs predominate at higher temperatures (20–30 °C). After settling, stable phosphorous removal at high temperatures (30 °C) could be achieved by selectively removing biomass from the top of the AGS bed because, according to FISH measurements, the lighter top fraction of the granules was more abundant in GAOs. The denser bottom fraction of the AGS bed, on the other hand, contained more PAOs. It has been discovered that the stability of the phosphorus removal is influenced by the carbon supply.

In a study, *Zoogloea, Acidovorax*, and *Thiothrix* dominated the propionate-fed reactor during the first 50 days of operation, whereas *Thiothrix* predominated the acetate-fed reactor. The propionate-fed reactors operated effectively when the biological phosphorus removal activity began, whereas the acetate-fed reactor was unsteady. *Accumulbacter, Competibacter, Chloroflexi,* and *Acidivorax* were present in both reactors. The two reactors included various varieties of Accumulibacter, as determined by T-RFLP based on the ppk1 gene (a polyphosphate gene that can be used as a phylogenetic marker) (Wilén et al. 2018).

4 Types of Activated Sludge Process Plants

Numerous variations have been made to the activated sludge process (ASP). ASP modifications have frequently been applied to improve process biochemistry or other aspects to generate a more stable operation at a lower cost. Many factors influence the type of process modification used in a given project, including treatment objectives, site constraints, operational constraints, client preferences, and design engineer preference and experience (Water Environment Federation 1994). Plant capacity and function are likely among the most critical aspects to consider while deciding on variations (Eckenfelder et al. 2020). Loading rates, reactor design, feeding and aeration patterns, and other criteria, such as multiple biological nutrient removal processes, can be used to classify the process. Volumetric loading rate, mean cell residence time (MCRT), or F: M are all ways to express the loading rate (Water Environment Federation 1994). Reactors in series are often used with aerobic, anoxic, and anaerobic conditions (Metcalf and Eddy et al. 2013). Hydraulic properties of the process are the primary focus of reactor design (Water Environment Federation 1994). Plug-flow (PF), complete mix (CM), and sequencing batch reactors (SBRs) are the most common types of reactors used in activated sludge processes (Metcalf and Eddy et al. 2013). Several variants of the completely mixed and PF systems are frequently utilized in practice (Tolliver 2016). The most popular choices for small plants about 3 million gallons per day (MGD) or less are oxidation ditches and SBRs. With these plants, no primary treatment is usually used. Some extended aeration and contact stabilization plants are also considered to be small plants. The conventional activated sludge method appears to be the best option for medium plants with a capacity of (3–50 MGD). Over 50 MGD plants often employ the same designs as medium-sized plants, with step aeration being a popular choice (Eckenfelder et al. 2020).

4.1 Plug-Flow

PF reactors are very lengthy biological reactors with narrow aeration tanks (Riffat 2012; Water Environment Federation 1994). The fluid particle entering the system in an ideal plug-flow system will circulate uniformly throughout the reactor without spreading in the fluid (Water Environment Federation 1994). PF can also be replicated by folding rectangular reactors more significantly than a 10:1 length-to-width ratio

(Metcalf and Eddy et al. 2013; Water Environment Federation 1994). The oxygen utilization rate is high at the start of the basin and decreases with aeration duration. These conditions generally encourage good floc formation and filamentous growth control (Eckenfelder et al. 2020). When industrial wastes were introduced, the usage of a plug-flow technique became problematic due to the hazardous consequences of certain discharges (Metcalf and Eddy et al. 2013). In fact, due to longitudinal dispersion induced by aeration and mixing, a pure PF regime is impossible to maintain (Riffat 2012).

4.2 Complete Mix

The CM process is also a completely mixed stirred tank reactor (CSTR) (Peters 2011). In the completely mixed reactor, mixing happens immediately and is finished (van Haandel 2012). Completely mixed reactors are typically single tanks with a square cross-section and a flat or slightly sloping base in the center. Typically, the tank has a liquid depth of 3.0–6.0 m and a freeboard of 1.0–1.5 m to confine spray (Gray 1992). This method is beneficial for wastewaters that contain hazardous or bio inhibitory chemicals or have highly variable loading patterns. Another benefit of a thorough mix procedure is that the oxygen uptake rate in the basin is equalized (Eckenfelder et al. 2020). However, the mixing regimes and tank shape of the PF and CM activated sludge processes are considerably different (Metcalf and Eddy et al. 2013). The PF system will be more efficient if the same substrate removal in a PF reactor and a CM reactor. In addition, the quality of the effluent from these two systems may be the same because they both make compounds that break down slowly (Droste and Gehr 1997). Table 1 shows typical parameters for the design and operation of several activated sludge processes.

4.3 Extended Aeration

Prefabricated units, often known as package plants, are used in extended aeration systems (Ong 2007). Sequencing batch reactors, contact stabilization plants, complete mix, and rotating biological contactors are also considered package plants (U.S. EPA 2000; Kerri 2008). Package plants are prefabricated wastewater treatment facilities used in small towns or on private properties to treat wastewater. Package plants can be constructed to treat flows as low as 0.002 MGD or as high as 0.5 MGD. According to manufacturers, they are most typically used to treat flows between 0.01 and 0.25 MGD (Spellman 2013; U.S. EPA 2000). The extended aeration process is a variation of the activated sludge process that offers biological treatment to remove biodegradable organic wastes under aerobic conditions (U.S. EPA 2000). The process is a continuous flow process in which the aeration basin's contents are CM (Droste

Table 1 Typical design and operation parameters for activated sludge process variations (Metcalf and Eddy et al. 2013; Mines 2014; Tolliver 2016; Qasim and Zhu 2017)

Process variation	Reactor	SRT (d)	F: M, kg BOD/kg MLVSS.d	Volumetric loading		MLSS, mg/L	HRT (θ), h	RAS ratio	Air supplied m^3/Kg BOD_5	BOD_5 removal efficiency (%)
				lb BOD/ 1000 $feet^3$.d	kg BOD/ m^3.d					
Conventional	PF	3–15	0.2–0.4	20–40	0.3–0.7	1,000–3,000	4–8	0.25–0.75	45–90	85–90
Complete mix	CM	3–15	0.2–0.6	20–100	0.3–1.6	1,500–4,000	3–6	0.25–1.00	45–90	85–90
Step feed	PF or staged	3–15	0.2–0.4	40–60	0.7–1.0	1,500–4,000	3–5	0.25–0.75	45–90	85–90
Extended aeration	CM or PF	20–40	0.04–0.1	5–15	0.1–0.3	2,000–4,000	20–30	0.50–1.50	90–125	75–90
Contact stabilization	CM or PF	5–10	0.2–0.6	60–75	1.0–1.3	1,000–3,000[a] 6,000–10,000[b]	0.5–1[a] 2–4[b]	0.50–1.50	45–90	80–90
Oxidation ditch	CM + PF	15–30	0.04–0.1	5–15	0.1–0.3	3,000–5,000	15–30	0.75–1.50	–	–
Deep shaft	PF	2–4	0.75–1.25	–	5.6–8.0	7,000–12,000	0.5–0.75	0.2–0.5	–	–

Remarks, a-MLSS and hydraulic retention time in Contact basin; b-MLSS and hydraulic retention time in Stabilization basin

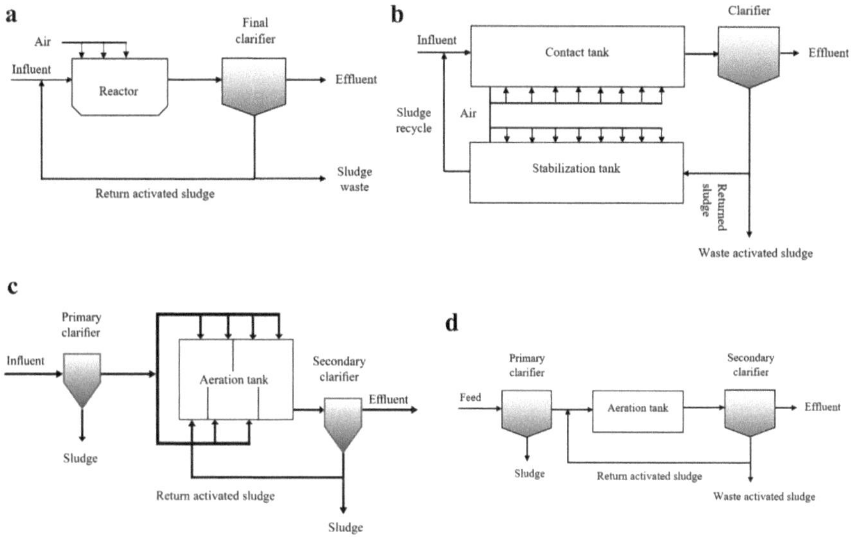

Fig. 3 Schematic diagram of **a** Extended Aeration; **b** Contact Stabilization; **c** Step Aeration; **d** Conventional Activated Sludge

and Gehr 1997). In extended aeration-activated sludge systems, low organic loadings, high biological solids concentrations, and long aeration periods are standard. A shallow earthen oval ditch with a horizontal brush aerator provided aeration and circular mixing in the plants (Mccarty and Brodersen 1962; Fuller 1927). Set concrete aeration basins are the most common. However, tanks made of steel or concrete can also be used. Mechanical aerators or diffused air are the two most common methods of achieving continuous complete mixing (Hammer and Hammer 2013). In most cases, primary clarification is avoided. Secondary clarifiers are designed to operate at lower hydraulic loading rates than activated sludge clarifiers (Metcalf and Eddy et al. 2013). The method is adaptable and can be utilized in circumstances where nitrification is required (Ong 2007). A schematic flow diagram of the extended aeration system is shown in (Fig. 3a).

4.4 *Contact Stabilization*

The development of the Contact Stabilization is also known as the biosorption process (Kayser 2008). The adsorptive properties of activated sludge were exploited in the development of biosorption (Ong 2007). The contact stabilization modification of the activated sludge process was made to help treat wastewater with many complex colloidal or soluble substrates (Richard O. Mines 2014). Contact stabilization includes two distinct tanks or compartments and a secondary clarifier to treat wastewater and stabilize activated sludge (Fig. 3b) (Fränzle et al. 2012; Metcalf and

Eddy et al. 2013; Mines 2014). The first step occurs in the "contact" tank, while the second takes place in the "stabilization" tank (Arceivala 1986). The adsorption and oxidation processes are carried out in separate tanks, respectively (Gray 1992). In a contact zone, the stabilized activated sludge is combined with the influent, which can be either raw or settled wastewater (Metcalf and Eddy et al. 2013; Fränzle et al. 2012). The contact zone is responsible for the rapid removal of soluble BOD. In contrast, colloidal and particle organics are collected in the activated sludge floc and degraded subsequently in the stabilization zone, according to the manufacturer. Return activated sludge (RAS) is aerated in the stabilization zone, and the detention time is in the order of 1–2 hours to maintain an adequate solids retention time (SRT) for sludge stabilization (Metcalf and Eddy et al. 2013). The contact process is comparable to the conventional activated sludge process in terms of performance efficiency but requires significantly less reactor volume (Arceivala 1986). Contact stabilization was first used at large coastal plants to remove 60–75% of BOD_5 from the water. It's made to expand current systems and package plants and is used in industrial waste application systems (Ong 2007).

4.5 Step Aeration

Step or step-feed aeration depicted in (Fig. 3c) is a traditional plug-flow procedure variation. Settled wastewater is fed at 3–4 feed locations in the aeration tank to equalize the F/M ratio and reduce peak oxygen demand (Metcalf and Eddy et al. 2013). This type of variation is referred to as a high-rate process (Droste and Gehr 1997). Most of the time, three or more parallel channels are used. A vital characteristic of this process is its adaptability, as the wastewater feed allocation may be adjusted to fit operational conditions (Metcalf and Eddy et al. 2013). Step aeration provides uniform air supply, while wastewater is fed at intervals or steps along the tank's first section (Hammer and Hammer 2013). This process shortens the time for the activated sludge to be in the mixed liquor. The process is used for a broad spectrum of applications (Ong 2007). Kraus process is a variation of the step aeration process that is used for waste that doesn't have a lot of nitrogen in it (Qasim and Zhu 2017).

4.6 Conventional Activated Sludge

Most conventional activated sludge (CAS) systems now run continuously, and they can be classified as either plug-flow or complete mixed systems according to their mixing regimes (Fuller 1927; Gray 1992). A primary clarifier and an aeration tank with air diffusers mix the activated sludge and wastewater with dissolved oxygen in a CAS process. There is also a secondary clarifier for solids removal and a sludge-return line from the clarifier bottom back to the aeration tank's inlet (Okafor 2011; Ong 2007). The regular aeration tank width is 8–10 m. Due to the space constraints imposed by the narrow width and the usage of liquid depths of 5–6 m (16–20 feet), tanks had to be at least 100 m long (330 feet). There usually are 3–4 channels in the aeration tank where the settled wastewater and RAS enter before flowing to the secondary clarifier (Metcalf and Eddy et al. 2013). Wastewater is aerated in long rectangular aeration basins in CAS systems for six to eight hours. For every m^3 of treated wastewater, 8 m^3 of air is provided. The sludge is kept suspended with the help of plenty of air. Adsorption, flocculation, and oxidation of organic materials occur during the aeration period (Fränzle et al. 2012). Figure 3d is depicted in the CAS process flow diagram.

4.7 Complete Mix Activated Sludge

Complete mix or completely mixed activated sludge (CMAS) techniques are the most commonly used, as the concentration of constituents inside the reactor is identical to the attention in the effluent. Primary clarifiers are often removed from the process treatment train (Mines 2014). The organic load on the aeration tank, the concentration of mixed liquor suspended solids (MLSS), and the oxygen demand is the same throughout the tank. The CMAS procedure is pretty simple to implement and maintain (Metcalf and Eddy et al. 2013). With careful selection of aeration and mixing equipment, the process should almost thoroughly mix the food. The oxygen demand is also the same across the tank (Ong 2007). To accomplish the CMAS, diffuse aeration or mechanical aerators might be used (Mines 2014). Complete mix techniques mitigate the effects of hydraulic and biological shock loadings that are frequently experienced during the process. However, the complete mix process has become more popular for cleaning industrial wastewaters. The process can be used for many things, but it can be harmed by filamentous growths (Ong 2007).

4.8 Oxidation Ditch

The oxidation ditch was invented in the 1950s at the Netherlands Research Institute for Public Health Engineering as a low-cost and straightforward technique for treating raw sewage from local communities and industries. However, in 1954 the first oxidation ditch plant was apparently constructed in Voorshoten, Holland (Ettlich and Evans 1978; USEPA 2000). When water flow is 2,000–20,000 m^3/d, oxidation ditches are also known as continuous loop reactors (CLRs) (Peters 2011). The oxidation ditch is a modification of the practice of extended aeration (Mines 2014). Although most oxidation ditches are complete mix systems, they can be changed to approach plug-flow conditions. Typical oxidation ditch treatment systems have single or multiple channel arrangements within a ring, oval, or horseshoe-shaped basin. Subsequently, oxidation ditch reactors are called “racetrack type” reactors (Spellman 2013; USEPA 2000). Oxidation ditches are usually made of reinforced concrete, but gunite, asphalt, butyl rubber, and clay have also been used (USEPA 2000). Wastes move around an oval channel 1.5–2 m deep at about 0.3–0.4 m/s with 45-degree sloping sidewalls (Cairncross and Feachem 2018; Ong 2007). The wastewater is placed into the ditch and aerated with mechanical or surface aerators (brush or disk type) that are mounted across the channel for a long time with a long hydraulic retention time (HRT) (Ong 2007; Peters 2011). Aerators are used to mix and circulate the water in the ditch, as well as to transport oxygen (Ong 2007; USEPA 2000). Although primary settling before an oxidation ditch is sometimes used, it is not common in this configuration. Depending on the effluent needs, tertiary filters may be necessary after clarification. Before final discharge, disinfection, and re-aeration may be required (Spellman 2013). To achieve partial denitrification, an oxidation ditch could be used. The Modified Ludzack-Ettinger (MLE) process is one of the most popular design changes for increased nitrogen removal (Spellman 2013; USEPA 2000). Nowadays, the oxidation ditch is famous because it removes a lot of BOD_5 (85–95%). It is also effortless to operate (Ong 2007).

4.9 Deep Shaft

The Agricultural Division of Imperial Chemical Industries (ICI) in the United Kingdom designed the deep shaft wastewater treatment process (Okafor 2011). The deep shaft process, also known as the vertical shaft bioreactor, is a high-rate activated sludge process and bioreactor that is used in areas where land is expensive (Gerba and Pepper 2015; Ong 2007). It is an aerobic biological technique for the treatment of basement wastewater (Qasim and Zhu 2017). In Japan, there are some installations. It must have been produced in the United States and Canada (Ong 2007). The deep shaft reactor is made up of a vertical shaft that is about 120–150 m or 400–500 feet deep. It has a U-tube aeration system. In place of primary clarifiers and aeration tanks, a shaft is used. An annular reactor is formed by lining the deep shaft with a steel shell and equipping it with a concentric pipe. Air is driven down the middle of the shaft and allowed to rise upward through the annulus, returning activated sludge (Fränzle et al. 2012; Ong 2007). Although there is a high degree of mixing in the process, there is little dispersion or reverse mixing, resulting in a plug-flow process (Gray 1992). A flotation tank is used to separate the biomass from the liquid. Most of the solids are sent back to the deep shaft reactor. Some are sent to an aerobic digester (Ong 2007). The advantages and disadvantages of various activated sludge processes are shown in (Table 2).

Table 2 Advantages and disadvantages of various activated sludge process variations (Lakatos 2018; Spellman 2013; Metcalf and Eddy et al. 2013)

Process	Advantages	Disadvantages
Conventional	• Widely applicable in a variety of contexts • Proven method • Process and structural design are well-known • Adaptable to a wide range of operating schemes, such as step feed, selector design, and anoxic or aerobic processes	• In the initial pass, it may be difficult to match oxygen delivery to oxygen demand • Capital and operating costs are moderate • Moderate ability to settle sludge
Complete mix	• A well-established process • Suitable for many types of wastewater • Excellent resistance to toxic shock loads compared to others • All sorts of aeration devices can use it	• Susceptible to the growth of filamentous sludge • Capital and operating costs are moderate
Step Feed	• Peak wet weather flows might be pushed until the final pass to reduce high clarifier solids loading • Operational flexibility • In comparison to CAS, capital costs are lower	• More sophisticated process and aeration system design • In most cases, flow split isn't measured or known very well
Extended Aeration	• Plants are easy to run because many are even manned for 2–3 hours a day • Systems are odorless and can be placed in a wide variety of locations • The systems are simple to install	• Energy is needed for a longer aeration time • Large aeration tanks and space are needed
Contact Stabilization	• Requires a smaller volume of air • Flows through wet weather with no loss of MLSS	• The more complicated operation • Reduced efficiency of nitrification
Oxidation Ditch	• Produces less sludge than other types of biological treatment • Possibility of high-quality effluent • The process that is both trustworthy and easy to use • It consumes less energy than other sorts of processes	• Concentrations of suspended particles in the effluent are relatively high • Large structure and more space needed • More energy is required for aeration

5 Aeration System

The conventional activated sludge (CAS) process uses aeration to provide the dissolved oxygen (DO) needed for the microbiological communities to break down the organic materials in wastewater, assuring their maintenance and proliferation. Since membrane filtration has replaced gravitational sedimentation in MBRs, which are activated sludge systems, this makes aeration systems a crucial component of both CAS plants and membrane bioreactor (MBR) systems. In addition to keeping solids suspended, it also reduces membrane fouling and enhances membrane cleaning in MBRs. One of the most energy-intensive processes, aeration uses almost 50% to up to 90% of the electricity required by a wastewater treatment plant. Additionally, its price ranges from 15 to 49% of the plant's overall cost. The DO content in the aeration tank regulates the effectiveness of treatment in CAS. Low DO concentrations cause lower effluent quality since bacteria don't thrive as quickly. Low DO levels (less than 1.1 mg O_2 L^{-1}) also promote the growth of filaments, which links between the most prevalent filamentous microbes and the environment. At present, different kinds of aeration devices are being adopted in CAS to assure optimum access of microorganisms to oxygen, like—mechanical surface aeration (the most frequent aeration method), diffused aeration (coarse, medium, and fine bubbles), and high purity oxygen aeration (less prevalent) (Skouteris et al. 2020; Drewnowski et al. 2019).

5.1 *Surface Aerators*

Up to the 1990s, surface aeration using Kessener brushes or rotor aerators was very common. The most widely used aerators today are surface horizontal or vertical shaft aerators, air pump systems, or brush aerators. An essential component of these devices is the blades that produce turbulence on the liquid surface when using the activated sludge method. Certain biological contact oxidation ditches employ rotating blades, rotors, or brushes. Aeration and, ultimately, biological processes taking place in wastewater are made easier by creating a turbulent flow. Only when modernizing treatment plants with shallow tanks (depth up to 3.5 m) or circulation ditches are such options now used. Many studies revealed their significant energy demand, since the associated energy consumption frequently exceeds 0.7 kWh/m^3 of effluent.

5.2 *Diffusion Aeration*

The most common approach in practical applications is aeration using blowers (because to the accessibility, efficiency, and dependability of blowers). Pipes carry the compressed air from blowers to the diffusers, which are typically found close to the bottom of a biological reactor and are responsible for introducing air to wastewater. A crucial factor in this procedure is the size of the air bubbles. Their size determines how much oxygen will transfer to wastewater via which a bubble advances toward their surface, as demonstrated on several times. When a bubble is expelled from a diffuser, the smaller it is, the slower it will rise to the surface and the longer it will stay in the wastewater; as a result, it will have more time and more oxygen to pass.

5.3 *Pure Oxygen Aeration*

In CAS, PO was first suggested as an alternative to air in 1940, but it wasn't employed commercially in the United States till the 1970s. In 1968, the first PO activated sludge (POAS) plants for treating municipal sewage were established. Because PO has a partial pressure that is 4.7 times greater than that of oxygen in the atmosphere, it enhances the driving power for oxygen transfer and the level of oxygen saturation that is feasible. Even when high strength or toxic wastewaters need to be treated, it increases OTR and maintains high DO concentrations at decreased flowrates. Higher gas phase oxygen concentrations, improved biokinetics, and faster treatment rates at higher MLSS concentrations and shorter hydraulic residence periods are all provided by PO in comparison to air (HRTs). Systems using PO aeration are straightforward and small, and they make it simple to store and handle gas. They regulate foul condensates without draining, which lowers odor and volatile organic compound (VOC) emissions, lowers sludge generation as more complete oxidation to CO_2 is achieved, and minimizes bulking and biomass foaming issues associated with sludge. The features of the biomass and the layout of the aeration system have an impact on oxygen transport in wastewater. The three characteristics of biomass—particle concentration, particle size, and viscosity—as well as aeration are interconnected. Particle size and viscosity are impacted by aeration intensity. The concentration of solids modifies the impact of any viscosity increase on oxygen transport. The associated effects of granular size and concentration on oxygen transport are also present (Skouteris et al. 2020; Drewnowski et al. 2019).

6 Conclusion

Activated sludge process (ASP) is one of the oldest and most widely used process around the world due to its economical values, easy to operate, and feasibility. In this chapter, detailed mechanism of a basic ASP system with how the process can be optimized by tuning the process variables has been described thoroughly. Apart from these, from above discussion it can also confirm that performance of the ASP depends of the concentration of nutrients, microorganism types as well as type of system which has been adopted. Performance of the ASP also depends on the aeration system of the process which plays great role in enhancing the efficiency of the process. However, there are several limitations in each type of ASP system but with more research and advancement of modern technologies these problems can be solved.

References

Cloete TE, Steyn PL (1988) A combined membrane filter-immunofluorescent technique for the in situ identification and enumeration of acinetobacter in activated sludge. Water Res 22(8):961–969. https://doi.org/10.1016/0043-1354(88)90142-X

Drewnowski J, Remiszewska-Skwarek A, Duda S, Łagód G (2019) Aeration process in bioreactors as the main energy consumer in a wastewater treatment plant. Review of solutions and methods of process optimization. Processes 7(5):311. https://doi.org/10.3390/pr7050311

Eckenfelder WW (1980) Principles of water quality management. Springer

Fan XY, Gao JF, Pan KL, Li DC, Zhang LF, Wang SJ (2018) Shifts in bacterial community composition and abundance of nitrifiers during aerobic granulation in two nitrifying sequencing batch reactors. Biores Technol 251:99–107. https://doi.org/10.1016/j.biortech.2017.12.038

Gonzalez-Gil G, Holliger C (2014) Aerobic granules: microbial landscape and architecture, stages, and practical implications. Appl Environ Microbiol 80(11):3433–3441. https://doi.org/10.1128/AEM.00250-14

Grabínska-Łoniewska A (1991) Denitrification unit biocenosis. Water Res 25(12):1565–1573. https://doi.org/10.1016/0043-1354(91)90189-W

Hreiz R, Latifi MA, Roche N (2015) Optimal design and operation of activated sludge processes: state-of-the-art. Chem Eng J 281:900–920

Kocaturk I, Erguder TH (2016) Influent COD/TAN ratio affects the carbon and nitrogen removal efficiency and stability of aerobic granules. Ecol Eng 90:12–24. https://doi.org/10.1016/j.ecoleng.2016.01.077

Mccarty PL, Brodersen CF (1962) Theory of extended aeration activated sludge. J (water Pollut Control Fed) 34(11):1095–1103

Orhon D (2015) Evolution of the activated sludge process: the first 50 years. J Chem Technol Biotechnol 90(4):608–640. https://doi.org/10.1002/jctb.4565

Peters RW (2011) Water and wastewater engineering: design principles and practice, 1st edition. Environ Prog Sustainable Energy 30(3):266–267. https://doi.org/10.1002/ep.10602

Schneider DW (2014) Who invented activated sludge. Environ Eng Sci 1:8–11

Skouteris G, Rodriguez-Garcia G, Reinecke SF, Hampel U (2020) The use of pure oxygen for aeration in aerobic wastewater treatment: a review of its potential and limitations. Biores Technol 312:123595. https://doi.org/10.1016/j.biortech.2020.123595

Tan CH, Koh KS, Xie C, Tay M, Zhou Y, Williams R, Ng WJ, Rice SA, Kjelleberg S (2014) The role of quorum sensing signalling in EPS production and the assembly of a sludge community into aerobic granules. ISME J 8(6):1186–1197. https://doi.org/10.1038/ismej.2013.240

Wan C, Lee DJ, Yang X, Wang Y, Wang X, Liu X (2015) Calcium precipitate induced aerobic granulation. Biores Technol 176:32–37. https://doi.org/10.1016/j.biortech.2014.11.008

Water Environment Federation (1994) Operation of municipal wastewater treatment plants: manual of practice, vol. I-III. Sci Total Environ 142(3):227. https://doi.org/10.1016/0048-9697(94)90331-x

Wilén BM, Liébana R, Persson F, Modin O, Hermansson M (2018) The mechanisms of granulation of activated sludge in wastewater treatment, its optimization, and impact on effluent quality. Appl Microbiol Biotechnol 102(12):5005–5020. https://doi.org/10.1007/s00253-018-8990-9

Winkler MKH, Kleerebezem R, De Bruin LMM, Verheijen PJT, Abbas B, Habermacher J, Van Loosdrecht MCM (2013) Microbial diversity differences within aerobic granular sludge and activated sludge flocs. Appl Microbiol Biotechnol 97(16):7447–7458. https://doi.org/10.1007/s00253-012-4472-7

Zhou D, Niu S, Xiong Y, Yang Y, Dong S (2014) Microbial selection pressure is not a prerequisite for granulation: dynamic granulation and microbial community study in a complete mixing bioreactor. Biores Technol 161:102–108. https://doi.org/10.1016/j.biortech.2014.03.001

Arceivala SJ (1986) Wastewater treatment for pollution control. Rwanda journal of …, 3rd ed. Tata McGraw-Hill Education. http://www.ajol.info/index.php/rjhs/article/view/82339

Cairncross S, Feachem R (2018) Environmental health engineering in the tropics. Environmental health engineering in the tropics, 3rd ed. Routledge. https://doi.org/10.4324/9781315883946

Droste RL, Gehr RL (1997) Theory and practice of water and wastewater treatment. Choice reviews online, 2nd ed, vol 34. John Wiley & Sons. https://doi.org/10.5860/choice.34-4491

Eckenfelder WW, Eckenfelder, Inc. Nashville, TN, Musterman JL, Musterman & Associates, Nashville, TN (2020) Activated sludge treatment of industrial waters. In: Eckenfelder W (ed) Activated sludge. CRC Press, pp 181–346. https://doi.org/10.1201/9780203968567-10

Ettlich WF, Evans FL (1978) A comparison of oxidation ditch plants to competing processes for secondary and advanced treatment of municipal wastes. https://nepis.epa.gov/Exe/ZyNET.exe/3000057D.txt?ZyActionD=ZyDocument&Client=EPA&Index=1976Thru1980&Docs=&Query=&Time=&EndTime=&SearchMethod=1&TocRestrict=n&Toc=&TocEntry=&QField=&QFieldYear=&QFieldMonth=&QFieldDay=&UseQField=&IntQFieldOp=0&ExtQFiel

Fränzle S, Markert B, Wünschmann S (2012) Introduction to environmental engineering. Introduction to environmental engineering, 5th ed. McGraw-Hill Higher Education. https://doi.org/10.1002/9783527659487

Fuller GW (1927) The activated sludge process. In: Blackall l, Seviour RJ (eds) American Journal of Public Health, illustrate, vol 17. Springer Science & Business Media, pp 1061–1061. https://doi.org/10.2105/ajph.17.10.1061-a

Gerba CP, Pepper IL (2015) Municipal wastewater treatment. In: Lee YK (ed) Environmental microbiology, 3rd ed. World Scientific Publishing Company, pp 583–606. https://doi.org/10.1016/B978-0-12-394626-3.00025-9

Grau P, Sutton PM, Elmaleh S, Grady CPL, Gujer W, Henze M, Koller J (1983) Recommended notation for use in the description of biological wastewater treatment processes (provisional). Pure Appl Chem 55(6):1035–1040. https://doi.org/10.1351/pac198355061035

Gray NF (1992) Biology of wastewater treatment. In: Biology of Wastewater Treatment, p 1439

Hammer MJ, Sr, Hammer MJ, Jr (2013) Water and wastewater technology: pearson new international edition PDF eBook, 7th ed. Pearson Education. https://books.google.com.bd/books?id=ma6pBwAAQBAJ

Kayser R (2008) Activated sludge process. In: Wanner J, Jenkins D (ed) Biotechnology: second, completely revised edition, illustrate, vol 11–12. IWA Publishing, pp 253–283. https://doi.org/10.1002/9783527620999.ch13l

Kerri KD (2008) Water treatment plant operation (a field study training program), 6th ed, vol 1. California State University

Lakatos G (2018) Biological wastewater treatment. In: Wastewater and water contamination: sources, assessment and remediation, 3rd ed. CRC Press, pp 105–128. https://doi.org/10.1201/b18368-4

Lin Y, de Kreuk M, van Loosdrecht MCM, Adin A (2010) Characterization of alginate-like exopolysaccharides isolated from aerobic granular sludge in pilot-plant. Water Res 44(11):3355–3364.https://doi.org/10.1016/j.watres.2010.03.019

Liu Y, Tay JH (2002) The essential role of hydrodynamic shear force in the formation of biofilm and granular sludge. Water Res 36(7):1653–1665.https://doi.org/10.1016/S0043-1354(01)00379-7

Metcalf & Eddy Inc, Tchobanoglous G, Burton FL, Tsuchihashi R, Stensel HD (2013) Wastewater emgomeeromg: treatment and resource recovery. Wastewater engineering, 5th ed. McGraw-Hill Education. https://books.google.com.cu/books?id=6KVKMAEACAAJ.

Mines RO, Jr (2014) Design of wastewater treatment systems. In: Mines RO, Jr (ed) Environmental engineering: principles and practice. John Wiley & Sons, pp 331–447

Okafor N (2011). Environmental microbiology of aquatic and waste systems. Springer Science & Business Media. https://doi.org/10.1007/978-94-007-1460-1

Ong, SK (2007) Wastewater engineering. In: Lin SD, Lee CC (eds) Environmentally conscious materials and chemicals processing, 2nd ed. McGraw Hill Professional, pp 207–235. https://doi.org/10.1002/9780470168219.ch8

Porter JE (1921) The activated sludge process of sewage treatment: a bibliography of the subject, with brief abstracts, patents, news items, etc. comp. mainly from current literature. General Filtration Company, Incorporated

Qasim SR, Zhu G (2017) Wastewater treatment and reuse: theory and design examples: volume 1: principles and basic treatment. CRC Press. https://doi.org/10.1201/b22368

Randall CW, Barnard JL, Stensel HD (1992) Design and retrofit of wastewater treatment plants for biological nutrient removal, vol 5. CRC Press

Riffat R (2012) Fundamentals of wastewater treatment and engineering. CRC Press.https://doi.org/10.1201/b12746

Spellman FR (2013) Handbook of water and wastewater treatment plant operations, 4th ed. Taylor & Francis Group. https://doi.org/10.1201/b15579

Toerien DF, Gerber A, Lötter LH, Cloete TE (1990) Enhanced biological phosphorus removal in activated sludge systems. In: Advances in microbial ecology. Springer, pp 173–230. https://doi.org/10.1007/978-1-4684-7612-5_5

Tolliver R (2016) Environmental engineering. In: Using the engineering literature, 7th ed. McGraw-Hill, pp 277–315.

U.S. EPA (2000) Wastewater technology fact sheet package plants. In: United States environmental protection agency, pp 1–7. http://www3.epa.gov/npdes/pubs/final_sgrit_removal.pdf

USEPA (2000) Wastewater technology fact sheet dechlorination. In: Environmental protection agency, pp 1–7

van Haandel AC (2012) Handbook of biological wastewater treatment: design and optimisation of activated sludge systems. Water intelligence online, 2nd ed, vol 11. IWA Publishing. https://doi.org/10.2166/9781780400808

Wanner J (2020) Process theory: biochemistry, microbiology, kinetics, and activated sludge quality control. *Activated Sludge*, 15–72.https://doi.org/10.1201/9780203968567-7

Wu L, Li A, Hou B, Li M (2017) Exogenous addition of cellular extract N-acyl-homoserine-lactones accelerated the granulation of autotrophic nitrifying sludge. Int Biodeterior Biodegradation 118:119–125.https://doi.org/10.1016/j.ibiod.2017.01.035

Advanced Oxidation Processes for Industrial Wastewater Treatment

Md. Didarul Islam, Farzana Yeasmin, and M. Mehedi Hasan

1 Introduction

In last few decades, due to industrialization several environmental problems have raised and need to be resolved. Of them contamination of surface and groundwater by the discharge of organic and inorganic pollutants are the serious threat for the environment. Moreover, the presence of some pollutants at lower concentration level (e.g. ng to mg) in the water body may be harmful for the aquatic environment and or human health in the long-term exposure (Garcia-segura et al. 2018). Among those pollutants some are highly non-biodegradable, recalcitrant (e.g. pharmaceuticals, industrial dyes, pesticides) and cannot be removed or degraded by conventional wastewater treatment processes such as activated sludge, biological treatment process, fixed-bed reactor, and so on (Garcia-segura et al. 2018). In such cases, an effective alternative treatment process for the removal or degradation of contaminants from industrial actual wastewater is mandatory.

Nowadays, much progress has been achieved in the treatment of wastewater from effluents. Those techniques can be divided into three groups, (1) physical (e.g. precipitation, membrane filtration), (2) chemical (e.g. advanced oxidation process (AOP), adsorption, coagulation), and (3) biological methods (e.g. activated sludge, membrane bioreactor, biological filtration) (Luo et al. 2020). Among them physical treatment process possesses several limitations such as required huge chemical consumption, huge sludge production, handling and disposal problems, and so on

Md. D. Islam
National Institute of Textile Engineering and Research, Nayarhat, Savar, Dhaka, Bangladesh

F. Yeasmin · M. M. Hasan (✉)
Department of Applied Chemistry and Chemical Engineering, Faculty of Engineering, Bangabandhu Sheikh Mujibur Rahman Science and Technology University, Gopalganj-8100 Dhaka, Bangladesh
e-mail: mhasan.acdu@gmail.com

M. P. Shah (ed.), *Advanced and Innovative Approaches of Environmental Biotechnology in Industrial Wastewater Treatment*,
https://doi.org/10.1007/978-981-99-2598-8_3

(Crini and Lichtfouse 2020). On the other hand, biological treatment process cannot degrade several organic pollutants such as refractory pollutants, some pharmaceuticals, and pesticides. Most of those problems can be solved by advanced oxidation process-based treatment process. Advanced oxidation processes (AOPs) are versatile, efficient, cost-effective, easily automatable, and clean. Briefly, free radicals are generated during wastewater treatment process which can partially or completely degrade refractory organic pollutants and produce carbon dioxide, nitrogen oxides, sulfur oxides, water, and inorganic ions (Bagal and Gogate 2013; Zhao et al. 2017). However, several conventional AOP process shows several undesired by-products and or unexpected side reaction may occur which limits their application in large-scale industrial application. These problems can be solved by proper modification of the entire operating condition such as pH, addition of catalysts, and addition of oxidizing agents. Some study reviled that several activation methods such as heat treatment (Ji et al. 2016), alkaline addition (Lou et al. 2017), ultraviolet (Zhang et al. 2016), ultrasound (Su et al. 2012), and transition metals (Hao et al. 2019) can enhance free radicals generation and hence augment degradation efficiency.

The aim of this chapter is to understand the potential application of AOPs for the degradation of natural and synthetic wastewater at laboratory and pilot plant. This chapter also discusses the fundamentals of AOPs to better understand their operation, their application, advantages, and limitations obtained during industrial application.

2 Wastewater: Sources and Composition

Municipal wastewater and industrial wastewater (such as pharmaceutical, food processing, pulp and paper, textile, chemical, petroleum, tannery, and manufacturing industries) are the major sources of wastewater. While there is a huge difference in contamination levels between them. For example, industrial wastewater usually contains higher organic load, extreme physicochemical nature (such as pH, temperature, salinity, higher COD, BOD, suspended solids (SS), ammonium nitrogen (NH_4^+–N), heavy metals, pH, color, turbidity, and other biological parameters). The chemical and physical nature of wastewater strongly influences by types of industrial wastewater, types of operation used in industry during production. Table 1 summarizes the characteristics of several industrial wastewaters. In addition, several industrial wastewaters also contain a large variety of potentially toxic compounds such as aromatics, chlorinated or fluorinated compounds, phenols, volatile organic compounds, heavy metals, surfactants, biocides, and microorganisms.

Table 1 Composition of several industrial wastewaters

Parameter	Pharmaceutical Gadipelly et al. (2014)		Food processing Lin et al. (2012)	Pulp and paper Hermosilla et al. (2015), Ashra et al. (2015), Lin et al. (2012)	Textile Paździor et al. (2019), Lin et al. (2012)	Tannery Zhao and Chen (2019), Mannucci et al. (2010), Lofrano et al. (2013), Lin et al. (2012)	Oily and petrochemical Priyadarshini et al. (2021), Jain et al. (2020), Lin et al. (2012)
	Chemical processes	Fermentation processes					
COD (mg/L)	375–32,500	180–12,380	1,920–154,100	1,124–13,000	50–4,750	732–8,000	124–60,000
BOD_5 (mg/L)	200–6,000	25–6,000	1127–80,000	128–8,500	115–730	2,000–2,400	52–32,000
TOC (mg/L)	860–4,940	–	–	–	–	–	–
TKN (mg/L)	165–770	190–760	17–1,200	12–86	–	52–676	–
TP	–	–	–	2.3–36	103–239	–	102–227
NH_4^+–N (mg/L)	148–363	65.5–190	0.25–1,400	–	5–57	80–228	56–125
pH	3.9–9.2	3.3–11	3.5–6.3	2.5–10.5	2–11.2	2.5–9.7	2.5–8.79
TDS (mg/L)	–	1,300–28,000	–	72–850	–	–	–
TSS (mg/L)	–	57–7,130	–	37–6,095	50–880	330–2,820	20–950
Conductivity (μS/cm)	–	1600–44,850	–	–	–	–	–
Cl^- (mg/L)	–	182–2,800	–	–	–	95–3,260	–
SO_4^{2-} (mg/L)	–	160–9,000	–	–	–	900–2,270	–
PO_4^{3-}–P	–	–	20–830	–	–	–	–
Cr^{3+}	–	–	–	–	–	120	–
Oil	–	–	–	–	–	–	20–1,000

3 Advanced Oxidation Processes for Industrial Wastewater Treatment

3.1 Fenton and Fenton-Like Processes

The combination of ferrous ion and H_2O_2 in acidic conditions results in the generation of hydroxyl free radicals known as Fenton process, have been extensively used in advanced oxidation processes in industrial wastewater treatment. The Fenton reaction must be operated under acidic conditions to prevent the precipitation of iron. Like catalytic ozonation process, in Fenton-based AOP process iron act as a catalyst with catalytic activity at pH 3, especially due to the precipitation of ferric oxyhydroxide higher pH level. Moreover, the addition of H_2O_2 may lead to a reduction of ferric ion to ferrous ion which enhances the reaction rate. To get maximum reactor performance, reactors should be designed in such a way which allow optimum mixing of ferrous ion and H_2O_2 to generate maximum free radical generation. Complete mineralization of organic contaminants may take place by the reaction of $^{\bullet}HO$ shown in the following reactions 1 and 2.

$$Fe^{2+} + H_2O_2 = Fe^{3+} + {}^{\bullet}OH + OH^- \tag{1}$$

$$\text{Pollutant} + OH^{\bullet} = [\text{Pollutant}]^{\bullet} + H_2O = CO_x, NO_x, SO_x \tag{2}$$

Study showed that the degradation rate can further improved by operating under irradiation of UV–Vis light at wavelengths greater than 300 nm this process is known as the photo-Fenton process. Fenton reaction under UV–Vis results regeneration of ferrous ion along with more free radical formation as shown in the equation:

$$Fe(OH)^{2+} + h\nu = Fe^{2+} + {}^{\bullet}OH + OH^- \tag{3}$$

Regeneration of ferrous ion can further be increased by ultrasound irradiation which is known as Sono-Fenton process. The advantage of using ultrasound with Fenton process is combined treatment process can generate more hydroxyl free radicals from water without H_2O_2 consumption (Ammar 2016). Furthermore, during sono-Fenton process, ultrasonic irradiation can generate sulfate radicals from peroxydisulfate ion which can produce more hydroxyl free radicals into aquatic environment and lead to degrade organic pollutants more rapidly and efficiently (Chen and Su 2012; Ammar 2016).

$$S_2O_8^{2-} + h\nu = SO_4^{\bullet -} \tag{4}$$

$$SO_4^{\bullet -} + H_2O = H^+ + SO_4^{2-} + {}^{\bullet}OH \tag{5}$$

The Fenton and Fenton-based method is an attractive AOP process, as the availability of H_2O_2 and ferrous salts, environmentally safe compared to other chemical-based treatment processes, easy to handle, easy separation process (magnetic separation) of residual iron, and the reaction took place under ambient pressure and temperature (Khan and Yadav 2021). Until now, those methods have been used for the treatment of wide range of toxic organic pollutants such as herbicides, pesticides, dye compounds, phenolic compounds, and other highly carcinogenic organic compounds. Moreover, Fenton-based process can also be used with the biological treatment process to augment treatment efficiency.

3.2 Photocatalytic Oxidation Processes

In photocatalytic oxidation process (PCO), organic pollutants are generally immobilized in the semiconductor surface (catalyst) which is activated by the absorption of irradiation energy. The selection of catalyst is a major concern for the degradation of organic pollutants. The absorption irradiation must be equal or higher than the catalyst band gap. In such case, an electron from valance band is transported to the conduction band by absorbing solar radiation and generates electron hole pair. This electron hole pair is responsible for the generation of hydroxyl free radicals and degradation of organic pollutants as shown in the following Eqs. 6–13 (Batuira et al. 2019).

$$\mathrm{TiO_2 + h\nu = h^+_{VB} + e^-_{CB}} \tag{6}$$

$$\mathrm{e^-_{CB\ (untrapped)} + h^+_{VB} = e^-_{VB} + heat\left(e^-/h^+ recombination\right)} \tag{7}$$

$$\mathrm{h^+_{VB} + H_2O = {}^\bullet OH + H^+} \tag{8}$$

$$\mathrm{O_2 + e^-_{CB} = O_2^{\bullet -}} \tag{9}$$

$$\mathrm{O_2^{\bullet -} + H^+ = HO_2^\bullet} \tag{10}$$

$$\mathrm{2HO_2^\bullet = H_2O_2 + O_2} \tag{11}$$

$$\mathrm{H_2O_2 + O_2^{\bullet -} = {}^\bullet OH + OH^- + O_2} \tag{12}$$

$$\mathrm{Pollutant + OH^\bullet = [Pollutant]^\bullet + H_2O = CO_{\mathit{x}}, NO_{\mathit{x}}, SO_{\mathit{x}}} \tag{13}$$

Two types of photocatalysis processes have been used in water treatment plant (1) homogeneous photocatalysis and (2) heterogeneous photocatalysis. Among them, heterogeneous photocatalysis is currently considered the most promising and used method in water purification systems as it can fragment complex and long organic molecules into smaller fragments or completely mineralized in some cases. Titanium dioxide and zinc oxide are the most commonly used photocatalyst because of their high chemical stability and photocatalytic activity. However, cadmium selenide (CdS) and zinc selenide (ZnS) and colloids are also used after surface modification as some of them are highly toxic and may contaminate aquatic environment (Joshi and Shrivastava 2011; Sadegh et al. 2019).

Until now, although PCO systems have been used for the treatment of wastewater from synthetic wastewater in laboratory pilot plant but their application in real ETP-based wastewater plant is limited. Furthermore, PCO system can also be hybridized with other biological (e.g. membrane bioreactor) and oxidation process (e.g. ozonation) and shows synergistic effects. For example, a hybrid wet scrubber with UV/PMS system has been reported for the removal of volatile organic compounds and exhibited promising removal efficiency (Xie et al. 2019).

3.3 *Ozonation Processes*

Ozone is one of the most powerful oxidants that have been used for the treatment of wastewater. Ozone is highly selective and attacks primarily electron rich functional groups such as double bonds, amines, oxygen containing compounds, and aromatic rings containing compounds. In this treatment processes free radicals such as hydroxyl radical ($^{\bullet}HO$) and superoxide radical ($^{\bullet}O_2{}^{-}$) are formed in water through a series of reactions which is shown in the following reactions 3 to 8 (Parsons and Williams 2004).

$$O_3 + H_2O + h\nu = 2^{\bullet}OH + O_2 \tag{14}$$

$$O_3 + {}^{\bullet}OH = HO_2^{\bullet} + O_2 \tag{15}$$

$$O_3 + HO_2^{\bullet} = {}^{\bullet}OH + 2O_2 \tag{16}$$

$$2^{\bullet}OH + HO_2^{\bullet} = H_2O + O_2 \tag{17}$$

$$2^{\bullet}OH = H_2O_2 \tag{18}$$

$$\text{Pollutant} + OH^{\bullet} = [\text{Pollutant}]^{\bullet} + H_2O = CO_x, NO_x, SO_x \tag{19}$$

These free radicals may degrade organic contaminants in wastewater forming produce carbon dioxide, nitrogen oxides, sulfur oxides, water, and inorganic ions (Bagal and Gogate 2013; Zhao et al. 2017). Free radical generation rate depends on several factors such as pH of the solution, contact time, ozone dose, etc., several studies reviled that at higher pH levels disintegration rate is maximum and degradation rate of organic pollutants will be higher than expected. In recent year ozone-based process such as ozone/hydrogen peroxide, ozone/UV, ozone/ultrasound, and ozone/titanium dioxide photocatalysis has been used for the management of industrial wastewater. Among all ozone-based process, H_2O_2/O_3 system is considered to be the most established and used process. In single O_3 treatment process generation of $^{\bullet}OH$ by the reaction of O_3 and water molecules is not significant, which can be solved by the addition of H_2O_2 with O_3 in a ratio of 1:2 (Merényi et al. 2010; Miklos et al. 2018). The advantages of using H_2O_2/O_3 treatment process are requirement of lower dose of O_3 to degrade organic pollutants, ability to degrade non-biodegradable organic compounds with lower or bromate ion in the final effluents, and lower operating and installation costs (Khan and Yadav 2021). The ozone/catalyst system can be subdivided into two groups: (1) homogeneous catalytic ozonation, and (2) heterogeneous catalytic ozonation. In homogeneous catalytic ozonation process decomposition of ozone is occurred in the presence of transition metal ions. Homogeneous catalytic oxidation processes the following two major mechanisms: (a) decomposition of ozone by transition metal ions that lead to generate hydroxyl free radicals (Sauleda and Brillas 2001), and (b) complexes formation between organic molecules and catalyst (transition metal ions) and oxidation of this complex (Rivas and Beltra 2005). Mn(II), Fe(III), Fe(II), Co(II), Cu(II), Zn(II), and Cr(III) are the most commonly used transition metal ions used as catalyst during catalytic ozonation process (Nawrocki and Kasprzyk-Hordern 2010). In heterogeneous catalytic ozonation process uses both ozone and catalyst to generate hydroxyl free radicals. Metal oxides (e.g. MnO_2, TiO_2, Al_2O_3, CeO_2, etc.), metals (e.g. Cu, Ru, Pt, Co), activated carbon, zeolites are the most widely used catalyst in heterogeneous catalytic oxidation process (Nawrocki and Kasprzyk-Hordern 2010). This type of ozonation system is more effective for the reduction of chemical oxygen demand (COD) and total organic carbon (TOC) from raw wastewater compared to other ozonation treatment process at similar operating parameter (same ozone dose, same pH etc.). In 2019, Liya et al. studied several ozone-based advanced oxidation processes (single ozone, ozone/H_2O_2, ozone/catalyst (TiO_2) for the removal of TOC from secondary petrochemical wastewater (Fu et al. 2019). This study showed better TOC removal performance by TiO_2 catalyst-based ozonation process and is the other ozone-based treatment processes at same ozone dose (35 mg/L), ozonation time (1 hour).

3.4 Sonolysis Processes

Sonolysis is the best environmentally friendly advanced oxidation process (AOP), in which fine bubbles are formed and explode and generate higher temperature (5000 K) and pressure (1000 atm) and producing highly reactive species (e.g. H•, OH•) at that condition. The decomposition of organic pollutants occurs by thermal decomposition or by reactions with reactive free radicals. In sonolysis process bubbles are formed in the aquatic system by applying ultrasound in the range of 20–50 kHz. However, this treatment process requires more energy consumption and less heat transfer is exhibited compared to UV irradiation. Moreover, the number of free radicals generated during sonolysis process is not sufficient. For that reason, the researcher has been trying to couple sonolysis techniques with other systems such as ultrasound with UV irradiation (sonophotolysis); oxidants (O_3/H_2O_2); catalysts (TiO_2) (sonocatalysis); or ultrasound with UV irradiation and catalyst (sonophotocatalysis). The reaction involved during sonolysis process is shown in the equations:

$$H_2O + \text{Ultrasound} = {}^{\bullet}OH + {}^{\bullet}H \tag{20}$$

$$\text{Pollutant} + OH^{\bullet} = [\text{Pollutant}]^{\bullet} + H_2O = CO_x, NO_x, SO_x \tag{21}$$

$$\text{Pollutant} + \Delta H = CO_x, NO_x, SO_x \tag{22}$$

In last few decades, sonolytic-based processes have been used for the removal of bio-refractory organic pollutants such as m-nitrotoluene, pentachlorophenol, textile dyes, trinitrotoluene (TNT), cyclotrimethylene-trinitramine (RDX), methyl tert-butyl ether, cyclohexane, and so on (Wang et al. 2012). Destaillats et al. used a combined sonolysis/O_3 to estimate the degradation of azobenzene and methyl orange and compared with a single US and O_3 process (Destaillats et al. 2000). The study showed that TOC removal efficiency significantly increased with the treatment with US/O_3 almost 80% while only 20 and 30% were found to be removed by US and O_3 process. In 2011, Wang et al. use US/O_3 for the removal of tetracycline at optimum operating conditions (Wang et al. 2012). More than 91% of COD was removed from wastewater by the treatment of US/O_3, while only 76% and 0% of COD removal efficiency were associated with the treatment of O_3 and US systems.

3.5 Electrochemical Oxidation (EO) Processes

Electrochemical oxidation (EO) is an environment-friendly advanced oxidation process (AOP) which is activated by the transfer of electrons and generating hydroxyl free radicals leading to the removal of organic and inorganic pollutants. This process is versatile, efficient, cost-effective, easily automatable, and clean. Briefly, in electrochemical oxidation process, a strong oxidizing agent is used which allows the complete degradation of organic pollutants. In direct EO process, electron transfer occurs at the electrode surface without presence of any other chemical reagent while in indirect EO process, some electroactive species known as mediators (e.g. HClO, $H_2S_2O_8$ and catalytic mediators such as Ag^{2+}, Co^{3+}, Fe^{3+} etc.) are generated at anode or cathode surface which oxidize organic pollutants (Martínez-Huitle and Ferro 2006; Martínez-Huitle and Panizza 2018).

EO process can be applied for the treatment of turbid and colored wastewater containing refractory organic contaminants, which are very difficult to reduce by the conventional treatment process. However, like other processes, EO process possesses several drawbacks such as polarization, passivation, and corrosion of electrodes may occur for long-time wastewater treatment. Polarization results poor mass transfer and the accumulation of gases on the electrode surface leading reduction of electroactive species in the surface of the electrode (Shestakova and Sillanpää 2017). On the other hand, in passivation process polymer or oligomer compounds (from pollutants) are augmented near the surface of the electrode leading reduction of degradation capacity of the EO process (Lee et al. 2016). During the oxidation process several corrosive products may generate such as chlorine gas which may corrode anodic materials which can be solved by developing new electrode materials such as coating inert materials on the anode surface or by using inert material. However, powerful agitation by rapid stirring or turbulence promoter can significantly reduce the polarization and passivation. Another drawback of EO process is it cannot be applicable to the removal of metal-based contaminants. This problem can be solved by combined treatment of wastewater that can be used for the separation of those metal-based pollutants.

3.6 *Ionizing Radiation Processes*

The ionizing radiation (including gamma ray and electron beam) is an efficient and promising method to degrade organic pollutants in the aquatic environment. The ionizing or emitted radiation energy is transferred in surrounding media, which excites the atoms by striping their electrons or breakdown the chemical bonds within the molecules leading to the degradation of molecules. In electron beam system, an electron source generates electron and exposes it to the surroundings. The electrons then penetrate the water surface and lead to the formation of electrically excited free radicals including free radicals and ionic species. The penetration depth of the excited electron depends on the energy of the incident electrons. Furthermore, degradation can be improved by increasing the surface area of contaminated wastewater sources such as spraying by using nozzle. This process showed higher oxidizing ability and very little interferences by the water matrix with electrical efficiency (Miklos et al. 2018) (Fig. 1; Table 2).

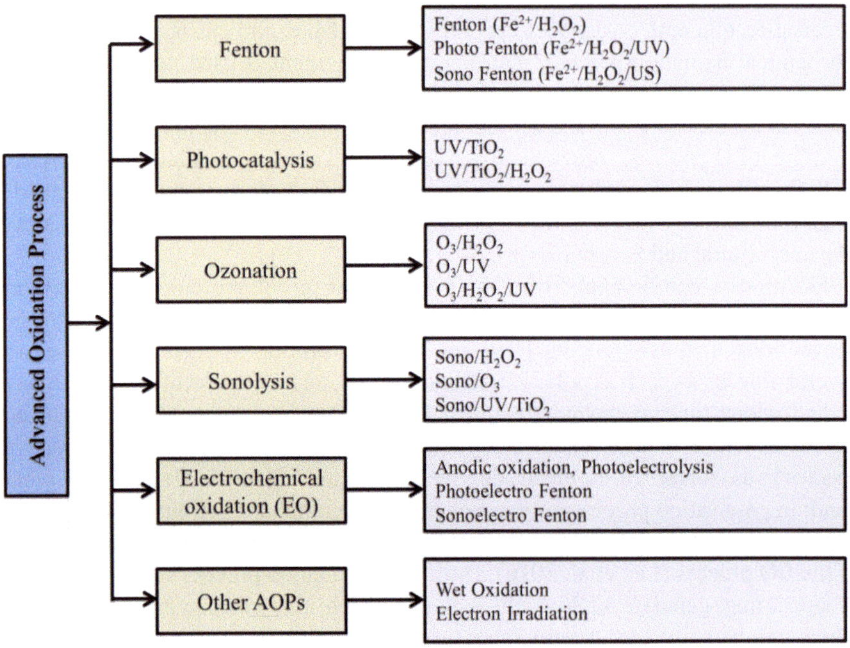

Fig. 1 General classification of various advanced oxidation processes (modified from [Khan and Yadav 2021])

4 Factors Influencing the Performance of AOPs

4.1 Characteristics of Catalysts

In AOP, catalysts influence the performance of the system significantly. Catalyst either directly or indirectly influences the parameters to control the rate of formation of radical. For example, in case of photocatalytic AOPs photocatalyst controls directly the formation of hydroxyl formation in wastewater while for ozone-based AOPs catalysts promotes the formation of radicals by enhancing the degradation of ozone molecules (Deng and Zhao 2015; Khan and Yadav 2021). Commonly, a catalyst should have economic feasibility, good dispersibility, non-toxicity, and resistance to corrosion. Both homogeneous and heterogeneous catalysis processes have been utilized by many studies. Both of these methods have their own advantages and disadvantages. Thus, the selection of catalysis method depends on number of factors such as rate of formation of active species, possibility of secondary pollutants formation, precipitation, or inactivation of catalysts at operating conditions. For example, though the homogeneous Fenton process exhibits great performances against numerous organic pollutants, low utilization of auxiliary chemical H_2O_2 makes it non-feasible for practical applications. Moreover, a constant pH of around 3 is mandatory for homogeneous Fenton process to avoid precipitation which increase

Table 2 Selected results reported for the treatment of industrial wastewater by several advanced oxidation treatments

Treatment process	Wastewater types	Operating conditions	Pollutants (mg/L)	Removal efficiency (%)	References
US	Petrochemical wastewater	pH: 5.5 ± 0.5 US Power: 200 W Treatment time: 60 min Rotation speed: 300 rpm	COD: 680 mg/L	9.3	Kakavandi and Ahmadi (2019)
UV		pH: 5.5 ± 0.5 UV light source Treatment time: 60 min Rotation speed: 300 rpm		7.8	
H_2O_2		pH: 5.5 ± 0.5 US Power: 200 W H_2O_2: 6.0 mmol/L Treatment time: 60 min Rotation speed: 300 rpm		6.9	
US/UV		pH: 5.5 ± 0.5 US Power: 200 W UV light source Treatment time: 60 min Rotation speed: 300 rpm		16.4	
US/H_2O_2		pH: 5.5 ± 0.5 US Power: 200 W H_2O_2: 6.0 mmol/L Treatment time: 60 min Rotation speed: 300 rpm		17.8	

(continued)

Table 2 (continued)

Treatment process	Wastewater types	Operating conditions	Pollutants (mg/L)	Removal efficiency (%)	References
UV/H_2O_2		pH: 5.5 ± 0.5 UV light source H_2O_2: 6.0 mmol/L Treatment time: 60 min Rotation speed: 300 rpm		21.7	
US/UV/H_2O_2		pH: 5.5 ± 0.5 US Power: 200 W UV light source H_2O_2: 6.0 mmol/L Treatment time: 60 min Rotation speed: 300 rpm		26.7	
MPAC		pH: 5.5 ± 0.5 MPAC: 0.3 g/L Treatment time: 60 min Rotation speed: 300 rpm		12.6	
MPAC/H_2O_2		pH: 5.5 ± 0.5 MPAC: 0.3 g/L H_2O_2: 6.0 mmol/L Treatment time: 60 min Rotation speed: 300 rpm		35.2	
MPAC/US/H_2O_2		pH: 5.5 ± 0.5 MPAC: 0.3 g/L US Power: 200 W H_2O_2: 6.0 mmol/L Treatment time: 60 min		40.8	

(continued)

Table 2 (continued)

Treatment process	Wastewater types	Operating conditions	Pollutants (mg/L)	Removal efficiency (%)	References
MPAC/UV/H_2O_2		pH: 5.5 ± 0.5 MPAC: 0.3 g/L UV light source H_2O_2: 6.0 mmol/L Treatment time: 60 min		45.6	
MPAC/UV/US		pH: 5.5 ± 0.5 MPAC: 0.3 g/L UV light source US Power: 200 W Treatment time: 60 min		36.7	
MNPs/US/UV/H_2O_2		pH: 5.5 ± 0.5 H_2O_2: 6.0 mmol/L MNPs: 0.3 g/L US Power: 200 W Treatment time: 60 min		45.3	
PAC/US/UV/H_2O_2		pH: 5.5 ± 0.5 H_2O_2: 6.0 mmol/L PAC: 0.3 g/L US Power: 200 W Treatment time: 60 min		39.8	
MPAC/US/UV/H_2O_2		pH: 5.5 ± 0.5 H_2O_2: 6.0 mmol/L MPAC: 0.3 g/L US Power: 200 W Treatment time: 60 min		58.3	

(continued)

Table 2 (continued)

Treatment process	Wastewater types	Operating conditions	Pollutants (mg/L)	Removal efficiency (%)	References
Fenton	Synthetic wastewater	Fe^{2+}: 30 mg/L H_2O_2: 375 mg/L pH: 3.0	Amoxicillin: 10 mg/L	12.0	Verma and Haritash (2019)
Photo-Fenton		Fe^{2+}: 30 mg/L H_2O_2: 375 mg/L pH: 3.0 UV (365 nm) intensity: 672 W/m^2		3.5	
Sono photo-Fenton		Fe^{2+}: 30 mg/L H_2O_2: 375 mg/L pH: 3.0 Ultrasound: 40 kHz UV (365 nm) intensity: 672 W/m^2		6.0	
Solar photo-Fenton		Fe^{2+}: 30 mg/L H_2O_2: 375 mg/L pH: 3.0 Light source: Sunlight		9.0	
Sono-Fenton		Fe^{2+}: 30 mg/L H_2O_2: 375 mg/L pH: 3.0 Ultrasound: 40 kHz		20.0	

(continued)

Table 2 (continued)

Treatment process	Wastewater types	Operating conditions	Pollutants (mg/L)	Removal efficiency (%)	References
Sonochemical process	Hospital wastewater	Volume: 350 ml Ultrasound: 375 kHz Ultrasonic power density: 88.0 W/L pH: 7.9	Diclofenac: 0.04 μg/L Carbamazepine: 1.9 μg/L Venlafaxine: 0.0015 μg/L Loratadine: 8.1 μg/L Sulfamethoxazole: 0.0001 μg/L Trimethoprim: 0.03 μg/L Norfloxacin: 3.9 μg/L Ciprofloxacin: 10.7 μg/L Irbesartan: 0.05 μg/L Azithromycin: 27.9 μg/L Clarithromycin: 23.4 μg/L Clindamycin: 25.4 μg/L	Diclofenac: 67.4 Carbamazepine: 100.0 Venlafaxine: 100.0 Loratadine: 49.7 Sulfamethoxazole: 100.0 Trimethoprim: 100.0 Norfloxacin: 89.4 Ciprofloxacin: 63.3 Irbesartan: 100.0 Azithromycin: 74.9 Clarithromycin: 74.0 Clindamycin: 89.5	Serna-galvis et al. (2019)
Sono photo-Fenton		Fe^{2+}: 90 μmol/L Volume: 350 ml Ultrasound: 375 kHz Ultrasonic power density: 88.0 W/L UV: 4 W pH: 7.9		Diclofenac: 100.0 Carbamazepine: 100.0 Venlafaxine: 100.0 Loratadine: 99.3 Sulfamethoxazole: 100.0 Trimethoprim: 100.0 Norfloxacin: 89.0 Ciprofloxacin: 69.9 Irbesartan: 100.0 Azithromycin: 87.4 Clarithromycin: 82.2 Clindamycin: 98.7	

(continued)

Table 2 (continued)

Treatment process	Wastewater types	Operating conditions	Pollutants (mg/L)	Removal efficiency (%)	References
Electrochemical oxidation	Paper mill effluents	Anode: Pb Current density: 2.2 mA/cm^2 pH: 8 Temperature: 25 °C Electrolyte (NaCl): 1 g/L Rotation speed: 100 rpm Contact time: 60 min	COD: 5500 mg/L Color: Brown	COD: 97 Color: 100	Garcia-segura et al. (2018)
Electro-oxidation	Tannery	Anode: Boron doped diamond Fe^{2+}: 3.0 mM Electrolyte: Raw Current density: 111 mA/cm^2 Volume of wastewater: 250 mL pH: 4 Temperature: 25 ± 1 °C Rotation speed: 400 rpm Contact time: 180 min	TOC: 1800 mg/L	65	Isarain-chávez et al. (2014)

(continued)

Table 2 (continued)

Treatment process	Wastewater types	Operating conditions	Pollutants (mg/L)	Removal efficiency (%)	References
Electro-Fenton (EC)		Anode: Boron-doped diamond Fe^{2+}: 3.0 mM Electrolyte: Raw Current density: 111 mA/cm^2 Volume of wastewater: 250 mL pH: 3 Temperature: 25 ± 1 °C Rotation speed: 400 rpm Contact time: 180 min		72	
Photoelectro-Fenton (PEF)		Anode: Boron-doped diamond Fe^{2+}: 3.0 mM Electrolyte: Raw UVA light source: 6 W Current density: 111 mA/cm^2 Volume of wastewater: 250 mL pH: 3 Temperature: 25 ± 1 °C Rotation speed: 400 rpm Contact time: 180 min		80	

(continued)

Table 2 (continued)

Treatment process	Wastewater types	Operating conditions	Pollutants (mg/L)	Removal efficiency (%)	References
Electro-Fenton (EC)/ Photoelectro-Fenton (PEF)		Anode: Boron-doped diamond Fe^{2+}: 3.0 mM Electrolyte: Raw UVA light source: 6 W Current density: 111 mA/cm^2 Volume of wastewater: 250 mL pH: 3 Temperature: 25 ± 1 °C Rotation speed: 400 rpm Contact time: 180 min		90	

US: Ultrasound, UV: Ultraviolet, MNPs: Magnetic nanoparticles, PAC: Powdered activated carbon, MPAC: Magnetic powdered activated carbon the operating cost even further (Deng and Zhao 2015). On the other hand, in the case of heterogeneous Fenton-like process, solid state catalysts can be designed which can work on a wide range of pH and can utilize H_2O_2 more than homogeneous catalysis system. Furthermore, as catalysts are used in solid forms, hence they can be used several times which reduces the overall cost of the operation in practical applications (Joshi and Shrivastava 2011). In addition, sometimes catalysts with special properties are needed to consider for some specific AOPs. For instance, in case of photocatalytic AOPs catalyst should have a low band gap. Along with above-mentioned properties of heterogeneous catalysts low band is needed for photocatalysts to harvest visible light to reduce cost and enhance performance (Khan and Yadav 2021).

4.2 Concentration of Main Variable and Catalysts

The main variable varies for different AOP systems, for instance, in case of ozone-based AOPs main variable is ozone while in case of irradiation AOPs main variable is the intensity of the irradiation on the other hand photocatalyst itself is the main variable for photocatalytic AOPs. In general, with the increase of main variables degradation performance is increased as their higher concentration results in the formation of higher number of active species. However, excessive dosages of main variable can impact the performance in negative way. For instance, an excessive amount of photocatalysts results in blocking the penetration of the photons and causes the loss of light energy by shielding, reflection, and scattering. This ultimately lowers the harvesting of light which reduces the efficiency of the system. Additionally, the performance of the main variables significantly increases by using proper catalysts in proper dosages. In theory, at higher catalysts concentration more activity becomes available which should be beneficial for the adsorption of pollutants and thus degradation. However, using high amount of catalyst doesn't always provide benefits to the system. It has been reported that though degradation efficiency increases with the increase of dosages of catalysts at the initial state, however, degradation efficiency doesn't increase significantly after reaching a certain threshold. Hence using additional amount of catalysts after threshold limit becomes pointless and adds additional cost. Moreover, excessive catalysts can also act as a scavenger for the active radicals which results in lowering the efficiency even further. Hence, optimization of both concentration of the main variables and catalyst dosages is needed for practical applications (Miklos et al. 2018; Deng and Zhao 2015).

4.3 pH

Another important variable which greatly influences the degradation performances of the pollutants is the pH of the wastewater. pH affects the performance of AOP systems by influencing the precipitation of the catalytically active ions, controlling the amount of available surface charges on the catalysts, providing hydroxyl radicals, altering the charged form of substrate, and also by controlling the number of active species for degradation (Deng and Zhao 2015; Miklos et al. 2018; Shukla et al. 2021; Joshi and Shrivastava 2011).

Precipitation problem is common in case of homogeneous catalysis process where catalyst ions are mixed with wastewater in the same state. If the applied catalysts are prone to precipitate by altering the pH of the matrix hence operation must be done in the range of pH where the catalysts are non-precipitated. For instance, ferrous ions are used as catalysts in Fenton process, in this case, ferrous ions are found to precipitate at pH higher than 3. Hence to obtain good efficiency operation must be done at pH lower than 3. This problem limits the practical application of homogeneous catalysis process in wastewater treatment. However, by using the composite of iron supported with other oxides or materials, wastewater's wide range of pH can be treated effortlessly. For example, while in case of zero-valent iron catalysts Fenton oxidation efficacy decreased by rising pH from 3.0 to 5.8, efficacy was possible to enhance at the same pH range for Fe/CeO_2 catalysts (Zhang et al. 2019).

The nature of surface charge can be controlled by lowering or increasing the pH of the matrix from the point of zero charges of the catalysts. On the other hand, charged form of the substrates is also changed when the pH is higher or lower compared to pKa value of the substrate (Mehrjouei et al. 2015). At pH higher than pKa and point of zero charges both catalysts and substrate act as negatively charged compounds while in lower pH act as positively charged compounds. As similar charged particles repel one another both scenarios are not suitable for the degradation. Hence, operation must be conducted at optimized pH where catalysts and substrates are in an opposite charged state.

Furthermore, pH also influences the number of active radicals in the system. In case of ozone-based AOP process, based on pH, contamination degradation can be done in either direct or indirect methods. At lower pH, only ozone acts directly to degrade the contaminates while at higher pH additional hydroxyl radicals enhance the degradation rate of the process in significant scale. Moreover, at higher pH, more ozone molecules are found to transform into hydroxyl radicals which augments the oxidation degradation process. pH can also alter the number of active species by promoting the side reactions where active radicals react with unwanted compounds and losses their potential. Such as, at low pH, H_2O_2 reacts with proton and forms peroxonium ion which decreases its reactivity with Fe^{2+} ion Fenton reaction. On the other hand, at higher pH, precipitation of Fe^{3+} occurs which results in the formation of lower hydroxyl radicals (El-Ghenymy et al. 2013).

4.4 Role of Auxiliary Chemicals or Materials

Apart from above-mentioned parameters, the presence of auxiliary chemicals or unwanted materials also influence the efficiency of the AOPs. H_2O_2 is used with ferrous ion in case of Fenton and Fenton-like AOPs. Studies confirmed that the amount of H_2O_2 plays a great role in degradation efficiency as the presence of active hydroxyl radical is highly influenced by the amount of H_2O_2. Though a low concentration of H_2O_2 results in lack of hydroxyl radicals, higher concentration may rise of undesirable conditions for the degradation of pollutants. Thus, optimum amount of H_2O_2 is required to obtain the best performance and it varies for each pollutant and catalysts (Wang and Zhuan 2019). On the other hand, the presence of inorganic anions and organic matters has also been found to affect the performance of the degradation in an AOP system. Because inorganic anions interfere with active ions or radiations which decreases the efficacy of the main variables thus results in lower efficiency. Moreover, these unwanted species react with active free radicals and results in unwanted side reaction which lowers the number of available radicals for degradation. For example, HSO_4^- can scavenge the required hydroxyl radicals which is crucial for the degradation of pollutants. A similar, scenarios can also be witnessed in case of HCO_3^-, NO_2^-, etc., inorganic ions (J. Wang and Zhuan 2019).

5 Advantages of AOP Process

For the removal of resistant organic substances from municipal and industrial wastewater, advanced oxidation processes (AOPs) are extensively employed as a pre-treatment approach, accompanied by biological processes for cost-effectivity and economic viability. Chemical wastewater treatment with AOPs can lead in either the thorough mineralization of pollutants to water, CO_2, and inorganic compounds with little/no sludge production, or at the very least the transformation of pollutants into more benign products along with partial degradation of non-biodegradable organic contaminants into biodegradable intermediates (Canizares et al. 2009). Since it completely processes the distillery wastewater with a high rate of AOP reactions, extensive filtration of the sample is not necessary (Kumar and Shah 2021). The application and benefits of major AOPs are highlighted as follows.

5.1 *Ozonation or Catalytic Ozonation Processes*

- With the standard redox potential of 2.07 V, O_3 reveals high oxidation performance along with an elongated endurance period under low- and medium-temperature conditions (<270 °C) producing non-toxic O after decomposition. For the removal of colorant from molasses-derived wastewater or synthetic distillery effluent, the ozonation/ozonolysis process is reported to be a potential oxidation treatment technology. This process is also commonly used for disinfection because of its significant cell lytic activity, which allows it to kill bacteria in wastewater (Martínez et al. 2011). Up to 80% decolorization of biologically treated wasted wash can be attained by O_3 oxidation along with simultaneously reducing COD by 15–25% and enriching the effluent's biodegradability.

- Once dissolved in H_2O, O_3 reacts with a wide range of organic molecules by direct oxidation as molecular O_3 at pH lower than 7 or indirect oxidation as secondary oxidants like free radicals, resulting in $^{\bullet}OH$ at pH higher than 7. Both O_3 and $^{\bullet}OH$ can rigorously oxidize a variety of organic contaminants, lowering the value of COD in wastewater up to 27% (Rice 1996). Preozonation was also found effective to improve the nitrogen removal and settling qualities of sludge in subsequent aerobic biological processes, in addition to lowering COD and TOC.
- The inclusion of UV radiation, solar radiation, or catalysts (heterogeneous and/or homogeneous) boosts the ozone action. In this way, the presence of external methods can aid in the disintegration of molecular ozone into hydroxyl radicals or other reactive oxidative species like ozonide and/or hydroperoxyl radicals. When these adjuvants are combined with ozone, organic matter decomposition can be accelerated.
- In acid or neutral conditions, even low concentrations of ozone can selectively attack the unsaturated bonds and functional groups like $-CH_3$, $-OH$, and $-CH_3$. So, melanoidins, polyphenols, and other distillery effluent pollutants having C=C and N≡C and other functional groups in their structure can be specifically broken up by O_3 without oxidizing the bulk organic matter (Mantzavinos and Psillakis 2004).
- The strong odor of livestock wastewater is one of the most serious issues with this type of effluent, causing significant distress to the local community and posing health and environmental threats. Different chemicals, such as ammonia (NH_3), hydrogen sulfide (H_2S) methane (CH_4), phenols, and indoles are found in swine wastewater air emissions which are accountable for these adverse consequences. Ozonolysis is a powerful option to degrade the refractory compounds by oxidation (Domingues et al. 2021).

5.2 *Fenton's and Photo-Fenton's Process*

- Fenton's reagent is a potent oxidant for organic pollutants indulging Fenton and photo-Fenton process effective for the treatment of wastewater produced in chemical manufacture, dye manufacture, pulp bleaching, and agricultural processing, as well as supporting biological wastewater treatment as Fenton pre-treatment step. The photo-Fenton reaction produces higher rates and degrees of mineralization. Solar irradiation can also be used to trigger this process because low-energy photons can drive the reactions. Because of this, the treatment's operational costs can be greatly reduced (Stasinakis 2008).
- To treat complex distillery effluent, Fenton oxidation is an efficient and modern technology, which improves the biodegradability of the effluent and removes residual COD, toxicity, and color from the wastewater. Biogas recovery from biomethane waste wash can be promoted by Fenton oxidation and adsorption pre-treatment. In an acidic medium (pH 3) at ambient temperature (30 °C), a 54%

reduction in COD can also be achieved within an hour using a ferrous sulfate catalyst for diluted wastewater (Kumar and Shah 2021).

- For the removal of hydrocarbons from saline wastewater, the photo-Fenton technique is effective. The starting concentration of ferric ion (Fe^{2+}) and salinity had been reported to show a major impact on the degradation process, with photodegradation of hydrocarbons greater than 81%. Up to 86% reduction of COD of oil recovery industry wastewater by using Fenton process and more than 90% of color removal from olive mill wastewater by photo-Fenton reactions are so far reported (Stasinakis 2008).
- Under the optimal conditions for Fenton and photo-Fenton reactions, lower iron sludge production and higher COD and fecal coliform removal had been reported while treating wastewater from animal farms, including cattle, swine, and poultry farms (Domingues et al. 2021).
- Nitrophenols are very poisonous organic compounds, are widely employed in the chemical industry to synthesize medicines, insecticides, and synthetic dyes. The breakdown of 4-nitrophenol by electro-Fenton technique in batch recirculation mode was investigated. Increasing Fenton's reagent dosage, Fe(II)-to-H_2O_2 molar ratio, and electrical current at the beginning, followed by continuous hydrogen peroxide addition improved the COD removal efficiency. Electro-Fenton oxidation method radiated with UV light efficiently can also decompose 2,4-dichlorophenol (a chemical precursor for the manufacture of herbicides) by 93% and mineralize by 78% in an hour. Electro-Fenton methods are also capable of degrading dyes such as—methyl red and alizarin red (Poyatos et al. 2010).

5.3 *Fenton-Like Oxidation Processes*

- In homogeneous Fenton-like system, additives viz- Ce^{3+}, Cu^{2+}, Co^{2+}, and Mn^{2+} showed a synergistic impact with Fe^{2+} to enhance the Fenton system's·OH release appreciably and can drag oxidation efficiency of NO above 85% under the optimal reaction conditions. Simultaneous injection of Fe^{2+} and H_2O_2 solution into the bubble bed reactor following the in situ Fenton process evidently revealed improved denitrification effect and better oxidative removal of coal-fire flue gas pollutants due to enhanced gas–liquid contact area (upon addition of ascorbic acid) which accelerate the shifting of ferric iron into divalent iron (Wang et al. 2021).
- In the Heterogeneous Fenton-like systems, solid Fe-based catalysts can be utilized to replace Fe^{2+} in the solution to promote the breakdown of H_2O_2 and produce highly active $^{\bullet}OH$ radicals, broadening the pH window of the reaction. Some Fe-containing solid materials have distinctive ferromagnetic characteristics, which makes it easier to recycle the catalyst. In the UV/heterogeneous Fenton system, UV light enriched the catalyst surface by Fe^{2+}/Fe^{3+} redox pairing to produce $^{\bullet}OH$ in the system or activate surface heteropoly acid (PMO) to produce $^{\bullet}OH$ and $^{\bullet}O$

for oxidation and absorption of pollutants, by which upto 84% NO and 100% SO_2 removal could be achieved. In the H_2O_2 steam catalytic oxidation reaction system, converting H_2O_2 into steam to raise the reaction temperature of the denitrification and desulfurization processes boosted the rate of H_2O_2 breakdown and pollutant molecule oxidation, resulting coexistence of $^{\bullet}OH$ and $^{\bullet}NO_x$ in the gas phase. As a result, the reaction efficiency was increased allowing a significant reduction of H_2O_2 usage and operating costs (Wang et al. 2021).

5.4 *Photocatalytic Oxidation Processes*

- Photocatalytic degradation is a cost-effective and efficient wastewater treatment method for increasing the biodegradability of hazardous and non-biodegradable contaminants including persistent organic pollutants. In this process, chemical contaminants can be converted into less dangerous compounds with more readily biodegradable structural characteristics, which would be particularly useful for treating effluent having highly hazardous chemicals and for which biological processes aren't feasible (Poyatos et al. 2010).
- TiO_2 is a low-cost, commonly available substance with highly oxidizing photogenerated holes. TiO_2 may also oxidize a wide spectrum of organic molecules, converting them to innocuous chemicals like CO_2 and H_2O. Because of the ambient operating conditions, the lack of mass transfer constraints when nanoparticles are used as photocatalysts, and the accessibility of solar irradiation, the TiO_2/UV light method has been widely used for wastewater treatment. The effects of starting phenol concentration, amount of catalyst, solution pH, and light intensity on the oxidation of phenol-containing wastewater were investigated using photocatalysis with TiO_2. After 4 hours of irradiation, almost all of the target chemicals had been eliminated (Stasinakis 2008). In comparison to the synergetic activity that occurs when H_2O_2 and TiO_2 are used together, the TiO_2/H_2O_2 combination appears to be more efficient.
- Solar light-assisted photocatalytic degradation of contaminants has been successfully exploited as an economically viable approach that can replace expensive and harmful artificial light sources. This approach is also effective for removing EDCs (estrone, 17β-estradiol). At photocatalytic degradation of C.I. Acid Red 88, Ag-doped TiO_2 is more efficient than undoped TiO_2. Silver's potential to trap electrons explains its beneficial influence on dye degradation (Stasinakis 2008).
- For dye removal, the UV/H_2O_2 method is commonly utilized. In a study of exploring C. I Acid Orange 7 (a commonly identified color in textile effluent), increases in H_2O_2 concentration and pH at basic levels improved dye removal efficiency. The impact of operating parameters such as H_2O_2 dosage, initial dye concentration, pH, and UV dosage on decolorization of Reactive Blue 19 was studied, revealing the dye solution being entirely decolorized in less than 30 minutes using optimal H_2O_2 dosage and low-pressure mercury UV-C lamps. The rate of decolorization increased in a nonlinear relationship with UV dose.

- In order to improve color removal during dyeing wastewater treatment, researchers have combined UV/H_2O_2 with ultrasonic vibrations (US). The efficiency of UV/ H_2O_2 and US/UV/H_2O_2 processes on the removal of malachite green as a model contaminant from the textile industry was investigated in a recent study, showing US/UV/H_2O_2 emerging as the most effective process for the degradation of malachite green, where ultrasound combined with photolysis contributed to H_2O_2 bond cleavage. The removal rate of malachite green obeyed first-order kinetics, and the reaction rate rose directly with the temperature (294–307 K), power density (0.049–1.163 W mL^{-1}), and inversely with an initial concentration of the target chemical (10–2 mg L^{-1}) (Stasinakis 2008).

5.5 *Electrochemical Oxidation Processes*

- Due to the minimal equipment required, and safety, electrochemical oxidation involves great energy economy, automation, and ease of handling.
- The distillery contaminants are eliminated by either direct or indirect oxidation in the electrochemical process. The contaminants are adsorbed on the anode surface before being eliminated by the anodic electron transfer reaction in a direct anodic oxidation process. On the other hand, Strong oxidants such as hypochlorite/chlorine, O_3, or H_2O_2 are electrochemically created in an indirect oxidation process by either the anodic or cathodic process. The organic contaminants are then eliminated due to oxidation in the bulk solution by the resultant oxidants (Martínez-Huitle and Panizza 2018).
- A wide range of contaminants can be found in the industrial effluents of ink producing operations. The effects of using a Boron-doped diamond (BDD) anode and stainless steel (AISI 304) cathode in an electrochemical oxidation process on various dyes, solvents, and surfactants were investigated. This form of oxidation removed nearly all COD from the effluent with high current performance for all of the chemicals examined. The treatment of dyes was promoted at a chloride media. The application of sulfates or phosphates, on the other hand, improved the elimination of the aliphatic pollutants tested (solvents) (Poyatos et al. 2010).
- In a study, electrochemical oxidation with an activated carbon fiber (ACF) electrode was used to degrade another dye (amaranth), which resulted in 95.4% of color, 35% of COD, and 30% of TOC removal. An electrochemical treatment of anaerobic digestion effluents from purine wastewater treatment with a Ti/Pt–IrO_2 electrode is reported to result in TOC, inorganic carbon, and turbidity reduction percentages of 51.4, 73.8, and 95.5, respectively (Poyatos et al. 2010).

6 Limitations and Challenges of AOP Process

Despite having a number of advantageous applications in industrial effluent treatment, AOPs have several drawbacks. The development of effective and affordable materials to support adequate treatment, the use of renewable energy sources, the adoption of strategies for process integration, the targeting of new classes of pollutants, and the commercialization of processes that have thus far only been used in the laboratory may be the main challenges facing the use of AOPs in wastewater treatment in the future. Limitations of major AOPs due to the above-mentioned challenges are briefly discussed below:

- In ozonolysis of gaseous pollutants by O_3 and chlorine dioxide, both are strong oxidizers and could harm personnel's respiratory systems, putting their health at risk. The amount of energy required to produce O_3 is also quite significant. These issues limit the process's industrial use. Although sodium hypochlorite and hydrogen peroxide have inexpensive costs, their aqueous solutions have poor oxidizing capacity and have a limited ability to remove NO. In an acid environment, sodium chlorite decomposes to produce harmful chlorine. As a result, the storage and application conditions for sodium chlorite are relatively harsh, limiting the process's industrial use (Wang et al. 2021). The hefty installation and operation expenses limit its application. If the operation releases large volumes of effluents on a daily basis, the problem gets even more challenging. Another drawback of the ozonation method is the low COD removal, which is particularly troublesome when the feed stream has a high COD level (Kumar and Shah 2021).
  - The majority of investigations on the Fenton method's integrated removal of flue gas contaminants are still not at the pilot stage. The related primary issues are addressed below:
  - The system typically features a powerful oxidizing gas or liquid, which makes transportation and storage difficult during operation and demands stabilizers to prevent explosions induced by shaking and crashes during shipment. Likewise, preserving hydrogen peroxide in a cold, well-ventilated warehouse and avoiding direct sunlight or heat during storage and transportation unquestionably raise operating expenses and safety concerns.
  - Anti-corrosion coating protection on the downstream flue and equipment is required during the operation process because of the use of high levels of strong oxidants. Meanwhile, the impact of the escaping oxidant on the absorption of oxidation products and the environment, as well as the service life of flue and equipment after coating, must be assessed quickly.
  - The oxidation absorption products could produce hazardous sludge due to the introduction of metal ions in the catalytic process while producing high-value-added fertilizer, mandating additional treatment which poses significant environmental and economic concerns.

  - If the catalyst is sprayed into the system, the breakage induced by the catalysts striking could raise the pressure drop of the subsequent capture device, diminish catalytic oxidation efficacy, and increase the risk of excessive particulate matter and gaseous pollutant emissions. Meanwhile, due to collision loss, the catalyst's cycle efficiency is diminished, necessitating further catalyst replenishment, which raises the cost of operation. A velocity gradient and a concentration gradient of the gas pollutants are dispersed in front of the first catalyst layer if the arrangement is in fixed-bed mode. The catalyst is partially deactivated as a result of this circumstance (Wang et al. 2021).

- The reaction process in the H_2O_2 oxidation method had to ensure that the temperature was over 500 °C, which consequently hindered the implementation of this technology in practical engineering. So a combination of any of the auxiliary means, like homogeneous Fenton, heterogeneous Fenton, UV light, and their combination, are to be opted for low-temperature conditions, which have other compromising difficulties like for UV/H_2O_2 process, high manufacture and operative costs, safety concerns, and short pH range. The UV/Fenton combined process involves high energy consumption, high investment cost, and the difficult recycling of catalyst where, H_2O_2 steam catalytic oxidation has poor catalyst stability issues (Wang et al. 2021). In cases where the Fenton process is followed by biological oxidation, the toxicity of H_2O_2 to various microorganisms and the usage of excessive levels of H_2O_2 could potentially degrade the overall degradation effectiveness.
- For removing the high organic content of molasses effluent, photocatalytic degradation is inadequate and is viable only as post-treatment for anaerobic digestion.
- In the case of sonolysis, the time scale and expended power required to achieve thorough mineralization of contaminants are not economically feasible. In addition, the use of sonolysis in the treatment of complicated effluents is constrained by substantial installation/process costs and low COD removal.
- The generation of significant amounts of sludge, the consumption of chemicals, and the requirement of extreme pH are all disadvantages of electrochemical oxidation (Kumar and Shah 2021).

7 Conclusion

Day by day numerous types of different contaminate are being added in the list of already existing water contaminates around the world. The rise of population and uncontrolled management are the primary causes of this situation. Thus, awareness among the mass population can be a solution and government should act on enforcing the strict regulations on companies to prohibit the dumping of effluent without treatment. However, along with the awareness, huge research on developing feasible methods to remove already existed contaminates in wastewater should also be conducted. In past decades, numerous physical, biological, and chemical processes have been extensively analyzed by different research communities around the globe. Among them, advance oxidation process is one of the most prominent types of technology that can be efficiently used for the removal of contaminates from wastewater. In this process with the help of active radicals, water contaminates are oxidized into non-harmful compounds. AOP can be established by using different types of active radical producers such as Fenton process, Photocatalysis, Ozonolysis, sonolysis, etc. Whatever the process is selected for the radical production all follow the same path of oxidizing the contaminates. Because of this, by using these processes, reluctant contaminates can also be removed from wastewater which may seem impossible to get rid of by other processes. Studies confirmed that this process exhibits higher efficiency of contaminated degradation compared to other traditional processes. However, some limitations are also being associated with these processes as mentioned above. By using a hybrid system of multiple AOPs instead of single system some of the limitations have been mitigated. Hence, keeping in mind the huge potential of these processes, more research must be done. AOP process with high efficiency, having reusability and economically feasible must be the priority while developing new future AOP process.

References

Ammar HB (2016) Sono-Fenton process for metronidazole degradation in aqueous solution: effect of acoustic cavitation and peroxydisulfate anion. Ultrason Sonochem 33:164–169

Ashra O, Yerushalmi L, Haghighat F (2015) Wastewater treatment in the pulp-and-paper industry: a review of treatment processes and the associated greenhouse gas emission. J Environ Manage 158:146–157

Bagal MV, Gogate PR (2013) Wastewater treatment using hybrid treatment schemes based on cavitation and fenton chemistry: a review. Ultrason Sonochem 21(1):1–14

Batuira M, Filho C, Vilar VJP (2019) Strategies for the intensification of photocatalytic oxidation processes towards air streams decontamination : a review laboratory of separation and reaction engineering - laboratory of catalysis and corresponding authors. Chem Eng J 123531. https://doi.org/10.1016/j.cej.2019.123531

Canizares P, Paz R, Sáez C, Rodrigo MA (2009) Costs of the electrochemical oxidation of wastewaters: a comparison with ozonation and fenton oxidation processes. J Environ Manage 90(1):410–420

Chen W-S, Su Y-C (2012) Removal of dinitrotoluenes in wastewater by sono-activated persulfate. Ultrason Sonochem 19(4):921–927

Crini G, Lichtfouse E (2020) Advantages and disadvantages of techniques used for wastewater treatment. Environ Chem Lett (0123456789). https://doi.org/10.1007/s10311-018-0785-9

Deng Y, Zhao R (2015) Advanced oxidation processes (AOPs) in wastewater treatment. Curr Pollut Rep 1(3):167–176. https://doi.org/10.1007/s40726-015-0015-z

Destaillats H, Colussi AJ, Joseph JM, Hoffmann MR (2000) Synergistic effects of sonolysis combined with ozonolysis for the oxidation of azobenzene and methyl orange. J Phys Chem A 104(39):8930–8935

Domingues E, Fernandes E, Gomes J, Martins RC (2021) Advanced oxidation processes perspective regarding swine wastewater treatment. Sci Total Environ 776:145958

El-Ghenymy A, Rodríguez RM, Arias C, Centellas F, Garrido JA, Cabot PL, Brillas E (2013) Electro-Fenton and Photoelectro-Fenton degradation of the antimicrobial sulfamethazine using a boron-doped diamond anode and an air-diffusion cathode. J Electroanal Chem 701:7–13

Fu L, Wu C, Zhou Y, ZuoJ, Song G (2019) Chemosphere ozonation reactivity characteristics of dissolved organic matter in secondary petrochemical wastewater by single ozone, ozone/ H_2O_2, and ozone/catalyst. Chemosphere 233:34–43. https://doi.org/10.1016/j.chemosphere.2019.05.207

Gadipelly C, Pérez-gonzález A, Yadav GD, Ibañez R, Rathod VK (2014) Pharmaceutical industry wastewater: review of the technologies for water treatment and reuse. Ind Eng Chem Res 53(29):11571–11592

Garcia-segura S, Ocon JD, Chong MN (2018) Electrochemical oxidation remediation of real wastewater effluents—a review. Process Saf Environ Prot 113:48–67

Hao S-M, Yu M-Y, Zhang Y-J, Abdelkrim Y, Qu J (2019) Hierarchical mesoporous cobalt silicate architectures as high-performance sulfate-radical-based advanced oxidization catalysts. J Colloid Interface Sci 545:128–137

Hermosilla D, Merayo N, Gascó A (2015) The application of advanced oxidation technologies to the treatment of effluents from the pulp and paper industry: a review. Environ Sci Pollut Res 22(1):168–191

Isarain-chávez E, De C, Godínez LA, Brillas E, Peralta-hernández JM (2014) Comparative study of electrochemical water treatment processes for a tannery wastewater effluent. J Electroanalyt Chem 713:62–69. https://doi.org/10.1016/j.jelechem.2013.11.016

Jain M, Majumder A, Ghosal PS, Gupta AK (2020) A review on treatment of petroleum refinery and petrochemical plant wastewater: a special emphasis on constructed wetlands. J Environ Manage 272:111057

Ji Y, Xie W, Fan Y, Shi Y, Kong D, Lu J (2016) Degradation of trimethoprim by thermo-activated persulfate oxidation: reaction kinetics and transformation mechanisms. Chem Eng J 286:16–24

Joshi KM, Shrivastava VS (2011) Photocatalytic degradation of chromium (VI) from wastewater using nanomaterials like TiO_2, ZnO, and CdS. Appl Nanosci 1(3):147–155. https://doi.org/10.1007/s13204-011-0023-2

Kakavandi B, Ahmadi M (2019) Efficient treatment of saline recalcitrant petrochemical wastewater using heterogeneous UV-assisted Sono-Fenton process. Ultrason Sonochem. https://doi.org/10.1016/j.ultsonch.2019.03.005

Khan SH, Yadav VK (2021) Advanced oxidation processes for wastewater remediation: an overview. In: Shah MP (ed) Removal of emerging contaminants through microbial processes. Singapore: Springer Singapore, pp 71–93. https://doi.org/10.1007/978-981-15-5901-3_4

Kumar V, Shah MP (2021) Advanced oxidation processes for complex wastewater treatment. In: Advanced oxidation processes for effluent treatment plants. Elsevier, pp 1–31

Lee B-S, Park H-Y, Choi I, Cho MK, Kim H-J, Yoo SJ, Henkensmeier D et al (2016). Polarization characteristics of a low catalyst loading PEM water electrolyzer operating at elevated temperature. J Power Sour 309:127–134. https://doi.org/10.1016/j.jpowsour.2015.12.139

Lin H, Gao W, Meng F, Liao B-Q, Leung K-T, Zhao L, Hong, Chen J, Huachang (2012) Membrane bioreactors for industrial wastewater treatment: a critical review. Crit Rev Environ Sci Technol 42(7):677–740

Lofrano G, Meriç S, Zengin GE, Orhon D (2013) Chemical and biological treatment technologies for leather tannery chemicals and wastewaters: a review. Sci Total Environ 461:265–281

Lou X, Fang C, Geng Z, Jin Y, Xiao D, Wang Z, Liu J, Guo Y (2017) Significantly enhanced base activation of peroxymonosulfate by polyphosphates: kinetics and mechanism. Chemosphere 173:529–534

Luo H, Zeng Y, He D, Pan X (2020) Application of iron-based materials in heterogeneous advanced oxidation processes for wastewater treatment: a review. Chem Eng J 127191. https://doi.org/10.1016/j.cej.2020.127191

Mannucci A, Munz G, Mori G, Lubello C (2010) Anaerobic treatment of vegetable tannery wastewaters: a review. Desalination 264(1–2):1–8

Mantzavinos D, Psillakis E (2004) Enhancement of biodegradability of industrial wastewaters by chemical oxidation pre-treatment. J Chem Technol Biotechnol 79(5):431–454

Martínez-Huitle CA, Ferro S (2006) Electrochemical oxidation of organic pollutants for the wastewater treatment: direct and indirect processes. Chem Soc Rev 35(12):1324–1340. https://doi.org/10.1039/B517632H

Martínez-Huitle CA, Panizza M (2018) Electrochemical oxidation of organic pollutants for wastewater treatment. Curr Opin Electrochem 11:62–71. https://doi.org/10.1016/j.coelec.2018.07.010

Martínez SB, Pérez-Parra J, Suay R (2011) Use of ozone in wastewater treatment to produce water suitable for irrigation. Water Resour Manage 25(9):2109–2124

Mehrjouei M, Müller S, Möller D (2015) A review on photocatalytic ozonation used for the treatment of water and wastewater. Chem Eng J 263:209–219

Merényi G, Lind J, Naumov S, Von Sonntag C (2010) Reaction of ozone with hydrogen peroxide (peroxone process): a revision of current mechanistic concepts based on thermokinetic and quantum-chemical considerations. Environ Sci Technol 44(9):3505–3507. https://doi.org/10.1021/es100277d

Miklos DB, Remy C, Jekel M, Linden KG, Drewes JE (2018) Evaluation of advanced oxidation processes for water and wastewater treatment – a critical review. Water Res. https://doi.org/10.1016/j.watres.2018.03.042

Nawrocki J, Kasprzyk-Hordern B (2010) The efficiency and mechanisms of catalytic ozonation. Appl Catal B 99(1–2):27–42

Parsons S, Williams M (2004) Advanced oxidation processes for water ans wastewater treatment

Paździor K, Bilińska L, Ledakowicz S (2019) A review of the existing and emerging technologies in the combination of AOPs and biological processes in industrial textile wastewater treatment. Chem Eng J 376:120597

Poyatos JM, Muñio MM, Almecija MC, Torres JC, Hontoria E, Osorio F (2010) Advanced oxidation processes for wastewater treatment: state of the art. Water Air Soil Pollut 205(1):187–204

Priyadarshini M, Ahmad A, Das S, Ghangrekar MM (2021) Application of microbial electrochemical technologies for the treatment of petrochemical wastewater with concomitant valuable recovery: a review. Environ Sci Pollut Res 1–20

Rice RG (1996) Applications of ozone for industrial wastewater treatment—a review. Ozone Sci Eng 18(6):477–515

Rivas FJ, Beltra FJ (2005) Iron type catalysts for the ozonation of oxalic acid in water. Water Res 39:3553–3564. https://doi.org/10.1016/j.watres.2005.06.018

Sadegh H, Ali GAM, Nia HJ, Mahmoodi Z (2019) Nanomaterial surface modifications for enhancement of the pollutant adsorption from wastewater: adsorption of nanomaterials. Nanotechnol Appl Environ Eng 143–170

Sauleda R, Brillas E (2001) Mineralization of aniline and 4-chlorophenol in acidic solution by ozonation catalyzed with Fe^{2+} and UVA Light. Appl Catal B 29:135–145

Serna-galvis EA, Silva-agredo J, Botero-coy AM, Moncayo-lasso A, Hernández F, Torres-palma RA (2019) Science of the total environment effective elimination of fifteen relevant pharmaceuticals in hospital wastewater from colombia by combination of a biological system with a sonochemical process. Sci Total Environ 670:623–632. https://doi.org/10.1016/j.scitotenv.2019.03.153

Shestakova M, Sillanpää M (2017) Electrode materials used for electrochemical oxidation of organic compounds in wastewater. Rev Environ Sci Biotechnol 16(2):223–238. https://doi.org/10.1007/s11157-017-9426-1

Shukla S, Khan R, Daverey A (2021) Synthesis and characterization of magnetic nanoparticles, and their applications in wastewater treatment: a review. Environ Technol Innov 24:101924. https://doi.org/10.1016/j.eti.2021.101924

Stasinakis AS (2008) Use of selected advanced oxidation processes (AOPs) for wastewater treatment–a mini review. Global NEST J 10(3):376–385

Su S, Guo W, Yi C, Leng Y, Ma Z (2012) Degradation of amoxicillin in aqueous solution using sulphate radicals under ultrasound irradiation. Ultrason Sonochem 19(3):469–474

Verma M, Haritash AK (2019) Journal of environmental chemical engineering degradation of amoxicillin by Fenton and Fenton-integrated hybrid oxidation processes. J Environ Chem Eng 7(1):102886. https://doi.org/10.1016/j.jece.2019.102886

Wang B, Song Z, Sun L (2021) A review: comparison of multi-air-pollutant removal by advanced oxidation processes-industrial implementation for catalytic oxidation processes. Chem Eng J 409:128136

Wang J, Zhuan R (2019) Degradation of antibiotics by advanced oxidation processes : an overview. Sci Total Enviro 135023. https://doi.org/10.1016/j.scitotenv.2019.135023

Wang Y, Zhang H, Chen L, Wang S, Zhang D (2012) Ozonation combined with ultrasound for the degradation of tetracycline in a rectangular air-lift reactor. Separat Purificat Technol 84:138–146

Xie R, Ji J, Guo K, Lei D, Fan Q, Leung DYC, Huang H (2019) Wet scrubber coupled with UV/PMS process for efficient removal of gaseous VOCs: roles of sulfate and hydroxyl radicals. Chem Eng J 356:632–640

Zhang N, Chen J, Fang Z, Tsang EP (2019) Ceria accelerated nanoscale zerovalent iron assisted heterogenous Fenton oxidation of tetracycline. Chem Eng J 369:588–599

Zhang R, Yang Y, Huang C-H, Li N, Liu H, Zhao L, Sun P (2016) UV/H_2O_2 and UV/PDS treatment of trimethoprim and sulfamethoxazole in synthetic human urine: transformation products and toxicity. Environ Sci Technol 50(5):2573–2583

Zhao C, Chen W (2019) A review for tannery wastewater treatment: some thoughts under stricter discharge requirements. Environ Sci Pollut Res 26(25):26102–26111. https://doi.org/10.1007/s11356-019-05699-6

Zhao Q, Mao Q, Zhou Y, Wei J, Liu X (2017) Metal-free carbon materials-catalyzed sulfate radical-based advanced oxidation processes: a review on heterogeneous catalysts and applications. Chemophere 189:224–238

Microbial Biofilms in the Treatment of Textile Effluents

Anitha Thulasisingh, Shivani Kumar, Suparna Perumal, and Sathishkumar Kannaiyan

1 Introduction

The fabric enterprise is one in every of the most important water pollution with inside the global in phrases of the quantity of chemical compounds produced, the quantity of chemical compounds discharged, including the quantity of wastewater released (Tchobanoglous et al. 2003). The enterprise produces about seventy billion heaps of effluent every year. This shows how it's not safe to release the wastewater without being treated, for it's too unsafe. Textile companies' factories use custom water treatment stations because the aquatic flora cannot thrive treat textile companies' wastewater. The different techniques used by textile companies (dyeing, printing, bleaching, scrubbing, degreasing, hydrolysis, etc.) lead to different types of stains. The flow of these wastes differs greatly in terms of the chemicals they contain and their concentrations, but they can change significantly over time, even in the same factory. Laws on what textile companies can and cannot dispose of in nature vary from one country to another. Crucial way for evaluating textile waste companies is its chemical composition and contamination level. Aquatic medicine technology needs to catch up with the rapidly evolving substance companies (in the phrase of substances and methods). Prompt response to water treatment technology is paramount to stabilizing the need for criminal pollutants in the sustainability of textile companies in your price range. The largest common abundant pollutions caused by textile companies are dyes, sulphides, enzymes, starches and various other insecticides plus peroxides.

A. Thulasisingh (✉) · S. Kumar · S. Perumal
Department of Biotechnology, Rajalakshmi Engineering College,
Chennai, TamilNadu, India 602105
e-mail: anitha.t@rajalakshmi.edu.in

S. Kannaiyan
Department of Chemical Engineering, Sri Sivasubramaniya Nadar College of Engineering,
Kalavakkam, Chennai, TamilNadu 603110, India

M. P. Shah (ed.), *Advanced and Innovative Approaches of Environmental Biotechnology in Industrial Wastewater Treatment*,
https://doi.org/10.1007/978-981-99-2598-8_4

Azo dyes are the most important and most frequently used elegance of the less simple industrial dyes in the textile industry as well as in the printing, paper, imaging and tanning industries (Pazdzior et al. 2019). If all the untreated effluents were disposed improperly, they can reduce the oxygen level in aquatic life that is photosynthesis of aquatic plants and algae is affected moreover they are toxic, mutagenic and carcinogenic. These can cause anilines (Pazdzior et al. 2017; Zhu et al. 2018). Therefore, the removal of chemicals from industrial wastewater is important for the environment and public health. In general, wastewater treatment in factories consists of physical and chemical steps, usually coagulation/flocculation and tapping plus organic processes (Manekar et al. 2014). It is confirmed that within azo dye drugs, biofilms developed into stronger cells than loose cells, resulting in superior dye discolouration (Anjaneya et al. 2013). Compared to biomass loss, the use of biofilm can reduce the amount of sludge, reducing installation and operating costs. Most herbal wastewater plants include number of algae in water body at a given time and wastewater stabilization ponds, which are powerful and inexpensive techniques worldwide (Babu et al. 2010; Craggs et al. 2015).

This chapter talks on fabric effluents, its additives, and their want for treating them. The chapter additionally supplies significance to biofilms, the function it performs in fabric effluent remedy, why biofilm is favoured over others, how diverse strategies are employed, benefits and drawbacks of biofilms in treating fabric effluent. This chapter concluded with the aid of using thinking about dangerous dye additives in fabric effluent, its current research and development and destiny components in effluent remedy. The most recognized of the chapter is on superior physicochemical and organic strategies with biofilms for the remedy of fabric dye wastewaters.

2 Textile Effluents

Textile company produces the world's most polluted wastewater. Textile mill wastewater is also very rich and contains many pollutants. This makes fabric turbine wastewater treatment difficult, unique and expensive. Although many possible treatment methods are currently used by organized companies, these methods are often a way to create a blueprint from one power source to another if they are not critical or sustainable. As organizational companies evolve, research on hydrotherapy must further keep pace with the growing demand. The main reason for wastewater treatment by fabric companies is to maximize the efficiency of decontamination even when wastewater treatment is judged to be eco-friendly or harmless to the local community. During the 10-year break, there was a significant change in the cap's ability to reduce the demand for natural oxygen and ammonium in wastewater. These trends reinforce the following questions: Could the significant use of this technology be more effective for your industry? It analyses the use of cardiac biomineralization, the use of durable floor-mounted microbial biofilms, and batches in a mobile mattress bioreactor (Moga et al. 2018) (Fig. 1).

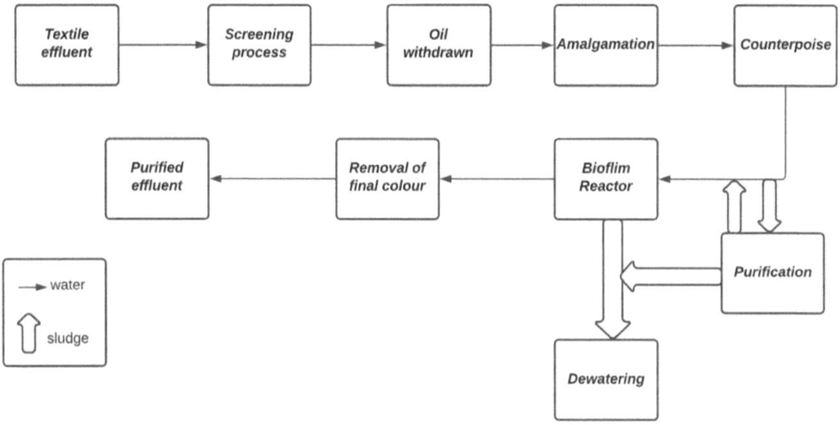

Fig. 1 Treatment methodology of textile effluents

3 Biofilms

Biofilms are created by microbes when they are interconnected with each other and surfaces to form aggregates that stick together by the slime microbes secrete (Gross et al. 2007) (Wimpenny et al. 2000). These are used in biofilm-based reactions as biocatalysts (Li et al. 2007). Spontaneously biofilms are produced. Several steps are present in the growth of biofilms, cell attachment on surface of solids, extracellular polymeric substances (EPS) for adhesion to surface, proliferation maturation of biofilms at last cell dispersion from EPS (Stoodley et al. 2002). Matrix of biofilm approx contains 97% water either as solvent or bound to microbial cell capsule. The solute that is dissolved with it determines physical properties such as viscosity, and water binding capacity (Sutherland 2001). Matrix also contains nutrients, metabolites that get absorbed, polymers that are secreted and compounds for cell lysis along with materials present in the environment where it is attached (Rummel et al. 2017). External environment contains many macromolecules of different classes such as proteins, DNA, RNA and polysaccharides along with lipids, peptidoglycan, phospholipids and some cell particles (Sutherland 2001). The organization of biofilms is altered or affected by the mass transport event of biofilm, where substrate availability plays a role. Convective transport for solute movement occurs between water channel and pores while for denser clumps diffusion takes place (Horn and Morgenroth 2006) (Fig. 2).

The availability of moist environment with required nutrients, aids in the growth of biofilms. A particular species of bacteria or multiple fungi, bacteria, protozoa and algae can form biofilms (Sutherland 2001). Bacterial biofilms are more frequent. There is a large concentration of micro-heterogeneity as multiple microenvironments co-exist together.

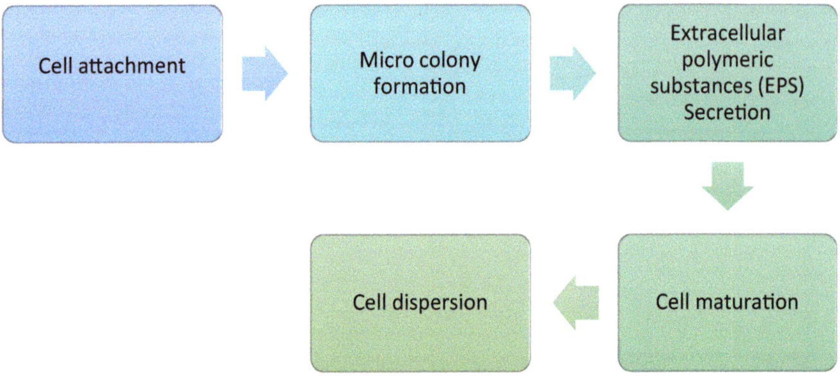

Fig. 2 Biofilm cycle

Table 1 Biofilm application for textile dye treatment

Method	Wastewater	Results	Reference
Anaerobic Biofilm method	Textile azo dyes wastewater	Usage of photo Fenton oxidization after biological treatment by eliminating toxic effluents from textile industry	Donkadokula et al. (2020)
Moving bed biofilm reactor	Textile dye wastewater	COD removed by the MBBR process for treating dye wastewater	Punzi et al. (2015)
Dye-degrading bacterial strain	Textile wastewater	Efficient textile wastewater treatment	Kee et al. (2014)

3.1 Applications of Biofilms

For the past few years biofilms are a great success and are used commercially as they are reliable and easy to work with (Qureshi et al. 2005). Multiple aspects of bioremediation use biofilms (Ahmad et al. 2020). Hetero-species biofilms are applied in wastewater treatment, soil contamination and off-gas treatment. Acetic acid bacteria in production of vinegar is an example of food industry where the process uses trickle bed reactor (Rosche et al. 2009). Biofilms used in catalytic reactions prevents degradation or spoilage of biocatalyst for many months, an industrial catalysis application (Liu et al. 2005). Major role of biofilms is acid removal for acid-mine drains in mining activities. Oxygen availability is increased for the remediation process with bacterial biofilms (Bang et al. 2000). Sugar, glycerol and starch are converted to alcohols or acid using single biofilm species is an agricultural application (Table 1).

3.2 *Characterization of Biofilms*

There are numerous methods that's being used to determine about the biofilm colonies, from the traditional methods and modern ways. Some of those methods are being discussed.

3.2.1 Weight Determination

This is a basic method where the dry and wet weight of biofilms is being compared by a digital weighing balance. The laminar flow chamber where the wet biofilm is obtained on soft rinsing using distilled water, is dried till the weight of polystyrene and polypropylene filter media is gained under aseptic condition (Naz et al. 2013).

3.2.2 Heterotrophic Plate Count (HPC)

The standard method of serial dilution is applied to find the heterotrophic plate count concentration (HPC/ml). Pure culture of biofilm is prepared by dissolving biofilms in 0.9% saline solution, diluted 10–5 times which is layered out on growth media plate. At 37 °C it's incubated for 24–48 h. Further differentiation of the incubated culture is done using microscopic, biochemical tests and colony morphology.

3.2.3 Microscopic Analysis

Visualizing biofilms is accurate by using microscopic methods which happens without any disturbance in structure. Electron microscopy (EM) and light microscopy (LM) are the old school methods for image analysis. Nevertheless, the morphology and topography of bacteria and the relation or network of biofilms are displayed by using scanning electron microscopy (SEM). There are also advanced technologies of magnetic resonance imaging (MRI), scanning transmission X-ray microscopy (STXM) and laser scanning microscopy (LSM) where shape, chemical and microbe's composition, and further in situ exploration could be carried out (Palmer and Sternberg 1999).

3.2.4 Microbial Finger Printing

A contrast of microbes and their colonies, i.e., biofilms overall profile in characterizing the constituents in a bio-molecule like DNA, RNA and phospholipids is done by methods of microbial fingerprinting (Ahmad et al. 2011). Terminal restriction fragment length polymorphism (T-RFLP), denaturing gradient gel electrophoresis (DGGE) and phospholipids fatty acid analysis (PLFA) are sub-methods of microbial printing (Thies 2007).

3.2.5 Biofilm Filter Media

For textile treatment filter or the packing media which indeed is fundamental of the biofilm growth on their surface. Common properties of size, chemical and erosion resistance, density and porosity are determined along with the media being insoluble, durable and has chemical resistance (Christensson and Welander 2004). In order to prevent clogging and boost ventilation for biofilm growth the ideal filter medium should have high porosity, high surface area and at a cheaper or cost-effective way. The hydrodynamic factors of reactor such as the geometry and surface area of support material influence the water treatment as they alter the reactor condition and affect the formation of biofilm (Yu et al. 2008). Nowadays various natural and synthetic substances are used as filters. Usage of pebbles, polypropylene, polystyrene and tire-derived rubber by researchers for fixed biofilm reactor in textile effluent treatment's bio-filter medium. The basic composition of the media and its chemical concentration, in par with media compatibility to the biofilms growing has to be assessed. The spectroscopic techniques of X-ray energy-dispersive spectroscopy (EDS or XEDS) and photoelectron X-ray spectroscopy (XPS) are mainly employed in quantifying and detecting the filter medium. Surface properties or chemicals of the medium are done with a chemical analysis technique XPS where the element's electronic state that is available within a material, empirical formula and chemical nature of the primary composition with parts per thousand range. Similarly, the chemical characters or the basic analysis of filter medium can be employed with the technique of EDS (Naz et al. 2015).

3.3 *Techniques for Textile Effluent Treatment Using Biofilm*

Treatment of textile effluent using biofilms is a prominent way compared to other biological methods (Sehar and Naz 2016; Shah 2020). We know that biofilms are microbial colonies where extracellular polymeric substance (EPS) binds them together, which is a five-step process (Naresh Yadav et al. 2021). Solid support media (SSM), suspended growth reactors contain them so as to make the biofilm attach and multiply. The pollutant disintegration and concentration of microbes have been increased as a result of SSM (Pal et al. 2010). The EPS matrix of biofilms contains compounds containing phosphorous, nitrogen, carbon and as well as confined microorganisms and wastewater nutrients that are broken down by the microbe colony present inside the biofilm.

3.3.1 Biofilm Reactors

Development of reactors for biofilms is a reason for the various applications of biofilms. The design of biofilm reactor needs to take productivity, product concentration stability, attachment of biofilms and aeration process into consideration

(Nicolella et al. 2000; Qureshi et al. 2005; Shah 2021a, b). Static surface biofilms are cultured in packed bed, trickle bed and submerged reactors where trickle bed can attain aeration easily. Flocculating biofilms or suspended particles use fluidized bed and airlift reactor where much productivity is attained as its functional mixing, abundant surface area and mass transfer properties. The major disadvantages that affect the production of reactor, would be biofilm's heterogeneity (Werner et al. 2004), genetic instability in cells of biofilms and active biomass concentration varies due to long usage. In spite of the disadvantages reactors are used for manufacturing catalysis as the productivity attained is comparatively large than suspension cell reactor (Liu et al. 2005).

3.3.2 Membrane Bioreactor (MRBs) for Biofilms

Provided with ultrafiltration (UF) and low-pressure microfilter membranes MBRs are integration between membrane filter and growth treatment of biofilms suspends. Used mainly for industrial (textile, food, beverage) and municipal wastewater treatment. On the basis of electric charge or the relative particle size, the two-dimensional membrane segregates the effluent fluids. The two categories of MRDs are

- Hollow flat or fiber membrane equipped with bioreactor or any membrane tank where gravity or vacuum-based systems are available
- On the bioreactor exterior, pipe cartridges or pressure-operated systems are present.

About 10 or 11 subsystems are mostly available in MBRs which consist of membrane zone, fine screening area and post disinfection procedure. The pollutants/effluent reach the membrane zone after passing through a sequence of membrane filters, the initial step, i.e., degradation of effluents by microorganisms/biofilms, a biological process takes place. Working membrane unit is known as a merged set of modules (single membrane unit). In order to facilitate aeration and agitation, air is inoculated using integral diffusers. Increased BOD and COD removal efficiency, for low sludge production there is a high mixed liquor suspended solids (MLSSs) concentration, and no need for clarification or disinfection or polishing are the major advantages of MBRs. The drawback would be fouling of membrane surface, clogging and high capital and maintaining cost (Judd 2008).

3.3.3 Biofilms Carrier Containing Zn Nanoparticles

For textile effluent this cost-effective and easy method is considered to be more apt (Li et al. 2015). Sometimes there might be problems with the low efficiency of treatment due to imbalance of micronutrients. Micronutrients like Zn, Ni, Cu, Fe and so on are required at lower concentration as cofactors of enzymes but at the same time they can be toxic or just inhibitory to microbes or the reactions, destructing the binding, structure or activity. For nitrite oxidizing bacteria (NOB), bacteria oxidizing ammonia and nitrifying bacteria the frequent micronutrient present is zinc (Zhang et al. 2017). Higher efficiency of removal in activated sludge action Zn can aid to the dye-degrading bacteria gathering. Controlled release of Zn should be carried out so that the required concentration can be given for the growth of microorganism (Bishara and Domb 2005). One method for controlled release would be using PLA (poly lactic acid), aliphatic polymer that has hydrolysable ester bond, and forms water and carbon-di-oxide on degradation. PLA is biodegradable and assists in Zn release during different time intervals for a long period. Proteins, peptides, narcotics and anesthetics are components that are administered (Kumar et al. 2017). In support of different industrial effluent treatment, moving bed biofilm reactor (MBBR) is being used as it has increased biomass attachment. Hard plastics like high-density polyethylene (HDPE) and ethylene (PE), are employed to produce the core of MBBR, i.e., biocarrier. Biocarrier aids in microbial adhesion and hence they can affect the quality of attachment, biofilm-forming ability and efficiency of treating effluents (Barwal and Chaudhary 2014). Bioreactors which has biocarrier incorporated with Zn Nanoparticles are developed with controlled release of zinc toenhance the bioactivity. Compounding technology of melting and mixing PLA, HDPE and Zn nanoparticles is used to form the biocarriers. The amount of Zn liberated is directly proportional to the amount of Zn NPs loaded. The optimal microbial stimulation for growth and efficiency of effluent treatment is for biocarrier of 20 wt % Zn NPs where 0.4 mg/L zinc ion is released at a time using controlled release (Wang et al. 2018a).

3.3.4 Algal Biofilm

Plant architectures have been studied in relation to microalgae in wastewater treatment agents (Santiago et al. 2013). Floating algae biomass and waste stabilization basins make up the majority of phytoplankton treatment structures, and they're widely used as a potent and low-cost technology around the world (Craggs et al. 2015).

The symbiotic movement of microalgae and microorganisms is the foundation of wastewater treatment (Aravantinou et al. 2017). Algae use CO_2 as a source of carbon, nutrient and as a mild source of electricity in order to provide O_2 for cardiac heterotrophic degradation of natural substances by microorganisms (Boelee et al. 2014). The proliferation of microalgae in wastewater and the necessity for algal cell extraction costs are the main disadvantages of these structures (Vergini et al. 2016; Wang et al. 2018b). Microalgae are estimated to account for 60–90% of the organic oxygen requirements in sewage pond drainage (Stoodley et al. 2002), whereas the biomass harvesting fee from me is 21% of the entire open pond. Algal biofilm culture has gained popularity in recent years. Harvesting and drainage costs can be greatly decreased because the biomass produced is much more (Schnurr and Allen 2015). Many parameters, including adjutants, low intensity, temperature, nutrients, substrates, extracellular polymer compounds and species interactions, keep the growth of algae biofilms at a minimum.

Important types of algal biofilms were discussed in the literature based on the motion of the helping material: rotating and desk-bound algal biofilms. The helping fabric is partially submerged in a pond in spinning biofilms, in such a way to form a contact between liquid and air medium. Scraping supportive fabric harvests the algal biomass (Blanken et al. 2014). The biofilm forms on any vertically, horizontally or even voluntarily positioned levelled surfaces in desk-bound reactors, with a thin layer of liquid flowing across it (Naumann et al. 2013). Another option for selling biofilm increase is by submerging helping cloth in a pond (Genin et al. 2014). To investigate biofilm formation, a spinning hollow drum has been developed for tertiary wastewater treatment and investigated the excellent growth of substrata on wire or sheet association. Cotton wire is said to be the most basic substratum. In order for the determination of the impact of natural carbon quantity of biomass in biofilm (Shi et al. 2014), used immobilized microalgae to create a double layer photo bioreactor on removing N and P from primary secondary effluent. Genin et al. (2014) created a parallel plate air elevate reactor to investigate the increase kinetics of integrated traditional biofilms by growing algae on a saline medium.

In phototrophic biofilm reactors, the effects of harvesting biofilm with biomass production and removal of nutrients were investigated. Artificial wastewater was treated in a vertical and horizontal reactor. The biofilm was produced in the vertical reactor using polyethylene-polymer completely geotextile, in turn positioned on a nonwoven geotextile, and each layer was held in place by a polypropylene assist plate. Evaluating the impact of the biofilm on nutrient removal in an open photo bioreactor over a long length of time and findings of reactor wall-connected biofilm supported high phosphorus and nitrogen removal. The goal of this research was to investigate about eliminating vitamins from secondary-treated effluent in lab-scale ponds, as well as algae biomass attachment to unusual surfaces. The pond featured vertically aligned geotextiles, polyethylene and cotton textiles sheets to aid in the improvement of biofilm using algae. The behaviour of pond either batch or non-stop waft studies, as well as the contentt of lipid present in the suspended and linked materials was aslo studied for algae based biofilms (Su et al. 2016). The table shows various methods and microbes used for textile effluent treatment (Table 2).

4 Other Methods

4.1 Enhanced Wastewater Oxidation Process

The inorganic and organic contaminants in wastewater can be eliminated using chemical treatment methods. Persistent organic pollutants in wastewater are rapidly degraded using strong oxidizing agents (Deng and Zhao 2015).

Table 2 Additional textile effluent removal method using biofilms

S. No	Textile effluent	Method	Biofilm	Reference
1.	Textile wastewater collected from a small-scale zipper factory	Core nitrogen cycle within biofoulant in full-scale anoxic and oxic biofilm membrane bioreactor	Anoxic and oxic biofilm	Zhou et al. (2021)
2.	Methyl orange [MO] containing effluent	Decolourization, degradation and detoxification of carcinogenic sulfonated azo dye methyl orange	Biofilm consortia e.g. *Escherichia coli, Enterobacter asburiae and E. ludwigii*	Haque et al. (2021)
3.	The neighbour-hood treatment plant supplied textile wastewater and sludge	The successful treatment of effluent, where bio-granules incorporated with specific decolourizing bacteria, with consecutive anaerobic–aerobic cycle	*Brevibacillus panacihumi strain, Bacillus pumilus strain, Lysinibacillus fusiformis strain, and Bacillus cereus strain*	Kee et al. (2014)

4.2 Nano Catalytic Applications of Wastewater Treatment

The advanced oxidation process (AOP) has received outstanding recognition for advances in wastewater treatment technology over 20 years. Several techniques, including fenton, cavitations, ozonation and photocatalytic oxidation, have been effectively used to decompose persistent organic pollutants at pilot scale. Catalysts like TiO_2, ZnO, UV, visible light, different oxidizing agents H_2O_2, O_3 were used for wastewater treatment implementing nanoforms. Several studies of various AOPs have demonstrated successful degradation of dye contaminants in aqueous media. The use of nanomaterials in AOPs for treatment of effluent in many previous studies has shown the use of micro catalytic catalysts to break up organic complexes in various AOPs (Geng et al. 2009). Amid the photocatalytic materials, ZnO is taken as one of the best materials in efficient decomposition of pollutants. Diverse study has shown ZnO is photochemically reactive and inexpensive compared to TiO_2 and also is non-toxic to decomposition of organic pollutants (Strunk et al. 2009).

4.3 Ceramic Membrane for Wastewater Filtration Purification

Recently, ceramic membrane has attracted a lot of attention due to its, long duty cycle, easy cleaning, recyclability, excellent chemical reliability and excellent contamination-free properties in water purification applications. Wastewater with the exception of reverse osmosis (2011), the remaining three membranes are used in treatment of industrial wastewater (Fig. 3).

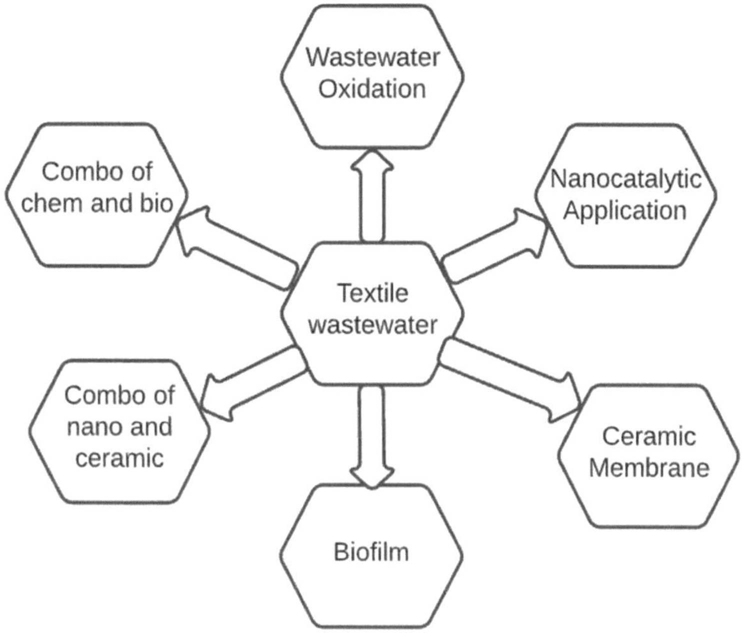

Fig. 3 Methods involved in the treatment of textile effluents

5 Future Prospects

This chapter concludes that the various techniques examined can be efficiently used to treat textile effluent. Different wastewater treatment methods must be integrated into an optimized polluted water treatment system and emission standards must be met. The technologies discussed in the present are the top choices for small-scale as well as large-scale real industrial wastewater treatment. In the present scenario, this treatment is of considerable interest to environmental organizers. These methods are considered viable because of their potentially longer life cycle, ease of maintenance, low capital cost and ability to purify wastewater from many organic pollutants. However, only a few studies have discussed the cost of treatment, and when cost data are not available, the best, non-restrictive technique becomes difficult. In the near future, researchers will need to focus on analysing the cost of processing, not just processing complex dyes. We have investigated and reviewed the development progress in photocatalytic material use and biofilm choice in textile effluent treatment (Donkadokula et al. 2020).

6 Conclusion

A large volume of water is utilized in textile industry leading to discharge of inorganic and organic effluents. To safeguard our environment these pollutants need to be treated and reused. This chapter focuses mainly on physicochemical and biological techniques in textile effluent treatment. Apart from being a significant economy contributor, textile companies have a harmful impact on environment. They consume lots of energy and at the same time pollutants release is also high. Xenobiotic compounds present in the effluent have dangerous impact on terrestrial and aquatic ecosystem. The advantage of applying biological methods, is that they are eco-sustainable and have the ability to degrade different xenobiotic compound. In this chapter we included about textile effluents, its components and the need for treating it, also about biofilms, the role it plays in eradicating textile effluent, techniques involved in exterminating hazardous dye components in textile effluent, its recent studies and development, how various methods are employed in this but why biofilm is superlative over others, advantages and disadvantages of biofilms in treating textile effluent, discussion and finally concluding it with future aspects.

References

Ahmad I, Ahmad F, Pichtel J (eds) (2011) Microbes and microbial technology: agricultural and environmental applications. Springer, London. https://doi.org/10.1007/978-1-4419-7931-5

Ahmad F, Zhu D, Sun J (2020) Bacterial chemotaxis: a way forward to aromatic compounds biodegradation. Environ Sci Eur. https://doi.org/10.1186/S12302-020-00329-2

Anjaneya O, Shrishailnath SS, Guruprasad K, Nayak AS, Mashetty SB, Karegoudar TB (2013) Decolourization of Amaranth dye by bacterial biofilm in batch and continuous packed bed bioreactor. Int Biodeter Biodegr 79:64–72. https://doi.org/10.1016/J.IBIOD.2013.01.006

Aravantinou AF, Barkonikou EF, Manariotis ID (2017) Microalgae biomass and lipid production using primary treated wastewater. Desalin Water Treat 91:228–234. https://doi.org/10.5004/DWT.2017.21069

Babu MA, Hes EMA, van der Steen NP, Hooijmans CM, Gijzen HJ (2010) Nitrification rates of algal-bacterial biofilms in wastewater stabilization ponds under light and dark conditions. Ecol Eng 36(12):1741–1746. https://doi.org/10.1016/J.ECOLENG.2010.07.023

Bang SW, Clark DS, Keasling JD (2000) Engineering hydrogen sulfide production and cadmium removal by expression of the thiosulfate reductase gene (phsABC) from Salmonella enterica serovar typhimurium in *Escherichia coli*. Appl Environ Microbiol 66(9):3939–3944. https://doi.org/10.1128/AEM.66.9.3939-3944.2000

Barwal A, Chaudhary R (2014) To study the performance of biocarriers in moving bed biofilm reactor (MBBR) technology and kinetics of biofilm for retrofitting the existing aerobic treatment systems: a review. Rev Environ Sci Bio Technol 13(3):285–299. https://doi.org/10.1007/S11157-014-9333-7

Bishara A, Domb AJ (2005) PLA stereocomplexes for controlled release of somatostatin analogue. J Control Release 107(3):474–483. https://doi.org/10.1016/J.JCONREL.2005.05.026

Blanken W, Janssen M, Cuaresma M, Libor Z, Bhaiji T, Wijffels RH (2014) Biofilm growth of Chlorella sorokiniana in a rotating biological contactor based photobioreactor. Biotechnol Bioeng 111(12):2436–2445. https://doi.org/10.1002/BIT.25301

Boelee NC, Temmink H, Janssen M, Buisman CJN, Wijffels RH (2014) Balancing the organic load and light supply in symbiotic microalgal-bacterial biofilm reactors treating synthetic municipal wastewater. Ecol Eng 64:213–221. https://doi.org/10.1016/J.ECOLENG.2013.12.035

Christensson M, Welander T (2004) Treatment of municipal wastewater in a hybrid process using a new suspended carrier with large surface area. Water Sci Technol 49(11–12):207–214. https://doi.org/10.2166/WST.2004.0843

Craggs R, Park J, Sutherland D, Heubeck S (2015) Economic construction and operation of hectare-scale wastewater treatment enhanced pond systems. J Appl Phycol 27(5):1913–1922. https://doi.org/10.1007/S10811-015-0658-6

Deng Y, Zhao R (2015) Advanced oxidation processes (AOPs) in wastewater treatment. Curr Pollut Rep 1(3):167–176. https://doi.org/10.1007/S40726-015-0015-Z/TABLES/2

Donkadokula NY, Kola AK, Naz I, Saroj D (2020) A review on advanced physico-chemical and biological textile dye wastewater treatment techniques. Rev Environ Sci Biotechnol 19(3):543–560. https://doi.org/10.1007/S11157-020-09543-Z/TABLES/5

Geng B, Jin Z, Li T, Qi X (2009) Kinetics of hexavalent chromium removal from water by chitosan-Fe0 nanoparticles. Chemosphere 75(6):825–830. https://doi.org/10.1016/J.CHEMOSPHERE.2009.01.009

Genin SN, Stewart Aitchison J, Grant Allen D (2014) Design of algal film photobioreactors: material surface energy effects on algal film productivity, colonization and lipid content. Bioresour Technol 155:136–143. https://doi.org/10.1016/J.BIORTECH.2013.12.060

Gross R, Hauer B, Otto K, Schmid A (2007) Microbial biofilms: new catalysts for maximizing productivity of long-term biotransformations. Biotechnol Bioeng 98(6):1123–1134. https://doi.org/10.1002/BIT.21547

Haque MM, Haque MA, Mosharaf MK, Marcus PK (2021) Decolorization, degradation and detoxification of carcinogenic sulfonated azo dye methyl orange by newly developed biofilm consortia. Saudi J Biol Sci 28(1):793–804. https://doi.org/10.1016/J.SJBS.2020.11.012

Horn H, Morgenroth E (2006) Transport of oxygen, sodium chloride, and sodium nitrate in biofilms. Chem Eng Sci 5(61):1347–1356. https://doi.org/10.1016/J.CES.2005.08.027

Judd S (2008) The status of membrane bioreactor technology. Trends Biotechnol 26(2):109–116. https://doi.org/10.1016/J.TIBTECH.2007.11.005

Kee TC, Bay HH, Lim CK, Muda K, Ibrahim Z (2014) Development of bio-granules using selected mixed culture of decolorizing bacteria for the treatment of textile wastewater. New Pub Balaban 54(1):132–139. https://doi.org/10.1080/19443994.2013.877853

Kumar S, Singh S, Senapati S, Singh AP, Ray B, Maiti P (2017) Controlled drug release through regulated biodegradation of poly(lactic acid) using inorganic salts. Int J Biol Macromol 104(Pt A):487–497. https://doi.org/10.1016/J.IJBIOMAC.2017.06.033

Li XZ, Hauer B, Rosche B (2007) Single-species microbial biofilm screening for industrial applications. Appl Microbiol Biotechnol 76(6):1255–1262. https://doi.org/10.1007/S00253-007-1108-4

Li C, Zhang Z, Li Y, Cao J (2015) Study on dyeing wastewater treatment at high temperature by MBBR and the thermotolerant mechanism based on its microbial analysis. Process Biochem 50(11):1934–1941. https://doi.org/10.1016/J.PROCBIO.2015.08.007

Liu X, Chung YK, Yang ST, Yousef AE (2005) Continuous nisin production in laboratory media and whey permeate by immobilized Lactococcus lactis. Process Biochem 40(1):13–24. https://doi.org/10.1016/J.PROCBIO.2003.11.032

Manekar P, Patkar G, Aswale P, Mahure M, Nandy T (2014) Detoxifying of high strength textile effluent through chemical and bio-oxidation processes. Bioresour Technol 157:44–51. https://doi.org/10.1016/J.BIORTECH.2014.01.046

Moga IC, Ardelean I, Petrescu G, Crăciun N, Popa R (2018) The potential of biofilms from moving bed bioreactors to increase the efficiency of textile industry wastewater treatment. Ind Textila 69(5):412–418. https://doi.org/10.35530/IT.069.05.1500

Naresh Yadav D, Naz I, Anand Kishore K, Saroj D (2021) Evaluation of tire derived rubber (TDR) fixed biofilm reactor (FBR) for remediation of Methylene blue dye from wastewater. Environ Technol 42(23):3627–3640. https://doi.org/10.1080/09593330.2020.1737736

Naumann T, Cebi Z, Podola B, Melkonian M (2013) Growing microalgae as aquaculture feeds on twin-layers: a novel solid-state photobioreactor. J Appl Phycol 25(5):1413–1420. https://doi.org/10.1007/S10811-012-9962-6

Naz I, Batool SAU, Ali N, Khatoon N, Atiq N, Hameed A, Ahmed S (2013) Monitoring of growth and physiological activities of biofilm during succession on polystyrene from activated sludge under aerobic and anaerobic conditions. Environ Monit Assess 185(8):6881–6892. https://doi.org/10.1007/S10661-013-3072-Z

Naz I, Seher S, Perveen I, Saroj DP, Ahmed S (2015) Physiological activities associated with biofilm growth in attached and suspended growth bioreactors under aerobic and anaerobic conditions. Environ Technol 36(13):1657–1671. https://doi.org/10.1080/09593330.2014.1003614

Nicolella C, Van Loosdrecht MCM, Heijnen JJ (2000) Wastewater treatment with particulate biofilm reactors. J Biotechnol 80(1):1–33. https://doi.org/10.1016/S0168-1656(00)00229-7

Pal S, Sarkar U, Dasgupta D (2010) Dynamic simulation of secondary treatment processes using trickling filters in a sewage treatment works in Howrah, west Bengal India. Desalination 253(1–3):135–140. https://doi.org/10.1016/J.DESAL.2009.11.019

Palmer RJ, Sternberg C (1999) Modern microscopy in biofilm research: confocal microscopy and other approaches. Curr Opin Biotechnol 10(3):263–268. https://doi.org/10.1016/S0958-1669(99)80046-9

Pazdzior K, Wrebiak J, Klepacz-Smołka A, Gmurek M, Bilinska L, Kos L, Sojka-Ledakowicz J, Ledakowicz S (2017) Influence of ozonation and biodegradation on toxicity of industrial textile wastewater. J Environ Manage 195(Pt 2):166–173. https://doi.org/10.1016/J.JENVMAN.2016.06.055

Pazdzior K, Bilinska L, Ledakowicz S (2019) A review of the existing and emerging technologies in the combination of AOPs and biological processes in industrial textile wastewater treatment. Chem Eng J 376:120597. https://doi.org/10.1016/J.CEJ.2018.12.057

Punzi M, Anbalagan A, Aragão Börner R, Svensson BM, Jonstrup M, Mattiasson B (2015) Degradation of a textile azo dye using biological treatment followed by photo-Fenton oxidation: evaluation of toxicity and microbial community structure. Chem Eng J 270:290–299. https://doi.org/10.1016/J.CEJ.2015.02.042

Qureshi N, Annous BA, Ezeji TC, Karcher P, Maddox IS (2005) Biofilm reactors for industrial bioconversion process: employing potential of enhanced reaction rates. Microb Cell Fact 4. https://doi.org/10.1186/1475-2859-4-24

Rosche B, Li XZ, Hauer B, Schmid A, Buehler K (2009) Microbial biofilms: a concept for industrial catalysis? Trends Biotechnol 27(11):636–643. https://doi.org/10.1016/J.TIBTECH.2009.08.001

Rummel CD, Jahnke A, Gorokhova E, Kühnel D, Schmitt-Jansen M (2017) Impacts of biofilm formation on the fate and potential effects of microplastic in the aquatic environment. Environ Sci Technol Lett 4(7):258–267. https://doi.org/10.1021/ACS.ESTLETT.7B00164/SUPPL_FILE/EZ7B00164_SI_001.PDF

Santiago AF, Calijuri ML, Assemany PP, Calijuri MDC, Dos RAJD (2013) Algal biomass production and wastewater treatment in high rate algal ponds receiving disinfected effluent. Environ Technol 34(13–16):1877–1885. https://doi.org/10.1080/09593330.2013.812670

Schnurr PJ, Allen DG (2015) Factors affecting algae biofilm growth and lipid production: a review. Renew Sustain Energy Rev 52:418–429. https://doi.org/10.1016/J.RSER.2015.07.090

Sehar S, Naz I (2016) Role of the Biofilms in wastewater treatment. Microb Biofilms Importance Appl. https://doi.org/10.5772/63499

Shah MP (2020) Microbial bioremediation & biodegradation. Springer

Shah MP (2021a) Removal of refractroty pollutants from wastewater treatment plants. CRC Press

Shah MP (2021b) Removal of emerging contaminants through microbial processes. Springer

Shi J, Podola B, Melkonian M (2014) Application of a prototype-scale Twin-Layer photobioreactor for effective N and P removal from different process stages of municipal wastewater by immobilized microalgae. Bioresour Technol 154:260–266. https://doi.org/10.1016/J.BIORTECH.2013.11.100

Stoodley P, Sauer K, Davies DG, Costerton JW (2002) Biofilms as complex differentiated communities. Annu Rev Microbiol 56:187–209. https://doi.org/10.1146/ANNUREV.MICRO.56.012302.160705

Strunk J, Kahler K, Xia X, Muhler M (2009) The surface chemistry of ZnO nanoparticles applied as heterogeneous catalysts in methanol synthesis. Surf Sci 603(10–12):1776–1783. https://doi.org/10.1016/J.SUSC.2008.09.063

Su Y, Mennerich A, Urban B (2016) The long-term effects of wall attached microalgal biofilm on algae-based wastewater treatment. Bioresour Technol 218:1249–1252. https://doi.org/10.1016/J.BIORTECH.2016.06.099

Sutherland IW (2001) The biofilm matrix – an immobilized but dynamic microbial environment. Trends Microbiol 9(5):222–227. https://doi.org/10.1016/S0966-842X(01)02012-1

Tchobanoglous G, Burton FL, Stensel HD (2003) Wastewater engineering treatment and reuse. *Wastewater Eng Treat Reuse*:1–24

Thies JE (2007) Soil microbial community analysis using terminal restriction fragment length polymorphisms. Soil Sci Soc Am J 71(2):579–591. https://doi.org/10.2136/SSSAJ2006.0318

Vergini S, Aravantinou AF, Manariotis ID (2016) Harvesting of freshwater and marine microalgae by common flocculants and magnetic microparticles. J Appl Phycol 2(28):1041–1049. https://doi.org/10.1007/S10811-015-0662-X

Wang F, Zhou L, Zhao J (2018a) The performance of biocarrier containing zinc nanoparticles in biofilm reactor for treating textile wastewater. Process Biochem 74:125–131. https://doi.org/10.1016/J.PROCBIO.2018.08.022

Wang JH, Zhuang LL, Xu XQ, Deantes-Espinosa VM, Wang XX, Hu HY (2018b) Microalgal attachment and attached systems for biomass production and wastewater treatment. Renew Sustain Energy Rev 92:331–342. https://doi.org/10.1016/J.RSER.2018.04.081

Werner E, Roe F, Bugnicourt A, Franklin MJ, Heydorn A, Molin S, Pitts B, Stewart PS (2004) Stratified growth in Pseudomonas aeruginosa biofilms. Appl Environ Microbiol 70(10):6188–6196. https://doi.org/10.1128/AEM.70.10.6188-6196.2004

Wimpenny J, Manz W, Szewzyk U (2000) Heterogeneity in biofilms: Table 1. FEMS Microbiol Rev 24(5):661–671. https://doi.org/10.1111/J.1574-6976.2000.TB00565.X

Yu Y, Feng Y, Qiu L, Han W, Guan L (2008) Effect of grain-slag media for the treatment of wastewater in a biological aerated filter. Bioresour Technol 99(10):4120–4123. https://doi.org/10.1016/J.BIORTECH.2007.09.001

Zhang X, Zhou Y, Zhang N, Zheng K, Wang L, Han G, Zhang H (2017) Short-term and long-term effects of Zn (II) on the microbial activity and sludge property of partial nitrification process. Bioresour Technol 228:315–321. https://doi.org/10.1016/J.BIORTECH.2016.12.099

Zhou L, Zhao B, Ou P, Zhang W, Li H, Yi S, Zhuang WQ (2021) Core nitrogen cycle of biofoulant in full-scale anoxic & oxic biofilm-membrane bioreactors treating textile wastewater. Bioresour Technol 325. https://doi.org/10.1016/J.BIORTECH.2021.124667

Zhu Y, Xu J, Cao X, Cheng Y, Zhu T (2018) Characterization of functional microbial communities involved in diazo dyes decolorization and mineralization stages. Int Biodeterior Biodegrad 132:166–177. https://doi.org/10.1016/J.IBIOD.2018.03.006

The Challenges of Wastewater and Wastewater Management

Sunita Kumari, Smita Dwivedi, Md E A Raghib Khan, Shreya Nayanam, Archna Dhasmana, and Sumira Malik

1 Introduction

Wastewater treatment and its management is a major ongoing challenge throughout the world. The lack of sanitation (Lofrano and Brown 2010) and various human activities led to hazardous situations. Recently, the United Nations published a report on the exploitation of wastewater around the world. It happens mainly in developing or underdeveloped countries like India and Japan, where less than 8% of the wastewater is treated. Untreated and unprocessed wastewater from household chores, toilets, kitchen, other chemicals from medicine and pharmaceuticals, agricultural waste, tannery industries, and dairy industries continue to harm the aquatic life and water quality on the globe. Wastewater consists of several pollutants and contaminants such as heavy metals, plant nutrients, pathogens, organic pollutants, and other microorganisms, which affect human and natural resources and eventually lead to economic impacts on society. It leads to several waterborne diseases and water quality depletion on the earth.

Historically, it has been estimated that wastewater management is a serious concern for people and governments with wide-ranging practical and political challenges (Lofrano and Brown 2010). Even though various economists and historians have recounted the growth of wastewater management for ages (Serneri 2007) but the biggest challenges of wastewater are the absence of sludge treatment facilities, lack of disposal of sludge on the land, ocean, and air, sediment runoff, microbial, nutrient and chemical overflow, improper use of energy resources, deadly diseases, and rainwater management. It requires genuine management for future generations

S. Kumari · S. Dwivedi · M. E. A. R. Khan · S. Nayanam · S. Malik (✉)
Amity Institute of Biotechnology, Amity University Jharkhand Ranchi, Jharkhand, India
e-mail: smalik@rnc.amity.edu

A. Dhasmana
Himalayan School of Biosciences, Swami Rama Himalayan University, Dehradun, India

M. P. Shah (ed.), *Advanced and Innovative Approaches of Environmental Biotechnology in Industrial Wastewater Treatment*,
https://doi.org/10.1007/978-981-99-2598-8_5

and a sustainable environment. An equilibrium of economic, technical, political, and social goals is required while fulfilling the demands of planners, environmental enthusiasts, environmental protection agencies, and customers (Burkhard et al. 2000). This paper intends to highlight those challenges and the management strategies with their future aspect.

2 Sources of Wastewater

Wastewater is the water generated by the impurities from households, hospitals, industries, etc. Wastewater generally contains heavy metals like nitrogen and phosphorus, organic and inorganic pollutants, microorganisms, dyes and colors, and toxic chemicals from the waste produced by soaps and detergents from the household and various industries further treated in wastewater management plants. Modern societies include a growing number of interests that create pollution and wastewater. While few contaminated water sources are more recognized and distinguished, other sources either appear or remain less neglected. A recent statistic by Eurostat relating to the origin of wastewater and volume emitted by the national level is rare and insufficient. For example, it is estimated that in Germany, a Western European country with a high IHDI, an annual waste of 1010 m^3 of wastewater is comparable to 125 m^3/per citizen. In India, many water sources are polluted by religious practices, unsustainable agricultural practices, urbanization, and industrialization.

Domestic and municipal waste is a common source of wastewater as every year, the average citizen of a developed country produces about half a ton of waste. Most of this domestic water use will be wastewater that carries a large set of chemicals such as drugs and their metabolites released after improper use or disposal by toilets (Mompelat et al. 2009). It can also be produced by household activities such as washing dishes, laundry, and showers. Biologically, domestic wastewater carries various pathogens and microorganisms from human feces infected with specific diseases.

Industries are one of the primary sources of wastewater, such as textile industries and dairy industries. The dairy industry is the most polluting in terms of the amount of production and properties of wastewater produced. Around 0.2–10 L of waste per liter of processed milk is produced at an average rate of 2.5 L of wastewater per liter of processed milk. Physically, dairy wastes contain suspended solid, soluble organics and are white and generally slightly alkaline. Due to the fermentation of milk to lactic acid, it becomes acidic (Shete and Shinkar 2013). All these organic components affect the environment in different ways depending on their biodegradabilities and their solubilities, such as BODs and river oxygen levels, the color and turbidity of receiving water, and the inorganic components which are toxic to humans and livestock (Shete and Shinkar 2013).

Pharmaceuticals and medicinal products are the well-known sources of wastewater. Pharmaceutical compounds and their components are one of the main content of pollution in the environment as it majorly affects terrestrial or aquatic life. In the twenty-first century, due to development in the pharmaceutical area, the mortality rate has cautiously decreased. The chemicals and their derivatives produced in the environment by medical and veterinary products become critical contaminants to the environment. Two thousand three hundred active pharmaceutical ingredients [API] in human medicine are toxic and bio accumulative. Veterinary components have played a leading role, remains of which are often detected in surface water, soils, etc. In a few cases ground water also is affected in large amounts (Pavithra et al. 2017), which is causing several waterborne diseases to humans.

Agricultural waste is another well-known contributor to the global amount of wastewater discharged yearly worldwide. Any unwanted waste produced from agricultural practices such as manure, silage plastics, pesticides and herbicides, fertilizers waste from poultry farms and slaughtered houses, and veterinary medicines can be considered wastewater (Ramírez-García 2019). Fertilizers are agricultural waste that contains nutrients including nitrogen, phosphorus, and potassium (N, P, K), which help plants to grow faster and increase yields, but only when exposed to open space leads to eutrophication of aquatic environment which is an utmost agricultural-based issue and results in the extraordinary richness of nutrients such as carbon, phosphorus, and nitrogen. Although, many cases of eutrophication are also caused by human activity or can occur naturally (Ramírez-García 2019). Additionally, many pesticides and herbicides dissolve in water. They can leak and get into the water and enter food products, which can lead to severe health issues in humans, including cancer, Alzheimer's disease, Parkinson's disease, reproductive disorders, and congenital disabilities (Mostafalou and Abdollahi 2013). On average, farms worldwide account for 70% of all water consumed annually. Of that 70% used by the farmers, 40% is lost to the environment due to poor irrigation systems evaporation and overall poor water management.

3 The Challenges of Wastewater

Fresh water is an essential resource for the environment, and providing clean and fresh water to the people is a concerning challenge facing the world, especially in an increasingly water-scarce generation. Wastewater from urban cities through industries and other sources is filled up directly in the water sources like rivers, lakes, ponds, and seas without proper treatment leading to the contamination of water and causing a threat to the life of beings and the ecosystem. From time immemorial, wastewater has been a significant environmental and public health problem, and its management has been an ongoing challenge ever since (Villarin and Merel 2020). It reflects deep concern for human and environmental health, keeping in mind the quantity and quality of wastewater produced by homes, hospitals, and industries. Therefore, proper wastewater management is essential, including collection and treatment, minimizing potential environmental impact (Villarin and Merel 2020).

Depending on regional factors, socio-economic conditions, and waste management laws, the challenges of purifying wastewater vary (M. Hosami 2016). Thus, it is difficult to identify the same challenge that applies to all situations. The lack of financial and technical resources in most developing countries makes widespread wastewater collection and treatment a long-term strategy for the future (M. Qadir et al. 2010).

3.1 Diseases Caused by Contaminated Water

Pathogen contamination is a severe threat to how we live our day-to-day life, making it essential to understand and recognize the threat it poses. It is such a dire issue that the United Nations has identified improving water quality as one of the eight Millennium Development Goals (MDGs). In 2006, Fenwick estimated that waterborne diseases are rampant in developing nations, such as in Africa, where the disease infected millions (Fenwick 2006). An estimate from WHO is that annually 3.4 million people and primarily children die from waterborne diseases (WHO 2014). An estimate from UNICEF is that almost 4000 children die from contaminated water. WHO accessed that almost 2.6 billion of the human population lack access to clean water. Improving the water quality of these nations can reduce the global disease burden by almost 4%. Though it is a massive burden for these developing nations, even the developed nations face a considerable crisis from water-contaminated diseases. A study by Arnone and Walling, who compiled data on outbreaks in the United States (1986–2000), reported 5,905 cases and 95 outbreaks associated with recreational water (Arnone and Walling 2007). Each year 42,000 cases of *salmonella* are reported in the United States.

All in all, the mortality caused by contaminated water is enormous and has to be controlled by improving clean water security in both developing and developed countries. Water contamination has plagued human life for generations, and we can find its presence in history, such as descriptions in the Sushruta Samhita about waterborne diseases resembling cholera in an Indian text written in Sanskrit early as 500–400 B.C. Cholera, typhoid fever, diarrhoea, ulcers, hepatitis, arsenicosis, kidney damage, viral gastroenteritis, and respiratory tract infection are a few diseases caused by contaminated water that are harmful and can lead to the death of individuals (Heinrich et al. 2017). Besides the treatment and proper medication, these diseases can be cured in various ways (Heinrich et al. 2017). Protection from these diseases should have a standard to measure water cleanliness safe for consumption. The United States has introduced something similar to this, known as MCL the maximum contaminant level which is safe to drink without any known risk to public health (Table 1)

Table 1 Different types of microorganisms cause different waterborne diseases (Ashbolt 2004)

Type of microorganism	Name of microorganisms/pathogen	Major diseases	Causes of primary sources
1. Viruses	Adenoviruses	Gastrointestinal and upper respiratory illness	Human feces
	Hepatitis E virus	Infectious hepatitis; miscarriage, and death	Human feces
	Hepatitis A virus	Infectious hepatitis	Human feces
	Polio viruses	Poliomyelitis	Human feces
	Echo viruses	Aseptic meningitis	Human feces
	Coxsackie virus A	Aseptic meningitis	Human feces
	Coxsackie virus B	Aseptic meningitis	Human feces
	Rotaviruses	Gastroenteritis	Human feces
	Norovirus	Gastroenteritis	Families and water
2. Bacteria	*Vibrio cholera*	Cholera	Human feces and freshwater zooplankton
	Salmonella typhi	Typhoid fever	Human feces
	Salmonella paratyphi	Paratyphoid fever	Human feces
	Other *Salmonella*	Salmonellosis	Human and animal feces
	Yersinia enterocolitica	Gastroenteritis	Human feces
	Shigella spp.	Bacillary dysentery	Human feces
	Campylobacter jejuni	Gastroenteritis	Human and animal feces
	Leptospira spp	Leptospirosis	Animal and human urine
	Various mycobacteria	Pulmonary illness	Soil and water
	Opportunistic bacteria	Variable	Natural water
	Enteropathogenic *E. coli*	Gastroenteritis	Human feces
3. Helminths	*Ascaris lumbricoides*	Ascariasis	Animal and Human feces
4. Protozoa	*Naegleria fowleri*	Primary amoebic meningoencephalitis	Warm water
	Giardia lamblia	Giardiasis (gastroenteritis)	Animal feces and water

(continued)

Table 1 (continued)

Type of microorganism	Name of microorganisms/ pathogen	Major diseases	Causes of primary sources
	Entamoeba histolytica	Amoebic dysentery	Animal and human feces
	Cryptosporidium hominis, C. parvum	Crystospordiosis (gastroenteritis)	Water, human and other mammal faeces
	Balantidium coli	Balantidosis (Dysentry)	Human and animal feces
	Acanthamoeba castellani	Amoebic meningoencephalitis	Human feces

3.2 Agricultural Wastewater Challenges

Agriculture is a vital part of our economy in India, accounting for 17% of GDP, and most of the population is employed by it. Agriculture also consumes 70% of the world's water and contributes significantly to water contamination. Agricultural wastewater is produced by various farm activities, including the processing of agricultural products and animal feeding operations, that may contribute to the contamination of surface and groundwater in lack of proper care. The generation of wastewater has grown in tandem with population growth, urbanization, better living circumstances, and economic progress. Agriculture, which contributes to the eutrophication of coastal and inland waters and the pollution of subsurface water with nitrates and salts, has surpassed contamination from towns and industries in many high-income nations. Pesticides, fertilizer, and unsustainable land use are the primary contributors to pollution in low-income nations, raising health and environmental concerns (Evans et al. 2019). Many countries cannot use dilution to control large-scale pollution (Bibby and Peccia 2013). Below are some of the elements that contribute to agricultural contamination of the environment:

Pesticides: Fungicides, herbicides, and insect sprays should all be used responsibly and deliberately since they can pollute water resources with cancer-causing compounds and other toxic substances that can harm humans. Weeds and creepy crawlies are also killed by pesticides, threatening our biodiversity. Rather than utilizing more expensive and well-established pesticides, greater emphasis may be placed on employing modern pesticides that are less toxic and more targeted. Acute pesticide poisoning from toxic pesticide formulations used by farmers in underdeveloped nations causes human morbidity and mortality globally (Sagasta et al. 2017).

Salts: As irrigation has expanded, so has the production of brackish drainage and leaching water. Salinization occurs when salts deposited in soil are mobilized and transferred by drainage water to receiving water bodies. The geochemical cycles of significant elements such as carbon, nitrogen, iron, phosphorus, silicon, and sulfur are affected by the salinity of water, which has an overall influence on ecosystems.

It also impacts the freshwater biota, producing changes in species and, as a result, biodiversity loss and migration. It is common knowledge that when water salinity rises, microbes, algae, and marine plants and animals' biodiversity diminish (Sagasta et al. 2017).

Sediments: Sediment is a complicated mixture of organic debris and minerals that may include physical and chemical contaminants. In agriculture, unsustainable land use, inappropriate tillage, and poor soil management increase erosion and sediment flow in water bodies, resulting in massive soil loss. Contaminants such as silt and clay particles, nutrients, heavy metals, and organic pollutants are primarily transferred to bodies of water by sedimentation. Sediments can obstruct fish spawning, jam fish gills, choke streams, and diminish the amount of storage space in useable reservoirs (Sagasta et al. 2017).

Organic matter: Organic matter degrades and contributes to hypoxia by consuming the dissolved oxygen in the water. Algal blooms and eutrophication are more likely to occur in lakes, riverbeds, reservoirs, and coastal locations (Sagasta et al. 2017).

Pathogens: Viruses, bacteria, fecal coliforms, helminth eggs. The parameters that harm human health can be seen in water and food. They can survive for a long time and cause several communicable diseases (Hussain 2002) (Table 2).

The detrimental environmental consequences of agricultural wastewater cannot be underestimated since they significantly harm human health and marine life. They necessitate careful monitoring and precautionary actions. Parameters, tolerance limits, frequency, indicators, sample points, and information regarding the amount of data should all be adequately defined. Farmers must avoid cultivating the neighboring water bed using pesticides, chemicals, and synthetic fertilizers. In addition, integrated aquaculture-agriculture and forestry should be implemented, in which cattle, fish, and crops may all be cultivated together, increasing output stability. This method of agriculture guarantees that another reuses trash from one culture. This can help reduce pollution and optimize resources (Sagasta et al. 2017).

3.3 Sludge Disposal Challenges

The rising population in urban cities has given rise to a new problem. It is estimated that almost half of the population will live in urban cities until 2050. Moreover, this rising population gives rise to extreme sludge disposal issues. Sewage sludges consist of natural organic matter and nutrients that can be utilized in agriculture. However, there is concern among the public about the potential of it contaminating the environment by spreading trace elements, toxic pollutants, and pathogens to the environment. To solve this issue, innovative and cost-effective sludge treatment is a prerequisite for the safe and effective disposal of sludge.

Sludge can be defined as a residual semi-solid material that is left from municipal or industrial wastewater or sewage treatment processes. Different challenges are faced in engineering and design concerning the reuse and disposal of sludge (Zhang et al. 2017). Wastewater is generally treated via sedimentation, flotation, flocculation,

Table 2 Some impacts of agriculture on water quality

SN. No	Agricultural activity	Impacts		References
		Surface water	**Groundwater**	
1.	Plant food nutrients	with an excessive amount of N and P cause eutrophication	Nitrogen leaches into groundwater, causing adverse effects on the environment and human health	(Hussain 2002)
2.	Biodegradable organics	Causes depletion of oxygen, unsuitable habitat, and environment inhibit pond-breeding amphibians		(Hussain 2002)
3.	Irrigation	Runoff of salts leads to soil salinization, runoff of fertilizers and pesticides leads to ecological harm and bioaccumulation in edible fish species and waterlogging	Salts, especially nitrate, contaminate groundwater, and water tables rise	(Zia et al. 2013)
4.	Residual chlorine in tertiary treated wastewater	Causes leaf-tip burn, greenhouse effect	causes contamination of groundwater by leaching	(Hussain 2002)
5.	Heavy metals	Toxic to plants and animals makes wastewater unsuitable for irrigation and accumulates in irrigated soils	Metals may leach into the groundwater and contaminate it	(Hussain 2002)
6.	Plowing/Tillage	Loss of spawning and home ground, siltation of river beds, phosphorus, and pesticides carried by the sediments, and decreased light penetration	Groundwater recharge is decreased	(Zia et al. 2013)
7.	Aquaculture	Eutrophication process through pesticide runoff, feed, and feces nutrients		(Zia et al. 2013)
8.	Livestock	Drinking water quality is degraded due to bacteria, viruses, and heavy metals	Leaching of metals and nitrogen toward groundwater	(Zia et al. 2013)

and microbial treatment. The main objectives of treatment are removing contaminants, managing by-products, and meeting quality standards regarding discharged water quality (Raheem et al. 2018). Generally, sludge is used for energy generation or resource recovery. Sludge includes 59–88% w/v biodegradable organic matter (OMs), composed of 50–55% Carbon, 25–30% Oxygen, 10–15% Nitrogen, 6–10% Hydrogen with a small amount of P and S (Tyagi et al. 2013). Bacteria such as proteins, lipids, and their decay by-products with cellulose and inorganic matter from the chemical structure of sludge.

Sludge has been under high vigilance due to large concentration of heavy metals like copper (Cu), arsenic (As), chromium (Cr), lead (Pb), zinc (Zn), cadmium (Cd), nickel (Ni), mercury (Hg), and selenium (Se) (Raheem et al. 2018). The presence of such metals in high concentrations can be a problem in the extraction of biochemical materials and energy recovery from sludge. Considering the repercussions of potentially harmful heavy metals during gasification, it should be mentioned that ash dust is one of the chief carriers of Cd and Pb in turbulent gasification routes. If gasification occurs under non-turbulent conditions, particulate ash emissions are almost eliminated. Save et al. have investigated metallic species during sludge gasification, indicating that heavy metals like Zn, Cu, and Zn were nearly retained in char. At the same time, Cadmium and Mercury were depleted from the sludge and got into a different downstream section like char residues, particles in filters, and condensate liquids. Cadmium was observed during the co-gasification of solid fuel and coal (H. Wu et al. 2013). The conversion of heavy metals (Cd, Cr, Zn, Cu, Pb, and Ni) from gasification of sludge to by-products and therefore the central accumulation in liquid and solid residues, i.e., char, tar, and ash, has been reported by several studies (S. Werle 2015). Mercury and arsenic can be entirely mobilized with aerosol particles due to high volatility at high temperatures (P. Manara and Zabaniotou 2012). The scheme linking sludge and biofuel production can be introduced as a promising approach. However, implementation of appropriate thermochemical processes and their optimum operating conditions are highly desirable based on the type and content of heavy metals in the sludge, reactor design, and probable pre-/post-treatment techniques (Y. Hu et al. 2016). Though, for clean bioenergy recovery from sludge contaminated with heavy metals thermochemical processes can be utilized as a promising way for clean bioenergy recovery from sludge contaminated with heavy metals, still, they are in the infancy stages in terms of implication of pre-/post-treatment toward the immobilization and minimization of heavy metals transfer to the products (X. Wang et al. 2016).

Apart from heavy metals, sludge also comprised a wide range of micropollutants. The severe environmental and human risks associated with the contamination of WWTP sludge (through biosolids-amended soils and groundwater) have been widely reported in the literature (L. Barron et al. 2010). These results are mainly caused because of the hydrophobicity or propensity of pollutants to be adsorbed on particles (sludge) during primary and biological treatments in wastewater sludge treatment (Raheem et al. 2018).

3.4 The Energy Challenges

Wastewater treatment plants are widely used in many companies and municipalities in order to lower the amount of harmful waste that is released into nearby receiving water bodies. It was pointed out that most wastewater treatment plants are designed to satisfy quality requirements without considering the sufficient energy requirements (Rojas and Zhelev 2012). The water-energy nexus is already becoming a topic in policies researching a series of studies on the relationship between energy and water for sustainable development (Hardy et al. 2012). The scarcity of water resources is increasing the use of energy water conveyance treatment systems and supply, and increasing attention on energy conservation has made the pursuit of sustainable development (Friedrich et al. 2012). The correlation between the quality of water and energy consumed is direct. As water quality increases, the amount of energy consumed increases. Wastewater treatment plants are generally the most intense energy consumers managed by municipalities. Most major stages in wastewater treatment plants, like the gathering and conveyance of wastewater, physical and chemical treatment, biological treatment, sludge treatment, and discharge, require considerable energy (Gu et al. 2016). In conventional wastewater treatment plants, 25–40% of operating costs are attributed to energy consumption (Panepinto et al. 2016).

All effective wastewater treatment and sludge disposal procedures require energy, primarily as electricity, gas, or other fuels for pumping, mixing, separating, and treating wastewater and sludge. The energy demand of a wastewater treatment plant depends on the plant location, plant size, type of treatment process and aeration system, effluent quality requirement, age of the plant, and knowledge of the operators (Bodik and Kubaská 2013). The energy requirements associated with wastewater treatment plants will likely grow with population growth and increased regulatory control of water quality discharge standards. A study in the United States has shown that the energy consumed by wastewater treatment plants accounts for about 20% of the municipality's total energy use. However, it is predicted that in the United States, this proportion will increase by 20% in the next 15 years because of the growth in contaminant load caused by the population growth and ever-increasing stricter regulations and the related environmental protection laws and standards for water quality and reclaimed water.

Due to increasing energy costs and environmental concerns, the potential for energy self-sufficient wastewater treatment plants has become a neighborhood of accelerating research and innovation. Energy self-sufficient wastewater treatment plants and carbon–neutral (zero greenhouse gas emissions) wastewater treatment plants are different, although carbon neutrality is often referred to by its narrow definition: energy neutrality. Total energy self-sufficient wastewater treatment is often achieved through energy recovery and improved energy efficiency. Nevertheless, few wastewater treatment plants satisfy both energy and carbon neutrality. Less than 10% of wastewater treatment plants in the United States produce surplus biogas for commercial use (Hao et al. 2015). There are currently no high-energy self-sufficient (>90%) wastewater treatment plants in developing countries, where a significant gap exists for energy self-sufficient wastewater treatment plants. Regarding technological barriers and environmental protection issues, establishing energy-self-sufficient wastewater treatment plants presents many limitations with ongoing research of technologies developed several decades ago.

Wastewater treatment plants of various sizes pose different problems. For instance, it may be difficult to understand complete energy independence in wastewater treatment plants with a small load. Cost is additionally a drag within the establishment of energy self-sufficient wastewater treatment plants because the upfront input in constructing an energy self-sufficient wastewater treatment plant is costlier than constructing a typical one, with home technologies especially, like CHP and photovoltaics, requiring a significant investment of development. In terms of environmental protection, inadequate anaerobic treatment may influence the adjacent environment, and CH4 and N2O leakages may contribute to heating and pollution (Nowak et al. 2011). The best combination of various P.E. sources must be determined to support the particular conditions in several areas (Hao X et al. 2015). Therefore, a good range of possible factors should be considered using state-of-the-art models, like LCA, to gauge the entire benefit and choose the foremost suitable technologies for energy self-sufficient wastewater treatment plants (Schaubroeck et al. 2015).

4 Wastewater Management

The most significant and severe environmental and public health concern is the production of large amounts of wastewater, whose management has been a constant challenge for ages. Large amounts of wastewater are generated daily from households, hospitals, and industries, which consist of components that are harmful to the environment and sanitation. Households are significant contributors to wastewater globally, with people using several liters of water per day to satisfy their basic needs of drinking, hygiene, food preparation, and sanitation. Hospitals are significant contributors to wastewater where patients use water for hygiene and to keep the other facilities clean (Dou et al. 2017). Wastewaters from industries and rainwater runoff, especially in urban areas, are other contributors to wastewater. Besides hospitals, households, and industries emerging activities like hydraulic fracturing also contribute to wastewater production (Villarín and Merel 2020).

Although wastewater management strategies and sanitation systems have been developed already, wastewater management is still an issue, and the strategies are evolving to protect public health and limit the impact on ecology. Many remediation methods have been modified, paying more heed to make the wastewater treatment techniques more and more sustainable (Ye et al. 2018). Wastewater management is considered a primacy that must be addressed globally rather than a local issue faced by each municipality individually. The advancement of wastewater management, their perception, and new requirements of modern society have led to the growth of research. One of the principal challenges in wastewater research is comprehensive chemical and toxicological analysis (Villarín and Merel 2020).

Wastewater treatment advancements have improved over the last years, and researchers continue fostering new processes. Presently, conventional wastewater treatment plants comprise a combination of chemical, physical, and biological processes that permit the elimination of organic matter, solids, and even some nutrients. The conventional wastewater techniques are activated sludge, trickling filters (costing more to build than an activated sludge system), and rotating biological contractor methods (Amoatey and Bani 2011). The rotating biological contractors and tricking filters are sensitive to temperature and remove less BOD. There are filtration methods, including processes like ultrafiltration, nanofiltration, and osmosis. Dealing phosphorous and nitrogen aggravates the eutrophication of receiving waters and harmful blooms. Therefore, wastewater elimination is a significant research interest (Amoatey and Bani 2011).

The activated sludge is a biological treatment process based on the principle that activated sludge or flocs of bacteria can be formed by intense wastewater aeration that degrades the organic matter and can be isolated by sedimentation. They require less space and have high effluent efficiency. The disadvantage it carries is that toward one end of the tank, the BOD is higher than the other end, which makes the microorganisms at that end physiologically active until the complete mixing activated sludge system process is utilized. On the other hand, the expulsion of suspended solids and the efficiency of trickling filters in BOD is high. However, it requires more space compared to other innovations and has the possibility of odor and filter flies. The rotating biological contractors (RBC) are easily expandable and have fewer power necessities. They can additionally be powered by compressed air, which circulates air through the system (Amoatey and Bani 2011).

Highly reactive ozone or radical species transforms the contaminants that are not removed by conventional treatments, which helps reduce the discharge's effect on receiving waters. Many approaches to AOPs have been made, including TiO2-based processes, Ozone-based cycles, and Fenton-related cycles, which have shown powerful outcomes for the constriction of drugs for a long time. Hence water contaminants are transformed into multiple unknown by-products instead of being removed (Amoatey and Bani 2011).

The non-conventional techniques include stabilization ponds, oxidation ditch, soil aquifer treatment, and constructed wetlands. They are low-cost, low-technology, and less sophisticated in maintenance and operation. Though these are land concentrated yet are more compelling in eliminating microbes. They are reliable and work continuously if the system is designed correctly and is not overloaded (Amoatey and Bani 2011). Waste stabilization pond is a low-technology treatment process comprising 4–5 ponds of different depths and biological activities. Treatment of wastewater is finished by sedimentation or transformation by biological or chemical processes. The anaerobic ponds are designed for the settlement and removal of suspended solids. The oxidation ditches consist of a single or multichannel arrangement inside an oval or a ring. They are easier to control and require more power than other mentioned technologies (Amoatey and Bani 2011).

Nature developed constructed wetlands (CW) for treating domestic wastewater over 15 years as Phragmifilter that can achieve high pollutant removal efficiency. Throughout the most recent 20 years, they have been widely adopted by small communities in France, which guarantees good integration of landscapes along with low operational expenses compared to conventional treatment plants. They are considered extensive systems or broad frameworks unsuitable for urban areas. There have been ongoing advancements in technology by the Syntea group that have resulted in the intensification of these systems for better treatment efficiency. A game changer for natural sanitation systems was the introduction of CW with constrained aeration technology which made the treatment of decentralized wastewater via constructed wetlands possible in urban areas. The classic constructed wetland systems demand huge spaces, whereas conventional wastewater treatment plants, majorly adopted by cities, require a high level of maintenance and exploitation along with high energy demand (Dou et al. 2017).

A combination of classic CWs with a constrained aeration framework called "Rhizosph Air" permits oxygen exchange by a factor of 3 which means that the plant surface can be divided by 2 to 3 (Dou et al. 2017). This can complete nitrification and denitrification, which is usually not that easy to be managed in natural sanitation frameworks. This innovation can also manage the inflow from the agro-food industry and other industries. The space requirement issue of traditionally constructed wetlands is likewise tackled by this technology which further improves the treatment efficiency and makes the urban treatment plant decentralized (Dou et al. 2017). The results had been proved satisfying because of the complete biological degradation of nitrogen.

The production of wastewater can never be stopped as living organisms cannot survive without water. When water is used for numerous human activities, wastewater will eventually be produced. These wastewaters should be treated to guarantee general well-being and a healthy environment. The choice of conventional and non-conventional methods should be according to the characteristics of wastewater, cost implications, power requirement, and technical expertise for maintenance and treatment. It has been seen that in developing countries like Ghana, non-conventional methods have been more successful than conventional methods. Though there are several challenges in treating and managing wastewaters, they cannot be ignored. Rather they ought to be overcome by giving sufficient consideration.

4.1 Rainwater Management

The natural circulation of rainwater guarantees the sustainability of surface and underground waters. For that reason, precipitation waters should be shielded from degradation and appropriately managed. Legitimate water management is vitally important for the environment and the design of sanitation systems. Because of expanded water deficiencies, flooding, and outrageous rain events, rainwater management has increased considerably. For rainwater management, various methods have been created, which include storage systems like stormwater tanks, settling tanks, and detention. Recent ones include flux reduction at valleys, vegetable roofs and walls, and porous roadways. At the same time, the broad frameworks include infiltration tanks, lagoons, and traditionally constructed wetlands (Dou et al. 2017).

Centralized wastewater treatment plants in metropolitan regions experience issues managing excessive rainwater flows. Most of the time, the additional rainwater flows are spilled straight over the natural water bodies instead of being treated. The issue of overflow or flood can be settled by decentralized constructed wetlands that will accomplish the treatment job. They also treat the city's natural green spaces, further developing the biodiversity and micro-climate and decreasing the urban heat island effects (Dou et al. 2017). In order to mitigate the climate crisis, the construction of eco-cities needs decentralized natural sanitation systems which are economical and have a simple operation. Their efficiency, suitability for shared spaces and environmental benefits are crucial. Through natural and alternative solutions are, water management carried out in urban and peri-urban areas, including the reuse of treated water in urban irrigation systems and stormwater management (Dou et al. 2017). Rainwater needs to be protected from degradation and managed correctly in urbanized areas where surface sealing causes the discharge of a substantial majority of rainwater into sewage systems. Traditional storm drainage system comprises catch basins and drainage pipes that transport water to the nearby outfall. This type of drainage system is still prevalent in many areas. This type of system can also include many other structures, like controllable structures, silt traps and storage tanks. The combined sewer systems convey storm runoff and domestic/industrial wastewater together, while the separate sewer system coveys it separately in two parallel pipes (Burkhard et al. 2000). Table 3 discusses different rainwater management techniques.

Traditional strategies include quick and effective interception of rainwater runoff and directing it to the receiving water, which causes lowering of groundwater table, and hydraulic overloading during excessive rainfall. In this manner, building owners

Table 3 Techniques of rainwater management

Techniques	Building conditions	References
Infiltration and collection systems		
Permeable pavements	It was constructed on residential roads and parking lots, pricy, and clogs after 1–3 years	(Burkhard et al. 2000)
Infiltration basins	Constructed in grassy areas flooded during rainfall, basins can be used for other purposes during dry weather	(Burkhard et al. 2000)
Swales	Grassed ditches take runoff from roads and parking lots and no stagnant water in properly designed ones	(Burkhard et al. 2000)
Infiltration trenches	Circular underground trenches filled with gravel medium	(Burkhard et al. 2000)
Detention systems		
Ponds	Rainfall-runoff is collected, can be integrated into the landscape surrounding, and serve as a habitat for wildlife	(Burkhard et al. 2000)
Constructed wetlands	Purification ponds clean spilled water through combined- sewer overflows	(Burkhard et al. 2000)
On-site retention	Grassed roofs and runoff are temporarily stored	(Burkhard et al. 2000)

of various nations are obliged to pay tariff charges for sewerage water treatment plants (Daniel Słyś et al. 2012).

Following sustainable development, the EU Water Framework Directive focused on managing precipitation waters by retaining rainwater through the natural surface and underground retention processes and water infiltration into the ground. Poland's "Water Law" coordinates an even management of water resources. Consequently, a sustainable water discharging system includes installing devices and facilities such as water dispersing boxes and chambers, sink basins, ditches, wells, and reservoirs. Urban regions, however, due to a shortage of free space, harvest and utilize rainwater through green roofs, which are classified under Sustainable Urban Drainage Systems as they can postpone the runoff from roof surfaces and decrease its total volume (Słyś et al. 2012).

Sustainable Urban Drainage Systems likewise include various systems for collecting and using rainwater in buildings, diminishing the volume of rainwater released into sewage systems, and reducing the consumption of consumable water. These establishments include systems for harvesting rainwater and utilizing it for watering gardens, laundry, washing cars, flushing toilets, and many more. Rainwater was observed to be able to fulfill up to 94% of water demands in some cases. Countries like Japan and Germany are equipped with large rainwater harvesting systems in sports facilities and airports, for watering plants, flushing toilets, and as a medium in the cooling water system. These analyses also test the feasibility of bringing such systems into dwelling houses (Słyś et al. 2012).

Human water demand within the Indian subcontinent is based on available water by exploiting natural resources like groundwater and rivers. The southern and central Indian regions with bedrock comprised of granite and basalt rocks have shallow aquifers with more limited residence time for groundwater, resulting in a significant fraction of rainfall flowing into the ocean. In Bangalore, monitoring the stable isotopic ratios in regional rainwater showed isotropic imprints of rainwater in bottled water. This proved that commercially available bottled water is produced by tapping followed by treating of monsoon-fed freshwater of the region (Rangarajan and Ghosh 2011).

The extent of monsoon rainfall harvesting to meet consumable water demands is critical. This approach can be carried out on a significant scale through community-based harvesting through augmentation into municipalities and systematic distribution of conserved water through proper channelized networks, which might circumvent drinking water deficiency at regional and community levels. Agriculture relies on irrigation water supplies provided from the surface of the groundwater blue water or through rainwater management (green water), both of which are rainfed. Climate-smart agro-technologies can improve the utilization of blue and green waters in farming (Gupta et al. 2020).

4.2 Domestic Wastewater Management

Water generated by people as a result of household activities like laundry, kitchen basins, washing of clothes, washing of vegetables, toilets, and restrooms is known as domestic wastewater. It contains solid particles like rags, plastic containers, feces, paper, vegetable peel, microscopic solids in colloidal suspension, and pollutants in a proper solution. Then comes sullage which also contains a wide variety of chemicals that are contributed by soaps, detergents, pesticides, sour milk, tea leaves, vegetable peelings, in short, anything that runs down the kitchen sink (Mara 2012). In many under-resourced regions, sullage is not considered a threat and is often discharged without treatment (Oladoja 2017). These pollutants appear not only objectionable and hazardous as they contain disease-causing organisms and cause several diseases. A small portion of human feces and urine contains fats, proteins, and carbohydrates. The carbohydrates and proteins resent in the organic fraction of human feces, and urine forms an excellent diet for bacteria. Along with the chemical compounds, urine, and human feces also contain several intestinal bacteria and other organisms, the majority of which are harmless and, to some extent, even beneficial, but some essential minorities are capable of causing human diseases (Mara 2012).

Domestic wastewater also contains a great deal of phosphorus which increases the risk of eutrophication in the receiving waters. The elimination of phosphorus from these wastewaters could be done by utilizing biological treatment, physio-chemical methods, or/and combinations of both. The physio-chemical processes have been used for many years to control phosphorus as they are effective and reliable. However, they come with limitations. They influence the effluent's pH and sometimes need some chemical addition before their final release or discharge. Often more steps are included in processing due to extraneous solid production at the treatment time. The processes consist of sorption precipitation or/and ion exchange mechanisms. The chemical techniques involve dosing metal salts to either conventional activated sludge reactors, the outlet from a secondary clarifier, or pre-treatment effluent (Bunce et al. 2018).

The higher the amount of organic matter, the more muscular the wastewater's strength. Often the strength is decided by its COD or BOD. These wastewaters are supplied with oxygen so that the bacteria utilize the contents of the food. Sewers, the underground pipes perform the collection of domestic wastewaters. It empties itself into a water body and affects the water by adding organic matter, intestinal bacteria, and adding hard detergents, which causes river foaming. This typical wastewater discharge technique is performed in under-resourced regions (Oladoja 2017). They should be treated to reduce the aquatic biota damage and the transmission of diseases they cause. Wastewater that is not treated causes significant damage to human health and the environment.

In well-resourced regions, this domestic wastewater management is done by several methods, including cesspools, septic tanks, on-site water treatment systems, and others. Watertight chambers, called cesspools, are meant to store wastewater where no treatment occurs. They are used for a small groups of houses or single dwellings. Septic tank is that domestic wastewater treatment facility which performs the primary treatment before discharging it through soak-away. This system allows the effluents to seep into the ground (Oladoja 2017) (Table 4).

The treatment systems should be cheap, odorless, hygienic, low maintenance, and easy to install. The on-site system of management is considered to be the ideal system. When such septic tanks are constructed and properly maintained, they are cost-effective, reliable, and safe. Constructed wetland is one such treatment system (Mara 2012). For natural management of domestic wastewater, vermicomposting, a process that converts the organic waste materials into soil conditioners with the help of microorganisms, has been used for decades. They are rich in plant nutrients and increase the water-holding capacity of the soil. The worms create burrows by ingesting and excreting food to increase the surface area allowing aeration (Bajsa et al. 2003). If these technologies are adopted, they will assist in saving money, protecting the investment of homeowners, ensuring a better solution for low-density communities, and providing excellent watershed management for an alternative for

Table 4 Different techniques of domestic wastewater management and their uses

Techniques	Uses	Reference
Centralized conventional sewage treatment works	It consists of primary, secondary, and tertiary stages. The screening takes place before the primary treatment and includes activated sludge, percolating filters, rotating biological contractors, and oxidation ponds	(Burkhard et al. 2000)
Decentralized conventional systems		
Non-biological treatment cesspools	Underground plastic tanks, water is collected and stored, pricy, water is transported for disposal, applicable where no option is available, includes septic tanks and settlement tanks	(Burkhard et al. 2000)
Package biological plants redirecting biological filter	Like percolating filters, the filter is contained in a plastic shell; wastewater passes through the filter, and regular cleaning is required	(Burkhard et al. 2000)
Leach fields	Used in remote areas, a bioactive layer of soil is required along with aquifer and infiltration pipes	(Burkhard et al. 2000)
Ecological treatment methods		
Constructed wetlands	Used for secondary treatment of domestic sewage, it consists of a watertight pool with aquatic plants	(Burkhard et al. 2000)
Living machines	It comprises a preliminary septic tank for sedimentation. Aerobic treatment takes place in a closed greenhouse tank. Nitrification and denitrification take place	(Burkhard et al. 2000)
Aquaculture	Large ponds with live fish and plants feed on secondary treated effluent or nutrients, and fish feed on plants	(Burkhard et al. 2000)
Sand filters	In vertical and pressure filters, wastewater applied is tertiary effluent	(Burkhard et al. 2000)

varying site conditions. Likewise, helpful solutions for ecologically sensitive areas are furnished (Oladoja 2017).

4.3 Management of Water Reuse

Wastewater treatment and its reuse have turned out to be an effective and attractive option in a condition where there is overcommitted water supply and cannot meet the demands of the growing population (Bahri et al. 2011). For a very long time, the lack of water has been overcome by traditional methods, including increasing resources through artificial canals and dams, as in Arizona. However, these alternatives turned out to be expensive, environmentally unsustainable, and insufficient. Hence reusing treated wastewater for proper water supply to the population is required (Villarín and Merel 2020).

Deteriorated water quality can be utilized for many non-potable purposes like landscape irrigation, agricultural irrigation, groundwater recharge, industrial recycling, and reuse, and urban and environmental uses till appropriate treatment are provided (Bahri et al. 2011). Potable water reuse can be differentiated into two classes, direct and indirect. Direct potable reuse (DPR) is directly discharged into the municipal water supply system after treatment or after utilizing an environmental buffer (like aquifer recharge). The very first DPR scheme was built in Windhoek in Nambia (1968), followed by Cloudcroft in New Mexico, USA (2007), Western Cape in South Africa (2010), and Big Spring in Texas (2014) (Villarín and Merel 2020).

Earlier reuse water was primarily used for ground recharge, but it has been used for agriculture (Ait-Mouheb et al. 2020). Now, in the last two decades, with the advancement in technology, water reuse has become a significant part of the world water cycle which is helping the human population in keeping ground recharge stable and also helping in a variety of applications like agriculture, industrial, aquaculture, fire brigade, road cleaning and many other (Pennell et al. 1996). Several countries are treating their wastewater and reusing it for their advantageous motive like in Jordan, where 87% of wastewater is treated and used in their work, correspondingly Israel with 40%, Tunisia with 25%, and many more countries reusing their wastewater on their excellent motive (Sun et al. 2011). All over the globe, domestic water contributes 8% of consumption, which unambiguously means there is 8% of the contribution from homes alone. A simple method called grey wastewater system has been used these days to save this wastewater. Many other ways like building a rain garden to save water in which a minor digging has to be done at any place besides the home and then connect the entire pipe to a single point and drain the water in it and then it can be used as irrigation, ground recharge purpose (Wurochekke et al. 2016).

Various advanced and innovative technologies have been included in the current design of wastewater treatment and their reuse in many locations, including ultraviolet disinfection, membrane bioreactors, and ozonation, along with advanced oxidation (Bahri et al. 2011). Reusing the treated wastewater helps protect the environment, especially the ecosystem, by decreasing the number of toxic contaminants and nutrients entering waterbeds. It also enables the use of treated water for applications that do not need high-quality drinking water (Bahri et al. 2011). However, there are many concerns regarding the use of DPR. Some wastewater contaminants like artificial sweeteners, pharmaceuticals, and personal care products are not completely removed even after the treatment. Since very little is known about the effects of exposure to low doses of a mixture of these components, therefore, they are not purely safe to be used directly. Another concern is to what extent the advanced treatments can protect the consumer's health. Due to this reason, most countries opt for indirect potable water reuse with an environmental buffer. This will limit human exposure and prevent health problems if the treatment is unsuccessful. However, many studies are being carried out to develop potable water reuse. population (Villarín and Merel 2020).

5 Social Aspects of Wastewater Management

Financial and technological aspects have always been given priority in wastewater treatment facilities, but social aspects have never been approached profoundly. However, the social issues are complicated to mention as they are not well defined but create connections between water and people. Institutions like culture, social compacts, and networks are involved in these aspects. Such social institutions are needed to be respected in decisions related to water management, but first, people must understand them. The right way society can meet its water needs is by participating in wastewater management processes, although that creates a social issue of responsibilities and roles (Padilla-Rivera et al. 2016). New methods to study social systems are becoming available as information technologies develop. In order to explain the social impacts, social impact assessment can be utilized in the same way as is used for environmental impact assessment.

Understanding and managing the wastewater problems is not that easy because social responses to decision-making in this case often result in unexpected other problems. It is challenging to predict the response of different stakeholders, especially in cases when the best decisions are being made. In order to have an excellent reasonable long half-life, the water management policies should be based on their affordability, sound science, and compatibility with the beliefs and values of individuals (Montanari et al. 2013).

Customs, history, education, and current social, natural, economic, cultural, and political environment impact the choices of the society. Steps taken by the social institutions to change the performance of the water system are giving valuable outcomes. A human can face water-societal problems by using approaches like vision planning, agent-based modeling, and innovative games, which involve the use of an interactive decision support system between the social system and the water system. The moment natural and social scientists start to work together, people can hypothesize about many different decisions society may take (Walker et al. 2015).

The social systems are complex, not fitting easily into different quantitative models of economists and natural scientists. Society's response to demands for change is difficult to predict, but the decision-making can be improved if we use the above approaches. This information on water-related cultural and social aspects is essential for achieving sustainable water management in this uncertain, unpredictable, and uncertain environment (Walker et al. 2015).

6 Conclusion and Future Perscpective

As we all know that with the increasing population, the issue of water scarcity is also increasing. Due to human activities, such wastewater is generated, which must be treated to meet the demands of water and guarantee a safe environment and public health. Many conventional and non-conventional methods are developed for treating wastewaters, and the choice of these methods should be based on factors like its source, expense, technical expertise for the maintenance and operation, and power requirement. Although the technological processes allow the conversion of wastewaters from low-quality to high-quality water suitable for human utilization, significant efforts in social research are needed for public acceptance. Besides facing many challenges due to its treatment and management, the biggest concern remains its impact on our environment and the ecosystem. Therefore, the government and every individual need to play their role in making wastewater treatment methods more efficient and effective.

References

Ait-Mouheb N, Mayaux P, Mateo-Sagasta J, Hartani T, Molle B (2020) Water reuse: a resource for Mediterranean agriculture. Water Resour Mediterrean Regn:107–136. https://doi.org/10.1016/b978-0-12-818086-0.00005-4

Arnone RD, Walling JP (2007) Waterborne pathogens in urban watersheds. J Water Health 5(1):149–162. https://doi.org/10.2166/wh.2006.001

Ashbolt NJ (2004) Microbial contamination of drinking water and disease outcomes in developing regions. Toxicology 198(1–3):229–238. https://doi.org/10.1016/j.tox.2004.01.030

Bajsa O, Nair J, Mathew K, Ho GE (2003) Vermiculture as a tool for domestic wastewater management. Water Sci Technol 48(11–12):125–132. https://doi.org/10.2166/wst.2004.0821. Retrieved January 30 2022

Barron L, Nesterenko E, Hart K, Power E, Quinn B, Kelleher B, Paull B (2010) Holistic visualization of the multimodal transport and fate of twelve pharmaceuticals in biosolid enriched topsoils. Anal Bioanal Chem 397(1):287–296. https://doi.org/10.1007/s00216-010-3494-1

Bibby K, Peccia J (2013) Identification of viral pathogen diversity in sewage sludge by metagenome analysis. Environ Sci Technol 47(4):1945–1951. https://doi.org/10.1021/es305181x

Bodík I, Kubaská M (2013) Energy and sustainability of operation of a wastewater treatment plant. Environ Prot Eng 39(2):15–24. https://doi.org/10.37190/epe130202

Bunce JT, Ndam E, Ofiteru ID, Moore A, Graham DW (2018) A review of phosphorus removal technologies and their applicability to small-scale domestic wastewater treatment systems. Front Environ Sci 6. https://doi.org/10.3389/fenvs.2018.00008
Burkhard R, Deletic A, Craig A (2000) Techniques for water and wastewater management: a review of techniques and their integration in planning. Urban Water 2(3):197–221. https://doi.org/10.1016/S1462-0758(00)00056-X
Dou T, Troesch S, Petitjean A, Gábor PT, Esser D (2017) Wastewater and rainwater management in urban areas: a role for constructed wetlands. Procedia Environ Sci 37:535–541. https://doi.org/10.1016/j.proenv.2017.03.036
Evans A, Mateo-Sagasta J, Qadir M, Boelee E, Ippolito A (2019) Agricultural water pollution: key knowledge gaps and research needs. Curr Opin Environ Sustain 36:20–27. https://doi.org/10.1016/j.cosust.2018.10.003
Fenwick A (2006) Waterborne infectious diseases—Could they be consigned to history? Science 313(5790):1077–1081. https://doi.org/10.1126/science.1127184
Friedrich E, Pillay S, Buckley C (2012) Environmental life cycle assessments for water treatment processes—A South African case study of an urban water cycle. Water SA 35(1). https://doi.org/10.4314/wsa.v35i1.76710
Gu Y, Dong YN, Wang H, Keller A, Xu J, Chiramba T, Li F (2016) Quantification of the water, energy and carbon footprints of wastewater treatment plants in China considering a water-energy nexus perspective. Ecol Ind 60:402–409. https://doi.org/10.1016/j.ecolind.2015.07.012
Gupta R, Tyagi NK, Abrol I (2020) Rainwater management and Indian agriculture: a call for a shift in focus from blue to green water. Agricultural Research 9(4):429–443. https://doi.org/10.1007/s40003-020-00467-2
Hardy L, Garrido A, Juana L (2012) Evaluation of Spain's water-energy nexus. Int J Water Resour Dev 28(1):151–170. https://doi.org/10.1080/07900627.2012.642240
Hao X, Batstone D, Guest JS (2015) Carbon neutrality: an ultimate goal towards sustainable wastewater treatment plants. Water Res 87:413–415. https://doi.org/10.1016/j.watres.2015.11.043
Heinrich K, Bach M, Breuer L (2017) Infectious disease research—What role is there for hydrologists? J Water Resour Prot 09(2):139–162. https://doi.org/10.4236/jwarp.2017.92011
Hosomi M (2016) New challenges on wastewater treatment. Clean Technol Environ Policy 18(3):627–628. https://doi.org/10.1007/s10098-016-1131-1
Hu Y, Chen G, Ma W, Yan M, Han L (2016) Distribution and contamination hazards of heavy metals in solid residues from the pyrolysis and gasification of wastewater sewage sludge. J Residuals Sci Technol 13(4):259–268. https://doi.org/10.12783/issn.1544-8053/13/4/3
Hussain I, (2002) Wastewater use in agriculture. Colombo, Sri Lanka: International Water Management Institute.
Lofrano G, Brown J (2010) Wastewater management through the ages: a history of humanity. Sci Total Environ 408(22):5254–5264. https://doi.org/10.1016/j.scitotenv.2010.07.062
Manara P, Zabaniotou A (2012) Towards sewage sludge based biofuels via thermochemical conversion—A review. Renew Sustain Energy Rev 16(5):2566–2582. https://doi.org/10.1016/j.rser.2012.01.074
Mompelat S, Le Bot B, Thomas O (2009) Occurrence and fate of pharmaceutical products and by-products, from resource to drinking water. Environ Int 35(5):803–814.https://doi.org/10.1016/j.envint.2008.10.008
Montanari A, Young G, Savenije HHG, Hughes D, Wagener T, Ren LL, Koutsoyiannis D, Cudennec C, Toth E, Grimaldi S, Blöschl G, Sivapalan M, Beven K, Gupta H, Hipsey M, Schaefli B, Arheimer B, Boegh E, Schymanski SJ, Belyaev V (2013) "Panta Rhei—everything flows": change in hydrology and society—The IAHS scientific decade 2013–2022. Hydrol Sci J 58(6):1256–1275. https://doi.org/10.1080/02626667.2013.809088
Mostafalou S, Abdollahi M (2013) Pesticides and human chronic diseases: evidences, mechanisms, and perspectives. Toxicol Appl Pharmacol 268(2):157–177. https://doi.org/10.1016/j.taap.2013.01.025

Nowak O, Keil S, Fimml C (2011) Examples of energy self-sufficient municipal nutrient removal plants. Water Sci Technol 64(1):1–6. https://doi.org/10.2166/wst.2011.625

Oladoja NA (2017) Appropriate technology for domestic wastewater management in under-resourced regions of the world. Appl Water Sci 7(7):3391–3406. https://doi.org/10.1007/s13201-016-0495-z

Padilla-Rivera A, Morgan-Sagastume JM, Noyola A, Güereca LP (2016) Addressing social aspects associated with wastewater treatment facilities. Environ Impact Assess Rev 57:101–113. https://doi.org/10.1016/j.eiar.2015.11.007

Panepinto D, Fiore S, Zappone M, Genon G, Meucci L (2016) Evaluation of the energy efficiency of a large wastewater treatment plant in Italy. Appl Energy 161:404–411. https://doi.org/10.1016/j.apenergy.2015.10.027

Pavithra KG, Senthil Kumar P, Sundar Rajan P, Saravanan A, Naushad M (2017) Sources and impacts of pharmaceutical components in wastewater and its treatment process: a review. Korean J Chem Eng 34(11):2787–2805. https://doi.org/10.1007/s11814-017-0255-2

Qadir M, Wichelns D, Raschid-Sally L, McCornick PG, Drechsel P, Bahri A, Minhas PS (2010) The challenges of wastewater irrigation in developing countries. Agric Water Manag 97(4):561–568. https://doi.org/10.1016/j.agwat.2008.11.004

Raheem A, Sikarwar VS, He J, Dastyar W, Dionysiou DD, Wang W, Zhao M (2018) Opportunities and challenges in sustainable treatment and resource reuse of sewage sludge: a review. Chem Eng J 337:616–641. https://doi.org/10.1016/j.cej.2017.12.149

Ramírez-García R, Gohil N, Singh V (2019) Recent advances, challenges, and opportunities in bioremediation of hazardous materials. Photo Manag Polluted Sites 517–568. https://doi.org/10.1016/b978-0-12-813912-7.00021-1

Rangarajan R, Ghosh P (2011) Rainwater management and harvesting strategies for human needs: an Indian perspective. Environ Sci Technol 45(22):9469–9470. https://doi.org/10.1021/es203308p

Rojas J, Zhelev T (2012) Energy efficiency optimization of wastewater treatment: study of ATAD. Comput Chem Eng 38:52–63. https://doi.org/10.1016/j.compchemeng.2011.11.016

Schaubroeck T, De Clippeleir H, Weissenbacher N, Dewulf J, Boeckx P, Vlaeminck SE, Wett B (2015) Environmental sustainability of an energy self-sufficient sewage treatment plant: improvements through DEMON and co-digestion. Water Res 74:166–179. https://doi.org/10.1016/j.watres.2015.02.013

Serneri SN (2007) The construction of the modern city and the management of water resources in Italy, 1880–1920. J Urban Hist 33(5):813–827. https://doi.org/10.1177/0096144207301452

Shete BS, Shinkar P (2013) Dairy industry wastewater sources, characteristics, and its effects on environment. Int J Curr Eng Technol 3(5):1611–1615

Słyś D, Stec A, Zeleňáková M (2012) A LCC analysis of rainwater management variants. Ecol Chem Eng S 19(3):359–372. https://doi.org/10.2478/v10216-011-0026-7

Sun F, Chen M, Chen J (2011) Integrated management of source water quantity and quality for human health in a changing world. Encycl Environ Health:254–265. https://doi.org/10.1016/b978-0-444-52272-6.00286-5

Tyagi VK, Lo SL (2013) Sludge: a waste or renewable source for energy and resources recovery? Renew Sustain Energy Rev 25:708–728. https://doi.org/10.1016/j.rser.2013.05.029

Villarín MC, Merel S (2020) Paradigm shifts and current challenges in wastewater management. J Hazard Mater 390:122139. https://doi.org/10.1016/j.jhazmat.2020.122139

Walker WE, Loucks DP, Carr G (2015) Social responses to water management decisions. Environmental Processes 2(3):485–509. https://doi.org/10.1007/s40710-015-0083-5

Wang X, Li C, Zhang B, Lin J, Chi Q, Wang Y (2016) Migration and risk assessment of heavy metals in sewage sludge during hydrothermal treatment combined with pyrolysis. Biores Technol 221:560–567. https://doi.org/10.1016/j.biortech.2016.09.069

Werle S (2015) Gasification of a dried sewage sludge in a laboratory scale fixed bed reactor. EnergS 8(8):8562–8572.https://doi.org/10.3390/en8088562

World Health Organization (WHO) (2014) Water quality and health. Drinking water chlorination—A review of disinfection practices and issues. http://www.waterandhealth.org/drinkingwater/wp.html. Retrieved April 28 2014

Wu H, Glarborg P, Frandsen FJ, Dam-Johansen K, Jensen PA, Sander B (2013) Trace elements in co-combustion of solid recovered fuel and coal. Fuel Process Technol 105:212–221. https://doi.org/10.1016/j.fuproc.2011.05.007

Wurochekke AA, Mohamed RMS, Al-Gheethi AA, Atiku H, Amir HM, Matias-Peralta HM (2016) Household greywater treatment methods using natural materials and their hybrid system. J Water Health 14(6):914–928. https://doi.org/10.2166/wh.2016.054

Ye Y, Ngo HH, Guo W, Liu Y, Chang SW, Nguyen DD, Liang H, Wang J (2018) A critical review on ammonium recovery from wastewater for sustainable wastewater management. Biores Technol 268:749–758. https://doi.org/10.1016/j.biortech.2018.07.111

Zhang J, Zuo W, Tian Y, Chen L, Yin L, Zhang J (2017) Sulfur transformation during microwave and conventional pyrolysis of sewage sludge. Environ Sci Technol 51(1):709–717. https://doi.org/10.1021/acs.est.6b03784

Zia H, Harris N, Merrett G, Rivers M, Coles N (2013) The impact of agricultural activities on water quality: a case for collaborative catchment-scale management using integrated wireless sensor networks. Comput Electron Agric 96:126–138. https://doi.org/10.1016/j.compag.2013.05.001

Application of Nanomaterials for the Removal of Heavy Metal from Wastewater

P. Priya, N. Nirmala, S. S. Dawn, Kanchan Soni, Bagaria Ashima, Syed Ali Abdur Rahman, and J. Arun

1 Introduction

One of the serious problems that countries are facing is the heavy metal pollution in water which affects the environment adversely. Heavy metals being highly toxic even at dilute concentration, non-biodegradable, and also accumulates through food chain destroying aquatic life and being serious threat to human life (Razzak et al. 2022). Heavy metals (Cr, Cd, Hg, As, Zn, Cu, Fe, Al, Ba, Ca, Mg, Pb, Mn, Ag, Na, and Se) contaminate drinking and irrigation water which are toxic for organisms that are present in excess amount. Nanomaterials are divided based on their role in adsorption applications which is dependent on their innate surface property (Beni et al. 2022). Broadly categorized, there are two essential approaches for nanoparticle blend: top-down approach and bottom-up approach. Top-down approach includes the break-down of bulk fabric into little particles. These methods are the extension of those that have been utilized to create micron-sized particles.

Top-down approaches are moderately less difficult and depend on expulsion or division of bulk fabric to its craved properties. A downside of this approach is the blemish of surface structure for case, nanowires made by lithography may contain numerous debasements and basic defects on the surface. A few illustrations of amalgamation procedures that use this approach are sol–gel amalgamation,

P. Priya · N. Nirmala · S. S. Dawn · J. Arun (✉)
Centre for Waste Management, Sathyabama Institute of Science and Technology, Tamil Nadu, Jeppiaar Nagar (OMR), Chennai 600 119, India
e-mail: arunjayaseelan93@gmail.com

K. Soni · B. Ashima
Department of Physics, Manipal University Jaipur, Dehmi Kalan, Jaipur, Rajasthan 303 007, India

S. A. A. Rahman
Department of Biotechnology, Sathyabama Institute of Science and Technology, Tamil Nadu, Jeppiaar Nagar (OMR), Chennai 600 119, India

M. P. Shah (ed.), *Advanced and Innovative Approaches of Environmental Biotechnology in Industrial Wastewater Treatment*,
https://doi.org/10.1007/978-981-99-2598-8_6

electron beam lithography, nuclear drive control, gas-phase condensation, etc. Bottom-up approach makes less squander and therefore is more conservative than the top-down approach This method refers to amalgamation procedures where build-up of the material occurs from the foot. It utilizes the concepts of atomic self-assembly and/or atomic acknowledgment. The preeminent advantage of the bottom-up approach is that it can synthesize homogenous nanostructures with culminated crystallographic and surface structures (Maitlo et al. 2019). A few illustrations are aqueous blend (Darr et al. 2017), format helped sol–gel, electrodeposition (Tonelli et al. 2019), etc. Among the two approaches, bottom-up is favored because it has numerous merits such as less surface defects, superior requesting and more homogenous chemical composition (Pareek et al. 2017).

The adsorption innovation on the other hand, due to its tall proficiency and effortlessness of operation, is respected as the foremost promising strategy to remove even follow sums of overwhelming metal particles from effluents. Even in spite of the fact that conventional sorbents are competent in expelling heavy metals from wastewater, their unobtrusive sorption capacities and efficiencies constrain their applications in concentrated arrangements (Haripriyan et al. 2022). Within the past decades, biodegradable polymeric nano adsorbents have been created as a potential differentiating choice to actuated carbon as distant as their boundless surface region, extraordinary mechanical inflexibility, pore measure dispersion, culminate surface chemistry, and regenerative capacity beneath gentle conditions. The transfer of utilized adsorbents containing overwhelming metal(s) may be done after recuperation of contaminants or straightforwardly without overwhelming metal recuperation, but in both the cases there will be auxiliary contamination from the used adsorbents and the chemicals utilized to treat the adsorbents for metal recuperation. However, metal-loaded adsorbents have poisonous impacts on people and environment. Hence, the utilized adsorbents ought to be released into the environment as it were after recuperation of the heavy metals totally.

Considering the requirement of metal desorption and recuperation, this paper summarizes the productivity of different recovering specialists utilized by different authors, proficiency of the adsorbents for evacuation of heavy metals, and recuperation of overwhelming metals. Overwhelming metal-induced poisonous quality and carcinogenicity includes numerous unthinking viewpoints, a few of which are not clearly illustrated or caught on. In any case, each metal is known to have special highlights and physicochemical properties that bestow to its particular toxicological components of activity. This audit gives an investigation of the natural event, generation and utilization, potential for human presentation, and atomic instruments of poisonous quality, genotoxicity, and carcinogenicity of arsenic, cadmium, chromium, lead, and mercury.

2 Water Pollution

Advancements in industries, natural disasters, human population, climate change, and modernization of livelihood have led to scarceness in access to safe drinking water in most of the developing countries. Textile dyes, heavy metals, metalloids, personal and pharmaceutical care products, and organic and inorganic pollutants are the most common pollutants in water ecosystem. Accumulation, segregation, and storage of industrial waste and household waste near the aquatic system damage the flora and fauna (Wadhawan et al. 2020). In this book chapter, we focus mainly on one of the promising toxic heavy metals. Heavy metals that are commonly reported in aquatic environment was lead (Pb), chromium (Cr), manganese (Mn), nickel (Ni), mercury (Hg), arsenic (As), cadmium (Cd), etc. These much variety of heavy metals are released into the water system due to the activity of metal fabrication, leather, batteries, electroplating, mining, fertilizers, paints, alloying, etc. (Yari et al. 2015).

Continuous exposure to toxic heavy metals leads to illness of nervous system, kidney, brain, liver, and other vital organs (Le et al. 2019). Most important point of this much organ toxicity of heavy metal was due to the carcinogenic activity. Exposure of lead (Pb) in kids, aged below six led to lesser IQ level, retarded growth, impaired hearing, and learning disabilities (Farghali et al. 2013). Considering the toxic nature of heavy metals, its need of hour to concentrate on development of cutting-edge techniques for the effective removal and degradation of heavy metals.

3 Heavy Metals: Types, Sources, and Toxicity

Heavy metals are described as metallic elements with a massively higher density when compared with water. Heavy metals include metalloids, such as arsenic, that can cause toxicity at even minimal level of concentration, based on the features of heaviness and toxicity (Tchounwou et al. 2012). On the basis of their toxicity, heavy metals were divided into two categories: essential heavy metals and non-essential heavy metals. At low concentrations, essential heavy metals are either nontoxic or less harmful, including cobalt, zinc, iron, and copper. Non-essential heavy metals are harmful even at low concentrations, including chromium, cadmium, arsenic, and mercury (Kim et al. 2019). Heavy metals are naturally occurring components that can be found in the earth's crust. The majority of heavy metal pollution and exposure is caused by human activities such as mining, smelting, industrial production, and the use of metals and metal-containing compounds at home and in agriculture (He et al. 2005; Herawati et al. 2000). Figure 1 provides a detailed note on the different release points of heavy metals into gas, soil, and water environment. Coal, petroleum, nuclear power, plastics, textiles, microelectronics, paper-processing, and wood preservation factories are all examples of industrial sources of heavy metal pollution (Sträter et al. 2010). Heavy metal contamination has also been correlated to weathering and volcanic eruptions (Nriagu 1989). Natural activities such as volcanic activity, metal corrosion, metal evaporation from soil and water and sediment re-suspension, geological weathering, soil erosion, and sediment re-suspension, among others, can contribute to heavy metal contamination (Briffa et al. 2020). Heavy metals are commonly defined as those with a density of more than 5 g per cubic centimeter. The maximum contamination level (MCL) guidelines for the most dangerous heavy metals are listed in Table 1.

Heavy metal deposition in the human body causes serious harm to a variety of organs, most notably the respiratory, neurological, and reproductive systems, as well as the digestive tract (Kim et al. 2019). Every metal has distinct physicochemical characteristics that determine its toxicological modes of action. Many studies have shown that arsenic toxicity is affected by the amount of exposure, the frequency and length of exposure, the kind of organism being exposed, the individual's age and gender, as well as genetic and dietary susceptibilities (Abernathy et al. 1999). Human lymphocytes and mouse leukocytes are both damaged by arsenic trioxide, which has been shown to cause DNA damage. Arsenic chemicals have also been demonstrated to cause gene amplification, cell mitotic arrest, and DNA repair inhibition (Banu et al. 2001). Toxicological effects of exposure to the metal form of chromium range from minimal to high depending on the specific oxidation state (Velma et al. 2009).

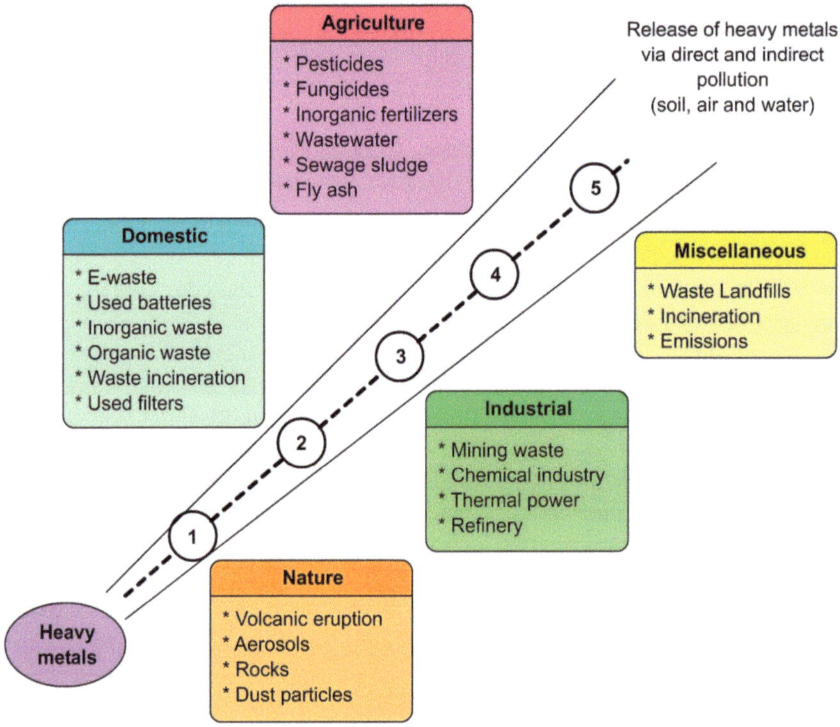

Fig. 1 Release points of heavy metals into gaseous, aqueous, and solid environments

Table 1 The MCL guidelines for the most dangerous heavy metals (Barakat 2011)

Heavy metals	Toxicity	MCL (mg/L)
Zinc	Neurological signs, continuous thirst, lethargy, and depression	0.80
Mercury	Nervous system problems, kidneys diseases, circulatory system, and rheumatoid arthritis	0.00003
Arsenic	Visceral cancers, skin manifestations, and vascular disease	0.050
Chromium	Carcinogenic, diarrhea, headache, vomiting and nausea	0.05
Cadmium	Carcinogen, kidney damage, and renal disorder	0.01
Lead	Circulatory system problems, fetal brain problems, kidney diseases, and nervous system diseases	0.006
Copper	Wilson disease, liver damage, and insomnia	0.25

Inhalation of less soluble or insoluble Cr (VI) compounds appears to be related to carcinogenicity. The elemental form of Cr (VI) has no toxicological properties. It varies substantially across various Cr (VI) complexes (Katz and Salem 1993). According to the results of epidemiological studies, Cr (VI) is a key factor in carcinogenesis. Chromium's solubility and other properties, including size, surface charge, crystal modification, and phagocytization, may be essential in predicting cancer risk (Tchounwou et al. 2012).

Lead's toxic and apoptotic effects on human cancer cells have been shown to involve numerous cell and molecular processes, including the induction of cell death and oxidative stress, DNA damage, the activation of stress genes, the externalization of phosphatidylserine, and activation of caspase 3 (Yedjou and Tchounwou 2007; Yedjou et al. 2010).

Lung is the most conclusive place where people get cancer from cadmium exposure. Other places where cadmium can cause cancer in animals are the injection site, testes, adrenals, and the hemopoietic system (Waalkes 1996). Cadmium, arsenic, chromium, and nickel, among other carcinogenic metals, have all been linked to DNA damage caused by mutation, deletion, or oxygen radical damage to DNA. Mercury in all forms is harmful, causing gastrointestinal neurotoxicity, toxicity, and nephrotoxicity (Tchounwou et al. 2003).

4 Nanoadsorbents: Synthesis Routes

There are two primary ways to synthesize nano adsorbents: top-down and bottom-up approaches. The top-down approach is the common method in which the process begins with bigger particles (macroscopic) and the size reduction to nano-scale is accomplished by the use of externally controlled energy sources such as sputtering, erosion, high-energy ball milling, mechanical alloying, and reactive milling.

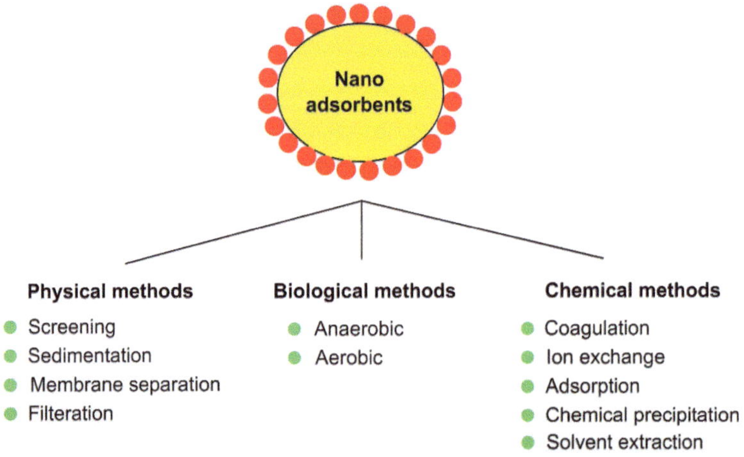

Fig. 2 Various synthesis routes of nano adsorbents for environmental applications

However, the fundamental issue with the top-down technique is the potential for crystallographic and surface structure damage during particle size reduction (El-sayed 2020; Singh et al. 2018). Bottom-up is the most recent method. It relies on how a substance is built from the bottom, either molecule by molecule or atom by atom, as in molecular self-assembly, physical/chemical vapor deposition and sol–gel method. Bottom-up methods have two major challenges: chemical purification is necessary, and large-scale manufacturing is challenging. However, the bottom-up approach is now the suitable technique since it may manufacture materials with precise qualities customized to the remediation needs based on the fabrication pathway chosen (Tulinski and Jurczyk 2017; Nik Abdul Ghani et al. 2021). Figure 2 showcases the various synthesis routes of nano adsorbents.

Furthermore, chemical and physical methods for the synthesis of nano sorbent materials frequently synthesize nano sorbent materials with a definite shape and size. However, these methods have been proven to be environmentally hazardous due to the use of harmful chemical substances and the elevated heat used during the synthesis methods (El-sayed 2020; Fosso-Kankeu 2019). The microbial production of nanomaterials can be accomplished by extracellular and intracellular methods, as evidenced in the biological entities that are high secretors of proteins and enzymes that are primarily responsible for metal ion reduction and nonabsorbent material control (El-sayed 2020; Tulinski and Jurczyk 2017). Biogenic synthesis method is a sustainable and green technology since it does not need the use of toxic chemicals for production methods (Fosso-Kankeu 2019).

Conventional adsorbents have limitations such as poor adsorption capabilities, a lack of functional tunability, recyclability, and reusability. To address such constraints, new nano-sized sorbents are being developed and used for water purification (Wadhawan et al. 2020). Nanomaterials have recently received a lot of interest as adsorbents in wastewater decontamination due to their huge specific surface area, lower flocculent generation, and availability of a lot of active groups for attaching heavy metals (Verma et al. 2017).

5 Nano Adsorbents in Heavy Metal Remediation

The increasing demand for clean water with low heavy metal concentrations necessitates the efficient removal of harmful heavy metals from industrial runoff prior to its discharge into the ecosystem (Tu et al. 2017). Ion exchange, reverse osmosis, solvent extraction, chemical precipitation, and other standard methods are available for purifying wastewater. Despite the fact that the aforementioned methods are effective and meet discharge criteria, the majority of them produce secondary pollutants. On the other hand, adsorption is seen as a promising option for the removal of toxic metals from waste water due to its accessibility, adaptability, and high proficiency. It is more suitable than other methods due to its simple design and minimal initial cost and space requirements. Furthermore, if the adsorbent employed is recyclable, the adsorption process becomes extremely cost-effective. Because of these qualities, researchers are paying close attention to the adsorption process in the treatment of industrial wastewater polluted by heavy metals (Wadhawan et al. 2020). There are several types of nano adsorbents used in wastewater treatment, and the most commonly employed nanomaterials are divided into four groups based on their shape, size, and chemical characteristics (Tulinski and Jurczyk 2017). They are carbon-based nano adsorbents, polymer-based nano adsorbents, metal oxide nano adsorbents, and magnetic nano adsorbents.

5.1 Carbon-Based Nano Adsorbents

Carbon-based nano adsorbents such as carbon nanotubes (CNT) and graphene are frequently used in wastewater treatment. These carbon nanotubes are classified into two types: single-walled carbon nanotubes (SWCNTs) and multi-walled carbon nanotubes (MWNTs). Nano nanotubes have unique properties such as surface areas, size, pore volumes, non-corrosiveness, nontoxicity, electrical conductivity, etc. These nano adsorbents have a high removal effectiveness of heavy metals including lead, copper, mercury, chromium, nickel, arsenic, and cadmium from wastewater (Šolić et al. 2020; Wadhawan et al. 2020). Several papers discuss the use of carbon nanotubes (CNTs) in heavy metal ion removal from wastewater. In a study, CNTs as adsorptive material showed the removal of ions with removal capabilities in the sequence Cu (II) > Pb (II) > Co (II) > Zn (II) > Mn (II) (Stafiej and Pyrzynska 2007). In another study, adsorption of Pb (II) from wastewater on CNTs showed 70.1 mg/g adsorption capacity (Rahbari and Goharrizi 2009).

Another carbon-based nanomaterial, graphene, has received a lot of interest in the field of environmental remediation. The presence of oxygen-containing functional groups on the surface of graphene oxide (GO) gives it a highly hydrophilic character, allowing for fine dispersion in water. Because of its diverse functional groups and large surface area, GO is a promising option for wastewater treatment. It possesses remarkable mechanical, electrical, and thermal properties. These nanoparticles are efficient against a wide range of contaminants found in wastewater (Wadhawan et al. 2020). Zhao et al. reported GO nanosheets for the removal and adsorption of Cd (II) and Co (II). Adsorption capacities of 106.3 and 68.2 mg/g for Cd (II) and Co (II) were achieved (Zhao et al. 2011).

5.2 Polymer-based Nano Adsorbents

Conventional adsorbents have disadvantages such as inadequate specificity, adsorption capacity, and recyclability. To address the problems of existing adsorbents, various organic–inorganic hybrid polymers with better adsorption capacity, improved thermal stability, high skeletal strength, and higher recyclability have been designed (Khajeh et al. 2013). Polymeric-based nano adsorbents with a large specific surface area, a porous structure, and functional groups on the surface have been shown to bind effectively to organic dyes and heavy metal ions such as arsenic, zinc, lead, and cadmium from wastewater. Polymer-based nano adsorbents are further classified based on the substance employed. They are chitosan, cellulose, and dendrimers (Nik Abdul Ghani et al. 2021).

Chitosan is a non-toxic, environmentally friendly, biocompatible, and hydrophilic polymer that may form complexes with a variety of metal ions. The existence of amino groups increases chelation interactions with metal ions, and chemical changes of chitosan improve its selectivity and sorption capability. Furthermore, to enhance the sorption ability and mechanical properties of the chitosan it can be processed with acidic medium (Wadhawan et al. 2020; Nik Abdul Ghani et al. 2021). In a study, chitosan-alginate nanoparticles were employed for mercury removal. The highest adsorption capacity of 217.39 mg/g was attained under ideal circumstances of pH 5 at 90 min with an initial ion concentration of 4 mg/L (Dubey et al. 2016).

Dendrimers are great indicators of polymeric nano adsorbents and their influence on eliminating organic and inorganic contaminants due to their repetitively branched molecules. Organic molecules can be adsorbed through the interior hydrophobic portions, whereas heavy metals can be adsorbed by the exterior branches (El-sayed 2020). To remove copper from the water, researchers used a combination of dendrimers with the addition of ultrafiltration, resulting in the removal of copper ions (Diallo et al. 2005).

The absorption of heavy metal ions by biopolymer-based nano adsorbents such as cellulose has been examined. Jamshaid et al. evaluated several cellulose-based nano adsorbents such as cellulose gels, nanocrystalline cellulose, and cellulose composites and derivatives, and the results demonstrated that the adsorption capacity of heavy metals was successfully increased (Jamshaid et al. 2017).

5.3 Metal Oxide Nano Adsorbents

Metal oxide nanoparticles have a high removal capacity, a large surface area, and a strong affinity for heavy metal adsorption, making them a promising nano adsorbent for wastewater treatment. The range of metal oxide nanoparticles in nanosized materials is 1–100 nm. It includes zinc oxide, manganese oxide, nickel oxide, aluminum oxide, iron oxide, zirconium oxide, titanium oxide, and magnesium oxide (Yang et al. 2019). Due to their high adsorption capacity and huge surface areas, these metal oxide-based nanoparticles are attractive nano adsorbents for heavy metal removal from aqueous systems. Metal oxide adsorption processes were regulated by complexation between dissolved metals and oxygen in two steps: first, metal ion adsorption on the external surface, and second, rate-restricted intra-particle diffusion along the porosities (El-sayed 2020; Wang et al. 2020).

5.4 Magnetic Nano Adsorbents

Magnetic nano adsorbents are extremely non-toxic, high selectivity, recyclable, and reusable (Mudhoo and Sillanpää 2021). They have magnetic properties that allow for ease of separation using an electromagnet. Magnetic nanoparticle modification has recently received increased interest as a consequence of magnetic nanoparticles' capacity to form bonds with various terminal groups of diverse molecules, especially ligands (Zhu et al. 2010). As a result, significant progress has been made in pollutant removal from wastewater using magnetic nanoparticles by reinforcing them onto other nanomaterials. Magnetic nanocomposite materials composed of two or more distinct nanoparticles have been used for water and wastewater treatment (Rebuttini 2014; Shah 2020, 2021a, b). For example, titanium dioxide (TiO_2) with powerful photocatalytic properties is strengthened into magnetic nanoparticles to boost the latter's photocatalytic properties for application in the removal of pollutants from wastewater (Lingamdinne et al. 2019; Khan et al. 2020).

6 Nano Adsorbents Recovery and Reutilization

Reutilization of spent adsorbent is the main aspect for successful real-time practical applications. Reusability of spent adsorbent helps to reduce the overall cost of the adsorption process as it reduces the need for new adsorbent synthesis and purchase. Reusability of spent adsorbents helps in reduced cost, toxicity, and large-scale usability of novel nanomaterials. Till date, plenty of research and review articles are published in desorption and utilization of adsorbents to enhance the overall viability of the process. Commonly nano adsorbents, activated carbon, graphene oxides, and biomaterials are utilized in adsorption process.

7 Conclusion and Future Directions

Nanomaterial, nanoscience, and nanotechnology have come across greater advancements in last years that have driven the way for synthesis, structuring of novel, economical, eco-friendly nano adsorbents. Higher selectivity and adsorption capacity of the adsorbent defines the electability and stand-alone candidate in the wastewater treatment technique. Even though adsorption of heavy metal is an easier physiochemical technique, in this book chapter we have focused mainly on heavy metal effects, nano adsorbents synthesis and types. In future, encapsulated and multi-dimensional nanomaterials can be used for achieving higher removal percentage of heavy metals from aqueous environment. This helps in achieving a comparative and comprehensive research methodology by identifying the research problem by various scientists and research groups globally.

Acknowledgements Authors wish to thank Sathyabama Institute of Science and Technology for their support.

Conflict of Interest None.

References

Abernathy CO, Liu Y-P, Longfellow D, Aposhian HV, Beck B, Fowler B, Goyer R, Menzer R, Rossman T, Thompson C et al (1999) Arsenic: health effects, mechanisms of actions, and research issues. Environ Health Perspect 107:593–597

Banu BS, Danadevi K, Jamil K, Ahuja YR, Rao KV, Ishaq M (2001) In vivo genotoxic effect of arsenic trioxide in mice using comet assay. Toxicology 162:171–177

Barakat MA (2011) New trends in removing heavy metals from industrial wastewater. Arab J Chem 4:361–377

Briffa J, Sinagra E, Blundell R (2020) Heavy metal pollution in the environment and their toxicological effects on humans. Heliyon 6:e04691

Darr JA, Zhang J, Makwana NM, Weng X (2017) Continuous hydrothermal synthesis of inorganic nanoparticles: applications and future directions. Chem Rev 117:11125–11238

De Beni E, Giurlani W, Fabbri L, Emanuele R, Santini S, Sarti C, Martellini T, Piciollo E, Cincinelli A, Innocenti M (2022) Graphene-based nanomaterials in the electroplating industry: a suitable choice for heavy metal removal from wastewater. Chemosphere 292:133448

Diallo MS, Christie S, Swaminathan P, Johnson JH, Goddard WA (2005) Dendrimer enhanced ultrafiltration. 1. Recovery of Cu (II) from aqueous solutions using PAMAM dendrimers with ethylene diamine core and terminal NH2 groups. Environ Sci Technol 39:1366–1377

Dubey R, Bajpai J, Bajpai AK (2016) Chitosan-alginate nanoparticles (CANPs) as potential nanosorbent for removal of Hg (II) ions. Environ Nanotechnol Monit Manag 6:32–44

El-sayed MEA (2020) Nanoadsorbents for water and wastewater remediation. Sci Total Environ 739:139903. https://doi.org/10.1016/j.scitotenv.2020.139903

Farghali AA, Bahgat M, Allah AE, Khedr MH (2013) Adsorption of Pb (II) ions from aqueous solutions using copper oxide nanostructures. Beni-Suef Univ J Basic Appl Sci 2:61–71

Fosso-Kankeu E (2019) Nano and bio-based technologies for wastewater treatment: prediction and control tools for the dispersion of pollutants in the environment. John Wiley & Sons

Haripriyan U, Gopinath KP, Arun J (2022) Chitosan based nano adsorbents and its types for heavy metal removal: a mini review. Mater Lett 131670

He ZL, Yang XE, Stoffella PJ (2005) Trace elements in agroecosystems and impacts on the environment. J Trace Elem Med Biol 19:125–140

Herawati N, Suzuki S, Hayashi K, Rivai IF, Koyama H et al (2000) Cadmium, copper, and zinc levels in rice and soil of Japan, Indonesia, and China by soil type. Bull Environ Contam Toxicol 64:33–39

Jamshaid A, Hamid A, Muhammad N, Naseer A, Ghauri M, Iqbal J, Rafiq S, Shah NS (2017) Cellulose-based materials for the removal of heavy metals from wastewater–an overview. ChemBioEng Rev 4:240–256

Katz SA, Salem H (1993) The toxicology of chromium with respect to its chemical speciation: a review. J Appl Toxicol 13:217–224

Khajeh M, Laurent S, Dastafkan K (2013) Nanoadsorbents: classification, preparation, and applications (with emphasis on aqueous media). Chem Rev 113:7728–7768. https://doi.org/10.1021/cr400086v

Khan FSA, Mubarak NM, Khalid M, Walvekar R, Abdullah EC, Mazari SA, Nizamuddin S, Karri RR (2020) Magnetic nanoadsorbents' potential route for heavy metals removal—a review. Environ Sci Pollut Res 27:24342–24356

Kim J-J, Kim Y-S, Kumar V (2019) Heavy metal toxicity: an update of chelating therapeutic strategies. J Trace Elem Med Biol 54:226–231

Le AT, Pung S-Y, Sreekantan S, Matsuda A et al (2019) Mechanisms of removal of heavy metal ions by ZnO particles. Heliyon 5:e01440

Lingamdinne LP, Koduru JR, Karri RR (2019) A comprehensive review of applications of magnetic graphene oxide based nanocomposites for sustainable water purification. J Environ Manage 231:622–634

Maitlo HA, Kim K-H, Kumar V, Kim S, Park J-W (2019) Nanomaterials-based treatment options for chromium in aqueous environments. Environ Int 130:104748

Mudhoo A, Sillanpää M (2021) Magnetic nanoadsorbents for micropollutant removal in real water treatment: a review. Environ Chem Lett: 1–21

Nik Abdul Ghani NR, Jami MS, Alam MZ (2021) The role of nanoadsorbents and nanocomposite adsorbents in the removal of heavy metals from wastewater: a review and prospect. Pollution 7:153–179

Nriagu JO (1989) A global assessment of natural sources of atmospheric trace metals. Nature 338:47–49

Pareek V, Bhargava A, Gupta R, Jain N, Panwar J (2017) Synthesis and applications of noble metal nanoparticles: a review. Adv Sci Eng Med 9:527–544

Rahbari M, Goharrizi AS (2009) Adsorption of lead (II) from water by carbon nanotubes: equilibrium, kinetics, and thermodynamics. Water Environ Res 81:598–607

Razzak SA, Farooque MO, Alsheikh Z, Alsheikhmohamad L, Alkuroud D, Alfayez A, Hossain SMZ, Hossain MM (2022) A comprehensive review on conventional and biological-driven heavy metals removal from industrial wastewater. Environ Adv:100168

Rebuttini V (2014) Functional iron oxide-based hybrid nanostructures

Shah MP (2020) Microbial bioremediation & biodegradation. Springer
Shah MP (2021a) Removal of refractory pollutants from wastewater treatment plants. CRC Press
Shah MP (2021b) Removal of emerging contaminants through microbial processes. Springer
Singh J, Dutta T, Kim K-H, Rawat M, Samddar P, Kumar P (2018) 'Green' synthesis of metals and their oxide nanoparticles: applications for environmental remediation. J. Nanobiotechnology. 16:1–24
Šolić M, Maletić S, Kragulj Isakovski M, Nikić J, Watson M, Kónya Z, Tričković J (2020) Comparing the adsorption performance of multiwalled carbon nanotubes oxidized by varying degrees for removal of low levels of copper, nickel and chromium (VI) from aqueous solutions. Water 12:723
Stafiej A, Pyrzynska K (2007) Adsorption of heavy metal ions with carbon nanotubes. Sep Purif Technol 58:49–52
Sträter E, Westbeld A, Klemm O (2010) Pollution in coastal fog at Alto Patache, northern Chile. Environ Sci Pollut Res 17:1563–1573
Tchounwou PB, Ayensu WK, Ninashvili N, Sutton D (2003) Environmental exposure to mercury and its toxicopathologic implications for public health. Environ Toxicol An 18:149–175
Tchounwou PB, Yedjou CG, Patlolla AK, Sutton DJ (2012) Heavy metal toxicity and the environment. Mol Clin Environ Toxicol:133–164
Tonelli D, Scavetta E, Gualandi I (2019) Electrochemical deposition of nanomaterials for electrochemical sensing. Sensors 19:1186
Tu Y-J, You C-F, Chen M-H, Duan Y-P (2017) Efficient removal/recovery of Pb onto environmentally friendly fabricated copper ferrite nanoparticles. J Taiwan Inst Chem Eng 71:197–205
Tulinski M, Jurczyk M (2017) Nanomaterials synthesis methods. Metrol Stand Nanotechnol Protoc Ind Innov:75–98
Velma V, Vutukuru SS, Tchounwou PB (2009) Ecotoxicology of hexavalent chromium in freshwater fish: a critical review. Rev Environ Health 24:129–146
Verma M, Tyagi I, Chandra R, Gupta VK (2017) Adsorptive removal of Pb (II) ions from aqueous solution using CuO nanoparticles synthesized by sputtering method. J Mol Liq 225:936–944
Waalkes MP (1996) Cadmium carcinogenicity and genotoxicity. Toxicol Met:231–244
Wadhawan S, Jain A, Nayyar J, Mehta SK (2020) Role of nanomaterials as adsorbents in heavy metal ion removal from waste water: a review. J Water Process Eng 33:101038
Wang L, Shi C, Pan L, Zhang X, Zou J-J (2020) Rational design, synthesis, adsorption principles and applications of metal oxide adsorbents: a review. Nanoscale 12:4790–4815
Yang J, Hou B, Wang J, Tian B, Bi J, Wang N, Li X, Huang X (2019) Nanomaterials for the removal of heavy metals from wastewater. Nanomaterials 9:424
Yari S, Abbasizadeh S, Mousavi SE, Moghaddam MS, Moghaddam AZ (2015) Adsorption of Pb (II) and Cu (II) ions from aqueous solution by an electrospun CeO2 nanofiber adsorbent functionalized with mercapto groups. Process Saf Environ Prot 94:159–171
Yedjou CG, Tchounwou PB (2007) N-acetyl-l-cysteine affords protection against lead-induced cytotoxicity and oxidative stress in human liver carcinoma (HepG2) cells. Int J Environ Res Public Health 4:132–137
Yedjou CG, Milner JN, Howard CB, Tchounwou PB (2010) Basic apoptotic mechanisms of lead toxicity in human leukemia (HL-60) cells. Int J Environ Res Public Health 7:2008–2017
Zhao G, Li J, Ren X, Chen C, Wang X (2011) Few-layered graphene oxide nanosheets as superior sorbents for heavy metal ion pollution management. Environ Sci & Technol 45:10454–10462
Zhu Y, Murali S, Cai W, Li X, Suk JW, Potts JR, Ruoff RS (2010) Graphene and graphene oxide: synthesis, properties, and applications. Adv Mater 22:3906–3924

Nanofiltration Applications for Potable Water, Treatment, and Reuse

Neha Patel, Archna Dhasmana, Shristi Kumari, Rajat Sharma, Shreya Nayanam, and Sumira Malik

1 Introduction

The world population is rapidly expanding, creating significant environmental stress and pressure. In the last decades, several human practices have utilized water as energy and food production resources, but the anthropogenic activities result in pollution, water, and food scarcity as an emerging threat to life in this ecosystem. In the recent studies by IWMI (International Water Management Institute), approximately 1.4 billion people face the water problem, which has continuously increased with population increase. Currently, around half of the water is used in households, and half is used for agricultural and industrial purposes.

Water will continue to be a critical and significant factor in human existence and activities. This is especially true in industrialized areas with a steadily growing population; however, industries will be under pressure to recapture, reclaim, and reuse some of their wastewater. Hence, to resolve the global environmental issues, many water treatment procedures or techniques have been developed and implemented, considering the source of pollution (Avlonitis et al. 2008).

The most promising wastewater approach is nanofiltration membrane technology for treating water and wastewater in various fields. It is a compression-compelled process that involves filter membranes acting as specifically annotated contaminants and pollutants, including organic metals, inorganic metals, nutrients, microorganisms, turbidity, and other oxygen-depleting pollutants, allowing comparatively pure

Neha Patel and Archna Dhasmana are contributed equally.

N. Patel · S. Kumari · R. Sharma · S. Nayanam · S. Malik (✉)
Amity Institute of Biotechnology, Amity University Jharkhand Ranch, Ranchi, Jharkhand, India
e-mail: smalik@rnc.amity.edu

A. Dhasmana
Himalayan School of Biosciences, Swami Rama Himalayan University, Dehradun, Uttarakhand, India

M. P. Shah (ed.), *Advanced and Innovative Approaches of Environmental Biotechnology in Industrial Wastewater Treatment*,
https://doi.org/10.1007/978-981-99-2598-8_7

and clear water to pass through. Nanofiltration technology can be a potential and intelligent solution for standardizing water quality and recycling (Shon et al. 2013). Compared to other technologies, common monovalent ion rejection, strong divalent ion rejection, and increased flux are vital differentiating features of nanofiltration membranes. Due to these qualities, nanofiltration has been applied in various applications for wastewater and industrial effluent treatment (Mohammad et al. 2015).

2 Nanofiltration

Nanofiltration was first discovered in the late 1980s; since then, it has come a long way. It is the most evolved and practiced method for liquid phase separation through the pressure-driven membrane technique. The high-rated flux and minimum energy intake in the nanofiltration have exceeded reverse osmosis in numerous applications (Mohammad et al. 2015). The nanofiltration membranes have intermediate properties to compact RO membranes having a solution diffusion mechanism that governs transport, and porous ultrafiltration (UF) membranes have size exclusion and charge effects separation properties (Shon et al. 2013). Nano filters separate a wide variety of organic and inorganic substances or chemicals from impure or contaminated liquid samples through diffusion at differentials pressure, i.e., lower than RO and higher than UF. Thus, the significant property of the NF to separate or differentiate and fractionate ionic and organic species, which has a relatively low molecular weight process, makes this system an effective technique (Sutherland 2008). In the aqueous solutions, the NF membranes get slightly charged and alienation of surface functional group for the prolonged adsorption of charge solutes particle, e.g., acid groups and anionic charged groups. Likewise, RO membrane and nanofiltration showed effective separation of the small organic molecule and inorganic salts (Mohammad. et al. 2015). Nanofiltration membranes have a pore size in the nanometre range (rim) (1 $\times$ 10^9 m). As a comparison, sodium and chloride ions have an atomic radius of about 0.97 nm (0.97 $\times$ 10^{-9} m) and 1.8 nm (1.8 $\times$ 10^9 m), respectively. This shows that nanofiltration membranes are capable or effective in removing small ions (Roth et al. 2009). Because of these characteristic features, NF membrane was used for the fractionation-specific remediation of the pollutants particles in the saturated samples by easy methods, and their application increased dramatically. New possibilities for producing drinking water have been discovered, bringing remedies or solutions to new-fangled challenges such as the remediation of toxic chemicals, heavy metals, and EDC from water bodies (Bruggen et al. 2008).

The utilization of membrane technology has been widely used during the last decade and has mainly emerged in wastewater treatment techniques and water purification for domestic and drinking usage. The growth is attributed to a combination of factors, including (1) increased demand for high-quality water. (2) increased or massive pressure to reuse wastewater. (3) improved membrane integrity and reliability. (4) lower prices of nanofiltration membranes due to widespread usage or increased demands. (5) more stringent standards like those in the drinking water industry. Other industries that might benefit from nanofiltration involve pharmaceuticals, food and dairy industries, and chemical processing industries. Membrane technology could save a considerable amount, and the environmental benefits of reduced energy consumption make nanofiltration even more appealing and desirable (Bruggen et al. 2008). NF system is the potential and promising for removing the charged compounds such as monovalent and multivalent ions and complex organic pollutants of variable dimensions. This possible technique results in significant application in industries effluents treatments such as pharmaceuticals, the paper industries, textile industries, dairy industries, hospital waste, and heavy metal retrieval. On the other hand, several challenges which need to be solved include membrane fouling, insufficient separation, concentrate treatment, chemical resistance and limited membrane lifetime, insufficient pollutant rejection in wastewater treatment, and the requirement for modeling and simulation tools are a few examples (Bruggen et al. 2008). NF systems come in various forms such as plate and frame, cylindrical, curved, capillary, and hollow fiber arrangements, consisting of natural, synthetic, semi-synthetic polymers, composites from inorganic material, organic or inorganic hybrids, and ceramics (Sutherland 2008).

3 Necessity of Nanofiltration Technology

Increased human activity in every sector, such as food, medicine, different industries, pharmaceuticals, fishing, etc., needs adequate or insufficient clean water. In these industries, nanofiltration is becoming more frequent. Limited fresh water supply, which comprises only 3% of the total water on this planet, necessitates or requires environmentally safe, efficient, clean output that can even be reused for numerous uses. Nanofiltration technology provides excellent selectivity, low energy consumption, cost-efficient, and environmental stability and can be used with different separation processes as it is chemically free and easy to scale up. Nanofiltration can be widely used in solving many issues, particularly in the treatment of wastewater (Mulyanti and Susanto 2018). The nanofiltration process is used to separate colors in the textile sector, recover metals, and clean effluent from olive mills. NF was also employed in treating coke effluent, paper industry, fuel/oil sea-shores, and the exclusion of mine industries in water bodies. The availability of raw materials and methodology for the fabrication of NF provide better scopes and research in the area.

4 Nanofiltration Mechanism

NF mode of action and processing is based on two parameters: electrical (Donnan) effect and steric (sieving) effects for the separation process. This combination enables NF membranes to separate effluents containing micron or small organic solutes or salts, either neutral or charged (Bowen and Mohammad 1998). The process includes passing a portion of the feedstock through the semipermeable membrane, in which the inlet is divided into two separate chambers: (i) Permeate having a filtered portion of the stream and (ii) Retentate having an unfiltered portion of the stream. Moreover, chlorine disinfection is used in the NF distribution system to remove microbial growth. NF membranes possess maximum and effectual removal of the organic component are eminence in reducing microbial growth.

NF has also confirmed the promising outcome of removing EDCs and pharmaceuticals (PACs). EDCs and PACs are extracted or removed with the help of membranes by the amalgamation of numerous processes, which includes screening by the size, charge, and adsorption mechanisms (Wang et al. 2021).

Size exclusion/screening—Physical blocking is the most common method of size exclusion/screening, which means if the molecular size is smaller than the membrane size, no effective removal occurs, and it cannot pass through the membrane. Researchers found that the removal rate is proportional to molecule size in many studies.

Charge exclusion—in this, the removal process is facilitated by electrostatic repulsion that occurs among the filter membrane and charged molecules in the effluents, which is a significant or non-negligible mechanism (Wang et al. 2021).

The adsorption mechanism is the critical process for the preliminary rejection process based on the distribution either at the fast or slow stage, along with the equilibrium time related to the molecular weight. Here, the chemical bond is used in the adsorption process (for example, hydrogen bonding) (Wang et al. 2021).

5 Nanofiltration for Wastewater

The wastewater treatment required for the effective treatment of removal of discharge of water and reuse involves different problems and challenges than the treatment of drinking water. Previously, wastewater processing only the removal of solids, organic compounds, and microorganisms. Nevertheless, the regulations may no longer be limited to only these contaminants (Roth et al. 2009). An innovative treatment combination was studied polishing wastewater from a municipality intending to ensure or maintain safe groundwater. The NF analysis indicated an appropriate option for treating tertiary effluent to decrease or lower dissolved organic carbon (DOC) and adsorbable organic halides (AOX) up to < 2–3 mg/l and 20 μg/l, respectively. Food industries widely use membrane processing technology to treat food industries' effluents. A single nanofiltration stage can reduce chemical oxygen demands instead of reverse osmosis or ultrafiltration, e.g., whey manufacturing effluent COD level was 100,000 mg^2/L^{-1}, and its filtrate with lower permeate COD level is 2787 mg^2/L. It was investigated that the practice of underwater NF sheets removes heavy aromatic organic pharmaceutical pollutants from the municipal wastewater treatment plant (MWTP) at comparable low pressure of 0.7 bar in Giessen, Germany. Recent research has engrossed on tertiary treatment, i.e., advanced oxidation processes (AOPs) for

MWTP to eliminate micropollutants (Mohammad et al. 2015). However, for the treatment of wastewater containing heavy metals, copper (Cu) and cadmium (Cd), NF membranes are widely used in different sectors. The throwaway Cu ion from highly acidic copper solutions of $CuSO_4$, 25 g/l in 8% H_2SO_4 at temperature 40 °C and pressure of 20–40 bar for two months showed NF membranes total Cu retention at 30 bar up to 88–95%. During the two weeks, the NF270 membranes retained 100% copper; after 27 days, they had essentially little Cu retention. NF's role in water softening, dye removal, industrial effluent handling, and reuse may be a better option than other technologies for removing hardness, color, and DBP precursors (Shahmansouri and Bellona 2015). Avlonitis et al. studied nanofiltration treatment on the effluents that come from textile industries containing high salts and organics concentration and can be reused, and it was found that a nanofiltration membrane named TRISEP (4040-XNA5-TSF) showed a fantastic result. These studies showed that the membrane could completely decolorize cotton dye effluents, lower overall salt concentration, and provide high-quality water that can be reused (Avlonitis et al. 2008). Moreover, various NF membranes are offered for wastewater treatment, each having its properties and applicability for the particular solutions to encounter discharge requirements. Nanofiltration membranes are designed to mark certain pollutants more effectively than RO in a single pass or after several passes (Roth et al. 2009).

Waste generated from anthropogenic human activity is a significant need but has many potentials to be utilized for various treatments. Generally, a large portion of waste is a recyclable and replenishable energy source. It can be counted as a new possibility since concerns about the environmental effects and impacts have been drastically minimized and an alternate remedy for a scarcity of clean water. The objective and implications of reusability of wastewater have reached the middle east nations and America, which is considered one of the finest in water reusability index. Wastewater reuse is increasing in European and Asian countries, and membrane technology can create and produce high effluents for reuse in various applications, including irrigation, processed water, tourism, public parks, and toilets (Mulyanti and Susanto 2018; Fig. 1).

6 Applications of Nanofiltration

NFS's innovative membrane technology has widespread uses together with wastewater treatment, as well as product purification in different sectors (food industry, chemical industry, textile industry, petroleum industry, etc.). Like any other procedure, nanofiltration is chosen based on technical and commercial contemplation.

NF operated as pressure-driven membrane eco-friendly separation technology shows extremely efficient and minimum-energy consumption processing for the Desalination, softening of seawater and brackish, and COD decrease.

The molecular weight limit for NF membranes is between reverse osmosis and ultrafiltration membranes, ranging from 200 to 1000 Da. The continuous population increase and manufactured eco-destructive activities result in global warming,

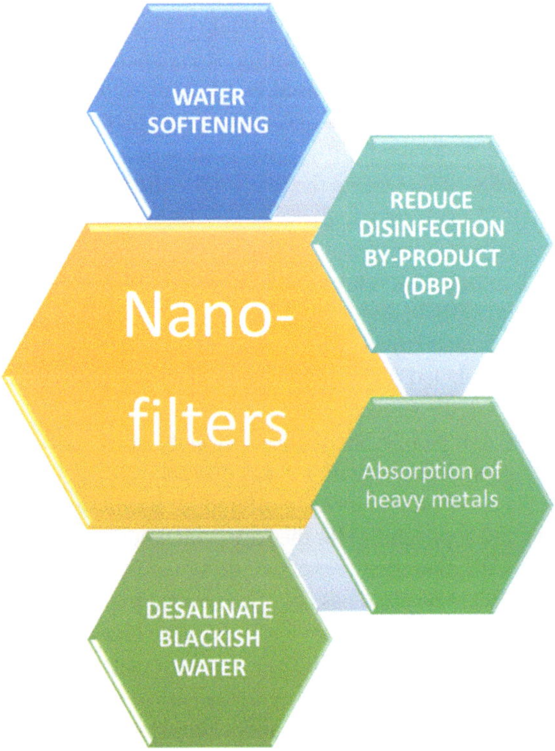

Fig. 1 Nanofilters application in wastewater treatment

water resources in diverse areas, and unprecedented pressures on the ecosystem. The imbalance of water demand and its availability is exacerbated by the concentration of people in metropolitan areas (Asano et al. 2007). Wastewater can be used as an alternate water source, reducing the need for fresh water. High salt concentrations in wastewater are a problem for some towns and industries because salts prevent biological treatment; saline effluents are often treated with physicochemical methods (Aloui et al. 2009). Most of these systems are biological agents, i.e., microbial conversion by following either anaerobic or aerobic pathways. Bioreactor systems are the utmost choice for treating organic and nitrogenous compounds in wastewater.

Consequently, maximum dissolved salts in wastewater have been shown to limit the efficacy of traditional treatments such as activated sludge, anaerobic digestion, nitrification, and denitrification processes. NF membrane technologies play an essential role in wastewater disposal, water recycling, retrieval, and recovering valued compounds from effluent streams to meet environmental regulations (Marcucci et al. 2001). NF and RO are two membranes that are the advanced treatment technologies used for the mixture of electrostatic and steric interacting groups. NF membranes were tested to study the pollutants removal that is not eliminated by biological treatment of municipal wastewater for the reusability in irrigation (Bunani et al. 2013).

6.1 Water Softening

Water softening is done by using a nanofiltration membrane. It is done when the refusal of monovalent salts is less and the preference of reverse osmosis over low pressure or energy of the membrane using the removal of color and DBP. Calcium (Ca^{2+} and magnesium (Mg^{2+}) ions are commonly found in high amounts in hard water (Ghizellaoui et al. 2005). Various compounds like nitrates, fluoride, carbons, arsenic, sulfate, phosphate, and other organic contaminants are removed using nanofiltration during water softening and wastewater treatment. Since the 1980s, NF has been an extensively researched technology in water purification, softening, and recycling. (Lee and Lee 2000).

NF stimulates ions separation processes in water softening and recycling industrial effluent by the high flux and discerning rejection of multivalent ions (Lee and Lee 2000; Tang et al. 2018). The membrane fouling caused by the scale formation on the surface of the membrane is one of the severe constrains in the operation of these processes (Lee and Lee 2000). Traditional water softening procedures include ash, calcium carbonate, zeolite, and ionic resin treatments. Unit water costs for three alternative processes for water softening and color removals are lime and soda ash, softening with ozone inoculation, granular-activated carbon (GAC) for decoloration, and NF membranes. The findings demonstrated that the water purification capacity of the NF membranes (NF 270 and NF 90) for the operating system at low pressures had been used to soften hard water and desalinate seawater. Despite these advancements, the complicated physiochemical interactions between the ionic solutes and NF membrane for the multi-ionic solutions are unpredictable (Silva et al. 2011). Before expanding NF softening powers model, we need to recognize or standardize the protocols for the separation process based on the sample to be transported and membrane functioning.

The previous research on NF modeling for water softening applications has specific difficulties and demonstrates the requirement for more research. In a study, the modified version of the two-dimensional capillary model of bipolar softening membranes showed solute rejection at variable pH and polyelectrolyte concentration of the treated samples. Donnan-Steric Pore Model Dielectric Exclusion (DSPM-DE) model was the first used to study the performance of the unique layer-by-layer LbL1.5C NF membrane fabricated multi-layer deposition and chemical crosslinking and used for water softening. The membrane was characterized using the DSPM-DE model and has a significant arrangement between trial and modeling samples. The stimulation and sensitivity examination of the LbL1.5C membrane was used to determine the membrane parameter, such as the strong selectivity for monovalent ions and multi-ionic interactions, and dictates the higher softening capacity (Labban et al. 2017).

6.2 Desalinate Blackish Water

The NF membranes used brackish water and seawater, indicating that they have the potential for the desalination process for water treatment (Tang et al. 2018). The most complicated use for cleaning water is seawater desalination. In recent years, many municipalities have built saltwater desalination plants employing traditional seawater ultrafiltration membrane techniques. Desalination with these traditional reverse osmosis membranes necessitates exceptionally high pressures, ranging from 55 to 83 bars. Furthermore, untreated saltwater and brine water are very corrosive and need iron alloy and stainless steel filters extensively. Thus, the process requires energy uptake, maintenance, and high cost. The desalination performance of chitosan (CS) and chitosan-piperazine (CIS-PIP) composite for NF membranes was investigated with seawater and brackish water under various compressions (Tang et al. 2018). Due to its limitations in removing monovalent ions, the NF desalination market is quite limited, but it has been proposed for some uses in the desalination sector and is an essential process in combination with other processes. For divalent ions such as SO_4 2, Mg^{2+} , and Ca^{2+} , NF membranes demonstrated improved rejection. The electrostatic repulsive effect and the sieving effect influence the separation properties of various ions.

Higher ion rejection is caused by a macro hydrodynamic ionic radius and a low diffusion coefficient. As a result, negative ions such as sodium hydroxide and chloride were repelled more strongly by CS and CS-PIP composite NF membranes (Tang et al. 2018). It studied the application of RO-NF and NF2 for future drinking water regulations. However, the NF membrane used as pretreatment of multi-stage flash desalination (MSF) and RO-MSF be significant eco-friendly and economic desalination processes. The combined effect of NF with RO and MSF for seawater desalination processes reduced the cost by up to 30%. According to the analysis, while NF can significantly boost desalination system recovery, the cost reductions are negligible. Although seawater desalination facilities can get their feed water from various sources and places along the coast, open seawater drinking is the most predominant.

In another approach, chemical compounds are used as disinfectants for the pretreatment, and subsequently, multi-media filtration performs to minimize bacterial development and biofouling in intake structures and to increase filter performance. The most used disinfectant for pretreatment and disinfection are free-chlorinated compounds (i.e., HOCl/OCl). Besides that, other disinfectants such as chloramines, ozone, and chlorine dioxide are routinely employed to deactivate pathogenic germs in the water bodies (Kim et al. 2015). Water resources with a mid-salinity, such as brackish groundwater or wastewater, are excellent sources of high-quality water.

Membrane bioreactors (MBR) or membranes combined with improved oxidation techniques is the high-quality water purification system from wastewater. Instead, NF and low-pressure RO have been presented as viable technologies for brackish groundwater purification. Fouling, on the other hand, reduces the lifetime of the membranes used in pressure-driven filtering, raising the overall cost of the system. As a result, current efforts are focused on evaluating novel membrane-based processes that utilize various driving forces (Giagnorio et al. 2019). The cost of desalinating brackish water is 60–70% less than desalinating seawater (Diawara 2008). Water desalination by RO makes water drinkable, but also for various industrial applications for ultra-pure water for sample testing and processing, e.g., electronics industry (Mohsen et al. 2003; Redondo 2001).

In case of water contamination due to the high salinity, NF and low-pressure RO membranes treat fluids of low- to high-salinity water. The limited power supply and energy in rural communities may limit the operation of such facilities. Solar energy, also known as photovoltaic energy, is an excellent form of renewable energy for addressing this issue. The most appropriate membrane for salt retention was pump energy designed the deliver a permeate flow of 400–1000 L/day from brackish wells. In the last few decades, science and technology in membrane designing have led to tremendous innovation in both procedures and products by rationalizing and optimizing new manufacturing processes (Matsuura 2001). The most exciting advancement in industrial membrane technology is combining multiple membrane activities in a single industrial cycle, resulting in significant improvements in sample quality, system compactness, environmental effect, and energy efficiency (Drioli and Romano 2001; Drioli et al. 2002).

6.3 Reduce Disinfection By-Product (Dbp)

Worldwide the desalination of seawater has become the technology used in many places as a supplementary for the water supply due to rising freshwater demand worldwide. Synthetic chemical disinfection for desalination facilities in the pretreatment and post-treatment is used to reduce biofouling and disinfect desalinated water. The study into the DBP precursor removal capacity of capacitive deionization (CDI) is significant for future disinfection in drinking water treatment. The ideal removal of bromide (Br) over natural organic matter, the minimizing Br substitution factor standards, and the lowered DBP formation potential were discovered to indicate that CDI might reduce the inorganic and organic DBP precursors (Liu et al. 2016). The most commonly used disinfectant Cl used in desalination plants produces DBPs, e.g., trihalomethanes [THMs], haloacetic acids [HAAs], and haloacetonitriles [HANs], having potential health risks to humans and animals. Hence, alternative oxidants are used as disinfectants, such as chloramines, to limit the expansion of chlorinated DBPs (Kim et al. 2015). Disinfection, as a necessary water treatment step for the prevention of waterborne infections, results in a wide range of DBPs in the final product (Xie et al. 2004). Even at low concentrations, many DBPs have been demonstrated to negatively influence health, e.g., THM and HAAs (Liu et al. 2016). Two factors are essential in DBP formation: chlorine dosage and contact time. DBP production is also influenced by organic carbon concentration, temperature, and pH. Interactions between disinfectants form the DBP precursor's natural organic matter (NOM) and bromide, and DBP precursors are one of the choices for successful DBP control. Electro-adsorption was found to be more critical than physical adsorption in the

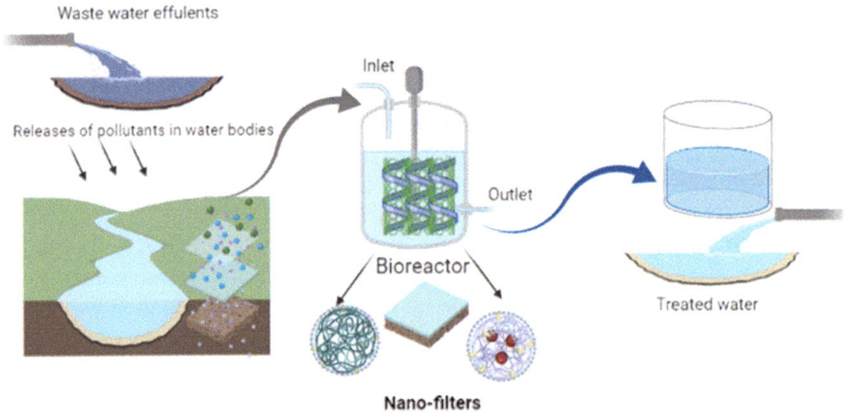

Fig. 2 Nanofiltration mechanism of wastewater treatment

removal of DBP production potential. CDI can be regarded as a viable technology for DBP control in drinking water handling due to its potential to reduce health hazards by less brominated DBPs after consequent chlorination (Liu et al. 2016).

7 Conclusion and Future Prospective

NF membrane processes are among the most operative methods for purification and desalination of water and wastewater. NF Membrane separation technology is a possible alternative to conventional filtration technologies. NF systems have a greater retrieval rate recompenses of lower operational costs and less ecological effect. NF is the extensively used membrane for water and wastewater handling such as softening, desalination, and diminished disinfection by-product (DBP) formation. Moreover, nanofiltration systems have a better recovery rate, providing the benefits of lower cost operations and less environmental effect. Hence, the NF applicable in various industries can or potentially give the best results in effluent treatment and improve the quality of the generated sewage water for reuse. The mechanism of nanofiltration techniques is summarized in Fig. 2.

References

Aloui F, Khoufi S, Loukil S, Sayadi S (2009) Performances of an activated sludge process for the treatment of fish processing saline wastewater. Desalination 246(1–3):389–396. https://doi.org/10.1016/j.desal.2008.03.062

Asano T, Burton FL, Leverenz HL, Tsuchihashi R, Tchobanoglous G (2007) Water reuse, issues, technologies, and applications. McGraw -Hill, New York

Avlonitis S, Poulios I, Sotiriou D, Pappas M, Moutesidis K (2008) Simulated cotton dye effluents treatment and reuse by nanofiltration. Desalination 221(1–3):259–267. https://doi.org/10.1016/j.desal.2007.01.082

Bowen W, Mohammad A (1998) Characterization and prediction of nanofiltration membrane performance—a general assessment. Chem Eng Res Des 76(8):885–893. https://doi.org/10.1205/026387698525685

Bunani S, Yörükoğlu E, Sert G, Yüksel Ü, Yüksel M, Kabay, Nalan (2013) Application of nanofiltration for reuse of municipal wastewater and quality analysis of product water. Desalination 315:33–36. https://doi.org/10.1016/j.desal.2012.11.015

Diawara CCK (2008) Nanofiltration process efficiency in water desalination. Sep Purif Rev 37(3):302–324. https://doi.org/10.1080/15422110802228770

Drioli E, Romano M (2001) Progress and new perspectives on integrated membrane operations for sustainable industrial growth. Ind Eng Chem Res 40(5):1277–1300. https://doi.org/10.1021/ie0006209.

Drioli E, Criscuoli A, CurcioaCurcio E (2002) Integrated membrane operations for seawater desalination. Desalination 147(1–3):77–81. https://doi.org/10.1016/S0011-9164(02)00579-9

Ghizellaoui S, Taha S, Dorange G, Chibani A, Gabon J (2005) Softening of Hamma drinking water by nanofiltration and lime in the presence of heavy metals. Desalination 171(2):133–138. https://doi.org/10.1016/j.desal.2004.05.001

Giagnorio M, Ricceri F, Tiraferri A (2019) Desalination of brackish groundwater and reuse of wastewater by forward osmosis coupled with nanofiltration for draw solution recovery. Water Res 153:134–143. https://doi.org/10.1016/j.watres.2019.01.014

Kim D, Amy GL, Karanfil T (2015) Disinfection by-product formation during seawater desalination: a review. Water Res 81:343–355. https://doi.org/10.1016/j.watres.2015.05.040

Labban O, Liu C, Chong TH, Lienhard JH (2017, Jan 1) Fundamentals of low-pressure nanofiltration: membrane characterization, modeling, and understanding the multi-ionic interactions in water softening. J Memb Sci 521:18–32

Lee S, Lee CH (2000) Effect of operating conditions on $CaSO_4$ scale formation mechanism in nanofiltration for water softening. Water Res 34(15):0–3866. https://doi.org/10.1016/S0043-1354(00)00142-1

Liu D, Wang X, Xie YF, Tang HL (2016) Effect of capacitive deionization on disinfection by-product precursors. Sci Total Environ 568:19–25. https://doi.org/10.1016/j.scitotenv.2016.05.219

Marcucci M, Nosenzo G, Capannelli G, Ciabatti I, Corrieri D, Ciardelli G (2001) Treatment and reuse of textile effluents based on new ultrafiltration and other membrane technologies. Desalination 138(1–3):75–82. https://doi.org/10.1016/S0011-9164(01)00247-8

Matsuura T (2001) Progress in membrane science and technology for seawater desalination a review. Desalination 134(1–3):47–54. https://doi.org/10.1016/S0011-9164(01)00114-X

Mohammad A, Teow Y, Ang W, Chung Y, Oatley-Radcliffe D, Hilal N (2015) Nanofiltration membranes review: recent advances and future prospects. Desalination 356:226–254. https://doi.org/10.1016/j.desal.2014.10.043

Mohsen MS, Jaber JO, Afonso MD (2003) Desalination of brackish water by nanofiltration and reverse osmosis. Desalination 157(1–3):167–168, 318. CK Diawara Downloaded by [Lulea University of Technology] at 20:49, 18 July 2013. https://doi.org/10.1016/S0011-9164(03)00397-7

Mulyanti R, Susanto H (2018) Wastewater treatment by nanofiltration membranes. IOP Conf Ser Earth Environ Sci 142:012017. https://doi.org/10.1088/1755-1315/142/1/012017

Redondo JA (2001) Brackish-, sea- and wastewater desalination. Desalination 138(1–3):29–40. https://doi.org/10.1016/S0011-9164(01)00241-7

Roth CD, Poh SC, Vuong DX (2009) Customization and multistage nanofiltration applications for potable water, treatment, and reuse. Nanotechnol Appl Clean Water 107–114. https://doi.org/10.1016/b978-0-8155-1578-4.50017-2

Shahmansouri A, Bellona C (2015) Nanofiltration technology in water treatment and reuse: applications and costs. Water Sci Technol 71(3):309–319. https://doi.org/10.2166/wst.2015.015

Shon H, Phuntsho S, Chaudhary D, Vigneswaran S, Cho J (2013) Nanofiltration for water and wastewater treatment—a mini-review. Drink Water Eng Sci 6(1):47–53. https://doi.org/10.5194/dwes-6-47-2013

Silva V, Geraldes V, Alves AB, Palacio L, Prádanos P, Hernández A (2011) Multi-ionic nanoltration of highly concentrated salt mixtures in the seawater range. Desalination 277(13):29–39. https://doi.org/10.1016/j.desal.2011.03.088. URL: http://www.sciencedirect.com/science/article/pii/S0011916411003195

Sutherland K (2008) Developments in filtration: what is nanofiltration? Filtr Sep 45(8):32–35. https://doi.org/10.1016/s0015-1882(08)70298-2

Tang YJ, Wang LJ, Xu ZL, Zhang HZ (2018) Novel chitosan-piperazine composite nanofiltration membranes for the desalination of brackish water and seawater. J Polym Res 25(5):118. https://doi.org/10.1007/s10965-018-1514-6

Van der Bruggen B, Mänttäri M, Nyström M (2008) Drawbacks of applying nanofiltration and how to avoid them: a review. Sep Purif Technol 63(2):251–263. https://doi.org/10.1016/j.seppur.2008.05.010

Wang S, Li L, Yu S, Dong B, Gao N, Wang X (2021) A review of advances in EDCs and PACs removal by nanofiltration: mechanisms, impact factors and the influence of organic matter. Chem Eng J 406:126722. https://doi.org/10.1016/j.cej.2020.126722

Xie L, Sapozhnikova Y, Bawardi O, Schlenk D (2004) Evaluation of wetland and tertiary wastewater treatments for estrogenicity using In Vivo and In Vitro assays. Arch Environ Contam Toxicol 48(1):81–86. https://doi.org/10.1007/s00244-004-0062-8

Sustainable Green Approaches for Wastewater Purification

Preeti Kumari, Archna Dhasmana, Shristi Kishore, Subham Preetam, Nobendu Mukherjee, and Sumira Malik

1 Introduction

Water impacts every element of human existence, including energy, health, food, and the economy. The demand for global water resources is growing due to fast industrialization and rapid population expansion, resulting in higher costs and harsher regulations. Consequently, natural resources are being depleted, and the environment degrades, posing a severe danger to economic development's long-term viability. Furthermore, the massive volume of wastewater produced by enterprises is not only a human health issue but also a severe concern for future calamities. The enormous rise in pollutants to hazardous levels has alarmed society with a slew of previously unknown environmental issues. Both surface and groundwater resources are depleted, resulting in pollution. Additionally, soil erosion, saltwater intrusion, fertilizers, pesticides, detergents, and heavy metals are causing aquifers to become less productive and contaminated globally (Yadav et al. 2019).

P. Kumari · S. Kishore · S. Malik (✉)
Amity Institute of Biotechnology, Amity University Jharkhand Ranch, Ranchi, Jharkhand 834001, India
e-mail: smalik@rnc.amity.edu

A. Dhasmana
Himalayan School of Biosciences, Swami Rama Himalayan University, Dehradun, Uttarakhand, India

S. Preetam
Centre for Biotechnology, Siksha O Anusandhan (SOA-DU), Bhubaneswar, Odisha, India

N. Mukherjee
Department of Microbiology, Ramakrishna Mission Vivekananda Centenary College, Kolkata, West Bengal, India

M. P. Shah (ed.), *Advanced and Innovative Approaches of Environmental Biotechnology in Industrial Wastewater Treatment*,
https://doi.org/10.1007/978-981-99-2598-8_8

Industrial waste poses some of the most severe risks owing to heavy metal toxicity, persistent organic chemicals, and other factors. Pathogens (viruses, bacteria, and parasites) and their removal are still essential aspects of wastewater treatment. Hormonal disruptors are one of several types of chemicals that offer a health risk to people and animals with unpredictable outcomes. These include pharmaceuticals, insecticides, plastic additives, bleaching, and cleaning agents. Heavy metals in wastewater have increased as industries and human activities have grown, which are non-biodegradable and may be carcinogenic. Thus their presence in water at excessive levels might cause serious health problems for living beings. Heavy metal-tainted garbage enters the environment, and the presence of these metals in water in excessive concentrations might cause serious health problems for living creatures (Qasem et al. 2021).

Natural water, and to a lesser degree, sewage, must be treated before use, reuse, or disposal to fulfill this requirement, and an essential technique for achieving a sustainable water supply is to convert wastewater into usable water by eliminating impurities. The phrase "green chemistry," first in 1991, was used to describe the elimination or reduction of harmful compounds to limit chemical exposure to humans and the environment. Green approaches are used in water purification, energy generation, and the manufacture of electronics, pharmaceuticals, polymers, and insecticides, among other things (Kharissova et al. 2019). For water purification, green and sustainable paths are utilized, and these pathways are eco-friendly.

2 Major Water Pollutants

Water is one of the primary necessities, and life without water anyone can never imagine, but water contamination has become a global issue, with emerging countries suffering the most due to their desire to grow. Pollution of the water bodies is a severe hazard to humanity and the aquatic ecology, and rapid population growth has accelerated climate changes. For example, different human activities and the emission of greenhouse gases by industry contribute significantly to global warming, a rise in global temperature and a drop in the quality of the atmosphere's air. Micropollutants in natural water can originate from a variety of sources. Industry and municipalities utilize over 30% of the world's renewable freshwater, resulting in massive amounts of wastewater containing various pollutants in various quantities (Schwarzenbach et al. 2010). Table 1 lists the many types of pollution.

3 Conventional Wastewater Treatment Process

Wastewater purification is a significant concern in removing organic pollutants, microorganisms, nutrients, suspended solids and sediments, inorganic pollutants, thermal pollutants and radioactive pollutants. Each pollutant has a specific ability to

Table 1 Different types of pollutants

S. N.	Types	Cause of pollution	Effect of pollutant	Reference
1.	Organic pollutants	Animals and humans	Increased number of bacteria, increased oxygen levels, deaths of aquatic life	Aksu (2005)
2.	Microorganism	Microorganisms include bacilli, viruses, and protozoa	Water that is unfit for consumption	Jung et al. (2014)
3.	Nutrients	Phosphates and nitrates	Algae blooms, water source eutrophication	Van Puijenbroek et al. (2019)
4.	Suspended solids and sediments	Particles suspended in the water settle	It warms the water, reduces the depth of water sources, and causes toxic deposits	Silva et al. (2011)
5.	Inorganic pollutants	Chemicals that are toxic and dangerous	It kills aquatic life, enters the human food chain, and causes birth deformities, infertility, cancer, and other ailments in people and animals	Cao and Li (2014)
6.	Thermal pollutants	Increased in temperature	Because of a decline in oxygen levels, fish and plants have died	Verones et al. (2010)
7.	Radioactive pollution	Isotopes of radioactivity	Aquatic creatures are killed, and people and other animals develop cancer and die	Bo et al. (2016)

resist through different purification processes, and the conventional wastewater treatment process is the only solution to all types of contaminates. In conventional wastewater treatment, various physical, chemical, and biological processes and operations are used to remove this contamination from wastewater. The prevalent concept used to describe distinct levels of treatment is preliminary, primary, secondary, tertiary, and advanced wastewater treatment in a succession of increasing treatment levels (Sonune and Ghate 2004).

3.1 Preliminary Treatment

The goal of preliminary treatment is to remove coarse particles and other heavy components commonly found in raw wastewater. Preliminary treatment helps to remove or reduce significant absorbed, suspended, and floating pollutants. These solid wastes include pieces of wood, fabric, paper, plastics, rubbish, and other solids, as well as some feces. Heavy inorganic materials like sand and gravel, as well as metal and glass, are also removed; grit and excessive levels of oils or greases are examples of these items (Sonune and Ghate 2004).

3.2 *Primary Method*

In the first treatment, organic and inorganic particles are removed using traditional sedimentation and flotation processes. Approximately 25–50% of the biochemical oxygen demand (BODs), 50–70% of total suspended solids (S.S.), and 65% of oil and grease are removed during the first treatment. Some organic pollutants like nitrogen, phosphorus, and heavy metals are also removed from solids during primary sedimentation (Kesari et al. 2021).

3.3 *Secondary Method*

This step aids in the removal of organic debris that has disintegrated and escaped the first treatment. Bacteria, protozoa, rotifers, fungus, algae, and other microorganisms are among those pollutants that play a complex role during the purification process. Microbes eat organic materials, converting them to carbon dioxide, water, and energy for their development and following the biological process, further settling is done to eliminate any remaining suspended particles. Secondary treatment can remove over 85% of suspended particles and biological oxygen demand (BOD) (Kesari et al. 2021; Jasim et al. 2020).

3.4 *Tertiary Method*

Tertiary water disinfection processes constitute coagulation, flocculation, sedimentation tanks, sand filters, and disc filtration. For chlorination or UV light disinfection process, suspended particles must be reduced, but the effectiveness of tertiary treatment phases varies significantly depending on the techniques and technology used (Kesari et al. 2021; Fig. 1).

Fig. 1 The conventional wastewater treatment process

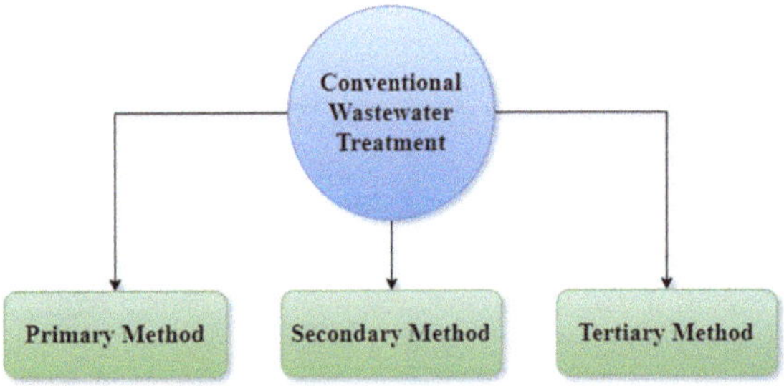

4 Green and Sustainable Wastewater Treatment Methods: Main Goal

Water quality is becoming more widely recognized as a significant problem worldwide, and consumers and communities must play a critical part in maintaining cleanliness near water sources to improve water quality. For a sustainable water-quality plan, it is essential to design a strategy that encompasses all components and prioritizes control measures according to the country's expectations (Sharma et al. 2017). "Provide universal access to and sustainable management of water and sanitation," says the sixth of the 17 Sustainable Development Goals. Given the importance of safe drinking water to overall socioeconomic progress and living standards, as well as health and environmental protection, even accomplishing a percentage of this objective would benefit humanity (Tortajada 2020). For that, bioremediation processes like bacterial bioremediation, active sludge method, membrane bioreactor(MBR), sequence batch reactor, up-flow anaerobic stage reactor, phytoremediation, final bioremediation, phytoremediation, and cyanobacterial bioremediation are used (Fig. 2).

4.1 Bacterial Bioremediation

Microorganisms are better suited to adapt to environmental changes and degradation and are vital to the preservation and sustainability of any ecosystem. Bacterial remediation has long been utilized in treating wastewater as they are eco-friendly and most effective. Using the microbial metabolic potential to eliminate environmental contaminants is a safe and cost-effective alternative to dumping them in landfills or using traditional physicochemical methods. Under experimental circumstances, microorganisms capable of mineralizing several hazardous chemicals have been identified (Azubuike et al. 2016).

Approaches based on microbial molecular ecology have primarily been used for wastewater treatment. To investigate the formation of flocs (active sludge) and biofilms in aerobic treatment systems (trickling filters), *Pseudomonas fluorescens*, *Pseudomonas putida*, and various *Bacillus* strains can be employed in biological wastewater systems. *Endophytes, Pseudomonas*, other bacteria strains, and species such as *B. subtilis* have all been used in the therapy procedure (Shah and Shah 2020). On the other hand, autochthonous bacteria's intrinsic metabolic variety is insufficient to safeguard the biosphere from anthropogenic contamination. In addition, since bacterial exopolysaccharides (EPS) are an excellent adsorption material, they might be employed in a novel approach to removing heavy metals (Kesari et al. 2021; Table 2).

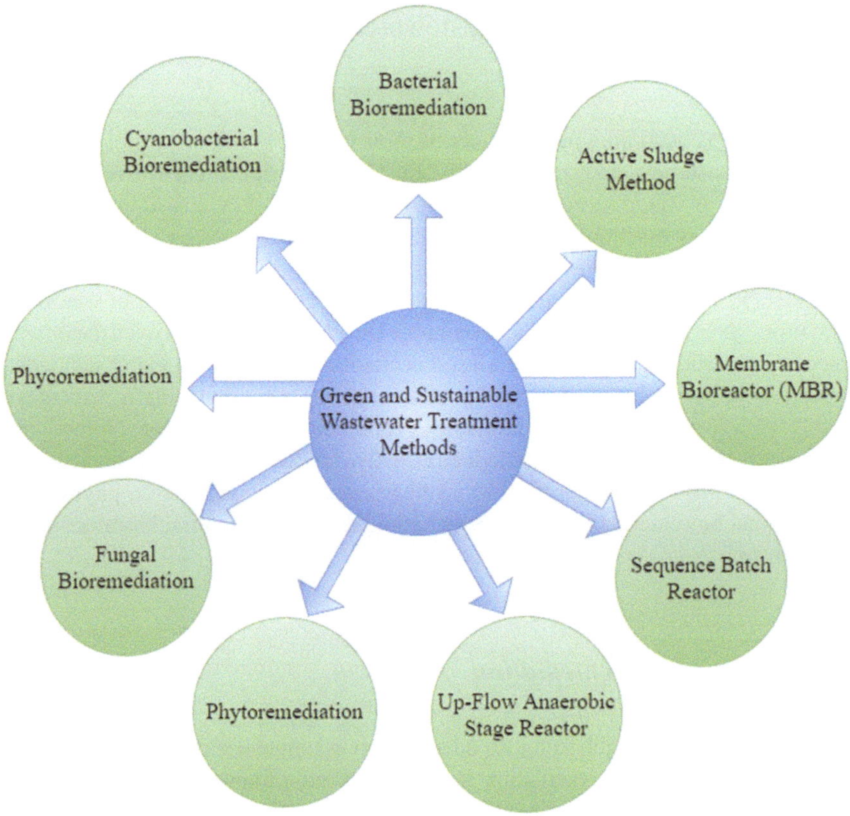

Fig. 2 Green and sustainable wastewater treatment

4.2 Active Sludge Method

The activated sludge technique, which has been around for over a century, is still and will continue to be at the forefront of wastewater treatment technology. Activated sludge refers to microbial cell aggregation in the form of flocs, granules, or biofilm. The characteristics and functions of activated sludge are influenced by the microorganisms and the extracellular polymeric substances (EPS) that surround them (Yu 2020). Most of the time, the design of these plants is based on facts. Since the 1960s, a more acceptable approach to the design of activated sludge has been devised. The latter method is based on the fact that when any form of urban or industrial wastewater is subjected to an aeration process over an extended length of time, it becomes more oxygenated. The amount of organic stuff in the sludge is reduced, resulting in a flocculent sludge. The mass load is a popular design parameter for activated sludge treatment systems (ML). The connection between the daily organic matter feed mass and the decaying biomass content is within the tank. The same tank is characterized as the mass load of an aeration tank (Cisterna 2017), but the drawback

Table 2 Selected wastewater-relevant microorganisms used in the treatment of wastewater (Daims et al. 2006)

S. N.	Microorganisms	Process
1.	*Nitrosomonas europaea* *Nitrosomonas eutropha* *Nitrosomonas oligotropha* *Nitrobacter hamburgensis* *Nitrobacter winogradskyi* *Nitrospira marina* 'Cand. *Nitrospira defluvii*' *Paracoccus denitrificans* 'Cand. *Kuenenia stuttgartiensis*' 'Cand. *Brocadia anammoxidans*'	Nitrogen removal Ammonia oxidation Nitrite oxidation Denitrification Anammox
2.	*Gemmatimonas aurantiaca* 'Cand. *Accumulibacter phosphatis*'	Phosphorus removal
3.	*Acidovorax temperance* (2 strains)	Floc/biofilm formation
4.	*Herpetosiphon aurantiacus*	Bulking

is that wastewater treatment plants generate a significant volume of wasted active sludge. Because of its large volume and toxic composition, sludge disposal becomes an issue.

4.3 Membrane Bioreactor (MBR)

Membrane bioreactor (MBR) technology has gained interest as an alternative to the activated sludge method for wastewater treatment (ASP) which was previously the primary municipal wastewater treatment method, throughout the last century. MBR is, in fact, one of the most critical advancements in wastewater treatment as it has been used in both municipal and industrial settings for wastewater treatment and reclamation because of its capacity to generate high-quality effluent. Membrane filtration combines traditional biological treatment with physical liquid–solid separation in an MBR. Biological and membrane filtration processes are also combined in MBR. The biomass breakdown occurs inside the bioreactor tank, while the separation of treated wastewater from microorganisms is completed in a membrane module (Al-Asheh et al. 2021).

4.4 Sequence Batch Reactor (SBR)

The SBR, on the other hand, is a time-oriented system that uses a preset, periodic operating strategy to regulate flow, energy input, and tank capacity. As a result, the SBR should be viewed as a time-based, repeatable procedure. However, it fits an activated sludge system with an unstable state that falls under this category. These technologies may establish a high level of control over organism selection. With this in mind, the SBR is described physically, mathematically, and biologically (Irvine et al. 1989). For activated sludge, the SBR is a fill-and-draw system. Each tank in the SBR system is loaded with wastewater for a specific time before being treated in batch mode. The clarified supernatant is taken from the tank after the mixed liquor has settled for a specific time after treatment. During treatment, sedimentation, and withdrawal, wastewater flows to another SBR tank in the system or to a storage tank in a single SBR tank arrangement, where it is pulled for treatment, and the supernatant withdrawal is made (Arrojo et al. 2004).

4.5 Up-Flow Anaerobic Stage Reactor

In the late 1970s, Lettinga and his colleagues created the UASB procedure. The UASB configuration allows for highly efficient mixing of biomass and wastewater, resulting in rapid anaerobic decomposition. The main component of a UASB reactor's functioning is its granular sludge bed, which expands when wastewater is forced vertically upwards through it. Since the microflora adhering to sludge particles eliminates contaminants from wastewater, biofilm quality and the closeness of sludge-wastewater contact are two of the most important parameters determining the performance of UASB reactors. The UASB reactor can effectively treat a variety of contaminants such as sugar, pulp and paper, dairy, chemical, potato starch, lentil balancing, fizzy drinks, and seafood processing industries, pasta processing, yeast production, butcher shop, and coffee processing industries. UASB technology can assist in the achievement of circular economy and environmental objectives. Such efficient biogas extraction meets renewable energy demand and lowers greenhouse gas emissions, especially in tropical locations, where reactor heating must be kept to a bare minimum (Mainardis et al. 2020).

4.6 Phytoremediation

Phytoremediation is an environmental cleansing approach that depends on plants and their accompanying bacteria to volatilize, stabilize, degrade, or remove pollutants. Potential pollutants including fertilizers, food spills, animal excrement, food waste, feces, metabolic by-products, residual biocides, and biostats are all sources of

Table 3 The impacts of various plant species used for wastewater remediation

S. N.	Plant species	Common name	Effects	Reference
1.	*Juncus effusus* L	Soft rush or common rush	BOD, COD, TSS, nitrogen, phosphate, and fecal coliforms had all reduced in concentration	Ren et al. (2021)
2.	*Azolla californiana*	Fairy moss	Turbidity reduction TSS, BOD, and COD	Carlozzi and Padovani (2016)
3.	*Oenanthe javanica*	Chinese celery, Indian pennywort, Japanese parsley,	Dissolved oxygen, pH, and temperature influence water filtration and nutrient absorption	Song et al. (2019)
4.	*Hydrocotyle vulgaris*	Marsh pennywort	Total nitrogen and NH4 nitrogen are removed	Duan et al. (2016)
5.	*Eichornia crassipes*	Water hyacinth	Heavy metal ions, ammonium, nitrates BOD, COD, TSS, sedimentation, and ammonia concentrations are all reduced	Zhang et al. (2015)

wastewater pollution. They are responsible for large amounts of disturbing aquatic ecology and the natural food chain. Explosives are just a few substances that may be found in the environment, including harmful gases are among the contaminated substances. In recent decades, phytoremediation has been extensively investigated as a powerful method for eliminating and decomposing these various hazardous chemicals (Mohd Nizam et al. 2020). Traditional cleanup methods are replaced by phytoremediation, a low-cost, non-invasive, ecologically friendly, and safe alternative. On the other hand, phytoremediation is seen as a more promising green remediation method because of its reduced cost and viability. More research and innovation are needed to improve and promote this technology in underdeveloped nations.

The various aquatic plants system also provides an ecologically beneficial and cost-effective solution. By consuming a part of contaminants as plant nutrients, these plants aid in removing pollutants (Zimmels et al. 2004). In Table 3, different plant species are used for wastewater remediation (Kesari et al. 2021).

4.7 Fungal Bioremediation

Fungi are microorganisms that are responsible for the breakdown of the majority of organic molecules in the environment. Since the 1980s, fungi belonging to the WRF (Water Research Foundation) have been used in water, and soil bioremediation approaches. The WRF (mostly basidiomycetes) is an ecophysiological category of fungi capable of lignin degradation. Lignin modifying enzymes (LME) are enzymatic machinery in this fungus responsible for lignin breakdown and wood disintegration (Akerman-Sanchez and Rojas-Jimenez 2021). Fungus, especially white rot fungi,

Table 4 Varying fungal strains have different capacities for eliminating metal ions

S. N.	Fungi	Metals
1.	*Aspergillus niger*	Pb, Cd, Cu, Ni
2.	*Aspergillus versicolour*	Cr, Cu, Ni
3.	*Aspergillus fumigatus*	Cu, Pb, Zn, Fe, Ni
4.	*Penicillum purpurogenum*	Pb, Cd, Hg, As
5.	*A. cylindrospora*	Cd
6.	*Aspergillus terreus*	Zn
7.	*Aspergillus sydowii*	Cr(VI)
8.	*Aspergillus terreus*	Azo dye

has long been known to break down a wide range of refractory substances through non-specific extracellular oxidative enzymes, including synthetic colors (Anastasi et al. 2012). Fungi contain potential metal ion binding sites in their cell walls, which are made up of deacetylated glucoseamine and polymers of N-acetyl, chitin, and chitosan potentially help them to be employed in wastewater treatment. Different fungi for eliminating metal ions are mentioned in Table 4 (Sharma et al. 2020).

4.8 Phytoremediation

In phytoremediation, the utilization of macroalgae takes place by eliminating nutrients (nitrogen and phosphorus) and xenobiotics from wastewater or biotransformation of contaminants by producing biomass. Microalgae have emerged as a popular wastewater treatment method in recent years due to its cost-effectiveness, resulting in the biodegradation of xenobiotics and hazardous chemicals. It is also an efficient alternative to traditional wastewater treatment technologies (Tripathi et al. 2019). Pollutants like nitrogen, phosphorus, sulfur, and minerals appear as "nutrients" to feed algae rather than "contaminants." Phosphates, nitrates, toxic metals, pesticides, hydrocarbons, nitrogen, and phosphorus are toxins that phytoremediation can remove (Kaloudas et al. 2021). Table 5 discusses various microalgae for wastewater treatment.

4.9 Cyanobacterial Bioremediation

Cyanobacteria have significant advantages over other living forms, allowing them to thrive in various environments. Their metabolic mechanisms and unique adaptations of some species, like desiccation resistance and nitrogen fixation, make them highly adaptable. At the same time, depending on the nature of the available nutritional

Table 5 Different microalgae for treating different types of wastewaters

S. N.	Microalgae	Types of wastewaters	Reference
1.	Scenedesmus	Ammonia removal from anaerobic digestion effluents with high ammonium and alkalinity values	Dickinson et al. (2014)
2.	*Botryococcus braunii*	Synthetic wastewater	Podder and Majumder (2016)
3.	*Chlorella sorokiniana*	Artificial wastewater	Chen et al. (2017)
4.	Ankistrodesmus and Scenedesmus *Scenedesmus quadricauda*	Olive oil mill wastewater and wastewater from the paper industry	Pinto et al. (2003)
5.	*Chlorella* sp.	Dairy manure that has been anaerobically digested	Lu et al. (2015)

medium, cyanobacteria can quickly transition to mixotrophic feeding (Zinicovscaia and Cepoi 2016).

Cyanobacteria are a diverse category of prokaryotic photosynthetic organisms, some of which have the unique capacity to fix nitrogen. They are widely distributed and may be found in practically any environment. Using cyanobacteria in wastewater treatment is an environmentally benign approach that produces no secondary pollution if the biomass generated is utilized and allows for efficient nutrient recycling (Sood et al. 2015). A live cyanobacterial strain (blue-green algae) was identified at an industrial wastewater treatment plant and named *Starria zimbabweensis*. It was utilized by Biswas et al. to remediate fluoride-contaminated simulated and actual wastewater (2018). In another study by Srimongkol et al. 2019, *Synechococcus sp.* VDW removes ammonium from contaminated brackish water and is also helpful for improving the quality of aquaculture wastewater. Cyanobacteria have recently been identified as phytochemically active chemicals with poisonous (cyanotoxins). It also has antibacterial, antiviral, antifungal, and anticancer properties. These organisms may create poisonous or antibacterial chemicals that influence bacterial populations once they are present in biological systems (Martins et al. 2011).

5 Conclusion

The demand for natural resources grows in lockstep with the world's population. Water resources are one of those that are in catastrophic condition, and their exploitation is at an all-time high. Finally, repurposing these resources is limited to meeting the population's continued needs. Various approaches have been utilized, resulting in the eventual discharge of waste treatment products into the environment. In reducing pollution, green and sustainable solutions have been created that are both environmentally benign and capable of meeting demand. Different bioremediation procedures are more successful, such as using plant and microbial species to treat these wastewaters.

References

Akerman-Sanchez G, Rojas-Jimenez K (2021) Fungi for the bioremediation of pharmaceutical-derived pollutants: a bioengineering approach to water treatment. Environ Adv 4:100071. https://doi.org/10.1016/j.envadv.2021.100071

Aksu Z (2005) Application of biosorption for the removal of organic pollutants: a review. Process Biochem 40(3–4):997–1026. https://doi.org/10.1016/j.procbio.2004.04.008

Al-Asheh S, Bagheri M, Aidan A (2021) Membrane bioreactor for wastewater treatment: a review. Case Stud Chem Environ Eng 4:100109. https://doi.org/10.1016/j.cscee.2021.100109

Anastasi A, Spina F, Romagnolo A, Tigini V, Prigione V, Varese G (2012) Integrated fungal biomass and activated sludge treatment for textile wastewaters bioremediation. Biores Technol 123:106–111. https://doi.org/10.1016/j.biortech.2012.07.026

Arrojo B, Mosquera-Corral A, Garrido J, Méndez R (2004) Aerobic granulation with industrial wastewater in sequencing batch reactors. Water Res 38(14–15):3389–3399. https://doi.org/10.1016/j.watres.2004.05.002

Azubuike C, Chikere C, Okpokwasili G (2016) Bioremediation techniques–classification based on site of application: principles, advantages, limitations and prospects. World J Microbiol Biotechnol 32(11). https://doi.org/10.1007/s11274-016-2137-x

Biswas G, Thakurta S, Chakrabarty J, Adhikari K, Dutta S (2018) Evaluation of fluoride bioremediation and production of biomolecules by living cyanobacteria under fluoride stress condition. Ecotoxicol Environ Saf 148:26–36. https://doi.org/10.1016/j.ecoenv.2017.10.019

Bo A, Sarina S, Liu H, Zheng Z, Xiao Q, Gu Y et al (2016) Efficient removal of cationic and anionic radioactive pollutants from water using hydrotalcite-based getters. ACS Appl Mater Interfaces 8(25):16503–16510. https://doi.org/10.1021/acsami.6b04632

Cao Y, Li X (2014) Adsorption of graphene for the removal of inorganic pollutants in water purification: a review. Adsorption 20(5–6):713–727. https://doi.org/10.1007/s10450-014-9615-y

Carlozzi P, Padovani G (2016) The aquatic fern Azolla as a natural plant-factory for ammonia removal from fish-breeding fresh wastewater. Environ Sci Poll Res 23(9):8749–8755. https://doi.org/10.1007/s11356-016-6120-8

Chen T, Zhao Q, Wang L, Xu Y, Wei W (2017) Comparative metabolomic analysis of the green microalga *Chlorella sorokiniana* cultivated in the single culture and a consortium with bacteria for wastewater remediation. Appl Biochem Biotechnol 183(3):1062–1075. https://doi.org/10.1007/s12010-017-2484-6

Cisterna P (2017) Biological treatment by active sludge with high biomass concentration at laboratory scale for mixed inflow of sunflower oil and saccharose. Environments 4(4):69. https://doi.org/10.3390/environments4040069

Daims H, Taylor M, Wagner M (2006) Wastewater treatment: a model system for microbial ecology. Trends Biotechnol 24(11):483–489. https://doi.org/10.1016/j.tibtech.2006.09.002

Dickinson K, Bjornsson W, Garrison L, Whitney C, Park K, Banskota A, McGinn P (2014) Simultaneous remediation of nutrients from liquid anaerobic digestate and municipal wastewater by the microalga *Scenedesmus* sp. AMDD grown in continuous chemostats. J Appl Microbiol 118(1):75–83. https://doi.org/10.1111/jam.12681

Duan J, Zhao J, Xue L, Yang L (2016) Nutrient removal of a floating plant system receiving low-pollution wastewater: effects of plant species and influent concentration. IOP Conf Ser Earth Environ Sci 41:012028. https://doi.org/10.1088/1755-1315/41/1/012028

Irvine R, Ketchum L, Asano T (1989) Sequencing batch reactors for biological wastewater treatment. Crit Rev Environ Control 18(4):255–294. https://doi.org/10.1080/10643388909388350

Jung A, Le Cann P, Roig B, Thomas O, Baurès E, Thomas M (2014) Microbial contamination detection in water resources: interest of current optical methods, trends and needs in the context of climate change. Int J Environ Res Public Health 11(4):4292–4310. https://doi.org/10.3390/ijerph110404292

Kaloudas D, Pavlova N, Penchovsky R (2021) Phytoremediation of wastewater by microalgae: a review. Environ Chem Lett 19(4):2905–2920. https://doi.org/10.1007/s10311-021-01203-0

Kesari K, Soni R, Jamal Q, Tripathi P, Lal J, Jha N et al (2021) Wastewater treatment and reuse: a review of its applications and health implications. Water Air Soil Poll 232(5). https://doi.org/10.1007/s11270-021-05154-8

Kharissova O, Kharisov B, Oliva González C, Méndez Y, López I (2019) Greener synthesis of chemical compounds and materials. Royal Soc Open Sci 6(11):191378. https://doi.org/10.1098/rsos.191378

Lu W, Wang Z, Wang X, Yuan Z (2015) Cultivation of *Chlorella* sp. using raw dairy wastewater for nutrient removal and biodiesel production: characteristics comparison of indoor bench-scale and outdoor pilot-scale cultures. Biores Technol 192:382–388. https://doi.org/10.1016/j.biortech.2015.05.094

Mainardis M, Buttazzoni M, Goi D (2020) Up-flow anaerobic sludge blanket (UASB) technology for energy recovery: a review on state-of-the-art and recent technological advances. Bioengineering 7(2):43. https://doi.org/10.3390/bioengineering7020043

Martins J, Peixe L, Vasconcelos V (2011) Unraveling cyanobacteria ecology in wastewater treatment plants (WWTP). Microb Ecol 62(2):241–256. https://doi.org/10.1007/s00248-011-9806-y

Mohd Nizam N, Mohd Hanafiah M, Mohd Noor I, Abd Karim H (2020) Efficiency of five selected aquatic plants in phytoremediation of aquaculture wastewater. Appl Sci 10(8):2712. https://doi.org/10.3390/app10082712

Pinto G, Pollio A, Previtera L, Stanzione M, Temussi F (2003) Biotech Lett 25(19):1657–1659. https://doi.org/10.1023/a:1025667429222

Podder M, Majumder C (2016) The use of artificial neural network for modelling of phytoremediation of toxic elements As(III) and As(V) from wastewater using *Botryococcus braunii*. Spectrochim Acta Part A Mol Biomol Spectrosc 155:130–145. https://doi.org/10.1016/j.saa.2015.11.011

Qasem N, Mohammed R, Lawal D (2021) Removal of heavy metal ions from wastewater: a comprehensive and critical review. Npj Clean Water 4(1). https://doi.org/10.1038/s41545-021-00127-0

Ren L, Zhao G, Pan L, Chen B, Chen Y, Zhang Q et al (2021) Efficient removal of dye from wastewater without selectivity using activated carbon-juncus effusus porous fibril composites. ACS Appl Mater Interfaces 13(16):19176–19186. https://doi.org/10.1021/acsami.0c22104

Schwarzenbach R, Egli T, Hofstetter T, von Gunten U, Wehrli B (2010) Global water pollution and human health. Ann Rev Environ Res 35(1):109–136. https://doi.org/10.1146/annurev-environ-100809-125342

Shah A, Shah M (2020) Characterization and bioremediation of wastewater: a review exploring bioremediation as a sustainable technique for pharmaceutical wastewater. Groundw Sustain Dev 11:100383. https://doi.org/10.1016/j.gsd.2020.100383

Sharma R, Yadav M, Gupta R (2017) Water quality and sustainability in India. Chemistry and Water 183–205. https://doi.org/10.1016/b978-0-12-809330-6.00005-2

Silva B, Jelic A, López-Serna R, Mozeto A, Petrovic M, Barceló D (2011) Occurrence and distribution of pharmaceuticals in surface water, suspended solids and sediments of the Ebro river basin, Spain. Chemosphere 85(8):1331–1339. https://doi.org/10.1016/j.chemosphere.2011.07.051

Song S, Wang P, Liu Y, Zhao D, An S (2019) Effects of *Oenanthe javanica* on nitrogen removal in free-water surface constructed wetlands under low-temperature conditions. Int J Environ Res Public Health 16(8):1420. https://doi.org/10.3390/ijerph16081420

Sonune A, Ghate R (2004) Developments in wastewater treatment methods. Desalination 167:55–63. https://doi.org/10.1016/j.desal.2004.06.113

Sood A, Renuka N, Prasanna R, Ahluwalia A (2015) Cyanobacteria as potential options for wastewater treatment. Phytoremediation 83–93. https://doi.org/10.1007/978-3-319-10969-5_8

Srimongkol P, Thongchul N, Phunpruch S, Karnchanatat A (2019) Ability of marine cyanobacterium *Synechococcus* sp. VDW to remove ammonium from brackish aquaculture wastewater. Agric Water Manag 212:155–161. https://doi.org/10.1016/j.agwat.2018.09.006

Tortajada C (2020) Contributions of recycled wastewater to clean water and sanitation sustainable development goals. Npj Clean Water 3(1). https://doi.org/10.1038/s41545-020-0069-3

Tripathi R, Gupta A, Thakur I (2019) An integrated approach for phytoremediation of wastewater and sustainable biodiesel production by green microalgae, *Scenedesmus* sp. ISTGA1. Renew Energy 135:617–625. https://doi.org/10.1016/j.renene.2018.12.056

van Puijenbroek P, Beusen A, Bouwman A (2019) Global nitrogen and phosphorus in urban wastewater based on the shared socio-economic pathways. J Environ Manag 231:446–456. https://doi.org/10.1016/j.jenvman.2018.10.048

Verones F, Hanafiah M, Pfister S, Huijbregts M, Pelletier G, Koehler A (2010) Characterization factors for thermal pollution in freshwater aquatic environments. Environ Sci Technol 44(24):9364–9369. https://doi.org/10.1021/es102260c

Yadav M, Gupta R, Sharma R (2019) Green and sustainable pathways for wastewater purification. Adv Water Purif Tech 355–383. https://doi.org/10.1016/b978-0-12-814790-0.00014-4

Yu H (2020) Molecular insights into extracellular polymeric substances in activated sludge. Environ Sci Technol 54(13):7742–7750. https://doi.org/10.1021/acs.est.0c00850

Zhang F, Wang X, Yin D, Peng B, Tan C, Liu Y et al (2015) Efficiency and mechanisms of Cd removal from aqueous solution by biochar derived from water hyacinth (*Eichornia crassipes*). J Environ Manag 153:68–73. https://doi.org/10.1016/j.jenvman.2015.01.043

Zimmels Y, Kirzhner F, Roitman S (2004) Use of naturally growing aquatic plants for wastewater purification. Water Environ Res 76(3):220–230. https://doi.org/10.2175/106143004x141762

Zinicovscaia I, Cepoi L (2016) Cyanobacteria for bioremediation of wastewaters. Basel, Switzerland: Springer International Publishing. https://doi.org/10.1007/978-3-319-26751-7

Contaminants of Emerging Concern and Hybrid Continuous Flow Treatment: A Promising Combination

Natalia Klanovicz, Thamarys Scapini, Fábio Spitza Stefanski, Priscila Hasse Palharim, Bruno Ramos, Shukra Raj Paudel, Helen Treichel, and Antonio Carlos Silva Costa Teixeira

Abbreviations

ABOPs	Advanced bio-oxidation processes
AMO	Ammonia monooxygenase
Anammox	Anaerobic ammonium oxidation
AOPs	Advanced oxidation processes
BOM	Biodegradable organic matter
BPA	Bisphenol A
CECs	Contaminants of emerging concern
COD	Chemical oxygen demand
DOM	Dissolved organic matter
EPS	Extracellular polymeric substance

N. Klanovicz · F. S. Stefanski · H. Treichel (✉)
Laboratory of Microbiology and Bioprocesses, Federal University of Fronteira Sul, Erechim, Brazil
e-mail: helentreichel@gmail.com

N. Klanovicz (✉) · P. H. Palharim · B. Ramos · A. C. S. C. Teixeira
Research Group in Advanced Oxidation Processes (AdOx), Department of Chemical Engineering, Escola Politécnica, University of São Paulo, São Paulo, Brazil
e-mail: nataliaklanovicz@gmail.com

T. Scapini
Department of Bioprocess Engineering and Biotechnology, Federal University of Paraná, Curitiba, PR, Brazil

S. R. Paudel
Department of Civil Engineering, Institute of Engineering, Pulchowk Campus, Tribhuwan University, Pulchowk, Lalitpur, Nepal

Department of Environmental Engineering, College of Science and Technology, Korea University, Sejong, Republic of Korea

M. P. Shah (ed.), *Advanced and Innovative Approaches of Environmental Biotechnology in Industrial Wastewater Treatment*,
https://doi.org/10.1007/978-981-99-2598-8_9

HA	Humic acid
HOMF	Hybrid ozone membrane filtration
HRP	Horseradish peroxidase
HRT	Hydraulic retention time
IR	Industrial revolution
MBR	Membrane bioreactor
PhACs	Pharmaceutical active compounds
PMR	Photocatalytic membrane reactor
RBCs	Rotating biological contactors
ROS	Reactive oxygen species
SBR	Sequencing batch reactor
SRT	Sludge retention time
T&O	Taste and odor
TAN	Total ammonia nitrogen
TMF	Transmembrane flux
TOC	Total organic carbon
TP	Total phosphorus
US	Ultrasound
UV	Ultraviolet
WTP	Water treatment plant
WWTPs	Wastewater treatment plants

1 Introduction

In recent decades, advances in technologies and research in human and animal health, associated with exponential population growth and industrialization, have generated concerns related to basic sanitation, environmental conservation, and the availability of freshwater resources. The identification in the environment of trace concentrations of widely used chemical compounds of high persistence and potential toxicity, often denominated contaminants of emerging concern (CECs), has led to worldwide discussions for the development of detection and treatment technologies (Tran et al. 2013; Tiwari et al. 2017; Parida et al. 2021).

CECs are considered a large class of chemical compounds. They include pharmaceuticals (antibiotics, anti-inflammatories, analgesics, diuretics, antiseptics, etc.), disinfectants, detergents, industrial chemicals, and personal care products (Geissen et al. 2015; Viancelli et al. 2020). These compounds are widely produced and used worldwide, associated with different points and diffuse sources of entry into the environment, e.g. leaching, volatilization, surface runoff, and inappropriate disposal. Consequently, their occurrence in wastewater treatment plants (WWTPs) has increased in recent years (Parida et al. 2021).

Pharmaceutical active compounds (PhACs) are the class of CECs whose predominant entry pathway into the environment is wastewater due to direct drug disposal and indirect residual excretion (Rout et al. 2021; Parida et al. 2021). The high consumption of PhACs has boosted the development of analytical methods capable of efficiently detecting their presence in waters and minimizing their discharge in WWTPs treated effluents. The potential for contamination is aggravated because most PhACs are not wholly assimilated or metabolized. Therefore, the unmodified drug or metabolites are excreted through feces and urine, entering the environment due to the absence of basic sanitation or the inefficiency of conventional wastewater treatment strategies (Tiwari et al. 2017; Saidulu et al. 2021; Parida et al. 2021).

Conventional WWTPs generally have limited efficiency for CECs removal, mainly due to the variability of physicochemical properties of the wastewater, and to the concentrations, chemical structure, and reactivity of the CECs molecules. This adds to the typical efficiency fluctuation observed in WWTPs due to the constant adjustment of operating conditions, such as the hydraulic retention time (HRT), mixing rate, or the dosage of additives. Another critical factor is the seasonality of CECs in wastewaters: higher concentrations of PhACs are often detected in winter due to increased consumption of cold and flu medicines (Mohapatra et al. 2016). Conventional treatment has limited efficacy if the operation is not regularized with demand. Removal efficiency between 60 and 90% of the total CECs concentration in a WWTP in India was measured over time, with a greater removal in the summer associated with greater biological activity of the processes (facultative aeration lagoon and cyclic activated sludge technology followed by chlorination) (Mohapatra et al. 2016). The aerobic granular sludge membrane bioreactor demonstrated a variation of anti-inflammatory drugs removal (from 63 to 98%), and did not exhibit sufficient effect on antibiotics, mainly associated with these compounds' chemical structure and characteristics (Zhao et al. 2014).

The removal of CECs from wastewater by hybrid treatment systems has been discussed as indispensable for efficient WWTPs (Rout et al. 2021; Saidulu et al. 2021). Given the impact of these compounds, actions focused mainly on reducing wastewater discharge with a potentially harmful effect on aquatic life and human health are urgently required. Considering the multiple sources, pathways of entry into the environment, and the physicochemical characteristics of each compound (e.g. solubility, volatility, recalcitrance, photodegradation, and biodegradability), studies have directed efforts to develop different strategies to integrate treatment approaches and minimize CECs discharge. The combination of two or more processes can be a viable and sustainable strategy for CECs degradation, promising to (i) improve the overall system performance, (ii) reduce discharge into the environment, and (iii) simultaneously degrade different contaminants.

The complexity and heterogeneity of wastewaters containing multiple CECs must be monitored and adapted in WWTPs using flexible and upgraded processes. Applying a single method limits the removal efficiency because the degradation mechanism must have a specific affinity for the target compounds. Hybrid systems have demonstrated improved removals (above 95%) for some CECs and can be incorporated straightforwardly in conventional wastewater treatment plants (Saidulu et al. 2021). Given this scenario, particular emphasis has been given to biological processes (whole-cells or enzymatic methods) with the potential to biodegrade a range of compounds via sustainable and straightforward techniques. These processes can be associated with physical, chemical, or physicochemical strategies, such as chemical precipitation and particularly advanced oxidation processes (AOPs) (Saidulu et al. 2021).

Following this trend, this chapter aims to contribute to research developments involving CECs removal from wastewater, addressing the possibility of using hybrid continuous flow treatments and considering the perspective of chemical-biological hybridization and their degradation mechanisms, operational and technical aspects.

2 Selecting the Treatment Approach

The enormous variety of CECs and the constant alert about their dangerous effects put the necessary societal pressure on researchers and regulatory agencies to find a feasible treatment approach to remove them from wastewater. The success of processes for micropollutant degradation and removal from wastewater is closely linked to the chemical and physical properties of the compounds and how the agents involved in the process interact with them.

CECs can have a wide range of water solubility (from 17 to 1,110 mg L^{-1}) and half-life time (from 2 to 120 h), and several countries have reported their occurrence in groundwater, surface water, and WWTPs effluents; that is, they are persistent in aquatic environments and cannot be entirely removed by the current technologies implemented in the WWTPs (Verlicchi et al. 2012; Klanovicz et al. 2021). These findings also show that the processes involved in CECs degradation have more impact on the success of the treatment than the physicochemical properties of the CECs.

Therefore, the choice of a wastewater treatment strategy is highly relevant, and there is a wide variety of processes and mechanisms to degrade CECs, as presented in Sects. 2.1 and 2.2.

2.1 Biological Systems: Fundamentals

Several microbial groups are associated with biological systems that biotransform CECs through metabolic and co-metabolic pathways (Saidulu et al. 2021). In this sense, this topic will explain how microorganisms and their metabolites interact with CECs, resulting in their degradation.

In the metabolic pathway, CECs are used as the only carbon source to maintain biomass and induce oxidative/reducing enzymes involved in biodegradation. Heterotrophic bacteria and fungi predominate as agents in the metabolic transformation of CECs (De Gusseme et al. 2011; Tran et al. 2013). In turn, in the co-metabolic pathway, a carbon source is critically needed by supporting substrates or another biodegradable compound. That is because many CECs are toxic and resistant to microorganisms, preventing their use as the unique carbon source in the catabolic and anabolic pathways of microbial cells. Both biotransformation pathways can occur under aerobic and anaerobic conditions and, in some cases, may yield degradation products with increased toxicity concerning the parent compounds (Tran et al. 2013).

In aerobic co-metabolic biodegradation, CECs are oxidized during the microbial metabolism of another growth substrate with oxygen by an enzyme or cofactor. The aerobic process is more prevalent and observed in autotrophic microorganisms that use inorganic carbon and ammonia as substrates. In this case, there is the induction of non-specific enzymes, such as ammonia monooxygenase (AMO) and the cofactors [NADH] and [NADPH]. In anaerobic co-metabolic biodegradation, CECs are assimilated by a reducing enzyme or cofactor generated from the microbial metabolism of another primary substrate in an oxygen-free environment (Tran et al. 2013).

In biological systems, CECs can be biodegraded by a microbial consortium such as activated sludge or by pure strains. In addition to biodegradation, CECs sorption can also occur in treatment plants. The mechanisms involved in this case are (i) the hydrophobic interaction of the aliphatic and aromatic group with the cell membrane of microorganisms, and (ii) the electrostatic interaction of a positively charged compound with negatively charged microorganism biomass, dead or alive (Tiwari et al. 2017).

Ammonia-oxidizing bacteria play an essential role in the biological degradation of many CECs, e.g. trimethoprim, acetaminophen, ibuprofen, and bisphenol. These bacteria can form associative communities with oxidizing ammonia archaea and act synergistically for the same purpose. Ammonia-containing wastewater can be a growth substrate, providing reducers and inducing the AMO enzyme for oxidative metabolism (Tran et al. 2013; Bilal et al. 2019).

With fungi species, CECs degradation is associated with the excretion of some enzymes: lignin peroxidases, known for degrading polycyclic aromatic and phenolic compounds under hydrogen peroxide (H_2O_2) presence; manganese-dependent peroxidases (acting in the oxidation of monoaromatic phenols and aromatic dyes under H_2O_2 presence); and laccases (oxidation of aromatic amines and aliphatic, and diphenols, under oxygen presence). The best-known species reported to biodegrade CECs is the white-rot fungi, which can remove up to 100% of persistent drugs such as diclofenac, naproxen, carbamazepine, and 17a-ethinylestradiol (Rodarte-Morales et al. 2012; Tiwari et al. 2017; Viancelli et al. 2020).

In membrane bioreactor (MBR) systems, which integrate the membrane technique with biological treatment, the removal of CECs results from a combination of physical retention by the membrane, biodegradation, biosorption, bioaccumulation, and volatilization. The microorganisms drive the biological treatment, while the physical retention is determined by the type of membrane used (microfiltration or ultrafiltration). The CECs not biotransformed by the microorganisms can be retained on the membranes if their molar weight is greater than the molecular cut-off. In general, the MBR system efficiently removes CECs following the order: endocrine-disrupting compounds > beta-blockers > PhACs > pesticides (Goswami et al. 2018; Zhao et al. 2021). MBRs have been widely applied for industrial wastewater treatment and water reuse at a large scale. In China, for example, over 180 MBRs facilities have been operating and removing CECs, high chemical oxygen demand (COD), total ammonia nitrogen (TAN), and total phosphorus (TP) (Zhang et al. 2021).

The biological processes in MBR systems involve whole cells (usually fungi and bacteria) or bioactive molecules (oxidative enzymes). In both cases, the biological agents can be adhered to the membrane surface and may favor CECs removal by stimulating specific metabolic pathways usually mediated by enzymes (even in whole cells systems) (Akkoyunlu et al. 2021). Therefore, CECs removal in MBR is strongly dependent on the microbiological community and their enzymatic cycles. In turn, the microbial consortium present in MBR processes is diverse and can depend on wastewater characteristics, with the predominance of microorganisms that produce oxidative enzymes and have the capacity to absorb contaminants, as discussed by Tiwari et al. (2017). Whole-cell MBR systems are more used than immobilized bioactive compounds, given the lower cost for microorganisms inoculation, and the natural adaptive process to the MBR and wastewater environment.

The molecular micropollutants properties are also important to understand the removal efficiency by MBR: there may be a relationship between biotransformation and bioadsorption simultaneously occurring for CECs removal. For hydrophobic compounds, bioadsorption tends to be the dominant pathway, whereas biodegradation is for highly hydrophobic compounds. Compounds with moderate hydrophobicity (e.g. carbamazepine), for instance, can accumulate in the solid phase (Wijekoon et al. 2013).

Biological systems for CECs bioremediation can also be developed with microalgae. The degradation occurs by three main mechanisms: bioadsorption, bioaccumulation, and biodegradation, among which biodegradation is the most discussed pathway because it transforms contaminants into smaller and, in most cases, less toxic compounds. This process can occur intracellularly or/and extracellularly. The enzymatic mechanism is a complex process that involves multiple intra- and extracellular enzymes, so it is not yet fully understood and can differ substantially between microalgae species. *Chlorella sorokiniana, Scenedesmus obliquus*, and *Chlorella vulgaris* are the species most frequently reported to remove CECs, which can degrade up to twenty compounds concomitantly (Rempel et al. 2021; Ricky and Shanthakumar 2022).

It is also important to mention that, in many cases, CECs can negatively affect biological treatments and compromise reactor performance. Antibiotics, metals, nanomaterials, and microplastics are reported as agents of concern and bring uncertain perspectives in the operation of the anaerobic ammonium oxidation (Anammox) process, for example. This process involves a chemolithoautotrophic consortium of microorganisms with low oxygen and carbon demand. The CEC's impact on this process can inhibit protein synthesis because of their high toxicity, resulting in DNA, lipids, and proteins damage. For example, the functional activity of some key enzymes can be affected due to the binding of microplastics with biomolecules on the cell surface. Another case is the generation of reactive oxygen species (ROS) triggered by the transport of antibiotics into the cell, impacting the tricarboxylic acid cycle and protein metabolism. Therefore, a comprehensive understanding of these and other obstacles to the maintenance and success of biological systems is needed (Wang et al. 2021).

2.2 Advanced Oxidation Processes: Fundamentals

Advanced oxidation processes (AOPs) are reaction systems that use oxidizing agents, mainly hydroxyl radicals ($^{\bullet}OH$), to oxidize organic compounds. Their main advantage is the possibility of achieving complete oxidation or mineralization of the contaminants through a process at near ambient pressure and temperature (Matilainen and Sillanpää 2010). AOPs are divided into homogeneous and heterogeneous, with or without external energy input, such as ultraviolet (UV) irradiation, electric current or ultrasound (US).

The processes and oxidizing agents can be carried out individually, e.g. ozone (O_3), H_2O_2, Fenton, persulfate, UV, sonication, and electron beam; or can be enhanced by combining some processes, e.g. O_3/H_2O_2, O_3/UV, UV/H_2O_2, catalyst/UV, photo-Fenton, and persulfate/UV systems (Malik et al. 2020; Teixeira 2021). Each system works differently regarding target contaminants and comprises various interacting parameters that affect their performance towards specific classes of contaminants of emerging concern.

UV/H_2O_2, one of the most applied processes in large scale, produces $^{\bullet}OH$ radicals *in situ*, following the reaction described by Eq. 1. At high H_2O_2 concentrations, side reactions may occur, such as the ones shown in Eqs. 2–3, leading to the formation of less reactive hydroperoxyl ($HO_2^{\bullet}$) radicals. Hydroxyl radicals are one of the most powerful oxidizing species, mainly due to their strong standard reduction potential (E_0) of 2.7 V SHE in acidic media and 1.8 V SHE in alkaline media. They can react with organic contaminants either by hydrogen abstraction (Eq. 4) from C–H, N–H or O–H groups; or by radical–radical interactions, such as molecular O_2 addition, forming an organic peroxyl ($RO_2^{\bullet}$) radical (Eq. 5); or by direct electron transfer (Eq. 6) leading to the formation of oxidized intermediates or CO_2, H_2O, and inorganic acids if complete mineralization is achieved (Kanakaraju et al. 2018; Mierzwa et al. 2018).

$$H_2O_2 + hv \rightarrow 2^{\bullet}OH \quad (1)$$

$$^{\bullet}OH + H_2O_2 \rightarrow H_2O + HO_2^{\bullet} \quad (2)$$

$$^{\bullet}OH + HO_2^{-} \rightarrow HO_2^{\bullet} + OH^{-} \quad (3)$$

$$^{\bullet}OH + RH \rightarrow H_2O + R^{\bullet} \quad (4)$$

$$R^{\bullet} + O_2 \rightarrow RO_2^{\bullet} \quad (5)$$

$$^{\bullet}OH + RX \rightarrow RX^{\bullet +} + HO^- \quad (6)$$

The UV/H_2O_2 system can promote the complete removal of different contaminants with high mineralization rates. Nevertheless, it is essential to note that byproducts may be formed from side reactions of reactive organic intermediates with the target pollutants or other species present in wastewater, such as nitrite ions and dissolved organic matter (Stefan 2017).

Several UV/H_2O_2 systems are currently operating at water utilities worldwide, remediating micropollutants and disinfecting drinking water sources concomitantly. For example, in the Netherlands, an UV/H_2O_2-based system was implemented to treat water from Ijssel Lake and achieve at least 80% removal of atrazine, an important pesticide. In Lincoln (UK), one water supply company installed the UV/H_2O_2 system to treat some new pesticides detected in surface water, which could not be removed by O_3, were poorly removed by O_3/H_2O_2, and very poorly reduced by adsorption. Among the contaminants, the system was able to reduce at least 65% of metaldehyde, 80% of atrazine, and 45% of clopyralid. Another case is in Mississauga, Canada, with one of the largest UV/H_2O_2 facilities in the world for taste and odor (T&O) treatment, designed to reduce 95% of geosmin and 90% of 2-methylisoborneol contents with a throughput of 200,000 m^3 day^{-1} (Stefan 2017).

Ozonation, another large-scale applied AOP, uses the very powerful oxidizing agent O_3, with oxidation potential of 2.07 V SHE (Wang and Zhuan 2020). The oxidation of organic pollutants by ozonation can take place in two ways: (i) indirectly, through the generation of secondary oxidants, such as hydroxyl radicals (Eqs. 7–11); or (ii) directly, by the reaction of molecular ozone with organic species in a very selective way, with reaction rates ranging from 1.0 to 10^6 L mol^{-1} s^{-1}. Due to its polar structure, ozone will react with unsaturated bonds, promoting their scission. The degree of the nucleophilicity of organic compounds, or electron density, determines how quickly the reaction proceeds (Wang and Zhuan 2020; Malik et al. 2020).

$$O_3 + H_2O \rightarrow 2^{\bullet}OH + O_2 \quad (7)$$

$$O_3 + OH^- \rightarrow O_2^{-\bullet} + HO_2^{\bullet} \quad (8)$$

$$O_3 + HO_2^{\bullet} \rightarrow 2O_2 + {}^{\bullet}OH \quad (9)$$

$$O_3 + {}^{\bullet}OH \rightarrow O_2 + HO_2^{\bullet} \quad (10)$$

$$2HO_2^{\bullet} \rightarrow O_2 + H_2O_2 \quad (11)$$

In most cases, organic contaminants react slower with ozone than with $^{\bullet}OH$ radicals, limiting the process efficiency through direct reactions; nevertheless, these compounds will still be removed by $^{\bullet}OH$ reactions (Stefan 2017). Although high removal can be achieved with ozonation, several studies have reported that this process cannot completely mineralize contaminants of emerging concern, usually reaching maximum mineralization of 50%. Moreover, during ozonation, carbonate, and bicarbonate ions present in wastewater can act as $^{\bullet}OH$ scavengers, inhibiting the removal of organic contaminants. Additionally, the wastewater pH is likely to decrease as the ozonation reaction proceeds, which is also unfavorable to the generation of $^{\bullet}OH$. At the same time, wastewater biodegradability can be enhanced due to the formation of low molar weight and readily biodegradable byproducts (Wang and Zhuan 2020). In this context, H_2O_2 and/or irradiation can be coupled to ozonation to improve the mineralization degree.

The O_3/H_2O_2 system, also known as peroxone, is advantageous for removing organic contaminants because H_2O_2 speeds the decomposition of ozone to generate oxidative species via an electron transfer mechanism, represented in Eqs. 12–13, and 15. This "radical booster" occurs in addition to the direct and indirect ozone reactions (Malik et al. 2020), improving the process performance. In contrast, in an O_3/UV system, O_3 photodecomposition generates H_2O_2 and $^{\bullet}OH$ under UVB and UVC irradiation (<310 nm), leading to the reactions given by Eqs. 14–15.

$$H_2O_2 \leftrightarrow HO_2^- + H^+ \tag{12}$$

$$HO_2^- + O_3 \rightarrow HO_2^{\bullet} + O_3^{\bullet -} \tag{13}$$

$$O_3 + H_2O + h\nu \rightarrow H_2O_2 + O_2 \tag{14}$$

$$O_3 + H_2O_2 \rightarrow HO_2^{\bullet} + {}^{\bullet}OH + O_2 \tag{15}$$

A full-scale ozonation treatment plant in Sweden was evaluated to remove several antibiotics, antimycotics, and biocides. With a dose of 0.55 g O_3 g^{-1} total organic carbon (TOC), most of the compounds studied were efficiently removed (>90%), except benzotriazoles and fluconazole, with removals lower than 50% (Östman et al. 2019). A full-scale application of the O_3/H_2O_2 system is also well reported, such as in the Sung-Nam Water Treatment Plant (WTP) in South Korea, where T&O compounds (geosmin and 2-methylisoborneol) can be efficiently removed (>70%). Another WTP in Austria uses the O_3/H_2O_2 system to remove tetrachloroethene and trichloroethene from groundwater, achieving at least 90% degradation for both contaminants. In Spain, one of the water suppliers in Andalucia implemented O_3/H_2O_2 to comply with the European Union drinking water regulations on trihalomethanes and pesticides. However, the O_3/UV system is not yet applied at pilot- or full-scale, although various lab-scale studies use different water matrices demonstrating its high efficiency (Stefan 2017).

Other AOPs have already shown promising results for possible large-scale implementation. The Fenton process, for instance, is based on the use of an oxidizing agent (H_2O_2) and a catalyst (a metal oxide or a divalent salt, usually of Fe), generating

$^{\bullet}OH$ under mildly acid conditions (Eqs. 16–17) (Matilainen and Sillanpää 2010; Kanakaraju et al. 2018). In the photo-Fenton process, irradiation from an artificial light source or sunlight is used, promoting the generation of additional $^{\bullet}OH$ radicals, thus increasing the degradation efficiency of the target pollutants (Matilainen and Sillanpää 2010). Although the Fenton reaction does not require as much energy as other AOPs, such as UV/H_2O_2 and O_3, it needs to be carried out under acidic pH (3–5), which is the main drawback for its large-scale application (Kanakaraju et al. 2018).

$$Fe^{2+} + H_2O_2 \rightarrow Fe^{3+} + OH^- + {}^{\bullet}OH \quad (16)$$

$$Fe^{3+} + H_2O_2 \rightarrow Fe^{3+} + HO_2^{\bullet} + H^+ \quad (17)$$

Photocatalytic oxidation, another extensively studied AOP, uses semiconductor materials as photocatalysts, being TiO_2 the most common material. The process involves a series of mechanisms: (i) the excitation of the photocatalysts by photons of adequate energy to generate electrons in the conduction band (e^-) and holes in the valence band (h^+), (ii) the reduction of surface-adsorbed O_2 by the excited e^- to produce superoxide radical anions ($O_2^{\bullet-}$), (iii) the migration of h^+ to the surface of the catalyst, and (iv) the oxidation of H_2O to form $^{\bullet}OH$ radicals, according to Eqs. 18–24. Thus, CECs are usually removed by $^{\bullet}OH$, $O_2^{\bullet-}$, or h^+ (Wang and Zhuan 2020).

$$TiO_2 + hv \rightarrow TiO_2 + h^+ + e^- \quad (18)$$

$$h^+ + OH^- \rightarrow {}^{\bullet}OH \quad (19)$$

$$h^+ + H_2O + O_2 \rightarrow {}^{\bullet}OH + H^+ + O_2^{\bullet-} \quad (20)$$

$$O_2 + e^- \rightarrow O_2^{\bullet-} \quad (21)$$

$$O_2^{\bullet-} + H^+ \leftrightarrows HO_2^{\bullet}O_2^{\bullet-} + H^+ \leftrightarrows HO_2^{\bullet}O_2^{\bullet-} + H^+ \leftrightarrows HO_2^{\bullet} \quad (22)$$

$$2HO_2^- \rightarrow O_2 + H_2O_2 \quad (23)$$

$$H_2O_2 + O_2^{\bullet-} \rightarrow {}^{\bullet}OH + OH^- + O_2 \quad (24)$$

Oxidation processes based on sulfate radicals ($SO_4^{\bullet-}$) have also gained great attention. The precursor, usually peroximonosulphate (HSO_5^-) or persulphate ($S_2O_8^{2-}$), can be activated by light, ultrasound, temperature, or a metal catalyst (e.g. Fe, Cu, Co), according to Eqs. 25–26. $SO_4^{\bullet-}$ radicals exhibit a pH-independent redox potential of

2.5–3.1 V SHE, allowing the effective oxidation of a wide range of CECs. Moreover, compared to $^{\bullet}OH$ radicals, they have longer lifetime ($t_{1/2(SO_4^{\bullet-})} = 30–40\ \mu s$; $t_{1/2\,(^{\bullet}OH)} = 20$ ns), less significant self-scavenging effect, and better selectivity. Studies have confirmed that $SO_4^{\bullet-}$ may lead to a better mineralization rate than $^{\bullet}OH$ for several contaminants. Although $SO_4^{\bullet-}$ generally plays the most important role in the degradation process, $^{\bullet}OH$ radicals are produced by side reactions and contribute to CECs removal (Dewil et al. 2017; Teixeira 2021). Information regarding continuous flow sulfate-based processes is still unclear, and most of the results so far are from lab-scale batch experiments.

$$S_2O_8^{2-} \xrightarrow{UV/heat/US} 2SO_4^{\bullet-} \tag{25}$$

$$S_2O_8^{2-} + M^{n+} \rightarrow M^{n+1} + SO_4^{\bullet-} + SO_4^{2-} \tag{26}$$

CECs can also be removed by electrochemical oxidation, an AOP based on applying electric current to the wastewater to oxidize and reduce the contaminants in anode and cathode, respectively. Organic pollutants can be converted to low molecular weight biodegradable substances or, in some cases, completely mineralized (Qiao and Xiong 2021). $^{\bullet}OH$ radicals physically adsorbed onto the anode are formed from the oxidation of water (Eq. 27) and can interact with the surface of this electrode, producing MO_{x+1} sites (Eq. 28). Thus, these active forms of oxygen can oxidize organic pollutants adsorbed on the anode surface. In indirect electrochemical oxidation, oxidants such as chlorine, $SO_4^{\bullet-}$, O_3, H_2O_2, and $^{\bullet}OH$ are electrochemically generated and subsequently react with CECs (Qiao and Xiong 2021; Teixeira 2021). The large-scale application of electrochemical processes is limited mainly because of high energy consumption and the high cost of electrodes (Qiao and Xiong 2021).

$$MO_x + H_2O \rightarrow MO_x(HO^{\bullet}) + H^+ + e^- \tag{27}$$

$$MO_x(HO^{\bullet}) \rightarrow MO_{x+1} + H^+ + e^- \tag{28}$$

3 How Can Process Hybridization Help with Micropollutant Degradation?

Several technologies have succeeded in degrading or mineralizing CECs, as presented in Sects. 2.1 and 2.2 of this chapter. With different mechanisms, biological, and chemical systems can significantly reduce the imminent danger of some micropollutant classes. Yet the central question remains: how can process hybridization help, considering that biological or chemical systems alone can efficiently remove CECs? The answer to this question is in the weakness of each technique.

Biological systems, for example, have enzymatic degradation and biosorption as central mechanisms in CECs removal. They are critically influenced by HRT or sludge retention time (SRT) and nutrient sources to microorganisms or cofactors to enzymes (Saidulu et al. 2021). These process factors are more easily controlled in batch mode operation than in continuous flow systems. In fact little has been done in the continuous mode operation of the aforementioned biological systems (Sect. 2.1), especially in a full-scale application.

In contrast, industrial and urban activities commonly generate wastewater containing multiple CECs and in continuous discharge flow. In this sense, the current treatment approaches to treating one CEC at a time and in batch mode do not fit the WWTPs demand. This emphasizes the importance of optimizing treatment systems by combining two or more processes (hybridization) and thus potentializing the removal mechanism of each. Additionally, the choice for a continuous flow regime is the future of WWTPs. It allows for automating the treatment, reducing labor costs and risks, achieving constant removal efficiency, and being more productive.

To encourage the hybridization and operation of WWTPs in continuous flow regime, this section will bring knowledge and insights into two or more processes combined, exploring the chemical and biochemical phenomena in Sect. 3.1 and the operational and technical aspects in Sect. 3.2, with particular emphasis on chemical–biological arrangements.

3.1 Combined Chemical and Biochemical Action on CECs Degradation

The hybridization of biological systems by combining with chemical processes, such as AOPs, is a feasible strategy to enhance overall treatment performance, as pointed out by Saidulu et al. (2021). Hybrid systems integrating bioprocesses with photo-Fenton and O_3, for example, have shown increased removal efficiency and better suitability for treating high organic loads. Combining chemical-biological techniques has shown better performance than biological-physical systems for removing ibuprofen, diclofenac, naproxen, ketoprofen, ciprofloxacin, atenolol, mecoprop, triclosan, and 2,4-dichlorophenoxyacetic acid. Chemical-biological systems have promising scale-up potential, achieving complete removal of CECs in full-scale systems (Saidulu et al. 2021).

The development of systems composed of AOPs and biological reactions coupled in a single equipment is an exciting perspective for the degradation of CECs and has already received a designation: advanced bio-oxidation processes (ABOPs). Although their combined action still lacks understanding, these systems take advantage of the ability of microorganisms and their metabolism to generate ROS, such as hydroxyl radicals, already generated in conventional AOPs. Thus, ABOPs can increase ROS levels, resulting in a system with a high oxidation–reduction capacity.

In biological processes, the metabolic route of •OH radicals production has been mostly described for ligninolytic fungi, which generate these species through quinone redox cycling. This cycle occurs, in turn, due to the action of oxidoreductase enzymes in the presence of chelated ferric ions (Gómez-Toribio et al. 2009; Marco-Urrea et al. 2010; Xie et al. 2020). The quinone redox cycle is induced by a combination of intracellular and extracellular enzymes and occurs in the presence of mediators, e.g. lignin-derived quinone and ferric ions. Under these conditions, the fungi can convert quinone into hydroquinone through the intracellular enzyme quinone reductase. Hydroquinone, in turn, is oxidized to a semiquinone radical via an extracellular enzymatic system (laccases and peroxidases). In addition to hydroxyl radicals, H_2O_2 and Fe^{2+}/Fe^{3+} molecules are generated in this redox cycle. This catalytic route resembles what occurs in Fenton reactions (Marco-Urrea et al. 2010; del Álamo et al. 2018; Vasiliadou et al. 2019; Xie et al. 2020).

In addition to being little known, the redox cycle of fungi and other microorganisms is complex and challenging to control. Therefore, the mechanisms of ABOPs are only superficially explained in studies on CEC degradation. Furthermore, systems on continuous flow and large scale generally operate with a microbial consortium, making the task of explaining biochemical and chemical phenomena even more difficult.

Aiming to bring insights and perspectives on WWTPs hybridization, Table 1 presents the state-of-the-art of hybrid continuous flow treatment systems. The following paragraphs detail their degradation mechanisms; the studies were selected based on the system operation (continuous mode) and the processes (only biological and AOPs mentioned in Sects. 2.1 and 2.2).

The MBR hybridization has been successfully applied to a broad range of CECs, especially from the pharmaceutical class, achieving excellent removal performance. The AOP-MBR systems are predominantly arranged in separate reactors, but the order of the arrangement can be reversed, as in Nguyen et al. (2013). The treatment strategy was to combine UV oxidation as a post-treatment of MBR permeate and take advantage of their different degradation mechanisms. The laboratory-scale MBR system was seeded with activated sludge from a municipal WWTP, and a UVC system was coupled in series. The MBR-UVC system was fed with synthetic wastewater with values of TOC, total nitrogen, and COD of about 180, 25, and 600 mg L^{-1}, respectively, and a set of twenty-two trace organic contaminants (see Table 1).

Nguyen et al. (2013) confirmed the synergy between MBR mechanisms (CECs adsorption to the suspended solids and biodegradation) and UVC mechanism (oxidation) by measuring the removal efficiency for separate and hybrid systems. Each process efficiently removed groups of compounds with specific characteristics: the MBR could effectively remove hydrophobic and readily biodegradable compounds; UVC readily degraded contaminants with a phenolic or chlorine group. When combined, the system removed all groups efficiently (above 85%). A remarkable finding was the removal efficiency for carbamazepine, a well-known biologically persistent compound. In addition to being poorly removed by MBR (32%) or UVC (30%), the hybrid arrangement exponentially increased carbamazepine degradation (96% of removal).

Table 1 Examples of successful hybrid and continuous flow chemical-biological systems for the degradation of contaminants of emerging concern

Processes	Mechanisms engaged	Wastewater composition		System parameters				Removal efficiency (%)	References
		CECs	Initial concentration of each CEC ($\mu g\ L^{-1}$)	Temperature (°C)	pH	HRT (h)	Flow rate (mL min^{-1})		
MBR - UVC	Adsorption to the suspended solids, biodegradation, and oxidation	A mixture of twenty-two synthetic compounds, including pharmaceuticals, personal care products, pesticides, steroid hormones, and industrial chemicals and their metabolites	5	22	7.2–7.5	24	7	>85	Nguyen et al. (2013)
O_3/US - MBR	Physical and thermal decomposition, hydroxylation, adsorption, and biodegradation	Synthetic compounds simulating real municipal wastewater containing diclofenac, sulfamethoxazole, and carbamazepine	4	n.r.[a]	5.0–7.5	2	20	80–84	Prado et al. (2017)
UVA/TiO_2 - MBR	Pre-oxidation and biodegradation	Synthetic wastewater containing diclofenac, ibuprofen, carbamazepine, furosemide, levetiracetam, amylmetacresol, and paracetamol	40	n.r.	7.2	48	5–7	3–98	Joannis-Cassan et al. (2021)

(continued)

Table 1 (continued)

Processes	Mechanisms engaged	Wastewater composition		System parameters				Removal efficiency (%)	References
		CECs	Initial concentration of each CEC ($\mu g\ L^{-1}$)	Temperature (°C)	pH	HRT (h)	Flow rate (mL min^{-1})		
Electrochemical - aerobic activated sludge	Adsorption, mass transfer, water oxidation, generation of hydroxyl radicals and active chlorine, and biodegradation	Synthetic municipal wastewater containing bezafibrate, gemfibrozil, indomethacin, and sulfamethoxazole	200	25	7	0.33 and 32	1200 and 0.52	100	Rodríguez-Nava et al. (2016)
Aerobic activated sludge - Fenton	Biodegradation, chelating action, and generation of hydroxyl radicals	Real swine wastewater containing ten synthetic sulfonamide antibiotics	95.6–108.7	n.r.	6.0–8.9	72–144	5.5	80–95	Qian et al. (2020)
Fungi/Fenton	Intra- and extracellular enzymatic system, H_2O_2 formation by fungi, and generation of non-selective hydroxyl radicals	Real urban wastewater containing amoxicillin, metronidazole, sulfamethoxazole, carbamazepine, caffeine, 4-acetamidoantipyrine, gemfibrozil, hydrochlorothiazide, and iohexol	0.011–89	n.r.	4.5	24	7	23–89	Cruz del Álamo et al. (2020)

(continued)

Table 1 (continued)

Processes	Mechanisms engaged	Wastewater composition		System parameters				Removal efficiency (%)	References
		CECs	Initial concentration of each CEC (μg L^{-1})	Temperature (°C)	pH	HRT (h)	Flow rate (mL min^{-1})		
Electrochemical/immobilized peroxidase/electrocoagulation	Cathode: O_2 reduction, H_2O_2 *in situ* generation, enzymatic biocatalysis, phenoxy radical generation, polymerization, and/or adsorption. Anode: Al^{3+} production via electrolysis, precipitation of polymerized molecules by Al coagulation	Real wastewater containing bisphenol A	4.1	30	7	0.08–0.33	3.5–14	94	Zhao et al. (2015)

Note In the first column, the hyphen symbol (-) represents processes occurring in separate equipment, and the slash symbol (/) represents processes coupled in a single piece of equipment

[a]Not reported

More than one advanced oxidation process is often combined in single reactors as a pre-treatment for the MBR system. Prado et al. (2017) successfully implemented this hybrid concept to degrade three selected drugs (see Table 1) by pre-treating the synthetic wastewater with combined ozonation and ultrasound in a single reactor. Then, the MBR inoculated with activated sludge from a full-scale urban WWTP was fed with pre-treated wastewater. Some remarkable findings can be highlighted from this hybrid system: (i) the AOP pre-treatment did not induce any significant changes in biomass growth and biodegradation performance; (ii) membrane fouling and extracellular polymeric substance (EPS) concentration was reduced; (iii) AOPs alone induce the formation of intermediate toxic compounds, and MBR biologically oxidizes them, reducing the toxicological potential.

The strategies behind CECs degradation by the O_3/US—MBR system are to pre-oxidize the selected drugs into smaller compounds, contributing to partial chemical oxidation and enhancing their biodegradability by making the organic substances more readily available to microorganisms. The O_3/US combination also demonstrates promising mechanisms: ultrasound improved the cleavage of S–N bonds so that CECs were more easily attacked by ozone. The AOPs experimental results suggest a radical attack in the bulk solution, leading to hydroxylation as the primary degradation mechanism for the selected pharmaceuticals. The extent of degradation can be attributed to the compound hydrophobicity, resulting in a higher concentration in the interfacial region than in bulk (Prado et al. 2017).

Sludge adsorption played a minor role in the hybrid system studied by Prado et al. (2017), indicating that biodegradation was the primary degradation mechanism in MBR treatment. Since pre-treatment by AOPSs was able to enhance the breakdown of macromolecular organics, there was a decrease in EPS concentration (reduced by 50%). The consequence of this treatment hybridization was a modification of the microbial consortia, a reduction in the MBR fouling propensity, and most important: a significant increase in contaminants removal. Similar consequences were found in the hybrid system investigated by Joannis-Cassan et al. (2021), especially in microorganism metabolism modifications when wastewater was pre-treated by UVA/TiO_2 before MBR treatment. The results were an improvement in the degradation of some molecules and maintenance of the removal performance of others already well biodegraded by MBR alone. Only carbamazepine maintained low removal efficiency (3%), not improved by hybridization. The authors suggest that the hybrid system operating under continuous recirculation can improve the removal efficiency for poorly degraded molecules because the photo-oxidation step can induce several oxidation/ biodegradation cycles.

Another possibility of biological process hybridization is the use of aerobic activated sludge, containing a microbial consortium, coupled with electrochemical pre-treatment (Rodríguez-Nava et al. 2016) or Fenton post-treatment (Qian et al. 2020). Both studies developed the processes in separate equipment and operated a continuous flow reactor that removed up to 100% of drugs. In the electrochemical reactions, Rodríguez-Nava et al. (2016) observed the simultaneous adsorption of organic molecules and oxidant species for CECs oxidation on boron-doped diamond (anodic oxidation). The current density of this process was essential to produce lower quantities of hydroxyl radicals and thus prevent the degradation of biodegradable organic matter (BOM). The authors reported that higher current densities formed an excess of chlorine and modified degradation to a more complex mechanism without increasing degradation performance.

The minimal transformation of BOM by the electrochemical process is important to the subsequent biological process, avoiding sludge bulking (which occurred without the pre-treatment step). The electrochemical process with an activated sludge reactor inoculated with biomass from an oil refinery WWTP removed 83% of BOM and 100% of pharmaceuticals. Scanning electron microscopy images of activated sludge showed filamentous microorganism growth in the system without AOP and bacillus when the chemical-biological system was operated (Rodríguez-Nava et al. 2016) indicating the influence of hybridization in the microbial consortium.

In the treatment strategy to degrade drugs from real swine wastewater discussed by Qian et al. (2020), aerobic-activated sludge with high biological activity from a municipal WWTP was inoculated in a sequencing batch reactor (SBR), previously to Fenton-like batch experiments conducted with Fe^{2+} and citric acid. In this hybrid and partially continuous flow system, the biological process needed an acclimatization period, but after several days the sludge was adapted, and the removal stabilized at >95%. Only the removal efficiency of trimethoprim remained ca. 10% and the authors attributed this to the absence of an S–N bond in its molecular structure compared to other sulfonamide-based antibiotics. Its removal efficiency was improved by hybridization with a Fenton-like oxidation process, enabling 80% of trimethoprim degradation.

Regarding the Fenton reaction, the addition of citric acid as a chelating agent significantly improved Fe^{2+}/Fe^{3+} solubility under nearly neutral conditions, contributing to the higher utilization efficiency of Fe^{2+}/Fe^{3+} by H_2O_2. The initial H_2O_2 concentration also was an essential factor on Fenton efficiency: the increase in H_2O_2 positively impacted the average CEC removal efficiency. Overall, the proposed system showed remarkable advances by combining aerobic biodegradation—which removed COD and biodegradable trace contaminants and inhibited the nitrification process—with Fenton-like reaction to further decompose nonbiodegradable or residual trace contaminants (Qian et al. 2020).

In addition to the arrangements presented in Table 1 for AOPs-activated sludge systems, relevant results have been reported in batch mode operation involving this biological process, e.g. treatment of a municipal wastewater effluent containing thirty-two CECs by activated sludge followed by UVC or neutral photo-Fenton reactions (Fe^{2+}/H_2O_2/UVC) (De la Cruz et al. 2012); treatment of primary municipal wastewater effluent containing nine pharmaceuticals by aerobic biodegradation followed by different AOPs (heterogeneous solar photocatalysis with TiO_2, solar photo-Fenton, and ozonation) (Gimeno et al. 2016); and treatment of pharmaceutically active compounds from real hospital wastewater by UVC/H_2O_2 before and after activated sludge treatment or fungal treatment (Mir-Tutusaus et al. 2021).

Continuous hybrid systems combining processes in separate reactors seem to be the predominant type of biological processes with a microbial consortium, and different AOPs can be part of these systems. For example, Cruz del Álamo et al. (2020) immobilized the white-rot fungi *Trametes versicolor* in a rotating biological contactors (RBCs) reactor to develop a fungi/Fenton continuous process to treat real urban wastewaters. The Fenton reaction and quinone redox cycle was promoted by adding to the reactor lignin-derived quinone-type mediator (gallic acid) and complexed metal sources (iron oxalate and manganese nitrate), inducing the occurrence of an advanced bio-oxidation process. In turn, for isolated microorganisms, the trend seems to be chemical-biological hybridization in a single reactor, as presented in Table 1.

The fungi/Fenton reactions promote the generation of highly oxidizing hydroxyl radicals and increase the degradation of CECs. The ubiquitous formation of H_2O_2 by *T. versicolor* enzymatic system enables the generation of non-selective and highly oxidizing species through Fenton-like reactions. Cruz del Álamo et al. (2020) found a relationship between CECs removal and its loading rate: higher CECs concentration led to lower overall removal efficiency. Likewise, there is remarkable fungal tolerance for the range of pharmaceutical loadings (see Table 1). The promising removals (up to 89%) meet the feasible operation characteristics: non-addition of nutrients source, non-fungal biomass refreshment, and non-external aeration. Although fungal communities dominated the degradation of CECs, the authors highlight the proliferation of bacteria in RBCs biofilm because of the continuous operation under non-sterile conditions.

Another successful hybrid fungi/Fenton system was developed by Vasiliadou et al. (2019) in batch mode to degrade thirteen pharmaceuticals from hospital wastewater. The system was evaluated using two ligninolytic fungi, *T. versicolor* and *Ganoderma lucidum*, and Fenton-like reaction was induced using four different substances (1,4-benzoquinone, 2,6-dimethoxy-1,4-benzoquinone, 2-methyl-1,4-naphthoquinone, and gallic acid). Removals of up to 100% were achieved under non-sterilized conditions and through the same catalytic cycle described earlier for fungi (ABOPs). The authors highlight the importance of gallic acid since its phenolic form is equivalent to that of hydroquinone and is considered a critical parameter in Fe^{3+} reduction.

The last work presented in Table 1 was also developed in a single piece of equipment, but used the commercial horseradish peroxidase (HRP) enzyme as the biological part of the system. Zhao et al. (2015) explored an electrochemical/biocatalysis/electrocoagulation process operated in continuous flow to degrade bisphenol A (BPA) from real wastewater. The treatment system consisted of a cathodic cell with a titanium electrode containing immobilized HRP, and an anodic cell containing an aluminum (Al) plate as the anode. The cells were separated by a membrane promoting ion exchange. The wastewater containing humic acid (HA) was continuously pumped into the electroenzymatic cell (cathode), and then into the electrocoagulation cell (anode). Oxygen and carbon dioxide gases were provided at the bottom of the cathode cell to supply dissolved oxygen and replenish the carbonate buffer as the hydrogen ion was consumed.

The proposed hybrid system combined several synergistic degradation mechanisms described in detail by Zhao et al. (2015). In short, the electric current starts the process by continuously generating H_2O_2 *in situ* on the cathode through a two-electron reaction. H_2O_2 activates the enzyme and it biocatalyzes BPA following a classical peroxidase cycle to generate phenoxy radicals. Due to high enzymatic activity (25 U), the reaction rate is speedy. The enzymatically generated phenoxy radicals can further be coupled with each other, and the functional groups of BPA will change substantially. In the electrocoagulation cell, the Al anode produces Al^{3+} via electrolysis. The Al^{+3} ions immediately undergo spontaneous hydrolysis reactions to form hydrolytic products of aluminum. When the wastewater flows from the cathodic cell into the anodic cell, the organic polymers formed during the electroenzymatic process precipitate by Al species coagulation. During electrocoagulation in the anodic cell, the polymerized BPA and structurally altered HA are removed by the Al species dissolved from the anode via electrolysis and hydrolysis. Through all these synergistic steps, it was possible to simultaneously remove BPA and TOC in a system requiring 0.016 kWh m^{-3} (Zhao et al. 2015).

While the literature brings some examples and positive perspectives on hybrid continuous flow systems for AOPs combined with MBR, aerobic activated sludge, fungus, and enzymes, to our best knowledge, there is a lack of AOP-microalgae hybridization in continuous operation mode. Some studies have explored treatment strategies in batch mode, and their promising results can encourage researchers. For example, Khan et al. (2020) proposed microalgae-assisted oxidation as a quaternary treatment for antibiotics and refractory organics degradation from livestock wastewater. The authors investigated the combination of microalgal process by *Scenedesmus quadricauda* as pre-treatment, and subsequent oxidation by O_3, O_3/H_2O_2, UVC, or UVC/H_2O_2. The microalgae-AOP system achieved 99% removal for twenty antibiotics, and the main degradation mechanisms were photodegradation, hydrolysis, and biosorption during algal treatment, and degradation or mineralization during AOPs.

Other combinations of microalgae-AOP strategy and their mechanisms in degrading CECs are reported in the literature, e.g. Fenton pre-treatment (Fe^{2+}/H_2O_2) followed by *Chlorella* treatment (Li et al. 2015), UVC and *Chlorella* treatment in a single piece of equipment (Du et al. 2015), UVA and microalgal treatment (*Chlorella, Selenastrum*, and *Scenedesmus*) in a single piece of equipment (Yang et al. 2017), UVA irradiation by xenon lamp simulating natural light and *Chlamydomonas* treatment in a single piece of equipment (Liu et al. 2017), and several arrangements of UVA pre-treatment followed by microalgal treatment (*Chlorella*) and then activated sludge treatment (Jiang et al. 2019). The predominant mechanisms in these studies were adsorption, indirect photodegradation, and biodegradation, achieving remarkable removal percentages (73–100%).

3.2 *Operational and Technical Aspects*

As with any industrial process, the hybridization advantages mentioned in the previous section can only be as good as the overall design of the process. An efficient implementation of hybrid technologies must consider important operational and technical aspects of the combined process, including the complementarity of the treatments, the composition, and physicochemical properties upstream and downstream of each treatment stage, the throughput, and the required residence times in each piece of equipment, and need of pre- or post-treatment. Some of these issues were already addressed throughout this chapter, but two of them will be the focus of this section: the rationale of complementarity as a key aspect of hybridization, and the technical aspects of hybridization by combination in separate pieces of equipment.

When combining two or more treatments, the expected outcomes and limitations of each treatment should serve as the basis for the proposed hybridization. Take membranes as an example, which are usually highly efficient for removing pollutants selectively, especially bacteria and other microorganisms, due to their controllable pore sizes and chemical affinity. However, membranes alone do not degrade the contaminants, which eventually accumulate on their surface and block their pores. Therefore, any treatment combined with membranes should promote the degradation of the pollutants and the constant cleaning of the membrane pores. Some examples of hybrid processes using membranes include the previously discussed MBR, the Photocatalytic Membrane Reactor (PMR), and the Hybrid Ozone Membrane Filtration (HOMF) process.

In PMR and HOMF processes, different AOPs are used to remove the contaminants adhering to the surface of the membranes. PMR relies on the oxidizing power of UV-irradiated photocatalysts, while HOMF uses an ozone flow to degrade the fouling contaminants. In a prominent example, Wang et al. (2017) prepared a hollow fiber PMR made from photocatalytic TiO_2 with a maximum water flux of 7.9 L m^{-2} h^{-1}. They could increase the transmembrane flux (TMF) by ca. 86% by irradiating the membrane—likely a combined effect of removing the fouling contaminants and improving the hydrophilicity of the membrane.

Guo et al. (2016) presented an elaborate concept of coated ceramic membranes with internal channels through which both the contaminated wastewater and ozone feed flow. The system was operated at a flow rate of 3.0 m^3 h^{-1}. Clean water permeates the ceramic body, and the excluded contaminants are constrained on the inner walls of the channels, where they react with ozone (or derived radicals). They can also be mechanically removed by shear forces, enhanced by the hydrodynamics of the heterogeneous flow. The combination of these two techniques allowed a reduction of over 16% in wastewater toxicity and improved ca. 6% of contaminant removal over ozonation alone.

In another example of complementarity, Kim et al. (2019) investigated the use of activated biochar combined with ultrafiltration membranes—ultrafiltration-activated biochar hybrid system—with interesting findings operating at a normalized permeate flux of ca. 0.85. While both technologies are essentially used to remove contaminants without degrading them, the hydrophobic nature of biochar helped improve (by ca. 45%) the removal of selected pharmaceutical compounds (ibuprofen, 17 α-ethinyl estradiol, and carbamazepine). Nevertheless, the TMF of the membranes in wastew aters containing dissolved organic matter (DOM) stresses the need to widely assess the impact of the combination of treatment technologies.

These selected examples, among many other successful implementations reported in Table 1, illustrate the key strategy behind the design of a hybrid process: the complementarity of the selected techniques. Besides carefully considering the objectives and requirements of each treatment stage, it is essential to highlight that continuous flow systems require an arduous task: control of relevant parameters. For example, pH, temperature, and HRT are crucial for biological and chemical processes, but their range can differ for each method. In this sense, the strategy of combining processes in a single reactor is not commonplace, especially in large-scale implementations.

The alternative is hybridization by sequential treatments, which was particularly successful in the municipal WTP of Méry-sur-Oise (France). Tasked with providing on average 160,000 m^3 d^{-1} of potable water from surface waters with a high DOM content for over 800,000 people, this treatment plant uses an array of combined treatment technologies (Fig. 1a). The WTP was established in the 1920s and was gradually equipped with novel technologies to address the changing quality of the Oise River water and the growing population demands. It currently has two parallel treatment lines, a biological one and a membrane-based one. The incorporation of micro and nanofiltration in 1999, for example, ensured the removal of suspended and most dissolved matter, including pathogens and some contaminants of emerging concern—further enhanced by UV reactors downstream: the number of viable bacteria was reduced by one log after the implementation of this novel system (Peltier et al. 2003). A side-effect of nanofiltration is that the water exiting the membrane line is stripped of most of its salinity, significantly altering physical–chemical and organoleptic properties. To overcome that, the waters from the membrane and biological lines are mixed in a suitable proportion to restore some of its salt contents to an acceptable level.

Another successful case is the Ariake Water Reclamation Center in Tokyo Bay (Japan). In addition to being an example of a remarkable and modern WTP design, the Ariake center adopts a combination of biological treatments for nutrient removal (A^2O process) coupled with ozonation to convert local wastewater into reclaimed water for non-potable public use. The A^2O process (Fig. 1b) is an advanced biological treatment to remove phosphorus and nitrogen, designed to prevent eutrophication and the frequent occurrence of red tides in Tokyo Bay. The A^2O is a hybridization of anaerobic, anoxic, and aerobic treatments, achieved by an elaborate control of recirculating water and activated sludge to foster the growth of dedicated microorganisms. The anaerobic section promotes enhanced biological phosphorus removal based on an activated sludge mixture with nutrient-rich wastewater. Since phosphorus accumulating organisms compete with other heterotrophic microorganisms, the activated sludge should undergo a nitrification stage; thus, the return sludge should be collected after an oxic (aerobic) treatment stage. The effluent of the anaerobic section is transferred to an anoxic tank, where it is mixed with a recycle stream from the mixture of the aerobic tank. This stream is rich in NO_x, consumed in the anoxic tank to promote denitrification and reduce the amount of nitrate fed into the anaerobic tank with the activated sludge. The efficiency of this combined approach made the A^2O process widely adopted in China, where it is the second most frequent technology used after oxidation ditches, present in approximately 25% of the WWTPs (Jin et al. 2014).

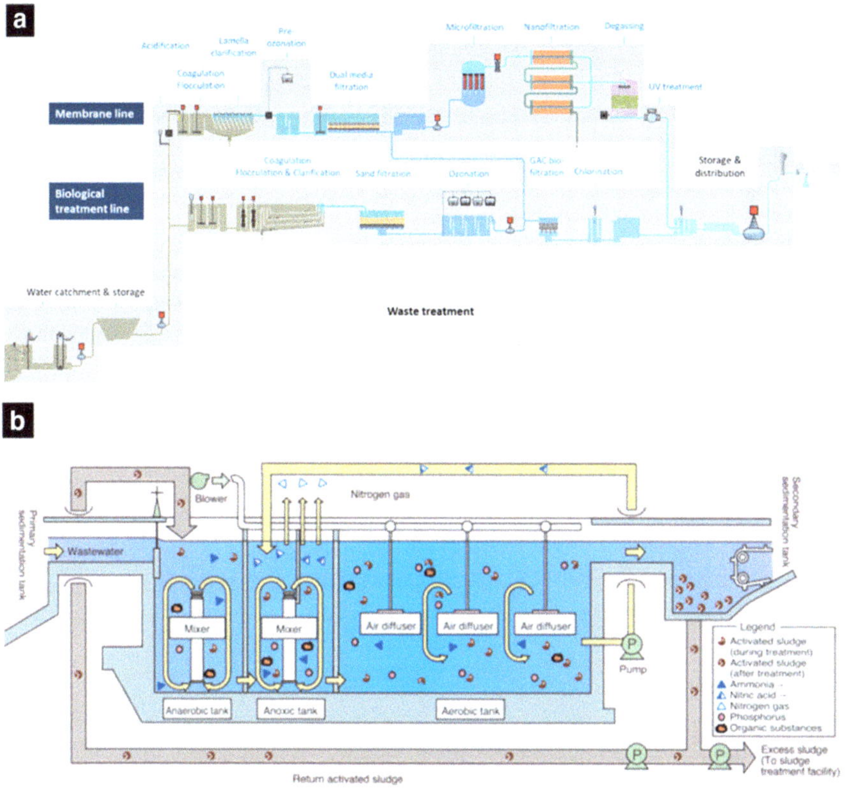

Fig. 1 Illustration of two full-scale hybrid chemical-biological systems. The system (a) refers to the Méry-sur-Oise Water Treatment Plant (France), and system (b) refers to the Ariake Water Reclamation Center (Japan). The illustrations were adapted from SEDIF (2015) and Ariake Water Reclamation Center (2008)

The flexibility of this system makes it prone to further combinations with membrane processes, such as the A^2O/MBR (Falahati-Marvast and Karimi-Jashni 2020), in which the effluent is separated from the activated sludge by membrane filtration instead of relying on gravitational settling. Additionally, this structure can be adapted to incorporate further treatment stages to conform to the properties of the wastewater. An example is the addition of a fermenter upstream of the A^2O tanks, as in the Kelowna WWTP (Rayne et al. 2005). This plant, located in the Okanagan Valley (Canada), was designed to treat the excessive nutrients present in local wastewater and serves an estimated population of 70,000, capable of delivering up to 40,000 $m^3\ d^{-1}$. The fermenter liquid, transferred to the first stage of the A^2O line, is rich in volatile fatty acids, enhancing the growth of phosphorus, accumulating microorganisms, and improving phosphorus removal.

In all the processes presented above, AOP stages—such as ozonation and UV irradiation—are adopted later to remove residual COD. This is normally the case, especially for UV irradiation treatments since the treated stream must be sufficiently clear to allow efficient UV light penetration. In addition, UV treatment can promote microorganism death; that is, it is feasible to concomitantly eliminate pathogens and treat the water, but it cannot be implemented in a single piece of equipment with a biological system involving living microorganisms.

Another attractive operational design is the use of upstream AOPs of a biological treatment line. Some wastewaters are rich in non-biodegradable organic matter, and others—particularly pharmaceutical industrial waters—can contain a high dose of antibiotics, significantly impairing the efficiency of a biological treatment. To overcome these limitations, AOPs can be used as pre-treatment technologies, improving the biodegradability of DOM contents, while also converting harmful pharmaceutical components into easily treatable intermediates (Nidheesh et al. 2021). Most of these combined approaches are still in a development stage, mainly due to the high capital and operational costs associated with installing AOP units; however, the results so far are quite promising (Shtepa et al. 2021).

4 Perspectives On Wastewater Treatment Hybridization for the Next Years

The worldwide WWTPs have several differences in resource availability and financial support, as each country deals with its current environmental problems in its own way. Still, the incoming adverse effects of CECs on human and wildlife health are undeniable. We see the future of WWTPs focusing on process sustainability and meeting local demands, but these aspects have different meanings to low-, middle- and high-income countries.

WWTP hybridization can be a suitable option for high-income countries, where environmental awareness and societal pressures are strong, and the market can support the operational and technical demands, such as increased energy and equipment costs. In these countries, the trend of hybridization can be considered in constructing new WWTPs or upgrading existing plants. Unfortunately, advanced wastewater treatment in low- and middle-income countries seems less likely to be implemented soon, mainly due to inadequate funding. As well highlighted by Prado et al. (2017), although several existing hybrid and continuous flow systems can altogether remove CECs, they do not match the social and economic realities of all regions.

Thinking of a prosperous scenario in countries interested in innovating and mitigating CECs problems, the WWTPs hybridization in a single reactor can be an excellent solution to reduce footprints and promote better utilization of public spaces. Reactor verticalization is also a feasible strategy to take better advantage of land use, given some biological processes need high residence times.

As seen in this chapter, the implementation of advanced bioprocesses is less frequent than AOPs application in full-scale WWTPs. We attribute this to the lack of interest of the environmental biotechnology field in addressing efforts towards CECs degradation, as this sector finds better market opportunities in the fine chemicals and waste2energy markets.

Another trend for the following years is WWTP automation and personalization, pressured by the upcoming fifth industrial revolution (IR). The industry is currently in the cyber-physical systems of the fourth IR, governed by globalization and digitalization, but, in the following decades, the fifth IR will introduce the cyber-physical cognitive concept, whereby personalization and cooperation between man and machine will be valorized (Ross and Maynard 2021). Since the industries will operate in this framework, wastewater treatment technologies have great potential to benefit from the tools of the fifth IR. For example, artificial intelligence connecting factories to their WWTPs can optimize time, costs, labor, and CEC removal performance.

It is worth highlighting that the current fourth IR is seen as an environmental, societal, and economic revolution and adopts the circular economy as a premise, governed by the non-competition for resources and the efficient use of the entire production chain. This concept can be applied to WWTPs since several resources can be generated or recovered during treatment, e.g. electricity by microbial fuel cells, phosphorus recovery by microbial consortium treatment, and potable water after advanced treatments.

The production of potable or reused water through advanced wastewater treatment strategies, such as the hybrid systems mentioned in this chapter, is particularly important in regions that already face severe water shortages. Brazil can be taken as an example, being a privileged country for its wide freshwater availability but experiencing localized seasonal water scarcity. In this context, the quantitative analysis of water availability in Brazil by Hespanhol (2017) indicated that the treatment and reuse of effluents could lead to a unit cost of water lower than that associated with the collection and treatment of surface water. In the industrial scenario, a significant reduction in water charges can be achieved by 60% of water reuse, a perfectly viable value in most industries.

In this sense, although wastewater treatment hybridization requires high initial investments, it is an essential tool for better environmental and economic management of water resources. It has the potential to be widely adopted in the coming years, especially in industries and high-income countries.

References

Akkoyunlu B, Daly S, Casey E (2021) Membrane bioreactors for the production of value-added products: recent developments, challenges and perspectives. Bioresour Technol 341:125793. https://doi.org/10.1016/j.biortech.2021.125793

Ariake Water Reclamation Center (2008) Bureau of sewerage of the Tokyo metropolitan government. https://www.gesui.metro.tokyo.lg.jp/business/b4/guide/sisc-list/03-12/index.html

Bilal M, Adeel M, Rasheed T, Zhao Y, Iqbal HMN (2019) Emerging contaminants of high concern and their enzyme-assisted biodegradation—a review. Environ Int 124:336–353. https://doi.org/10.1016/j.envint.2019.01.011

Cruz del Álamo A, Pariente MI, Martínez F, Molina R (2020) Trametes versicolor immobilized on rotating biological contactors as alternative biological treatment for the removal of emerging concern micropollutants. Water Res 170:115313. https://doi.org/10.1016/j.watres.2019.115313

De Gusseme B, Vanhaecke L, Verstraete W, Boon N (2011) Degradation of acetaminophen by Delftia tsuruhatensis and Pseudomonas aeruginosa in a membrane bioreactor. Water Res 45(4):1829–1837. https://doi.org/10.1016/j.watres.2010.11.040

De la Cruz N, Giménez J, Esplugas S, Grandjean D, de Alencastro LF, Pulgarín C (2012) Degradation of 32 emergent contaminants by UV and neutral photo-fenton in domestic wastewater effluent previously treated by activated sludge. Water Res 46(6):1947–1957. https://doi.org/10.1016/j.watres.2012.01.014

del Álamo AC, Pariente MI, Vasiliadou I, Padrino B, Puyol D, Molina R, Martínez F (2018) Removal of pharmaceutical compounds from urban wastewater by an advanced bio-oxidation process based on fungi Trametes versicolor immobilized in a continuous RBC system. Environ Sci Pollut Res 25(35):34884–34892. https://doi.org/10.1007/s11356-017-1053-4

Dewil R, Mantzavinos D, Poulios I, Rodrigo MA (2017) New perspectives for advanced oxidation processes. J Environ Manage 195:93–99. https://doi.org/10.1016/j.jenvman.2017.04.010

Du Y, Zhang S, Guo R, Chen J (2015) Understanding the algal contribution in combined UV-algae treatment to remove antibiotic cefradine. RSC Adv 5(74):59953–59959. https://doi.org/10.1039/C5RA10806C

Falahati-Marvast H, Karimi-Jashni A (2020) A new modified anoxic-anaerobic-membrane bioreactor for treatment of real wastewater with a low carbon/nutrient ratio and high nitrate. J Water Process Eng 33:101054. https://doi.org/10.1016/j.jwpe.2019.101054

Geissen V, Mol H, Klumpp E, Umlauf G, Nadal M, van der Ploeg M, van de Zee SEATM, Ritsema CJ (2015) Emerging pollutants in the environment: a challenge for water resource management. Int Soil Water Conserv Res 3(1):57–65. https://doi.org/10.1016/j.iswcr.2015.03.002

Gimeno O, García-Araya JF, Beltrán FJ, Rivas FJ, Espejo A (2016) Removal of emerging contaminants from a primary effluent of municipal wastewater by means of sequential biological degradation-solar photocatalytic oxidation processes. Chem Eng J 290:12–20. https://doi.org/10.1016/j.cej.2016.01.022

Gómez-Toribio V, García-Martín AB, Martínez MJ, Martínez AT, Guillén F (2009) Induction of extracellular hydroxyl radical production by white-rot fungi through quinone redox cycling. Appl Environ Microbiol 75(12):3944–3953. https://doi.org/10.1128/AEM.02137-08

Goswami L, Vinoth Kumar R, Borah SN, Arul Manikandan N, Pakshirajan K, Pugazhenthi G (2018) Membrane bioreactor and integrated membrane bioreactor systems for micropollutant removal from wastewater: a review. J Water Process Eng 26:314–328. https://doi.org/10.1016/j.jwpe.2018.10.024

Guo Y, Xu B, Qi F (2016) A novel ceramic membrane coated with MnO2–Co3O4 nanoparticles catalytic ozonation for benzophenone-3 degradation in aqueous solution: fabrication, characterization and performance. Chem Eng J 287:381–389. https://doi.org/10.1016/j.cej.2015.11.067

Hespanhol I (2017) Conservation and reuse as management tools to mitigate the billing charges for use of water in the industrial sector. In: de Mattos Bicudo CE, Tundisi JG, Scheuenstuhl MCB (eds) Waters of Brazil. Springer, Cham, pp 39–57. https://doi.org/10.1007/978-3-319-41372-3_4

Jiang R, Wei Y, Sun J, Wang J, Zhao Z, Liu Y, Li X, Cao J (2019) Degradation of cefradine in alga-containing water environment: a mechanism and kinetic study. Environ Sci Pollut Res 26(9):9184–9192. https://doi.org/10.1007/s11356-019-04279-y

Jin L, Zhang G, Tian H (2014) Current state of sewage treatment in China. Water Res 66:85–98. https://doi.org/10.1016/j.watres.2014.08.014

Joannis-Cassan C, Rodriguez Castillo AS, Dezani C, Gómez-Canela C, Reoyo-Prats B, Calas-Blanchard C, Barata C, Lacorte S, Plantard G (2021) Towards an innovative combined process coupling biodegradation and photo-oxidation for the removal of pharmaceutical residues. J Chem Technol Biotechnol 96(3):755–763. https://doi.org/10.1002/jctb.6589

Kanakaraju D, Glass BD, Oelgemöller M (2018) Advanced oxidation process-mediated removal of pharmaceuticals from water: a review. J Environ Manage 219:189–207. https://doi.org/10.1016/j.jenvman.2018.04.103

Khan W, Nam J-Y, Byun S, Kim S, Han C, Kim H-C (2020) Emerging investigator series: quaternary treatment with algae-assisted oxidation for antibiotics removal and refractory organics degradation in livestock wastewater effluent. Environ Sci Water Res Technol 6(12):3262–3275. https://doi.org/10.1039/D0EW00634C

Kim S, Park CM, Jang A, Jang M, Hernández-Maldonado AJ, Yu M, Heo J, Yoon Y (2019) Removal of selected pharmaceuticals in an ultrafiltration-activated biochar hybrid system. J Memb Sci 570–571:77–84. https://doi.org/10.1016/j.memsci.2018.10.036

Klanovicz N, Scapini T, Dalastra C, de Souza LP, de Jesus JMS, Treichel H, Teixeira ACSC (2021) Antiepileptic drugs: from public to environmental health problem. In: Thapar Kapoor R, Treichel H, Shah MP (eds) Biochar and its application in bioremediation. Springer, Singapore, pp 209–229. https://doi.org/10.1016/j.watres.2019.115313

Li H, Pan Y, Wang Z, Chen S, Guo R, Chen J (2015) An algal process treatment combined with the Fenton reaction for high concentrations of amoxicillin and cefradine. RSC Adv 5(122):100775–100782. https://doi.org/10.1039/C5RA21508K

Liu Y, Wang Z, Yan K, Wang Z, Torres OL, Guo R, Chen J (2017) A new disposal method for systematically processing of ceftazidime: the intimate coupling UV/algae-algae treatment. Chem Eng J 314:152–159. https://doi.org/10.1016/j.cej.2016.12.110

Malik SN, Ghosh PC, Vaidya AN, Mudliar SN (2020) Hybrid ozonation process for industrial wastewater treatment: principles and applications—a review. J Water Process Eng 35:101193. https://doi.org/10.1016/j.jwpe.2020.101193

Marco-Urrea E, Radjenović J, Caminal G, Petrović M, Vicent T, Barceló D (2010) Oxidation of atenolol, propranolol, carbamazepine and clofibric acid by a biological Fenton-like system mediated by the white-rot fungus Trametes versicolor. Water Res 44(2):521–532. https://doi.org/10.1016/j.watres.2009.09.049

Matilainen A, Sillanpää M (2010) Removal of natural organic matter from drinking water by advanced oxidation processes. Chemosphere 80(4):351–365. https://doi.org/10.1016/j.chemosphere.2010.04.067

Mierzwa JC, Rodrigues R, Teixeira ACSC (2018) UV-hydrogen peroxide processes. In: Suresh A, Rakshit A (eds) Advanced oxidation processes for waste water treatment. Elsevier, pp 13–48. https://doi.org/10.1016/B978-0-12-810499-6.00002-4

Mir-Tutusaus JA, Jaén-Gil A, Barceló D, Buttiglieri G, Gonzalez-Olmos R, Rodriguez-Mozaz S, Caminal G, Sarrà M (2021) Prospects on coupling UV/H2O2 with activated sludge or a fungal treatment for the removal of pharmaceutically active compounds in real hospital wastewater. Sci Total Environ 773:145374. https://doi.org/10.1016/j.scitotenv.2021.145374

Mohapatra S, Huang CH, Mukherji S, Padhye LP (2016) Occurrence and fate of pharmaceuticals in WWTPs in India and comparison with a similar study in the United States. Chemosphere 159:526–535. https://doi.org/10.1016/j.chemosphere.2016.06.047

Nguyen LN, Hai FI, Kang J, Price WE, Nghiem LD (2013) Removal of emerging trace organic contaminants by MBR-based hybrid treatment processes. Int Biodeterior Biodegradation 85:474–482. https://doi.org/10.1016/j.ibiod.2013.03.014

Nidheesh PV, Couras C, Karim A V., Nadais H (2021) A review of integrated advanced oxidation processes and biological processes for organic pollutant removal. Chem Eng Commun 1–43. https://doi.org/10.1080/00986445.2020.1864626

Östman M, Björlenius B, Fick J, Tysklind M (2019) Effect of full-scale ozonation and pilot-scale granular activated carbon on the removal of biocides, antimycotics and antibiotics in a sewage treatment plant. Sci Total Environ 649:1117–1123. https://doi.org/10.1016/j.scitotenv.2018.08.382

Parida VK, Saidulu D, Majumder A, Srivastava A, Gupta B, Gupta AK (2021) Emerging contaminants in wastewater: a critical review on occurrence, existing legislations, risk assessment, and sustainable treatment alternatives. J Environ Chem Eng 9(5):105966. https://doi.org/10.1016/j.jece.2021.105966

Peltier S, Cotte M, Gatel D, Herremans L, Cavard J (2003) Nanofiltration: improvements of water quality in a large distribution system. Water Supply 3(1–2):193–200. https://doi.org/10.2166/ws.2003.0103

Prado M, Borea L, Cesaro A, Liu H, Naddeo V, Belgiorno V, Ballesteros F (2017) Removal of emerging contaminant and fouling control in membrane bioreactors by combined ozonation and sonolysis. Int Biodeterior Biodegradation 119:577–586. https://doi.org/10.1016/j.ibiod.2016.10.044

Qian M, Yang L, Chen X, Li K, Xue W, Li Y, Zhao H, Cao G, Guan X, Shen G (2020) The treatment of veterinary antibiotics in swine wastewater by biodegradation and Fenton-like oxidation. Sci Total Environ 710:136299. https://doi.org/10.1016/j.scitotenv.2019.136299

Qiao J, Xiong Y (2021) Electrochemical oxidation technology: a review of its application in high-efficiency treatment of wastewater containing persistent organic pollutants. J Water Process Eng 44:102308. https://doi.org/10.1016/j.jwpe.2021.102308

Rayne S, Carey S, Forest K (2005) Evidence for tin inhibition of enhanced biological phosphorus removal at a municipal wastewater treatment plant. J Environ Sci Heal Part A 40(3):535–551. https://doi.org/10.1081/ESE-200046557

Rempel A, Gutkoski JP, Nazari MT, Biolchi GN, Cavanhi VAF, Treichel H, Colla LM (2021) Current advances in microalgae-based bioremediation and other technologies for emerging contaminants treatment. Sci Total Environ 772:144918. https://doi.org/10.1016/j.scitotenv.2020.144918

Ricky R, Shanthakumar S (2022) Phycoremediation integrated approach for the removal of pharmaceuticals and personal care products from wastewater—a review. J Environ Manage 302(Part A):113998. https://doi.org/10.1016/j.jenvman.2021.113998

Rodarte-Morales AI, Feijoo G, Moreira MT, Lema JM (2012) Biotransformation of three pharmaceutical active compounds by the fungus Phanerochaete chrysosporium in a fed batch stirred reactor under air and oxygen supply. Biodegradation 23:145–156. https://doi.org/10.1007/s10532-011-9494-9

Rodríguez-Nava O, Ramírez-Saad H, Loera O, González I (2016) Evaluation of the simultaneous removal of recalcitrant drugs (bezafibrate, gemfibrozil, indomethacin and sulfamethoxazole) and biodegradable organic matter from synthetic wastewater by electro-oxidation coupled with a biological system. Environ Technol 37(23):2964–2974. https://doi.org/10.1080/09593330.2016.1172669

Ross P, Maynard K (2021) Towards a 4th Industrial Revolution. Intell Build Int 13(3):159–161. https://doi.org/10.1080/17508975.2021.1873625

Rout PR, Zhang TC, Bhunia P, Surampalli RY (2021) Treatment technologies for emerging contaminants in wastewater treatment plants: a review. Sci Total Environ 753:141990. https://doi.org/10.1016/j.scitotenv.2020.141990

Saidulu D, Gupta B, Gupta AK, Ghosal PS (2021) A review on occurrences, eco-toxic effects, and remediation of emerging contaminants from wastewater: special emphasis on biological treatment based hybrid systems. J Environ Chem Eng 9(4):105282. https://doi.org/10.1016/j.jece.2021.105282

SEDIF (2015) Usine de production d'eau potable de Méry-sur-Oise. https://www.sedif.com/iso_album/brochure_mery_081015.pdf

Shtepa V, Balintova M, Chernysh Y, Chubur V, Demcak S, Gautier M (2021) Rationale for the combined use of biological processes and AOPs in wastewater treatment tasks. Appl Sci 11(16):7551. https://doi.org/10.3390/app11167551

Stefan MI (2017) Advanced oxidation processes for water treatment: fundamentals and applications. IWA Publishing

Teixeira ACSC (2021) Processos Oxidativos Avançados. In: Mancuso PCS, Mierzwa JC, Hespanhol A, Hespanhol I (eds) Reúso de água Potável como Estratégia para a Escassez. Manole, pp 224–243

Tiwari B, Sellamuthu B, Ouarda Y, Drogui P, Tyagi RD, Buelna G (2017) Review on fate and mechanism of removal of pharmaceutical pollutants from wastewater using biological approach. Bioresour Technol 224:1–12. https://doi.org/10.1016/j.biortech.2016.11.042

Tran NH, Urase T, Ngo HH, Hu J, Ong SL (2013) Insight into metabolic and cometabolic activities of autotrophic and heterotrophic microorganisms in the biodegradation of emerging trace organic contaminants. Bioresour Technol 146:721–731. https://doi.org/10.1016/j.biortech.2013.07.083

Vasiliadou IA, Molina R, Pariente MI, Christoforidis KC, Martinez F, Melero JA (2019) Understanding the role of mediators in the efficiency of advanced oxidation processes using white-rot fungi. Chem Eng J 359:1427–1435. https://doi.org/10.1016/j.cej.2018.11.035

Verlicchi P, Al Aukidy M, Galletti A, Petrovic M, Barceló D (2012) Hospital effluent: investigation of the concentrations and distribution of pharmaceuticals and environmental risk assessment. Sci Total Environ 430:109–118. https://doi.org/10.1016/j.scitotenv.2012.04.055

Viancelli A, Michelon W, Rogovski P, Cadamuro RD, de Souza EB, Fongaro G, Camargo AF, Stefanski FS, Venturin B, Scapini T, Bonatto C, Preczeski KP, Klanovicz N, de Oliveira D, Treichel H (2020) A review on alternative bioprocesses for removal of emerging contaminants. Bioprocess Biosyst Eng 43(12):2117–2129. https://doi.org/10.1007/s00449-020-02410-9

Wang J, Zhuan R (2020) Degradation of antibiotics by advanced oxidation processes: an overview. Sci Total Environ 701:135023. https://doi.org/10.1016/j.scitotenv.2019.135023

Wang DK, Elma M, Motuzas J, Hou W-C, Xie F, Zhang X (2017) Rational design and synthesis of molecular-sieving, photocatalytic, hollow fiber membranes for advanced water treatment applications. J Memb Sci 524:163–173. https://doi.org/10.1016/j.memsci.2016.10.052

Wang Y, Ji X-M, Jin R-C (2021) How anammox responds to the emerging contaminants: status and mechanisms. J Environ Manage 293:112906. https://doi.org/10.1016/j.jenvman.2021.112906

Wijekoon KC, Hai FI, Kang J, Price WE, Guo W, Ngo HH, Nghiem LD (2013) The fate of pharmaceuticals, steroid hormones, phytoestrogens, UV-filters and pesticides during MBR treatment. Bioresour Technol 144:247–254. https://doi.org/10.1016/j.biortech.2013.06.097

Xie Q, Xue C, Chen A, Shang C, Luo S (2020) Phanerochaete chrysosporium-driven quinone redox cycling promotes degradation of imidacloprid. Int Biodeterior Biodegradation 151:104965. https://doi.org/10.1016/j.ibiod.2020.104965

Yang K, Lu J, Jiang W, Jiang C, Chen J, Wang Z, Guo R (2017) An integrated view of the intimate coupling UV irradiation and algal treatment on antibiotic: Compatibility, efficiency and microbic impact assessment. J Environ Chem Eng 5(5):4262–4268. https://doi.org/10.1016/j.jece.2017.08.028

Zhang J, Xiao K, Liu Z, Gao T, Liang S, Huang X (2021) Large-scale membrane bioreactors for industrial wastewater treatment in China: technical and economic features, driving forces, and perspectives. Engineering 7(6):868–880. https://doi.org/10.1016/j.eng.2020.09.012

Zhao X, Chen ZL, Wang XC, Shen JM, Xu H (2014) PPCPs removal by aerobic granular sludge membrane bioreactor. Appl Microbiol Biotechnol 98(23):9843–9848. https://doi.org/10.1007/s00253-014-5923-0

Zhao H, Zhang D, Du P, Li H, Liu C, Li Y, Cao H, Crittenden JC, Huang Q (2015) A combination of electro-enzymatic catalysis and electrocoagulation for the removal of endocrine disrupting chemicals from water. J Hazard Mater 297:269–277. https://doi.org/10.1016/j.jhazmat.2015.04.063

Zhao Y, Qiu Y, Mamrol N, Ren L, Li X, Shao J, Yang X, van der Bruggen B (2021) Membrane bioreactors for hospital wastewater treatment: recent advancements in membranes and processes. Front Chem Sci Eng. https://doi.org/10.1007/s11705-021-2107-1

An Innovative and Effective Industrial Wastewater Treatments: A Brief History and Present Scenario

Pooja M. Patil, Rachna R. Ingavale, Abhijeet R. Matkar, Sangchul Hwang, Ranjit Gurav, and Maruti J. Dhanavade

1 Introduction

Earth is called a Blue Planet due to the presence of water that covers 71% of the globe and makes up 65% of human bodies. Water is essential to every living creature, and clean water is meant for drinking, local and industrial use, recreational activities, and many more. Polluted water loses its feasibility for the above-mentioned uses incurring economic and aesthetic damage. It may also lead to a heavy impact on the health of plants, animals, human beings, and aquatic life. The main water resources are either rivers, streams, lakes, and the industrial effluent contaminates these natural reservoirs (Sonune and Ghate 2004). The noticeable adverse impacts of releasing effluent from industrial processes into the ecosystems damages the health of the public, therefore, industrial wastewater is an urgent issue to be solved (Kong et al. 2019).

The original version of this chapter was revised: Author's proof corrections have been updated. The correction to this chapter is available at
https://doi.org/10.1007/978-981-99-2598-8_21

P. M. Patil · R. R. Ingavale
Department of Environment Management, Chhatrapati Shahu Institute of Business Education and Research, Kolhapur, India

A. R. Matkar
Department of Mechanical Engineering, D. Y. Patil College of Engineering and Technology, Kolhapur, India

S. Hwang · R. Gurav (✉)
Ingram School of Engineering, Texas State University, San Marcos, TX 78666, USA
e-mail: ranjit.g@txstate.edu

M. J. Dhanavade (✉)
Department of Microbiology, Bharati Vidyapeeth's Dr. Patangrao Kadam Mahavidyalaya, Sangli, India
e-mail: maruti.dhanavade@bharatividyapeeth.edu.in

M. P. Shah (ed.), *Advanced and Innovative Approaches of Environmental Biotechnology in Industrial Wastewater Treatment*,
https://doi.org/10.1007/978-981-99-2598-8_10

In developing countries growing population demands more manufactured products and the need is fulfilled by various types of industries. The backbone of every country is entirely dependent on industrialization. However, as the useful products are manufactured proportionate amount of waste is generated along with certain other harmful by-products which are polluting the environment. Industries having processing units contribute the maximum to water pollution and other issues (Aderibigbe et al. 2017). The contaminants set out from industrial liquid waste are the prime cause of many waterborne diseases. Surprisingly, some studies from developing countries have reported industrial liquid waste responsible for shorter life expectancy. Also, aquatic organisms are the most susceptible creatures who suffer if untreated or poorly treated waste water is released into any water reservoir. Liquid waste is released from a variety of sections like sanitation, manufacturing units releasing process waste, equipment section discharging wash water, and relatively less contaminated water from cooling, heating, and evaporation sections. In general, liquid waste from industry comprises both inorganic and organic components which are the main reason for high dissolved solids and high biochemical oxygen demand (BOD) (Barakat 2011). The exponential growth in industries, modernization, and urbanization are leading causes for the discharge of liquid waste from industrial processes. Industries such as chemical, foundry, petrochemical, sugar, textile, paper, and pulp contribute to the maximum pollutant release in water bodies and hence become the prime concern for global environmental issues (Khalaf 2016). Industrialization is also leading to the deterioration of the available freshwater resources in both quality and quantity values. Contrarily, industrialization is the foundation of the development and economic growth of society. But the problem associated is improper handling of waste produced from industries that are disposed without any proper treatments. In the context of low-income countries, the study reveals that only 8% of all the liquid waste generated is treated. Similarly, 80% of the overall liquid waste formed over the globe is sent into the water reservoirs without any adequate treatment (Fito et al. 2019).

In general, water pollution can be defined as an undesirable change in the physical, chemical, and biological properties of natural water bodies due to the discharge of various unwanted elements (Rathoure and Dhatwalia 2016). The formed mixed pollutants will hamper the normal functioning of aquatic plants, animals, and their habitats and ultimately the health of human beings. Sources of water pollution can be classified as point sources and non-point sources. A point source can be referred to as a single source of pollutant emission such as effluent discharge from industry into the water reservoir. The non-point source can be considered when the entry of pollutants is observed from several sources. The industrial sector is a key contributor to the water pollution. Enormous water is required in manufacturing units for production and cooling towers. The water used in the production process is ultimately turned into an effluent that contains several toxic elements (Sharma 2015). Hence, the treatment of such liquid waste has emerged as a prime issue to protect the available water resources. Various measures are taken to prevent the pollution of water reservoirs due to the release of industrial liquid waste. Prioritized methods include recycling wastewater which has gained global attention in the context of sustainable development. Improved water management, zero liquid discharge, and efficient treatment methods are some important steps undertaken for treating the water. Several physical, biological, and chemical methods have been reported for dealing with pollutants present in the effluent. These methods include flocculation, sedimentation, solvent extraction, ion exchange, carbon adsorption, ion exchange, precipitation, membrane filtration, phytoremediation, etc. Usually, these treatment methods are used in combination to achieve the desired water value before release in a possible economic way (Crini and Lichtfouse 2019). Along with the treatment of liquid waste, emphasis is given to contamination control measures. Three options can be followed for controlling industrial wastewater. Firstly, controlling the wastewater at the point itself i.e., at the generation site. The second is to treat the liquid water before disposing into the municipal sewer system. Lastly, completely treating the generated effluent and reusing it at the plant site itself or discharging it into the receiving water bodies (Sonune and Ghate 2004).

Industrial effluent released from industrial plants and manufacturing processes may signify, collectively, an imperative portion of municipal liquid waste and must be deliberated for effective liquid waste treatment plant operation. In some locations, industrial wastewater discharge is collected with other municipal liquid waste and the combined wastes are treated. In other instances, industries may provide some pre-treatment or partial treatment of their liquid waste preceding to the municipal septic tank. In the rest circumstances, the properties and volume of the industrial waste are such that separate collection and disposal are necessary. Industrial wastewaters vary widely in composition, strength, flow, and volume, depending on the specific industry or manufacturing establishment in the community (Crini and Lichtfouse 2018).

2 Crucial Contaminants and Probable Causes of Industrial Wastewater

Industrial waste is very difficult to generalize as compared to domestic sewage. The characteristics of industrial effluent vary considerably with the type of industry, e.g., the food industry has completely different characteristics of effluent from that of the petrochemical industry. Surprisingly, industries producing the same end product may have different plants, and thus generating effluent also does vary. The pollutants comprised the raw materials, process chemicals, process intermediates, final products, process by-products, and impurities in raw materials. These pollutants and their effects on the environment are broadly discussed as follows:

(a) **Organic pollutants**: These mainly include products of incomplete combustion (PCDD/Fs and PAHs) and pesticides like mirex, toxaphene, and hexachlorobenzene. When these organic pollutants enter into the water bodies it causes depletion in the oxygen content and imparts heavy loads on the treatment plants having a biological treatment section.
(b) **Inorganic pollutant**: These include elements such as chlorides, carbonates, nitrogen, etc. Such elements may sometimes make the water unfit for various desired use and also boost the growth of undesirable micro plants in the receiving water bodies.
(c) **Acids or alkalis**: The presence of such elements makes the water body unsuitable for the growth of fish and other aquatic organisms. It may also cause severe complications while treating wastewater.
(d) **Toxic elements**: Mixing toxic elements from industrial effluents into water bodies can have a hazardous impact on the fauna and flora. The workers engaged in the treatment of effluent consisting of toxic elements because it may pose a serious threat to health if it is untreated and mixed in natural water bodies.
(e) **Color**: Effluent may impart color due to substances like dyes which are used in many industries. Chemical dyes are highly toxic and their presence impacts the aesthetic view to the water body.
(f) **Oil and Grease**: A layer of oil and grease over the surface of effluent makes it difficult for further treatment. The self-purification system of the water body may also get hampered.
(g) **Other floating materials**: These floating materials also impact the effluent aesthetically as well as create difficulty in further processes (Meng et al. 2017).

There are various industrial sources releasing effluents and thus more focus is given to their treatment. Industries discharging a heavy load of pollutants include distilleries, petrochemical, coffee processing, food processing, meat and seafood processing, tanneries, sugar industry, paper, and pulp industry, acid mine drainage, dairy and supplementary units, textile, etc. Industrial effluent has varied components concerning biodegradability as well as toxicity whereas municipal wastewater has consistent composition. The main characteristics of industrial effluent are its high chemical oxygen demand (COD) and organic content, high acidity or alkalinity, salts, turbidity, color, nutrient load, colloids, total suspended solids, and specific toxic contaminants (Kataki et al. 2021).

Characteristics and quantity of industrial effluent are typically based on the type of industry from where the effluent originates. It may have biodegradable or non-biodegradable elements or may contain compounds recalcitrant to treatment or consist of organic synthetic elements or heavy metals. Effluent can also contain a heavy load of oxygen-demanding wastes, pathogens, nutrients that either stimulate or lead to excessive plant growth, inorganic chemicals, minerals, and sediments. The management of industrial effluent deals with all types of waste from industries during pre-production, on-site, or post-production activities and occasionally may lengthen up to end-users of customers. Industrial waste management examines the broader context of waste treatment directly or indirectly originating from industries and may include corporate sustainability, environmental impact, consideration of government policy and regulations, recycling, containment, handling, and transport, centralized compared to on-site treatment, technologies, economics, avoidance, and reduction. Industrial wastewater treatment deals with processes that include treating liquid wastes to remove undesirable by-products. After treatment, the treated industrial wastewater (or now called effluent) might be reused or discharged to a wholesome septic tank or in the environment water bodies either directly or through a water canal. Most industries produce large volumes of wastewater continuously which after treatment (effluent) is released into the aquatic environment. Recent trends have been to reduce such production (waste reduction) or to recycle treated wastewater (recycling or reuse) within the production process.

Areas with developing economies usually experience exhausted waste collection services and uncontrolled and inadequately managed dumpsites. Problems with governance make the situation more complicated. Waste management in these countries and their cities is an ongoing challenge because of weak institutions, rapid urbanization, and chronic under-resourcing. All these challenges, together with a lack of understanding of the various approaches that subsidize the hierarchy of waste minimization, affect waste treatment. On the other hand, industrial waste management is still a challenge in the developed world, although most of the developed countries and cities have well-established and functioning waste management systems and regulations. The treatment and management procedures of various industrial activities are essential. The related health problems and environmental impacts due to poor management of industrial waste have to be well understood and solved (Awuchi et al. 2020) (Table 1).

3 Characteristics of Various Industrial Wastewater

The characteristics and quantity of industrial effluent are typically based on the type of industry from where the effluent originates. The constituents present in the effluent may vary extremely in properties due to biodegradable or non-biodegradable elements or compounds recalcitrant to treatment. It may consist of organic synthetic elements or heavy metals, oxygen-demanding wastes, pathogens, and nutrients that either stimulate or lead to excessive plant growth, inorganic chemicals, minerals, and sediments.

Table 1 Effluent characteristics of various industries

Sr. No.	Type of industry	Effluent characteristics	Recommended treatment
1.	Textile Industry	High BOD, high suspended solids, dissolved solids and total solids, strong color, highly alkaline, Chlorides	Chemical methods: Equalization, neutralization, chemical precipitation, chemical, and biological oxidation, caustic recovery Biological methods: Trickling filters, activated sludge process, waste stabilization ponds, microbial, and phytoremediation Adsorption: Biochar, activated carbon, nanocomposites, biomass powder
2.	Food Industry	High BOD, COD, organic matter, suspended solids, color, flavoring and coloring agents, acids, alkalis	Biological treatments: Activated sludge process, trickling filters, anaerobic digesters, anaerobic lagoon, stabilization ponds
3.	Dairy Industry	Alkaline pH, high total dissolved solids, suspended solids, High BOD, traces of nitrogen, phosphorous, and chlorides, high oil, and grease	Grease trap Biological treatments: Trickling filters, activated sludge plants, oxidation ditch, aerated lagoons, waste stabilization pond
4.	Paper and pulp Industry	High BOD, high COD/BOD ratio, high sodium, strong color, highly alkaline, total solids, and suspended solids	Color removal techniques like activated carbon, lime treatment recovery of chemicals, clarification, flocculation Biological treatment: Stabilization ponds, aerated lagoons, anaerobic lagoons, activated sludge process
5.	Sugar Industry	Low pH, high BOD, and COD, high volatile solids, total solids, total suspended solids, total nitrogen, a strong odor	Biological treatments: activated sludge process, trickling filters, anaerobic digesters, anaerobic lagoon, stabilization ponds
6.	Distillery Industry	High COD and BOD, strong color, heavy load of sulphates and potassium, high suspended, dissolved solids, and total solids, total nitrogen	Biological treatments: the combination of aerobic and anaerobic lagoons, by-product recovery for nutrient-rich animal feed and potassium-rich fertilizers

(continued)

Table 1 (continued)

Sr. No.	Type of industry	Effluent characteristics	Recommended treatment
7.	Tannery Industry	High BOD, strong color and odor, presence of chromium, sulphides, and lime, high salt content, high dissolved solids	Chemical treatment: sedimentation, coagulation, neutralization Biological treatment: activated sludge process, chromium reduction, trickling filters, oxidation pond, anaerobic lagoons

3.1 Textile Industry

The textile industry wastewater differs with types of fibers viz. wool, cotton, synthetics, and regenerated. The composition and characteristics of wastewater generated through textile industries are highly dependent upon the type of fiber being used in the mill as each fiber used undergoes a different sequence of treatment during the process. Therefore, pollutants present in the textile industry effluent may be the result of certain chemicals used for the process or due to natural impurities from the fibers.

3.1.1 Cotton Textile Mill Waste

The liquid waste from cotton textile mills is generated during the process of sizing, weaving, desizing, scouring, bleaching, mercerizing, dyeing, printing, and finishing. In the slashing or sizing process, starch and other sizing chemicals are used to impart strength to the yarn. At this stage, a variable amount of BOD load is present in wastewater depending upon the chemicals used.

Weaving is the next step after sizing or slashing. The finished cloth is obtained after giving certain treatments to the manufactured cloth. In the initial stage, dilute H_2SO_4 or enzyme is used to perform the desizing process to remove the chemicals used during the sizing process. On complete washing, the acid is removed from the cloth that hydrolyses the starch. During this operation of desizing, a sizable amount of BOD is contributed to the generated effluent. In fact, at this stage, BOD is the highest that any other operation in the textile industry.

In the scouring process, the cloth is treated with 1–3% soda ash, caustic soda, sodium peroxide, sodium silicate, etc. Waxes, grease, natural fats, and certain impurities are removed in this operation. Hence a high amount of oil and grease is present in the effluent generated at this unit. The next stage is bleaching, where chemicals like hydrogen peroxide, alkaline hypochlorites, and chlorine are used for bleaching. High pH is a major characteristic of bleaching unit wastewater. After bleaching, mercerizing of clothes is done to increase the affinity for dyeing agents, surface luster, tensile strength, and reduce ability to shrink. Cold concentrated caustic soda is used for this treatment which is later followed by thorough washing with water and dilute acid to remove the remaining alkalies. Hence during this mercerizing, the effluent generated high levels of alkalinity. But usually, effluent from this unit is recycled.

Dyeing is the next step where color is imparted to the fiber. Various dyes are used for this process like vat, direct, naphthol, developed, sulfur or aniline black, and chemicals like chromium. The wastewater generated has high color and BOD along with suspended and dissolved solids. Therefore, this effluent has dyestuff of natural and artificial origin, wetting materials, gum thickeners, pH buffers, dye retardants, etc. Hence, the wastewater contains flocculants which are polymer-based, settling agents, BOD, color, COD, phenol, oil, grease, TSS, sulphides, and heavy metals like chromium, lead, zinc, copper, etc.

In textile industries, other processes include carpet manufacturing where yarn manufacturing, wool cleaning, and finishing and fabric finishing are common practices. Fabric finishing also includes waterproofing and flame-retardant clothes, resin treatment, etc. In wool processing industries, residues of insecticides in fleece are common while treating its effluent. Sometimes, animal fats are also present in such effluent which can be further recovered or recycled for tallow production (Vithanage et al. 2017).

3.2 Food Industry

Raw materials like fruit, vegetables, and milk are processed in food industries to make by-products and increase their accessibility to consumers. In the food industry, there is a requirement for a large amount of water for food processing. During food processing, water is mostly used for operations like production, cleaning, sanitizing, materials transport, and cooling. There are numerous food industries like dairy, beverages, distilleries, sweets, and snacks (Aderibigbe et al. 2017). In the food industries wastewater is found to be generated from the departments such as production sanitizing, cleaning, chilling, and constituents transport. The high organic content of this wastewater makes them biodegradable and non-toxic. This has resulted in high concentrations of BOD, COD, and suspended solids in the wastewater. Hence, wastewater generated from food industries is different from municipal sewage.

Slaughter-houses wastewater is also the most hazardous, harmful liquid waste released into the surroundings as stated by the United States Environmental Protection Agency (USEPA). In particular, the meat-processing industry uses 24% of the total freshwater consumed by the food and beverage industry and up to 29% of the consumed by the agricultural sector worldwide (Malollari et al. 2019). The wastewater from animal slaughter and meat processing units has very strong organic wastes from body fluids such as gut contents and blood. Ammonia, oil and grease, nitrogen, suspended solids, coliform bacteria, BOD, etc., are the common pollutants noted.

Effluents from the fruit, vegetable, and meat industries have a varied range of pH and BOD in composition. Seasonal variation in nature and post-harvesting of fruits, and vegetables add to the difference in the composition of generated effluent water. A huge amount of quality water is essential for the processing of food from raw materials. Ultimately, this water finds its way as an effluent. The washing of vegetables and fruits contributes to the maximum wastewater. It has a heavy load of particulate matter, dissolved organic matter, surfactants, etc. Some processed food industries have effluent rich in plant organic material, coloring agents, flavoring and preserving materials, salts, or alkalis, and sweetners used in beverages which have an important part during the processing unit. Fats and oils are also present in high quantities. Wastewater is also generated during the processes of plant clean up, conveying of materials, bottling, and washing of products.

The research revealed that effluents from food industries include extremely high levels of COD, BOD, phosphate, and nitrate. These parameters are above the limits set by WHO. They propose that these effluents are not appropriate to be discharged into a natural water body without any treatment. However, their rich organic content open the way for their biological treatment. As such, the purification performance of *Eichhornia crassipes* and *Panicum maximum* for the treatment of effluents showed good removal efficiency of PO_4^{3-}, COD, BOD and nitrate (Noukeu et al. 2016).

3.3 Dairy Industry

The dairy industry is vitally significant to India and subsidizes 35% of the total Asian milk. In various stages, clean water is used for various dairy operations, such as cleaning, milk processing, packaging, and washing of the milk tankers that is recognized as dairy effluent. Growing demand for milk products has led to the development of dairies of different sizes. Water used in the dairy industry is a ratio of 1:10 (water: milk) per liter of milk (Qasim and Mane 2013). Milk is collected from producers at the dairies and either it is supplied directly for consumption or other milk products are produced. In the dairy industry, liquid waste originates from units like receiving stations, bottling plants, butter plants, casein plants, cheese plants, dried milk plants, condensed milk plants, and ice cream plants.

At the receiving station, milk is collected usually from the farmers, and it is stored in a large container from where it is sent either for bottling or processed products. The empty cans and containers are rinsed, washed, and sterilized for further use. The bottling plant receives the milk from the receiving station and undergoes the processes such as cooling, clarification, filtration, pasteurization, and bottling. At both, these operations liquid waste is generated during the rinsing and washing of cans, bottles, and equipment and thus have milk residues and chemicals that are used for cleaning purposes.

In a cheese plant, either whole milk or skimmed milk is pasteurized, stored in a container, and then cooled where lactic acid-producing bacterial culture and rennet are added to it. This causes the separation of casein from the milk in the form of curd. The whey is then withdrawn and the formed curd is subjected to compression to drain out the extra whey from it. Eventually, some other ingredients are also added to the material, and a product called cheese is obtained. The liquid waste in this section consists primarily of whey and wash water used for cleaning and washing of vats and equipment used during cheese processing activity.

In a creamery process, whole milk is preheated at 30 °C to separate the cream from the milk. In a butter plant, the cream is pasteurized, and some selected acid and bacterial cultures are added for ripening. Later at 7–10 °C, it is churned to obtain the butter. The buttermilk is now drained off and obtained butter is also washed and then after standardization it is sent for packaging and sale. The liquid waste generated in this section includes buttermilk residues and wash water used to clean the churns and other equipment and a small amount of butter is also recorded in the generated liquid waste.

The skimmed milk obtained is sent either for human consumption after packaging or for further processing in the dairies to obtain non-fat milk powders. Milk powder is obtained on evaporation and then it is dried either by roller or spray method. The dry milk plant waste consists of wash water used to clean the containers and equipment. The soured or spoiled milk is also processed to obtain casein which is used in the production of plastics. Some mineral acids are added for the precipitation and coagulation of casein. The waste from this section contains whey, cleaning powder, and the chemicals utilized for the precipitation.

From all the above milk manufacturing units, some amount of uncontaminated cooling water comes as a waste. In general, the characteristics of dairy liquid waste have alkaline pH, high total dissolved solids and suspended solids, and high BOD and COD. It also contains total nitrogen, chlorides, and phosphorous, and some amount of oil and grease.

3.4 Paper and Pulp Industries

In Canada, pulp and paper industries contribute to producing 50% of pulp and paper wastes that are discarded into water bodies. Although the paper and pulp industries have a key role in the country's economy, the wastewater released from these industries needs to be properly treated to minimize their hazardous effects on the environment. More than 759 pulp and paper industries are established in India till date. It is well known that Indian pulp and paper industries are highly water-intensive, consuming about 100–250 m^3 freshwater/ton of paper for processing and generating a subsequent 75–225 m^3 wastewater/ton of paper produced (Zainith et al. 2019).

Paper is manufactured from pulp, and cellulosic materials like wood, bagasse, bamboo, recycled paper, and cotton are used to obtain pulp. Paper mills are usually integrated with pulp-generating units and paper-making units. The pollution load from pulp mills is generally high as compared to paper-making units. Pulp making is a unique process of digesting raw cellulosic material like wood, bagasse, etc., which is used for paper making. Digestion is carried under high temperatures and pressure to separate cellulose fibers from raw cellulosic material and lignin, and other non-cellulosic materials are dissolved. Following are the types of digestion viz. kraft process, sulphite process, and alkali process. In the kraft process, chemicals like sodium sulphate, sodium sulphide, and sodium hydroxide are used for digestion. Magnesium or calcium bisulphite and sulfurous acid are used for the digestion of raw material in the sulphite process. In the alkali process, agents like sodium hydroxide or lime is used for the making of pulp (Barbusiński and Filipek 2001). The waste generated from the pulping unit is termed "black liquor". This can be recovered for digesting chemicals as it contains a high amount of lignin and unused digesting chemicals. Thus, the cellulosic materials are recovered after digesting process from the "black liquor" and washed. These are later dewatered, and the effluent generated now is known as "brown stock wash". It is also called "unbleached decker waste" as this process is carried out in a cylindrical screen called a "decker". Further, a bleaching treatment is given to the washed cellulosic fibers. Chlorine, caustic, and, hypochlorite are used for the bleaching process in three successive stages. The wastewater released from the caustic treatment has a very strong color while light yellow colored wastewater is released from chlorine and hypochlorite treatment stages. The bleached pulp is added to chemical fillers and dyes, etc., to make it a refined pulp. This refined pulp is sent into a series of rollers to obtain a paper as a final product. The wastewater squeezed from the roller is called "white water" which is recycled for the wet chipping process. Wastewater from this unit may have dioxins (including 2, 3, 7, 8-TCDD), chloroform, furans, COD, and phenols.

The paper and pulp industry effluent have high color and high COD which is mainly due to the presence of lignin which is obtained from raw material. High BOD, suspended solids, high COD/BOD ratio, alkaline pH, and high total suspended solids are some other influencing parameters of paper and pulp industry effluent.

3.5 *Tannery Industry*

The tanning industry is one of the oldest industries, which involves the peculiar process of converting animal skins or hides into good-quality leather. According to statistical data from all over the globe, every year 300 million pieces of cowhide are handled in the world leather markets which consequences that 300 million tons of tannery wastewater annually generated. In China, the total annual water consumption of the tannery industry is about 1.4 hundred million tons, and 1.2 hundred million tons of water is discharged into the surrounding. Therefore, pollution is one of the most pressing problems, restricting the development of the tannery industry. In Brazil, the yearly water utilization of tanneries signifies the water depletion of 5.5 million residents where the water consumption of residents is 150 L/day (Zhao and Chen 2019).

In the modern tanning industry, this process undergoes three stages as follow:

(1) Hides preparation: Animal skin or hides are washed thoroughly with water to remove the dirt, and salt which is applied earlier as a preservative. Later this hide is soaked for the next few days in freshwater having sodium chloride and "Antimucin" as a preservative agent. Wastewater generated from the soaking of hides is rich in soluble proteins, dirt, and common salt. This spent liquor decomposes very easily thus generating a very strong and offensive odor.

Later this soaked hide is subjected to liming where a paste of lime and sulphide is applied over the hide. This helps in the easy mechanical removal of hairs and flesh from the skin in wooden containers called vats. This is followed by the processes of deliming and bating. Reduction in pH and swelling along with the removal of protein-degrading products is done in the bating process. The spent lime liquor and the spent bate liquor are highly loaded with dissolved and suspended lime, organic and ammonia nitrogen, proteins and their degradation products, sulphides, fatty contents, and lime sludge. Thus, it imparts high alkalinity and moderate BOD, and high ammonia–nitrogen content to the effluent from this process.

(2) Tanning: In this stage, the hide is prepared to make it soft after drying and non-decayable. Tanning agents used are either vegetable substance tanning or chrome tanning. In vegetable tanning, substances like bark, wood, nut, etc., having natural tannins are used while in chrome tanning, inorganic chromium salts are used as a tanning agent. For heavy leathers, vegetable tanning is preferred while chrome tanning is done for light leathers.

The wastewater from the tanning unit has a relatively strong color, high COD, and BOD load, organic acids, inorganic salts, mineral acids, and chromium salts.

(3) Finishing: It consists of stuffing and fat liquoring where tanned leather is incorporated with oil and grease and thus becomes soft, pliable, and resistant to tearing. Similarly, synthetic dyeing agents are used for the dyeing of leather (Miklos et al. 2018).

3.6 Sugar Industry

The sugar industry is considered to be the most significant type of industry in concern with production and effluent generation. Sugar is produced from raw materials like sugarcane or beetroots. Sugar industry requires enormous water for the production of sugar from these raw materials. Ultimately, the wastewater generation rate is also higher in the industry contributing to a lot of pollutants in the effluent. In India for one ton of sugarcane crushed about 1,000 L of wastewater is produced. The operations from typical cane milling consist of a variable load of pollutants including organic matter which leads to an offensive odor due to its putrefaction. Normally, the harvested sugarcanes are carried to the mill house for sugar production where they are cut into small pieces. A series of rollers are used to crush the cane to obtain juice. Lime is added to the extracted juice and is subjected to clarification by heating where the suspended and colloidal particles are coagulated. The coagulated juice is further clarified to remove the residual solids or sludge from it. The clarifier sludge is passed through filter press and the generated waste is called press mud. The next step is sulphitation where the clarified juice is bleached completely by passing it through SO_2 gas (Kushwaha 2015).

In the next stage, the clarified juice is preheated and concentrated in evaporators and vacuum pans leading to the crystallization of sugar. The partially crystallized juice called massecuite is sent for centrifugation to separate the sugar crystals. The spent liquor generated is called blackstrap molasses and can be used in the fermentation process to produce alcohol, yeasts, etc. The fibrous residue called bagasse is also generated as waste from mill houses. Bagasse is mostly used in boilers as a fuel or sent as a raw material in paper and pulp manufacturing units. In the sugar industry, almost all units generate wastewater. Various operations like floor washing, filter cloth washings, leakages, spillages, blow, etc., contribute to effluent generation. Maximum water is discharged from an operation called splash where enormous water is used to extract maximum juice from the cane while crushing, hence BOD of such effluent is usually high. Other characteristics include the heavy load of COD, high total solids, volatile solids, suspended solids, and total nitrogen (Wan et al. 2013). Cleaning of the milling floor and boiling house divisions like clarifiers, evaporators, vacuum pans, centrifugation, etc., produce a huge volume of liquid waste. Periodical cleaning of heat exchangers and evaporators with NaOH and HCl to eliminate the scales on the hose surface contributes to organic and inorganic pollutant loadings to wastewater. Leakages from pumps, pipelines, and centrifuging houses also contribute to wastewater produced. Besides this, wastewater is also produced from boiler blowdown, spray pond overflow, and from condenser cooling water which is discharged as wastewater when it gets contaminated with cane juice (Kushwaha 2015).

3.7 *Distillery Industry*

The distillery industry uses molasses as a raw material to obtain various industrial-valued products. The products obtained from the distillery industry include rectified spirit, industrial alcohol, absolute alcohol, silent spirit, beverage alcohol, etc. The products are generated by the application of the biochemical process of fermentation by using yeast which utilizes carbohydrates from molasses as a raw material. The process undergoes the following steps. In the first step, raw molasses is diluted with water and the sugar level is brought down to 15%. Later, the pH of this is adjusted to 4.0–4.5 to prevent bacterial growth. The last step is to add nutrients and yeast suspension in the above-diluted molasses in a unit called a fermenter to initiate the process of fermentation in a controlled condition (Fito et al. 2019).

The fermented liquor having alcohol contents is sent for distillation and the alcohol is separated. The waste generated is termed a "spent wash". The spent wash from the distillery is the major contributor to pollution. During this process, yeast sludge is also generated in large amount as a polluting agent. Sometimes the generated yeast sludge is mixed with a spent wash before disposal. The effluent, spent wash, from the distillery industry is primarily characterized by high COD and BOD and dark brown color. Strong color is imparted due to the caramelization of sugars as microbes hardly

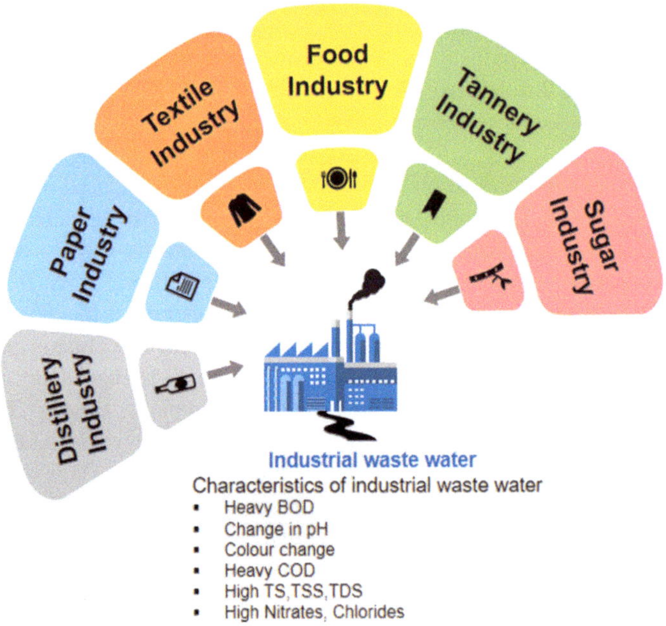

Fig. 1 Industrial wastewater sources, and characteristics

degraded it. Acidic pH, dissolved solids, suspended solids, substantial quantities of nitrogen, sulphates, and potassium are also characteristics of this effluent. Spent wash can be utilized as a good compost due to the abundant presence of nitrogen and potassium. On an average, 10–15 L of spent wash is generated for every liter of ethanol produced (Fito et al. 2019) which has the characteristics of BOD as 40,000–50,000 ppm and COD as 80,000–100,000 ppm (Fig. 1).

4 Effects of Industrial Wastewater on the Environment and Human Beings

Recently water pollution is becoming a serious concern due to the direct release of industrial and sewage water, industrial solid waste, and the mixing of numerous non-point sources into the various water reservoirs. Manufacturing discharge is a crucial cause of water contamination. Depending on the kind of industry, biological and chemical pollutants are straight or accidentally released into the liquid body which causes a severe threat to the atmosphere, and human health (Arihilam and Arihilam 2019). Pharmaceuticals, leather, paint, textile, agriculture, dairy, and dyes industry releases a huge number of organics, heavy metals, polychlorinated biphenyls (PCBs), pesticides, polycyclic aromatic hydrocarbons (PAHs), dioxins, phenolic components, and petrochemicals as effluent. The heavy metals (Cr, Cd, Pb, Hg, As, Ni, Zn, Cu, etc.) containing industrial effluent when reaching into the water body enters the food chain and bioaccumulate, bio magnifies in human bodies and result in organ damage, cancer, developmental retardation, immunological disorders, endocrine disruption, and even death (Crini et al. 2017).

Due to the concentration of biological matter, and contaminants in industrial wastewater, serious issues may arise that may affect the air, soil, and water. The discarded high organics into the aquatic body will degrade quickly by a microorganism which depletes the dissolved oxygen (DO) level and then the water body may get eutrophic. Due to low DO, and high suspended solids, the fauna will suffocate and cause gradual death (Rathoure et al. 2016). The light penetration will get reduced resulting in fish gills clogging. Contaminated water bodies become a proliferation place for hazardous diseases like yellow fever, malaria, chicken guinea, and dengue. A huge load of nitrate, ammonia, and nitrogen in industrial effluents is toxic to aquatic flora, and fauna and also causes methemoglobinemia. Moreover, this water body promotes numerous operational difficulties in water treatment units, like low settle ability of sludge, high BOD and COD demand, etc. Due to industrial wastewater, the physical, chemical, and organic properties of water and soil gets altered which decreases the quality and yield of a crop. The microbial population may also be impacted negatively due to the presence of a pollutant in the industrial wastewater (Patil and Bohara 2020) (Table 2).

5 Preventive Measures to Reduce Industrial Water Pollution

The contamination prevention is defined as a reduction or elimination of the pollutants through improved use of raw materials, water, energy, and other assets, or conservation, or protection of natural resources (Patil et al. 2019a, b). It is a demand for the careful use of capital through energy efficiency, source reduction, re-use of water and input substance during built-up, recycling of water and manufacturing products, and reducing the consumption of liquid. Two significant methods can be used in reducing industrial water pollution, i.e. process changes and product changes. These methods reduce the wastewater volume, toxicity, concentration, and treatment operational cost (Ahmed et al. 2017) (Fig. 2).

Water pollution prevention is the activity that decreases the quantity of hazardous, contaminants, pollutants, or substances in a wastewater course or released into the ecosystem preceding recycling treatment, and disposal. These preventive measures to reduce industrial water pollution reduce public health hazards and protect the environment from contaminants. Pollution prevention aims at high product design, eco-friendly products, and clean and improved technology for manufacturing, wastewater treatment, and disposal. These include recycling off-site, treating solid waste and converting it into a sustainable product, appropriate disposal of toxic or hazardous waste, and reducing the volume (Carboneras et al. 2018).

Table 2 Possible sources, major contaminants in industrial wastewater and their impact on humans and the environment

Possible sources	Pollutants	Impact on human	Effect on environmental
Paper, printing, leather, dyeing, oil refineries, and electroplating industry	Alkali	Dry skin, itching, stomach upset, etc.	Affects aquatic and terrestrial ecosystems, climate change, etc.
Mining, steel, chemical, electroplating, machinery industries, and non-ferrous metallurgy	Acid	Respiratory and lung problems, asthma, chronic bronchitis, etc.	Water pollution, damages crops, a negative impact on aquatic ecosystems
Smelting, electroplating, metal mining, batteries, metal processing, etc.	Cadmium and its compounds	Affect bones, kidneys, liver, etc.	Cadmium is toxic to plants, crops, and microorganisms. It disturbs the ecological balance of nature
Explosives, color alkali, mercury refining industry, mercury chemicals, pesticides, electroplating, instruments, etc.	Mercury and its compounds	Memory loss affects neuromuscular, headaches, etc.	Mercury gets bio accumulate and biomagnifies in the food chain. Contaminate crops, plants, and the environment
Pharmaceutical, metallurgical, pesticides, paint, ore processing, fertilizers, and other industrial	Arsenic, organic phosphorous, chlorine and its compounds	Cancer, nervous problems, skin lesion, even death	Inhibition in plant growth affects the photosynthesis process, and contaminate river, lake, and other water bodies
Glass industry, mining, smelting, electroplating, batteries, metal processing, etc.	Hexavalent chromium and its compounds	Nasal irritation, ulcer, cancer, etc.	Contaminate surface water, decrease the yield and growth of the crops
Coal gas, electroplating, coking, acrylic, acrylonitrile refining industry, metal cleaning, and, gold industry, etc.	Cyanide	Eye irritation vomiting, brain problems, heart failure, etc.	Hazardous to wildlife pollutes the water, kills the aquatic biota

(continued)

Table 2 (continued)

Possible sources	Pollutants	Impact on human	Effect on environmental
Synthetic resins, pharmaceuticals oil refining, dyes, and other industries	Phenol	Eye, skin, nose throat irritation, and damage to the nervous system	Death of plants, animals, birds, and fish. Polluted water affects the quality and quantity of crops
Machinery, food processing, and natural gas processing	Oil	Damage liver, respiratory system, reproductive system, etc.	Pollutes the aquatic ecosystem, stops oxygen receiving to the animals and plants
Leather, chemicals, dyeing, oil refining, and oil processing industry, etc.	Sulfide	Eye, skin, nose throat irritation, vomiting, coma, and even death	Risk to crops, plants, animals, and aquatic organisms. Pollutes the water
Pesticides, smelting, lead paint, gasoline explosion, enamel industry	Lead, organic phosphorus, chlorine, and its compounds	Anemia, brain and kidney damage, etc.	Pollutes the water and soil. Harmful to all living organisms
Textile bleaching and paper-making industry	Free chlorine	Eye, skin, nose throat irritation, affects nasal passage and the respiratory system	Contaminate water, crops, fish, and transfer in living organisms and other animals that consumes the fishes
Radioisotope laboratories, other weapons production, and nuclear industry	Radioactive substances	Affects the eye, skin, heart diseases, cancer, etc.	Bioaccumulation, biomagnification, damage to crops, and plants, contaminating soil water, etc.
Plastics, electricity, and lubricants	Polychlorinated biphenyls (PCBs)	Melanomas cause gall bladder, liver, brain cancer, etc.	High temperature produces dangerous dioxins which cause environmental pollution

In 1980 the U.S. legislation has progressively provided financial incentives to factories and industries which take preventive measures to reduce industrial pollution. They look over the cleanup, liability design, waste generation, resource utilization its efficient use, and disposal criteria (Castillo et al. 2008). Various industrial acts and industrial sections have been amended, and rules and regulations regarding the management and disposal of industrial wastewater have become strict as these factors raise the importance of pollution inhibition in the industrial sector. Prevention of industrial pollution supports nationwide environmental goals and with concurs

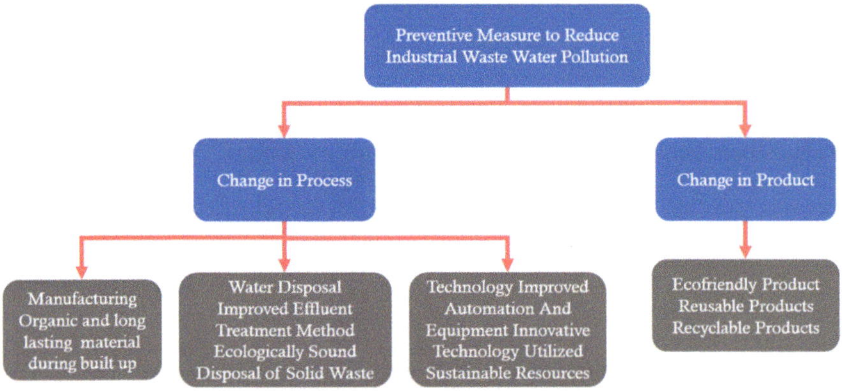

Fig. 2 Represents preventive measures to reduce industrial water pollution

industries' interests. Industries will also have more financial incentives for reducing the toxicity and volume of solid and liquid waste. Reducing wastewater pollution offers upstream benefits as it protects aquatic ecosystems, maintains the food chain and food web conserves water and protects the atmosphere. Furthermore, measures of wastewater effluence anticipation will boost the industry's image in public, improve public health, and overall ecological and conservational benefits (Patil et al. 2019a, b).

6 Conventional Industrial Wastewater Treatments in the Effluent Treatment Plant

6.1 Basic Treatments

For the removal of large particles and solid waste which regularly exist in unprocessed industrial wastewater, a preliminary treatment is done. In the preliminary treatment, the suspended elements, and floating substance that exists in wastewater are eliminated and reduced. Mostly solid particles like raw materials, plastics, glass, papers, etc., are separated by preliminary treatment. For solid waste separation, the industrial wastewater is passed in a filter, and screening is done. After preliminary screening of industrial wastewater then it is transferred to primary treatment (Chiam and Sarbatly 2011).

6.2 Primary Treatment

In primary treatment, physical and chemical treatments are used. In physical treatment flotation and sedimentation methods are used, whereas in chemical treatment flocculation, coagulation, and neutralization methods are carried out for industrial wastewater (Muruganandam et al. 2017). The foremost objective of primary treatment methods is to eliminate organic and inorganic substances that exist in wastewater. Almost 60–65% of oil, lubricant, and grease, 40–50% of biochemical oxygen demand, and 60–70% of total suspended solids are eliminated through primary treatment. In the sedimentation process, organic phosphorus, organic nitrogen, and heavy metals are eliminated. In the process of sedimentation, sludge is settled at the bottom of the chamber, and effluent is transmitted to a different tank. In wastewater treatment, primary treatment methods play a decisive role to reduce the contaminants present in effluent water (Choudhury et al. 2017). In primary treatment, to remove organic and inorganic substances, the clarifiers are applied primarily in industrial wastewater when more organic, inorganic, and colloidal substances are present. The primary treatment alone cannot achieve the superior removal of organic substances and cannot fix the water quality limits, through secondary treatment the organic matter which is present in wastewater can be effectively reduced (Maizel and Remucal 2017).

6.3 Secondary Treatment

Various secondary treatment methods are applied to industrial wastewater management. The treatments include the use of a trickling filter, sludge activation, a stabilization tank, aerated lagoons, and anaerobic lagoons. In secondary treatment, several types of microorganisms are used in a controlled environment. In secondary treatment, residual organic and suspended materials are removed from wastewater. After the secondary treatment methods approximately 30–35% suspended particles, 60–65% dissolved solids, and 85–95% of BOD are removed and only 5–10% colloidal particles are left in treated effluent (Varjani et al. 2020). Further, the effluent water can be processed in tertiary treatment to achieve maximum removal of contaminated pollutants.

6.4 Tertiary Treatment

Numerous procedures are carried out in tertiary treatment methods like denitrification, ion exchange, ultrafiltration, carbon absorption, reverse osmosis, and sand filtration. This tertiary treatment is also known as advanced treatment or the final method. In tertiary treatment, microbial contaminants are destroyed, and left-over residues are removed. After tertiary treatment, the treated water is used for irrigation,

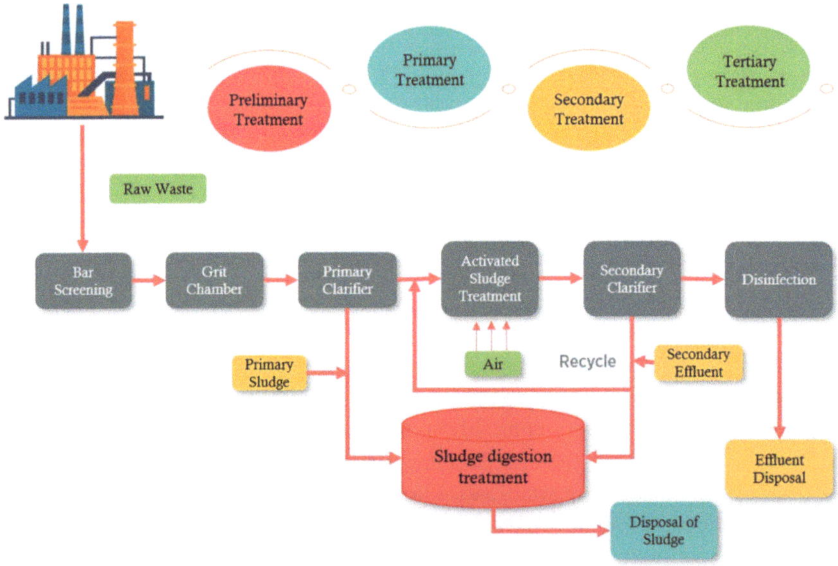

Fig. 3 Conventional treatments applied for industrial wastewater in effluent treatment plant

land application, or disposed of in ponds, rivers, lakes, etc. In tertiary treatment, all contaminants, infectious agents, and total dissolved solids are removed and before discarding in water bodies industries assure that all parameters of industrial treated water are within the permissible limit (Gaied et al. 2019; Arzate et al. 2019) (Fig. 3).

7 Innovative and Effective Technologies for Industrial Wastewater Treatment

7.1 Advanced Physical Treatment for Effluent

7.1.1 Membrane Filtration

The physical methods are mainly applied to remove undesired components from effluent water based on their particle size. In the filtration technique, the membranes provide a great level of systematization, less space is required, integrated configuration permits flexible design, and chemicals are not used. In membrane technology, the main challenge is about changes in characteristics between permeability and layer selectivity. But required energy is high for membrane utilization based on membrane pressure and process layer cleaning (Hube et al. 2020). Furthermore, the functional nanomaterials integration into membranes provides a great prospect that increases the fouling resistance of membrane, mechanical and thermal stability, and removes the pollutant with self-cleaning. In wastewater treatment plants, membrane filtration is mostly used depending on the types of target contaminants present in effluent water. During the effluent treatment, the filtration method can be executed and applied to reduce the pollutant load present in water. Depending on the cut-off size of the membrane the filtration type is used. For viruses, and bacteria removal nano filtrations membranes with aperture size between 10 and 30 μm are known to be ideal (Pronk et al. 2019).

7.1.2 Reverse Osmosis

Reverse osmosis is also known as hyper-filtration where the small particle ions separation is permitted, and it is also known to be the finest filtration. This treatment is used for water purification and enhances the property of water like color and taste. The membrane utilized is semipermeable, the liquid is permitted and purified through that membrane liquidating the pollutants. The wastewater passes, through the membrane, the contaminants are sieved out, and then it is evaporated. Further, cross-stream treatment is applied in reverse osmosis that consistently cleans the membrane (Zhao et al. 2020). In reverse osmosis, the chief thrust is to push the water over the membrane and the utmost pressure is applied by a pump that is seven to ten times greater than ultrafiltration. For microscopic organisms, sugars, salts, fats, proteins, and various particles more than 0.15–0.25 kDa atomic size the reverse osmosis treatment is satisfactory. The ions separation depends on the nature of the charge which interferes with the dissolved contaminants that have charge and will possibly be eliminated through the membrane than those which don't have charge. Reverse osmosis and nanofiltration comprise a diffusion solution mechanism (Anis et al. 2019).

7.2 Advanced Chemical Treatments

The important and most widely used treatments are advanced oxidation, Fenton, ozonation, coagulation, etc. Chemical treatments are usually coupled with membrane techniques or photo-catalysis. A few of the advanced chemical treatments are discussed below.

7.2.1 Advanced Oxidation Process

In the advanced oxidation process, the oxidizing agents are used to oxidize the contaminant or pathogens present in the effluent. For organic contaminant removal, sulfate radicals, and hydroxyl radicals are extensively used. This treatment is ecofriendly which kills the pathogens and converts them into a harmless substance that neither transfers the pathogens into the environment nor generates enormous amounts of sludge. Additionally, this process has numerous advantages, like a speedy rate of reaction and less retention time as compared to conventional methods. For specific contaminants, it has composite chemistry tailored which needs accomplished personnel for system designing. The advanced oxidation process is categorized into various types like ozone-based treatment, electrochemical advanced oxidation, ultraviolet treatment, advanced catalytic oxidation, and advanced photo-oxidation (Parsons 2004).

7.2.2 Advanced Photochemical Degradation

For industrial wastewater treatment and toxins degradation, the ultraviolet (UV) radiation are broadly used. Based on the wavelength of UV light it has been characterized into four categories, explicitly UV-A (315–400 nm), UV-B (280–315 nm), UV-C (180–280 nm), and Vacuum UV (10–180) having different wavelengths. The lethal effect is stronger when the wavelength of light is lower. For the treatment of industrial waste liquid, numerous UV lamps have been applied. For photochemical degradation, UV treatment has been widely used in combination with new technology. In the principle of photodegradation, light radiations are utilized as a source of energy (Ang et al. 2019). For the photochemical degradation process source of UV light is xenon, mercury lamp, or sunlight is used. When photons are absorbed from light energy by molecules, chemical, and physical changes happen in the species. The photochemical modification can be direct or indirect. In the direct method, the pollutants absorb photon energy and then go through homolytic cleavage to obtain the degraded material. In the indirect method, of photochemical, the photosensitive substances absorb photon energy, and then species are produced which interrelate with the target molecule to influence the degradation of contaminants. A direct and indirect method of photochemical comprises a sequence of oxidation and reductive reaction process which is known to be the redox process. The decomposition is instigated by a reduction or oxidation reaction (Cory and Kling 2018).

7.2.3 Ozonation

In the ozonation process, the ozone is used for the removal of contaminants either directly by the effect of ozone particles on pathogens or incidentally by the effect of oxidation on the free radicals resulting in the degradation of organic substances by ozone in water. Then chemical oxidants formed by free radicals are much more reactive and a bit selective. A lifetime of ozone gas is very short, so it is generated on the treatment sites from pure oxygen or dry air. The ozone formation is an endothermic reaction, this gas is thermodynamically unbalanced and hence readily relapses to oxygen. Ozone is extensively utilized in the treatment of industrial liquid waste based on its efficiency in the oxidation of contaminants, disinfection, and removal of color, taste, and odors. This method is more convenient because ozone is effortlessly produced from pure oxygen through electric discharge, with an organic and inorganic substance it reacts certainly, and extreme ozone in liquid decomposes certainly to oxygen without any hazard or without leaving any deposit. It is reported that through ozonation treatment, 23 pesticides have been removed from industrial wastewater. When ozonation treatment is combined with other treatments like activated carbon, nanomaterial, UV, and photodegradation much more efficient results for reducing contaminants were achieved (Habibi et al. 2017).

7.2.4 Advance Adsorption

This is a well-recognized, low cost and very effective industrial water treatment method that removes contaminants and pathogens present in a liquid. The adsorbent's effectiveness depends on the porosity, number of available sites, surface area, and probable interactions with the target contaminant. The adsorbents are grouped as bio-adsorbents, polymeric adsorbents, and inorganic adsorbents. Lately, research has been carried out to develop a faultless activated carbon that can remove numerous contaminants or pathogens present in effluent water. The granular and powder-activated carbon revealed high efficiency to absorb the waste which is present in effluent water (Habibi et al. 2017). Similarly, biochar is the carbon-rich adsorbent produced by pyrolysis (300–800 °C) of various waste biomass under an oxygen-free environment (Lyu et al. 2016; Vyavahare et al. 2019; Gurav et al. 2021a; Bhatia et al. 2020; Gurav et al. 2021c; Suryawanshi et al. 2023). Several pollutants like antibiotics, textile dyes, petroleum oil, cyanotoxins, etc., can be adsorbed onto pristine and modified biochar using different adsorption mechanisms depending on the type of pollutant and source of biochar (Zhang et al. 2016; Vyavahare et al. 2018, 2021; Song et al. 2021; Gurav et al. 2019, 2021b; Choi et al. 2020a, b; Kim et al. 2020). Artificial polymers like polymeric resins are also developed as an adsorbent. Bio-adsorption is a method of adherence of various pathogens to the cell wall of biotic agents, it doesn't involve oxidation via anaerobic or aerobic metabolism. Both dead and living cells assist as bio-adsorbent, but dead cells are more cost-effective than live-cell and don't require nutrients. Effective bio-adsorbents contain fermentation waste, seaweed, yeast, and sludge waste. Inorganic absorbents like graphene and clay are used widely to treat industrial wastewater because it is cost-effective. The organic, inorganic contaminant removal from industrial wastewater by adsorption seems to be a cost-effective, promising, relevant, and sustainable approach that can be used effectively (Burakov et al. 2018).

7.3 Biological Treatments

Numerous methods are designed to treat industrial wastewater using fungi, bacteria, and other microorganisms proficient in degrading contaminants. The biological methods are categorized as aerobic and anaerobic. Bio-mixtures are generally humic-rich materials, and they can be from exogenous or endogenous origins.

7.3.1 Pressurized Activated Sludge

The pressurized activated sludge technique is derived from activated sludge processes but this method has been modified and it has become innovative and advanced technology for industrial effluent treatment. The foremost objective of the pressurized method is to overcome the oxygen transfer hurdle by enhancing the level of dissolved oxygen. This method is an extremely efficient method it can be carried out in a small area, and it is a cost-effective sustainable, and eco-friendly approach as compared to chlorination. For disposal of sludge skilled and trained labor is needed in operation and maintenance. In activated sludge it is difficult to treat industrial waste liquid with more COD value due to the partial quantity of dissolved oxygen in atmospheric pressure and the process of aeration raises the cost of technology. In this method, the quantity of soluble oxygen rises with the increasing pressure of total air. The advanced pressurized activated sludge removes 85–92% COD from a discarded liquid by enhancing aeration time, sludge concentration, and operating pressure (Yang et al. 2018).

7.3.2 Membrane Bioreactor

In industrial wastewater technology, membrane bioreactor (MBR) is the most unique and latest treatment. The membrane filtration and biological treatments are combined in this method. It has high effectiveness, but it requires more energy, in supplementary with anti-fouling technique cost. The MBR system is known to be an integrated and complete membrane unit with associated constituents essential to consent the process to a role as desired. In this MBR system, often ten to eleven sub-processes are included like membrane zone, fine screening, and a few types of disinfection methods (Phommachanh et al. 2021). In MBR, microorganisms are exploited to decompose the contaminants and then the liquid waste is filtered by submerged membranes. The discrete membranes are retained in units known as cassettes, racks, modules, and a collective series of these units is known as an operational membrane unit. Oxygen is passed through central diffusers to persistently clean the surface of the membrane during facilitating mixing, filtration, and in specific cases, subsidizing oxygen to the biological methods. The MBR is advantageous because it includes a compact footprint, generally 30–50% lesser than a corresponding conventional activated sludge method with media tertiary filtration and secondary clarifiers. The MBR method gives excellent effluent quality that is proficient in meeting the most stringent water quality (Tan et al. 2019).

7.3.3 Bio-Augmented Treatment

In the treatment of bio-augmentation, additional microorganisms are incorporated into the wastewater system to improve the effectiveness of decomposition in an activated sludge process. The bio-augmented supports in the decomposition process remove the pollutants and disobedient compounds. Mostly, various bacterial strains survive naturally on industrial waste liquid, these bacterial species also help with the decomposition of contaminants over a period assuming a firm state (Habibi et al. 2017). Conversely, certain pollutants are proficient in resisting organic decomposition due to aspects like less water solubility, composite chemical composition, and high toxicity. Hence, an extensive species of microorganisms are essential in the biodegradation process and for this bio-augmented treatment is a prerequisite. Bio-augmented material includes multiple species or strains of fungi and bacteria which gradually create change in the microbial species. Due to bio-augmentation, the COD and BOD levels have been decreased drastically, this process enhances the settling proficiency of solids and improves nitrification (Phommachanh et al. 2021).

8 Conclusion

The existence of many chemicals and contaminants in industrial water is constant apprehension for the ecosystem, individual health, and the environment. The emerging pollutant and hazardous materials in freshwater resources are mainly constituted due to various industrial and anthropogenic activities. As mentioned above various treatments and technologies have emerged to treat these industrial liquid waste or to eradicate a huge concentration of contaminants from effluent. However, lacking significant skill and poor understanding still signifying the great challenge of reusing this treated wastewater for domestic purposes or consumption. The present book chapter evaluates the emerging contaminants and their adverse effects accompanied by the summarizing of several potential elimination methods. It is highlighted that the emerging contaminant removal by a single technological treatment seems not the distinguished approach demanding the coupled treatment implementation to overwhelm the insufficiencies of a single technology. The current book chapter encompasses various innovative methods and their recent advancements that are applicable to effectively reduce hazardous pollutants could be a good showcase for a new budding researcher. For imminent development the synergistic and sustainable approach and design should be applied to treat, degrade, or remove the contaminants from the industrial effluents.

References

Aderibigbe DO, Giwa ARA, Bello IA (2017) Characterization and treatment of wastewater from food processing industry: A review. Imam J Appl Sci 2(2):27

Ahmed MB, Zhou JL, Ngo HH, Guo W, Thomaidis NS, Xu J (2017) Progress in the biological and chemical treatment technologies for emerging contaminant removal from wastewater: a critical review. J Hazard Mater 323:274–298

Ang MBMY, Trilles CA, De Guzman MR, Pereira JM, Aquino RR, Huang SH, Lai JY (2019) Improved performance of thin-film nanocomposite nanofiltration membranes as induced by embedded polydopamine-coated silica nanoparticles. Sep Purif Technol 224:113–120

Anis SF, Hashaikeh R, Hilal N (2019) Reverse osmosis pretreatment technologies and future trends: A comprehensive review. Desalination 452:159–195

Arihilam N, Arihilam E (2019) Impact and control of anthropogenic pollution on the ecosystem–A review. J Biosci Biotechnol Discov 4(3):54–59

Arzate S, Pfister S, Oberschelp C, Sánchez-Pérez JA (2019) Environmental impacts of an advanced oxidation process as tertiary treatment in a wastewater treatment plant. Sci Total Environ 694:133572

Awuchi CG, Hannington T, Awuchi CG, Igwe VS, Amagwula IO (2020) Industrial waste management, treatment, and health issues: wastewater, solid, and electronic wastes. Eur Acad Res 8(2):1081–1119

Barakat MA (2011) New trends in removing heavy metals from industrial wastewater. Arab J Chem 4:361–377

Barbusiński K, Filipek K (2001) Use of Fenton's reagent for removal of pesticides from industrial wastewater. Pol J Environ Stud 10(4):207–212

Bhatia SK, Gurav R, Choi T-R, Kim HJ, Yang S-Y, Song H-S, Park JY, Park Y-L, Han Y-H, Choi Y-K, Kim S-H, Yoon J-J, Yang Y-H (2020) Conversion of waste cooking oil into biodiesel using heterogenous catalyst derived from cork biochar. Biores Technol 302:122872

Burakov AE, Galunin EV, Burakova IV, Kucherova AE, Agarwal S, Tkachev AG, Gupta VK (2018) Adsorption of heavy metals on conventional and nanostructured materials for wastewater treatment purposes: A review. Ecotoxicol Environ Saf 148:702–712

Carboneras MB, Villaseñor J, Fernández-Morales FJ, Rodrigo MA, Cañizares P (2018) Biological treatment of wastewater polluted with an oxyfluorfen-based commercial herbicide. Chemosphere 213:244–251

Castillo MDP, Torstensson L, Stenström J (2008) Biobeds for environmental protection from pesticide use: a review. J Agric Food Chem 56(15):6206–6219

Chiam CK, Sarbatly R (2011) Purification of aquacultural water: conventional and new membrane-based techniques. Sep Purif Rev 40(2):126–160

Choi Y-K, Choi T-R, Gurav R, Bhatia SK, Park Y-L, Kim HJ, Kan E, Yang Y-H (2020a) Adsorption behavior of tetracycline onto Spirulina sp. (microalgae)-derived biochars produced at different temperatures. Sci Total Environ 710:136282

Choi Y-K, Gurav R, Kim HJ, Yang Y-H, Bhatia SK (2020b) Evaluation for simultaneous removal of anionic and cationic dyes onto maple leaf-derived biochar using response surface methodology. Appl Sci 10(9)

Choudhury P, Uday USP, Mahata N, Tiwari ON, Ray RN, Bandyopadhyay TK, Bhunia B (2017) Performance improvement of microbial fuel cells for waste water treatment along with value addition: a review on past achievements and recent perspectives. Renew Sustain Energy Rev 79:372–389

Cory RM, Kling GW (2018) Interactions between sunlight and microorganisms influence dissolved organic matter degradation along the aquatic continuum. Limnol Oceanogr Lett 3(3):102–116

Crini G, Lichtfouse E (2018) Wastewater treatment: an overview, chapter 1. In: Crini G, Lichtfouse E (eds) Green adsorbents for pollutant removal—fundamentals and design, environmental chemistry for a sustainable world, vol 1.

Crini G, Lichtfouse E (2019) Advantages and disadvantages of techniques used for wastewater treatment. Environ Chem Lett 17(1):145–155

Crini G, Morin-Crini N, Fatin-Rouge N, Deon S, Fievet P (2017) Metal removal from aqueous media by polymer-assisted ultrafiltration with chitosan. Arab J Chem 10:S3826–S3839

Fito J, Tefera N, Kloos H, Van Hulle SW (2019) Physicochemical properties of the sugar industry and ethanol distillery wastewater and their impact on the environment. Sugar Tech 21(2):265–277

Gaied F, Louhichi B, Bali M, Jeday MR (2019) Tertiary treatment of wastewater by electrocoagulation, electro-Fenton and advanced electro-oxidation processes: comparative and economic study. Songklanakarin J Sci Tech 41(5)

Gurav R, Bhatia SK, Choi T-R, Choi Y-K, Kim HJ, Song H-S, Lee SM, Lee Park S, Lee HS, Koh J, Jeon J-M, Yoon J-J, Yang Y-H (2021a) Application of macroalgal biomass derived biochar and bioelectrochemical system with Shewanella for the adsorptive removal and biodegradation of toxic azo dye. Chemosphere 264:128539

Gurav R, Bhatia SK, Choi T-R, Choi Y-K, Kim HJ, Song H-S, Park SL, Lee HS, Lee SM, Choi K-Y, Yang Y-H (2021b) Adsorptive removal of crude petroleum oil from freshwater using floating pinewood biochar decorated with coconut oil-derived fatty acids. Sci Total Environ 146636

Gurav R, Bhatia SK, Choi T-R, Park Y-L, Park JY, Han Y-H, Vyavahare G, Jadhav J, Song H-S, Yang P, Yoon J-J, Bhatnagar A, Choi Y-K, Yang Y-H (2019) Treatment of furazolidone contaminated water using banana pseudostem biochar engineered with facile synthesized magnetic nanocomposites. Bioresour Technol 122472

Gurav RG, Bhatia SK, Jagtap UB, Yang Y-H, Choi Y-K, Tang J, Bhatnagar A (2021c) Utilization of invasive weed biomass for biochar production and its application in agriculture and environmental clean-up. In: Pant D, Bhatia SK, Patel AK, Giri A (eds) Bioremediation using weeds. Springer, Singapore, pp 207–224

Habibi M, Zinatizadeh AA, Akia M (2017) Advanced oxidation processes treating of Tire Cord production plant effluent: a comparative study. J Appl Res Water Wastewater 4(1):319–330

Hube S, Eskafi M, Hrafnkelsdóttir KF, Bjarnadóttir B, Bjarnadóttir MÁ, Axelsdóttir S, Wu B (2020) Direct membrane filtration for wastewater treatment and resource recovery: a review. Sci Total Environ 710:136375

Kataki S, Chatterjee S, Vairale MG, Dwivedi SK, Gupta DK (2021) Constructed wetland, an eco-technology for wastewater treatment: A review on types of wastewater treated and components of the technology (macrophyte, biolfilm and substrate). J Environ Manage 283:111986

Khalaf MN (ed) (2016) Green polymers and environmental pollution control. CRC Press, Oakville

Kim JE, Bhatia SK, Song HJ, Yoo E, Jeon HJ, Yoon J-Y, Yang Y, Gurav R, Yang Y-H, Kim HJ, Choi Y-K (2020) Adsorptive removal of tetracycline from aqueous solution by maple leaf-derived biochar. Biores Technol 306:123092

Kong Z, Li L, Xue Y, Yang M, Li YY (2019) Challenges and prospects for the anaerobic treatment of chemical-industrial organic wastewater: a review. J Clean Prod 231:913–927

Kushwaha JP (2015) A review on sugar industry wastewater: sources, treatment technologies, and reuse. Desalin Water Treat 53(2):309–318

Lyu H, Gong Y, Gurav R, Tang J (2016) Chapter 9—potential application of biochar for bioremediation of contaminated systems. In: Ralebitso-Senior TK, Orr CH (eds) Biochar application. Elsevier, pp 221–246

Maizel AC, Remucal CK (2017) The effect of advanced secondary municipal wastewater treatment on the molecular composition of dissolved organic matter. Water Res 122:42–52

Malollari I, Pinguli L, Buzo R, Lajqi V, Makolli S, Cani XH (2019) Actual situation of wastewater from food industry and a case study of their treatment. J Environ Prot Ecol 20(1):432–438

Meng Y, Luan F, Yuan H, Chen X, Li X (2017) Enhancing anaerobic digestion performance of crude lipid in food waste by enzymatic pretreatment. Biores Technol 224:48–55

Miklos DB, Remy C, Jekel M, Linden KG, Drewes JE, Hübner U (2018) Evaluation of advanced oxidation processes for water and wastewater treatment–A critical review. Water Res 139:118–131

Muruganandam L, Kumar MS, Jena A, Gulla S, Godhwani B (2017, November) Treatment of waste water by coagulation and flocculation using biomaterials. In: IOP conference series: materials science and engineering 263(3), 032006. IOP Publishing

Noukeu NA, Gouado I, Priso RJ, Ndongo D, Taffouo VD, Dibong SD, Ekodeck GE (2016) Characterization of effluent from food processing industries and stillage treatment trial with Eichhornia crassipes (Mart.) and Panicum maximum (Jacq.). Water Resour Ind 16:1–18

Parsons S (ed) (2004) Advanced oxidation processes for water and wastewater treatment. IWA Publishing
Patil PM, Bohara RA (2020) Nanoparticles impact in biomedical waste management. Waste Manage Res 38(11):1189–1203
Patil PM, Mahamuni PP, Abdel-Daim MM, Aleya L, Chougule RA, Shadija PG, Bohara RA (2019a) Conversion of organic biomedical waste into potential fertilizer using isolated organisms from cow dung for a cleaner environment. Environ Sci Pollut Res 26(27):27897–27904
Patil PM, Mahamuni PP, Shadija PG, Bohara RA (2019b) Conversion of organic biomedical waste into value added product using green approach. Environ Sci Pollut Res 26(7):6696–6705
Phommachanh V, Chiemchaisri W, Chiemchaisri C (2021) Co-digestion of greasy waste and mbr sludge using bio-augmented pseudomonas putida: effect of buffering agents. Asean Eng J 11(4):1–15
Pronk W, Ding A, Morgenroth E, Derlon N, Desmond P, Burkhardt M, Fane AG (2019) Gravity-driven membrane filtration for water and wastewater treatment: a review. Water Res 149:553–565
Qasim W, Mane AV (2013) Characterization and treatment of selected food industrial effluents by coagulation and adsorption techniques. Water Res Ind 4:1–12
Rathour R, Krishnan SG, Khanna A, Dhatwalia S, Kaachra A, Sharma TR, Singh AK (2016) Development and validation of co-dominant gene based markers for Pi9, a gene governing broad-spectrum resistance against blast disease in rice. Mol Breeding 36(12):1–8
Rathoure AK, Dhatwalia VK (eds) (2016) Toxicity and waste management using bioremediation. IGI Global, Hershey
Sharma SK (ed) (2015) Green chemistry for dyes removal from wastewater. Scrivener Publishing LLC Wiley, Beverley
Song HJ, Gurav R, Bhatia SK, Lee EB, Kim HJ, Yang Y-H, Kan E, Kim HH, Lee SH, Choi Y-K (2021) Treatment of microcystin-LR cyanotoxin contaminated water using Kentucky bluegrass-derived biochar. J Water Process Eng 41:102054
Sonune A, Ghate R (2004) Developments in wastewater treatment methods. Desalination 167:55–63
Suryawanshi SS, Kamble PP, Gurav R, Yang Y-H, Jadhav JP (2023) Statistical comparison of various agricultural and non-agricultural waste biomass-derived biochar for methylene blue dye sorption. Biomass Convers Biorefinery. 13:5353–5366
Tan X, Acquah I, Liu H, Li W, Tan S (2019) A critical review on saline wastewater treatment by membrane bioreactor (MBR) from a microbial perspective. Chemosphere 220:1150–1162
Varjani S, Joshi R, Srivastava VK, Ngo HH, Guo W (2020) Treatment of wastewater from petroleum industry: current practices and perspectives. Environ Sci Pollut Res 27(22):27172–27180
Vithanage M, Herath I, Joseph S, Bundschuh J, Bolan N, Ok YS, Rinklebe J (2017) Interaction of arsenic with biochar in soil and water: a critical review. Carbon 113:219–230
Vyavahare G, Gurav R, Patil R, Sutar S, Jadhav P, Patil D, Yang Y-H, Tang J, Chavan C, Kale S, Jadhav J (2021) Sorption of brilliant green dye using soybean straw-derived biochar: characterization, kinetics, thermodynamics and toxicity studies. Environ Geochem Health 43(8):2913–2926
Vyavahare G, Jadhav P, Jadhav J, Patil R, Aware C, Patil D, Gophane A, Yang Y-H, Gurav R (2019) Strategies for crystal violet dye sorption on biochar derived from mango leaves and evaluation of residual dye toxicity. J Clean Prod 207:296–305
Vyavahare GD, Gurav RG, Jadhav PP, Patil RR, Aware CB, Jadhav JP (2018) Response surface methodology optimization for sorption of malachite green dye on sugarcane bagasse biochar and evaluating the residual dye for phyto and cytogenotoxicity. Chemosphere 194:306–315
Wang Y, Zhang CH, Wang S, Shen LY, Ge Y (2013) Accumulation and transformation of different arsenic species in nonaxenic Dunaliella salina. Huan jing ke xue= Huanjing kexue 34(11):4257–4265
Yang Z, Lu X, Zhang S, Zhang K, Zhi S, Guo H, Li Q, Yu X (2018) Pressurized electro-dewatering of activated sludge: analysis of electrode configurations (anode). Waste Manage 81:157–167
Zainith S, Chowdhary P, Bharagava RN (2019) Recent advances in physico-chemical and biological techniques for the management of pulp and paper mill waste. Emerging and eco-friendly approaches for waste management, 271–297

Zhang H, Tang J, Wang L, Liu J, Gurav RG, Sun K (2016) A novel bioremediation strategy for petroleum hydrocarbon pollutants using salt tolerant Corynebacterium variabile HRJ4 and biochar. J Environ Sci 47:7–13
Zhao C, Chen W (2019) A review for tannery wastewater treatment: some thoughts under stricter discharge requirements. Environ Sci Pollut Res 26(25):26102–26111
Zhao DL, Japip S, Zhang Y, Weber M, Maletzko C, Chung TS (2020) Emerging thin-film nanocomposite (TFN) membranes for reverse osmosis: A review. Water Res 173:115557

Role of Lignocellulosic Waste in Biochar Production for Adsorptive Removal of Pollutants from Wastewater

K. Ankita Rao, Vaishakh Nair, G. Divyashri, T. P. Krishna Murthy, Priyadrashini Dey, K. Samrat, M. N. Chandraprabha, and R. Hari Krishna

1 Introduction

Environmental problems are continuously increasing due to the rapid increase in industrialization, urbanization and world population. Overconsumption of non-renewable energy resources is an important cause of releasing pollutants into water, air, soil and sediments. A large number of toxic and hazardous pollutants such as heavy metals, dyes, antibiotics, pesticides, polychlorinated biphenyls, endocrine disruptors and polycyclic aromatic hydrocarbons are posing a severe threat to the health of humans and the ecosystem. It is very crucial to develop efficient treatment methods to remove these pollutants for the safety of human health and the environment and to ensure sustainable development (Wang et al. 2019a, b). Among various natural resources, water is an important resource for the sustainable growth and smooth functioning of the ecosystem and society. Discharge of untreated effluent into freshwater sources leads to the addition of undesirable chemicals and deteriorates the quality of ground & surface water resources and causes water pollution (Ahmad and Danish 2018; Chakraborty and Mukhopadhyay 2014). It is critical to treat wastewater due to increasing wastewater disposal costs and strict regulations

The original version of this chapter was revised: Author's proof corrections have been updated. The correction to this chapter is available at
https://doi.org/10.1007/978-981-99-2598-8_21

K. Ankita Rao · V. Nair (✉)
Department of Chemical Engineering, National Institute of Technology Karnataka—Surathkal, Mangalore, Karnataka 575025, India
e-mail: vaishakhnair@nitk.edu.in

G. Divyashri (✉) · T. P. Krishna Murthy · P. Dey · K. Samrat · M. N. Chandraprabha · R. Hari Krishna
Department of Biotechnology, M S Ramaiah Institute of Technology, Bengaluru, Karnataka 560054, India
e-mail: gdivyashri@msrit.edu

M. P. Shah (ed.), *Advanced and Innovative Approaches of Environmental Biotechnology in Industrial Wastewater Treatment*,
https://doi.org/10.1007/978-981-99-2598-8_11

for permissible limits of pollutants in discharged effluent (Crini et al. 2018). To maintain the pollutants in permissible limits in the effluent several techniques such as chemical precipitation, reverse osmosis, ion exchange, electrochemical treatment, membrane processes (ultra-, micro- and nanofiltration), coagulation-flocculation, flotation, ozonation etc. are employed. Even though most of these methods are efficient in removal of many pollutants, they also possess some of the disadvantages viz., chemical and energy requirement, generation of toxic sludge and secondary pollution, high cost and partial removal of pollutants (Braghiroli et al. 2018). Among numerous treatment methodologies, adsorption is rapidly getting importance as a method of wastewater treatment. The adsorption method is chosen over other techniques as it is convenient and quick to remove toxic pollutants. It has also low initial costs, produces lesser secondary pollutants and is relatively simple concerning the design and operation of the treatment equipment. The adsorption process is based on mass transfer phenomena through which a solid material can selectively eliminate dissolved pollutants from the solution by attracting the dissolved pollutants to its surface. As adsorption does not require any special additional equipment, it makes it easy to run this process (Adeyemo et al. 2017; Singh et al. 2018).

Process parameters such as mode of adsorption, adsorbent nature, adsorbent dose, nature of wastewater (pollutant types, concentration, etc.), effluent pH, temperature, time, etc., significant influence on the efficiency of the adsorption process. Among these parameters to attain desired water quality, a careful selection of adsorbents is of prime importance. Adsorbents used in the amputation of pollutants from wastewater are of natural origin or obtained through the activation process. An ideal adsorbent utilized should be environmentally friendly, low cost and possess satisfactory adsorption properties (Kyzas et al. 2014; Sadegh et al. 2017). Adsorbents are solid materials that are insoluble porous and sponge-like substances with the capability of capturing and trapping pollutants/contaminants. A few adsorbents such as activated carbon, silica, zeolites and activated alumina are used commercially on an industrial scale to treat wastewater. But, due to high cost, low efficiency, lower adsorption rates and recycling difficulty of various adsorbents used, it is critical to develop more effective and low-cost adsorbents for the treatment of contaminants from industrial wastewater on large scale (Krishna Murthy et al. 2019).

Industrial production and various other human activities produce large amounts of biomass waste. The value-added utilization of biomass reduces the cost of waste and enhances its commercial value in application. Biomass is generally used biosorbent in three different forms (1) raw biomass (e.g., agricultural residues), (2) activated/synthetic biobased materials (e.g., activated carbon/fibres, biochar, magnetic and nanomaterial-biochar composites) and (3) nanomaterials. Recently, biochar has been extensively used as a sorbent for amputation of various contaminants owing to their superior properties, such as eco-friendly, availability of a large number of inorganic mineral species and functional groups, comprising of micro and/or mesoporous structures with high adsorption capacity. Also because of its improved physicochemical properties such as controllable porous structures, tunable surface areas and abundant oxygen-containing functional groups, biochar has shown widespread applications in multidisciplinary domains such as environment energy and agriculture. Also, biochar is a stable aromatic porous carbon-rich material that is obtained carbonization of biomass under high temperature and oxygen-limited conditions.

In the past few decades, biochar has fascinated researchers in applying them for pollutants management and soil improvement, wastewater treatment, photocatalytic degradation of organic pollutants and carbon sequestration (Singh et al. 2023).

2 Biochar

Lignocellulosic rich agricultural wastes are reported to be highly suitable for the production of biochar. Biochar is also known as biocarbon is a carbon-rich material obtained from the pyrolysis technique that decomposes organic matter at high temperatures (350 and 500 °C) in the presence of limited or no oxygen (Pradhan et al. 2020). Production of biochar from various ago wastes (e.g., wood waste, bagasse, green plant material, poultry litter, feedlot manure, etc.) has recently been attempted by various research groups due to its availability, applicability and smoother production abilities (Gabhane et al. 2020). Biochar preparation approaches are broadly categorized as traditional and modern methods based on their modernization and advancements (Table 1; Fig. 1). Primarily, it comprises 38–80% carbon (C) and the rest of nitrogen (N), hydrogen (H) and oxygen (O) (Murtaza et al. 2021).

Even though, pristine biochar possesses a wide range of environmental applications, "engineered biochar", i.e., biochar modified via physical, chemical or biological approaches has arisen as a new trend to meet specific purposes (Panahi et al. 2020). Many research efforts from the scientific community have been reported in preparation to activate biochar with the usage of steam, gas, acids, alkalis and oxidants (Panwar and Pawar 2020). In addition, the preparation of biochar composite with the incorporation of other materials also shows promising properties (Wang et al. 2021).

2.1 *Biochar Composites*

Biochar composites are classified into five groups viz., microorganism–biochar composites, carbonaceous engineering nano-composites, layered double hydroxide (LDH)–biochar composites, metal-biochar composites and mineral–biochar composites (Fig. 2). In comparison to pristine biochar, biochar composites display excellent properties for various environmental applications.

2.1.1 Microorganism–Biochar Composites

Microorganisms can improve biochar performances. Preparation of biochar composite using microorganisms having the ability to degrade organic contaminants thereby enhancing biodegradation properties. Modification of rice straw-biochar using engineered *Mycobacterium gilvum* resulted in effective degradation of soil polycyclic aromatic hydrocarbons (PAHs) in comparison to native biochar (Xiong et al. 2017). In line with this, rice husk-biochar inoculated with *Bacillus siamensis*

Table 1 Traditional and modern biochar production techniques

S. No	Biochar production technique	Production conditions	References
1.	Traditional Early Approach	Biomass burning in a pit covered by soil to limit oxygen supply	Thines et al. (2017)
2.	Traditional Slow Pyrolysis Technique	Temperature: 300–600 °C Heating rate: 5–7 °C min^{-1}	Lai et al. (2013)
3.	Traditional Fast Pyrolysis Technique	Temperature: >500 °C Heating rate: 300 °C min^{-1}	Huang et al. (2017a, b)
4.	Modern Gasification Technique	Temperature: >700 °C using steam as gasifying agent	Al-Rahbi and Williams (2017)
5.	Modern Torrefaction Technique	Temperature: 230–300 °C	Pentananunt et al. (1990)
6.	Modern Flash Pyrolysis Technique	Temperature: 900–1200 °C Heating rate: 800–1000 °C sec^{-1}	Li et al. (2013)
7.	Modern Vacuum Pyrolysis Technique	Pressure: 0.05–0.20 MPa Temperature: 450–600 °C Heating rate: 300 min^{-1}	Tripathi et al. (2015)
8.	Modern Microwave Pyrolysis Technique	12qwsd	Huang et al. (2017a, b)
9.	Modern Electro-modified Biochar Production Technique	Current supply: 0–100 V and 0–12 A for 5 min Pyrolysis temperature: 400–500 °C; Heating rate: 5 °C min^{-1}	Jung et al. (2015)
10.	Modern Hydrothermal Carbonization Technique	Pressure: 2–10 MPa Temperature: 220–240 °C	Gao et al. (2013)
11.	Modern Magnetic Biochar Production Technique	Temperature: 450–1,000 °C; material soaked with a solution of Fe_2O_3, $FeSO_4{\cdot}7H_2O$, or $FeCl_3{\cdot}6H_2O$ before pyrolysis	Thines et al. (2017)

showed the ability to degrade dibutyl phthalate (DBP) (Feng et al. 2020). Furthermore, the microorganism-biochar composite has shown an ability to fix nitrogen and release phosphate, thus can be employed to improve soil fertility (Wei et al. 2020).

2.1.2 Carbonaceous Engineering Nano-Composites

Carbonaceous-engineered nano-composites have found usefulness in the remedy of metals and adsorption of organic contaminants. Graphene oxide-biochar composites help sulfamethazine adsorption through π–π EDA interactions, ion exchange and hydrogen bonding (Huang et al. 2017a, b). Multiwalled carbon nanotube-biochar composites revealed a promising adsorption rate towards methylene blue via electrostatic interactions (Inyang et al. 2014; Shah Maulin 2020). Furthermore, multi-walled carbon nanotube-biochar composites were employed for simultaneous adsorption of Pb and sulphapyridine from the aqueous media (Inyang et al. 2015).

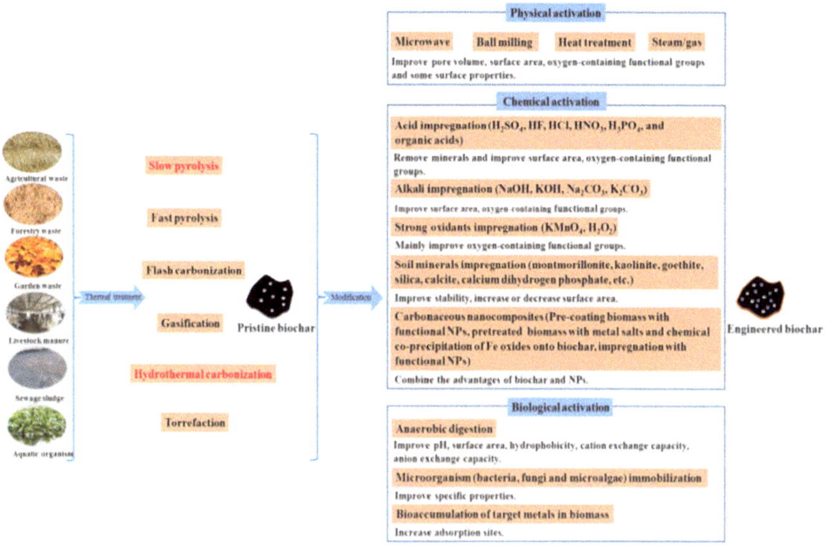

Fig. 1 Methods of preparation and modification of Biochar (Image courtesy: Wu et al. 2020)

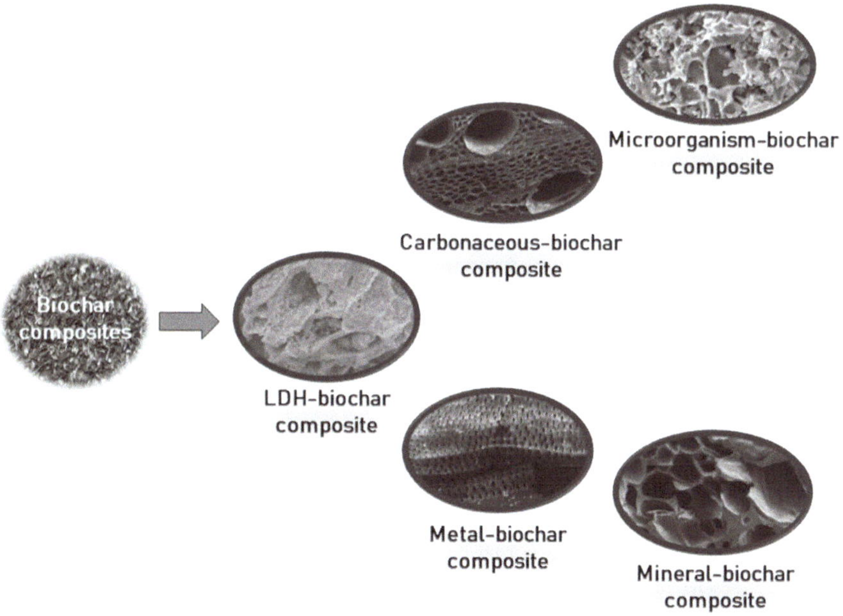

Fig. 2 Different types of engineered biochar composites

2.1.3 Layered Double Hydroxide (LDH)–Biochar Composites

LDH-biochar composites consist of metal hydroxide layers (positively charged) and are made up of anions in the interlayer space for charge neutralization (Ma et al. 2016). LDH-biochar composites that are applied for contaminant adsorption comprised of various divalent and trivalent metal cations such as Ni–Fe, Mg–Fe, Ca–Al, Mg–Al, Zn–Al. They found their usefulness in adsorption of particularly anionic contaminants, viz., phosphate (Yang et al. 2019), arsenic (Wang et al. 2016) and nitrate (Xue et al. 2016) through their anion-exchange mechanism. In addition to anionic contaminant removal, they are used for organic contaminant adsorption. The mechanism behind organic contaminants removal is attributed to pore-filling, hydrogen bonding and π–π EDA interactions (Zubair et al. 2020).

2.1.4 Metal-Biochar Composites

Iron incorporation in the preparation of metal-biochar composites has shown potential to improve biochar performances. Iron oxide-biochar, iron sulphide-biochar and nano zero-valent iron-biochar composites are the notable ones that have profound environmental applications (Lyu et al. 2020). These iron-biochar composites adsorb heavy metals and organic contaminants through a mechanism of improved surface complexation, electrostatic and precipitation interactions (Alam et al. 2020; Zhang et al. 2020). Other metal-biochar composites such as MnO-biochar, MoS_2-biochar and MgO-biochar composites are also developed for environmental applications (Wang et al. 2021). MgO-biochar composite was found to adsorb PB in clayey soil. Cation-π interaction is the mechanism behind Pb adsorption by biochar matrix, and MgO improved Pb immobilization through precipitation process (Shen et al. 2019). MnO-biochar composite can be prepared by soaking with $KMnO_4$ (Yu et al. 2017). The developed composite could adsorb arsenic from soil. In addition, the MoS_2-biochar composite showed a better ciprofloxacin adsorption rate in comparison to virgin biochar as they could provide more π electrons, thereby enhancing the π–π EDA interactions (Yang et al. 2020).

2.1.5 Mineral–Biochar Composites

The incorporation of natural minerals has shown the ability to improve biochar performances towards soil remediation, thereby enhancing soil fertility and assisting in wastewater treatment (Wang et al. 2021). Clay mineral, montmorillonite retains cationic nutrients and adsorbs metals through the cation exchange process (Wang et al. 2020; Shah Maulin 2021a, b). Herath et al. (2020) found that montmorillonite can be a silicon source to adsorb soil arsenic via silicone-ferrihydrite complex on Si-rich montmorillonite–biochar composite. Attapulgite, a fibrous clay material with a laminar structure has also attracted much attention due to its abundant hydroxyl groups. Attapulgite-biochar composites have shown promising results for the adsorption of arsenic and cadmium in river sediments due to improved surface complexation (Wang et al. 2019a, b).

3 Application of Biochar

The major reasons for which biochar has been utilized are environmentally friendly, inexpensive and the ease of preparation from diverse sources mainly, biomass (Manikandan, Giannakoudakis et al. 2023). The other characteristic properties which make biochar a desirable adsorbent for the remediation of various pollutants is its porous structure, large surface area and functional groups (Manikandan, Pallavi et al. 2023). Its absorptivity capacity can be attributed to these surface properties which can strongly adsorb different types of pollutants. Hence due to these reasons, there is a huge interest in researchers with the goal of sustainable usage of waste products for environmental applications.

3.1 Dyes

Dyes have been characterized as water or oil-soluble colour-imparting organic compounds; which are distinct from insoluble pigments. Industrial effluents are environmentally toxic and have high chemical oxygen demand and salinity, and will lead to a detrimental impact on people and wildlife health and on aquatic habitats due to their growing omnipresence in surface water (Singh et al. 2023). Due to their toxic properties, there has been a surge for new techniques for removing dye contaminants from water sources which has created a large amount of research interest. In comparison to all the techniques adsorption principle is the most widely used technique due to it being simple and high efficiency. In this paper, its main objective was to prepare a biochar composite made from corn straw having magnetic properties. This biochar (nZVI/BC) was made by pyrolysis with a temperature as high as 500 °C then converting the obtained BC into a magnetic composite was done by chemical reduction of ferric salt. The contaminant selected for this research was a cationic dye called Malachite green. The nZVI/BC composite was analysed by XRD, TEM, TEM-EDS, FTIR, VSM, TGA, XPS, zeta potential and BET surface area. It was found from isotherms and kinetic studies that the maximum adsorption capacity of nZVI/BC composite is 515.77 mg MG/g composite. The mechanism suggested in the remediation process is not just by adsorption but also through oxidative degradation mechanism. The present experiment is the most common source for lignocellulosic biomass which is available in nature abundantly namely, Rice husk and Rice Straw. The biochar was hydrothermally activated which was treated with $KMnO_4$ for MG removal. The optimized conditions for effective removal rates were 0.05 g of adsorbent to adsorb 50 mg·L^{-1} at ambient temperature for 120 min. At these conditions, the adsorption rate was seen to be over 90% (Table 2).

Another such cationic dye namely crystal violet was remediated by using a natural source, lignocellulosic biomass. CV when present even in the lowest quantity has deleterious effects on aquatic plants. The natural lignocellulosic source for this study was Date palm petioles. To obtain the DPP-BC direct pyrolysis was employed at 700 °C for 3 h. To confirm their surface anatomy and structure techniques like FTIR measurement, BET analyser, pH-pzc method and XPS technique. It was observed that the resultant BC was highly porous and carbonaceous in nature with a Carbon content of 86.61%. It also exhibited a high surface area for adsorption 640 m^2/g. From kinetic studies, it was seen that the process of adsorption reaches its equilibrium after

Table 2 Biochar in the removal of dye contaminant

Biochar source	Contaminant	Removal efficiency	Reference
NaOH-activated biochar prepared from peanut shell	Remazol orange RGB	Adsorption was found as 94.49%	Acemioğlu (2022)
Casuarina equisetifolia (also known as Jungli saru)	Methylene blue	Removal efficiencies of 81.5 and 89.1% in immobilized and free-cell reactor	Bharti et al. (2019)
Date palm petioles biochar	Crystal Violet	Fast adsorption equilibrium was established at approximately 15 min of contact	Chahinez et al. (2020)
Pinewood biomass activated by NaOH and DES (deep eutectic solvent)	Methylene blue and Rhodamine B	20% increase in adsorptive capacity when activated with DES	De Caprariis et al. (2018)
Wodyetia bifurcata biochar	Methylene blue	83% of methylene blue removal was achieved in 30 min of equilibrium time	Dos Santos et al. (2019)
Corn straw-derived biochar supported nZVI magnetic composite (nZVI/BC)	Malachite Green	Removal efficiency 99.9% after 20 min	Eltaweil et al. (2020)
Rice husk and rice stalk treated hydrothermally and activated by $KMnO_4$	Malachite green	Adsorption rate >90%	Li et al. (2019)
Walnut shell and Wood powder Acid- and Alkali-Modified Biochar	Methylene blue	Adsorption of wood is more than a shell. The adsorption capacity of treated biomass is found in this order $ZnCl_2 > KOH > H_3PO_4 > H_2SO_4$	Liu et al. (2020)
Switchgrass-biochar	Methylene Blue(MB), Orange G(OG) and Congo Red (CR)	Results showed that SB900 showed high adsorption capacity compared to SB600 due to the significant enhancement in surface area at the high pyrolysis temperature	Park et al. (2019)

15 min itself. The mechanism involved in this process was by π-π interaction, pore filling and hydrogen bonding formation, electrostatic attraction, cation exchange and van der Waals force. Another cationic dye namely Methylene blue was remediated using biochar from Seeds of *Casuarina equisetifolia* (also known as Jungli saru) and mainly the degradation was done by bacterial species for metabolizing dye molecules that were isolated from dye-contaminated water sources. The best bacterial strain was chosen by 16s DNA sequencing. Further, the removal efficiency for MB combining bio-film with biochar was estimated adsorption rate in packed bed bioreactor (PBBR) where the microbes were immobilized. The maximum MB efficiencies for 50 ppm

by batch reactors consisting of free and immobilized *A. faecalis* cells on the biochar surface are 81.5 and 89.1%. Similarly, for MB and Rhodamine removal an ingenious method was used for the activation of biochar was employed by using deep eutectic solvents (DES). Firstly thermal biochar (TB) was made by pyrolysis of the natural pine wood. Secondly, it was activated by NaOH and then with DES. The removal efficiency using this method of activation-based biochar showed about 480 mg/g for the MB adsorption. A comparison analysis was conducted for MB removal efficiency by wood powder biochar (WPC) and walnut shell biochar (WSC). It was produced by supplying a low concentration of oxygen during pyrolysis and a series of chemicals such as KOH, $ZnCl_2$, H_2SO_4 and H_3PO_4 were used to modify or activate them. It was observed that KOH treatment results in the highest specific surface area. Also, it was seen that the biomass showed better adsorption than the shell biomass. The treatments were seen to be effective in the following order $ZnCl_2 > KOH > H_3PO_4 > H_2SO_4$. The max adsorption capacities for each biomass treatment were 850.9 mg/g for WPC with $ZnCl_2$ treatment and 701.3 mg/g for WSC with KOH treatment respectively. For the first time, Wodyetia bifurcate (known as foxtail palm) biochar using vacuum pyrolysis was used for remediation purposes of MB dye. The activation was done by using H_2SO_4, H_3PO_4 and KOH and then subjecting it to thermal treatment. 83% of removal was seen within 30 min by using optimal conditions: pyrolysis temperature of 700 °C and activating reagent using H_3PO_4.

Other types of dyes on which work has been carried are reactive dyes called Remazol orange RGB. The removal was carried out by using NaOH-activated biochar. The source of the biomass was a peanut shell which was subjected to pyrolysis at 500 °C under nitrogen gas conditions. The results of batch adsorption show that about 94% of the dye molecule gats adsorbed onto the adsorbent (Acemioğlu 2022). A comparison study was conducted to check the behaviour of anionic and cationic dyes such as Orange G (OG), Methylene Blue (MB) and Congo Red (CR) on switchgrass biochar at two different carbon content at pyrolysis temperatures of 600 °C and 900 °C. From the kinetics, it was seen that both BC fit the pseudo-second-order model. To further understand the adsorption process intraparticle diffusion model enabled to understand that MB adsorption was chiefly done by inner pore diffusion. Two conclusions were drawn that the cationic dye (MB) was more likely to be adsorbed when compared to anionic. Secondly >85% of dye removal was observed with biochar pyrolyzed at 900 °C.

3.2 Heavy Metals

Heavy metal ion conversion to a lesser reactive form is quite difficult by nature since they persist longer leading to accumulation by enrichment. Due to this, they have been proven to be very injurious to humans and the environment. The most common source of producing hazardous pollutants from metal mining drainage electroplating water discharge, and dyeing industries. Long-term exposure to these contaminants causes defects in the human olfactory sense, in a more critical case may be injurious to the liver and kidney. For this reason, there is a dire need for technologies to remove or remediate these heavy metal-infested areas. There have been many techniques that can be used for this purpose. But the most inexpensive method is adsorption. Recently a study was conducted using Mg-loaded modified, different types of biochar namely, banana straw, cassava straw, Chinese fir straw, corn straw, camellia nutshells and taro straw were prepared by an impregnation process. The metal ions selected for the removal were Cu (II), Cd (II) and Pb (II). The different types of characterizations were carried out to differentiate the properties of each of the biomasses. The characterizations are zeta potential, FTIR, SEM–EDS, XRD, XPS and ICP–OES. The main objective of the paper was to study the mechanism involved in the process of adsorption which was found to be cation−π interactions, ion exchange, mineral precipitation and functional group interactions. Out of all the biochars BSB exhibited the highest adsorption capability (Li et al. 2020).

A mixture of heavy metals was removed by using four different types of biochar from three different categories (Table 3). Maize stalk and black gram (crop biomass), pine needle (tree) and *Lantana camara* (weed) biomass were selected for the following study. The metal of choice is Cadmium (Cd), lead (Pb), nickel (Ni) and zinc (Zn), copper (Cu) and arsenic (As)) for which the batch studies are carried out. Initially, it was found that all biochar has an effective capacity for the removal of six metal ions. It was seen that a maximal removal was found for arsenic and very little for nickel. The following result of the removal rate of heavy metal was obtained 49.5–66.1% (Cd), 47.3–60.0% (Pb), 45.5–60.6% (Ni), 46.6–60.8% (Zn), 49.3–63.2% (Cu) and 52.7–64.2% (As) for 4 biochars. Analysis of metal removal done by ICP-MS. It was concluded that the biochars in this experiment exhibit a more porous structure which allows them to behave as a bio-filter for wastewater treatment (Das et al. 2023). A study was conducted using single biochar but with variations to the pyrolysis temperature known as LEC200, LEC300, LEC400 and LEC500, respectively. The mixture of toxic ions chosen for this study are lead, zinc, copper and cadmium. It was observed from the batch studies that LEC500 was the foremost material to bind effectively to metals which can be supported by SEM, BET and elemental analysers. The maximum removal rates for Pb (II), Zn (II), Cu (II) and Cd (II) were as follows 39.09 mg g^{-1}, 45.40 mg g^{-1}, 48.20 mg g^{-1} and 44.04 mg g^{-1}.

Singular-based metal ions studies have also been carried out. One such research was to assess Pb (II) adsorption property of jujube pit biochar (JPB) which was pyrolyzed at 800 °C. From the kinetics, it was seen that the equilibrium was obtained within the first 30 min itself. Further for the recovery of the adsorbent desorption studies were carried out from which it was observed that even after 5 cycles JPB had 70% of its adsorption efficiency. Ultimately the maximum adsorption for lead ions was seen to be 137.1 mg/g. For understanding Cu (II) removal efficiency biochar was prepared from pine needles which were magnetic in nature. This BC is magnetized before and after the oxidation process which resulted in the production of PNMC: carbonized-magnetized pine needles and PNCOM: carbonized-oxidized-magnetized pine needles respectively. The results from the batch studies reveal that PNCOM significantly has higher adsorption capacity which is 1.0 mmol g^{-1}. It was also concluded that biochar magnetization after oxidation shows a higher adsorption rate due to carboxylic moieties. It is known that chromate ion [Cr (VI)] is toxic in nature and it becomes necessary to remove it from the source. Hence a novel method was developed by enhancing the surface by coating the biochar with MgO from sugarcane harvest residue (SHR). The mechanism proposed explained that the Cr (VI) ions were directly adsorbed onto the MgO-coated biochars by chemical interactions. The composite from sugarcane harvest residue was first pretreated with

Table 3 Biochar in the removal of heavy metal contaminant

Biochar description	Contaminant	Characteristics	Reference
Maize stalk& black gram, pine needle and *Lantana camara* (weed) biomass	Cadmium (Cd), lead (Pb), nickel (Ni), zinc (Zn), copper (Cu) and arsenic (As)	Four different biochars, the average removal rate of heavy metal from aqueous solution was 49.5–66.1% (Cd), 47.3–60.0% (Pb), 45.5–60.6% (Ni), 46.6–60.8% (Zn), 49.3–63.2% (Cu) and 52.7–64.2% (As)	Das et al. (2023)
Biochar from jujube pit	Pb (II)	The adsorption capacity of JPB for Pb (II) was calculated to be maximum for 137.1 mg/g at pH 6.0	Gao et al. (2020)
Cu-coated bamboo (*Acidosasa longiligula*) shoot shell) biochar composite	Re(VII)	The adsorption capacity of Cu-biochars was increased by 3–12 times at pH 3–6	Hu et al. (2018)
Long-root *Eichhornia crassipe* biochar	Lead, Zinc, Copper and Cadmium	Maximum removal capacities of Pb (II), Zn (II), Cu (II) and Cd (II) at 298 K were 39.09 mg g^{-1}, 45.40 mg g^{-1}, 48.20 mg g^{-1} and 44.04 mg g^{-1}	Li et al. (2018)
$MgCl_2$-Modified banana straw, cassava straw, Chinese fir straw, corn straw, camellia nut shells and taro straw Biochars	Cd (II), Pb (II), or Cu (II)	Banana straw biochar has the best adsorptive capacity	Li et al. (2020)
(3-Aminopropyl) triethoxysilane and iron rice straw biochar composite	Cr (VI) and Zn (II)	Fe and APTES biochar composites in the soil effectively reduced the metal toxicity and improved the soil physicochemical properties	Medha et al. (2021)
Carbonized-oxidized-magnetized pine needles biochar	Cu(II)	Biochar magnetization after oxidation results in significantly higher adsorption capacities	Nicolaou et al. (2019)
Sugarcane harvest residue biochar with diluted sulfuric acid-assisted MgO-coated composite	Cr (VI)	Better adsorptive capacity than pristine biochar	Xiao et al. (2018)
Corn straw biochar supported manganese sulfide	Cr (VI)	Equilibrium Cr (VI) adsorption amount was about 98.15 mg g^{-1} at pH 5.0–6.0	Zhang et al. (2019)
Pine wood zerovalent iron Biochar (ZVI/BC)	Cr (VI)	77% Cr (VI) detoxification BY ZVI/BC than BC	Zhou et al. (2020)

sulfuric acid which had a significant Cr (VI) removal efficiency. A one-step process was employed in the synthesis of a novel corn straw biochar supported on manganese sulfide. The Nano-sized MnS particles are strongly attached to the biochar matrix. Hence this adsorbent showed its best removal rate of Cr (VI) about 98.15 mg L^{-1}. The mechanism proposed was in between adsorption and reduction/precipitation for pH 5.0–6.0. Lignocellulosic biomass has 3 types of biopolymers which were pyrolyzed either individually or in a mixture to make zerovalent iron ZVI/BC biochar. The quantity of cellulose, hemicellulose and lignin present in pinewood were as follows 32.04, 22.10, and 35.50%. Characterization was carried out using TGA/GSC, XRD, Raman and BET for BCs and ZVI/BC. It was found that this new biochar was effective in the removal of Cr (VI) with 77% of Cr (VI) detoxification, which was done by ZVI/BC when compared to BC (Zhou et al. 2020). The presence of Cr (VI) in the environment can cause a major threat to plant and human health. In this research, an innovative adsorbent, iron and (3-Aminopropyl) triethoxysilane (APTES) biochar composites were synthesized from rice straw at a pyrolytic temperature of 400 and 600 °C. The pristine biochar was doped with Fe and NH_2 radicals to enhance the removal of Cr (VI) and Zn (II) heavy metals in the soil and solution. The maximum removal rate of Cr was 100.59 mg/g by APTES/SiBC 600 and for Zn^{2+} was 83.92 mg/g by Fe/BC 400. In the end, the results show that the amalgamation of Fe and APTES biochar composites in the soil strongly reduced the metal's toxic nature.

A new and novel Cu-coated bamboo (*Acidosasa longiligula*) shoot shell (BSS) biochar composite was synthesized for the removal of low quantity Re (VII). The reactivity of Cu-coated biochars in original pH solutions drastically reduced and it was observed that the adsorption ability was greater than 3–12 times in comparison with pristine biochar for the pH range of 3–6. The adsorption mechanism which was drawn from the batch and characterizations was electrostatic attraction and surface complexation (Hu et al. 2018).

3.3 Pharmaceutical Waste

Antibiotics are generally used in drugs for the treatment and prevention of diseases in humans and animals. But antibiotic residuals when released into the ecosystem have caused an upsurge in the growth of antibiotic-resistant bacteria (ARB) and antibiotic-resistance genes (ARGs) (Table 4). Due to these reasons, there is a surge in the number of research on remediation or removal aspects. For example, Daptomycin (DAP) is a cyclic lipopeptide antibiotic. The one of its kind batch analysis was done for the removal of DAP using two magnetic wood-based biochars from willow wood and pine wood. By employing the ball milling process the biochar was obtained, which was sent for the following analysis by laser particle size analyzer, elemental analyzer, ICP-OES, VSM, BET, TG-DTG, FTIR and SEM. Ultimately, max adsorption rate obtained as WSBC/Fe_3O_4 and PSBC/Fe_3O_4 was 217.39 and 212.77 mg/g which increased by 278–283% when compared to non-treated biochar (Ai et al. 2019).

In this analysis, the adsorption process was carried out by raw and steam-activated biochar from *Aegle marmelos* also called wood apple shell was used in the remediation of ibuprofen (IBP) from the model solution. Wood apple biochar (WAB) and Wood apple steam-activated biochar (WASAB) showed maximum removal around 90 and 95% respectively. The optimized IBP removal was seen at pH 2 for WASAB

Table 4 Biochar in the removal of pharmaceutical waste/drugs

Biochar description	Contaminant	Characteristics	Reference
Magnetic Willow Wood and pine wood-based biochar	Daptomycin	The maximum adsorption capacity was found to increase by 278–283% when compared to untreated biochars	Ai et al. (2019)
Steam activated *Aegle marmelos* (wood apple) fruit shell biochar	Ibuprofen	About 90–95% removal efficiency	Chakraborty, Banerjee et al. (2018)
Sugarcane bagasse biochar was chemically & physically activated by phosphoric acid and Steam activation	Ibuprofen	The chemically activated version showed 91% removal whereas the physically activated one exhibited 82%	Chakraborty, Show et al. (2018)
Syagrus coronate biochar supported with MgAl/layered double hydroxide	Diclofenac sodium	>82% removal efficiency seen	Dos Santos et al. (2020)
Biochar derived from luffa sponge	Norfloxacin (NOR)	99.86% removal efficiency	Feng et al. (2018)
Ultrafiltration-activated Pine bark biochar hybrid system	17 α-ethinyl estradiol (EE2), Ibuprofen (IBP) and carbamazepine (CBM)	Average retention: IBP > EE2 > CBM for the UF system alone and EE2 > IBP > CBM for the UF-ABC system	Kim et al. (2019)
Biochar derived from *Miscanthus floridulus* -supported by nanoscale zero-valent iron activated using H_2O_2	Ciprofloxacin	>70% removal of ciprofloxacin under optimal conditions	Mao et al. (2019)
Guayule bagasse and cotton gin waste-derived biochar	Sulfa pyridine-SPY, docusate-DCT and erythromycin-ETM	The adsorption capacity of SPY, DCT and ETM was dependent on contact time, pH biochar specific surface area and functional groups	Ndoun et al. (2021)
Plum kernels biochar was functionalized by chemical microwave by KOH	Naproxen, Carbamaxzepine	The strongest adsorption capacity was seen at pH 6	Paunovic et al. (2019)
Preparation of spherical biochar (derived from pure glucose) and non-spherical biochar (from pomelo peel wastes)	Paracetamol	For PRC maximum adsorption of spherical biochar was approximately double that of non-spherical biochar	Tran et al. (2020)

and 3 for WAB with an optimum dosage concentration of 0.33 g L^{-1} for WAB and 1 g L^{-1} for WASAB. Similarly, another analysis was conducted for the separation of but using different activated forms of sugarcane biochar as the adsorbent. Two activation methods were employed, i.e., chemical and physical activation using phosphoric acid giving out SCAB and by steam activation at 500 °C for 1 h SPAB. Therefore, SPAB and SCAB have had removal efficiency of 82% and 91% from an aqueous solution for a period of 18 and 12 h. Desorption analysis shows that the adsorbent can be reused after 4 cycles with methanol as desorbing agent. A mixture of antibiotic waste containing Ibuprofen (IBP), 17 α-ethinyl estradiol (EE2) and carbamazepine (CBM) was tried to remove by ultrafiltration using activated pine bark biochar hybrid system (UF-ABC) and powdered activated carbon (UF-PAC) because of it being a co-product in the combustion of waste, aromatization and porous properties. Three different conditioned systems considered for removal are UF, UF-ABC with/without humic acid (HA). Average retention was found to be as follows IBP > EE2 > CBM for individual UF and EE2 > IBP > CBM for UF-ABC system. It was also noted that the retention rate of UFPAC is a little higher than UF-ABC. The BC from cotton gin waste (CG) is the mixture of pharmaceuticals such as sulfa pyridine-SPY, docusate-DCT and erythromycin-ETM) from the model solution. It was seen that in basic conditions adsorption was mainly done by the development of strong negative charged H-bonding present among the sulfonamide moiety of SPY and surface carboxylic groups. It is known that many pharmaceuticals are weak acids or bases which make them reasonably hydrophobic in nature. This property is known to be effective in the removal of pollutants and further study of this biochar from the future perspective (Ndoun et al. 2021).

Another class of antibiotic addressed for its removal, i.e. norfloxacin (NOR) present in wastewater from luffa sponge biochar. NOR is a ubiquitous antibiotic that is generally used to treat enteritis dysentery. From batch analysis, it was seen that the adsorption of BC was 250 mg/g. The kinetic data suggest the following pseudo-second-order model, with a high $R^2 > 0.99$ value which indicates that chemisorption is the rate-limiting factor. From FTIR characterization it indicates that BC has a high number of acidic oxygen-containing groups like carboxyl, phenol hydroxyl, lactone and carbonyl. A novel composite called the Mg–Al/layered double hydroxide supported by *Syagrus coronata* biochar was synthesized for the removal of diclofenac sodium antibiotics. Mg Al/LDH was made by the co-precipitation method. From the batch experiments, it was seen that the adsorbent has a high rate of DS removal of about >82% (Dos Santos et al. 2020).

Ciprofloxacin antibiotic is presently one of the most frequent pollutants present in the aquatic system. Hence a biochar-supported nanoscale zero-valent iron (BC/nZVI) was proposed to be used as an adsorbent that was activated by hydrogen peroxide (H_2O_2). The biochar's source was *Miscanthus floridulus* and BC was subjected to liquid-phase reduction precipitation technique. It was observed that >70% of ciprofloxacin was adsorbed under the optimal condition: acidic condition whose pH ranges from 3 to 4, low doses of H_2O_2 and temperature of 298 K. Some other antibiotics such as naproxen and carbamazepine are known to have harmful effects on the environment. Hence a new functionalized biochar (WpOH) was fabricated from the source wild plum kernels by employing pyrolysis and microwave-assisted KOH treatment. In this study, naproxen was studied briefly than carbamazepine. WpOH exhibited its surface being domineered by micropore due to surface functionalities mostly with O_2 present group. The removal percentage of naproxen was heavily reliant on pH of the solution bearing a high adsorption capacity of 73.14 mg/g at pH 6. A novel adsorbent in the form of biochar spheres was synthesized with high porosity. This study uses the antibiotic paracetamol (PRC) for batch studies. The spheres have been prepared from pomelo peel wastes and pure glucose. A comparison is carried out for non-spherical biochar, which is made from dried parts of pomelo peels and was pyrolyzed at a different temperature range from 600 to 900 °C. The spherical biochar was synthesized by a two-phase method. Firstly, hydrothermal carbonization along with pyrolysis takes place. This step uses glucose in its pure form was used. The resultant spherical hydrochar precipitant is washed thoroughly. The second step is where the hydrochar is pyrolyzed for various carbonization temperatures to produce the spheres. It was revealed that results show that the adsorption was to some extent affected solution pH from 2 to 11 and it reached equilibrium within 120 min. The max adsorption rate of spherical biochar for PRC was doubled that of non-spherical biochar.

References

Acemioğlu B (2022) Removal of a reactive dye using NaOH-activated biochar prepared from peanut shell by pyrolysis process. Int J Coal Prep Util 42:671–693

Adeyemo AA, Adeoye IO, Bello OS (2017) Adsorption of dyes using different types of clay: a review. Appl Water Sci 7(2):543–568. https://doi.org/10.1007/s13201-015-0322-y

Ahmad T, Danish M (2018) Prospects of banana waste utilization in wastewater treatment: a review. J Environ Manag 206:330–348. https://doi.org/10.1016/J.JENVMAN.2017.10.061

Ai T, Jiang X, Liu Q, Lv L, Wu H (2019) Daptomycin adsorption on magnetic ultra-fine wood-based biochars from water: kinetics, isotherms, and mechanism studies. Bioresour Technol 273:8–15

Alam MS, Bishop B, Chen N, Safari S, Warter V, Byrne JM, Warchola T, Kappler A, Konhauser KO, Alessi DS (2020) Reusable magnetite nanoparticles–biochar composites for the efficient removal of chromate from water. Scientific Rep 10(1):1–12

Al-Rahbi AS, Williams PT (2017) Hydrogen-rich syngas production and tar removal from biomass gasification using sacrificial tyre pyrolysis char. Appl Energy 190:501–509

Bharti V, Vikrant K, Goswami M, Tiwari H, Sonwani RK, Lee J, Tsang DCW, Kim K, Saeed M, Kumar S, Rai BN, Giri BS, Singh RS (2019) Biodegradation of methylene blue dye in a batch and continuous mode using biochar as packing media. Environ Res 171:356–364

Braghiroli FL, Bouafif H, Neculita CM, Koubaa A (2018) Activated Biochar as an effective sorbent for organic and inorganic contaminants in water. Water Air Soil Pollut 229(7):1–22. https://doi.org/10.1007/s11270-018-3889-8

Chahinez H, Abdelkader Q, Leila Y, Tran HN (2020) One-stage preparation of palm petiole-derived biochar: characterization and application for adsorption of crystal violet dye in water. Environ Technol Innov 19:100872

Chakraborty D, Mukhopadhyay K (2014) Introduction (D. Chakraborty & K. Mukhopadhyay (eds.); 1st ed., pp. 1–21). Springer Berlin Heidelberg. https://doi.org/10.1007/978-94-017-8929-5_1

Chakraborty P, Banerjee S, Kumar S, Sadhukhan S, Halder G (2018) Elucidation of ibuprofen uptake capability of raw and steam activated biochar of Aegle marmelos shell: isotherm, kinetics, thermodynamics and cost estimation. Process Saf Environ Prot 118:10–23

Chakraborty P, Show S, Banerjee S, Halder G (2018) Mechanistic insight into sorptive elimination of ibuprofen employing bi-directional activated biochar from sugarcane bagasse: performance evaluation and cost estimation. J Environ Chem Eng 6:5287–5300

Crini G, Lichtfouse E, Wilson LD, Morin-Crini N (2018) Conventional and non-conventional adsorbents for wastewater treatment. Environ Chem Lett 1–19. https://doi.org/10.1007/s10311-018-0786-8

Das SK, Ghosh GK, Avasthe R (2023) Conversion of crop, weed and tree biomass into biochar for heavy metal removal and wastewater treatment. Biomass Conv Bioref 13:4901–4914

De Caprariis B, De Filippis P, Petrucci E, Scarsella M (2018) Activated biochars used as adsorbents for dyes removal. Chem Eng Trans 65:103–108

Dos Santos KJL, Dos Santos GES, De Sá IMGL, Ide AH, Duarte JLS, De Carvalho SHV, Soletti JI, Meili L (2019) Wodyetia bifurcata biochar for methylene blue removal from aqueousmatrix. Bioresour Technol 293:122093

Eltaweil AS, Mohamed HA, El-Monaem EMA, El-Subruiti GM (2020) Mesoporous magnetic biochar composite for enhanced adsorption of malachite green dye: characterization,adsorption kinetics, thermodynamics and isotherms. Adv Powder Technol 31:1253

Feng F, Chen X, Wang Q, Xu W, Long L, El-Masry GN, Wan Q, Yan H, Cheng J, Yu X (2020) Use of Bacillus-siamensis-inoculated biochar to decrease uptake of dibutyl phthalate in leafy vegetables. J Environ Manag 253:109636

Feng Y, Liu Q, Yu Y, Kong Q, Zhou LL, Du YD, Wang XF (2018) Norfloxacin removal from aqueous solution using biochar derived from luffa sponge. J Water Supply: Res Technol—AQUA 67:703–714

Gabhane JW, Bhange VP, Patil PD, Bankar ST, Kumar S (2020) Recent trends in biochar production methods and its application as a soil health conditioner: a review. SN Appl Sci 2(7):1–21

Gao J, Liu Y, Li X, Yang M, Wang J, Chen Y (2020) A promising and cost-effective biochar adsorbent derived from jujube pit for the removal of Pb(II) from aqueous solution. Sci Rep10:7473

Gao Y, Wang X, Wang J, Li X, Cheng J, Yang H, Chen H (2013) Effect of residence time on chemical and structural properties of hydrochar obtained by hydrothermal carbonization of water hyacinth. Energy 58:376–383

Herath I, Zhao FJ, Bundschuh J, Wang P, Wang J, Ok YS, Palansooriya KN, Vithanage M (2020) Microbe mediated immobilization of arsenic in the rice rhizosphere after incorporation of silica impregnated biochar composites. J Hazard Mater 398:123096

Hu H, Sun L, Jiang B, Wu H, Huang Q, Chen X (2018) Low concentration Re(VII) recovery from acidic solution by Cu-biochar composite prepared from bamboo (Acidosasa longiligula) shoot shell. Miner Eng 124:123–136

Huang D, Liu L, Zeng G, Xu P, Huang C, Deng L, Wang R, Wan J (2017a) The effects of rice straw biochar on indigenous microbial community and enzymes activity in heavy metal-contaminated sediment. Chemosphere 174:545–553

Huang D, Wang X, Zhang C, Zeng G, Peng Z, Zhou J, Cheng M, Wang R, Hu Z, Qin X (2017b) Sorptive removal of ionizable antibiotic sulfamethazine from aqueous solution by graphene oxide-coated biochar nanocomposites: influencing factors and mechanism. Chemosphere 186:414–421

Inyang M, Gao B, Zimmerman A, Zhang M, Chen H (2014) Synthesis, characterization, and dye sorption ability of carbon nanotube–biochar nanocomposites. Chem Eng J 236:39–46

Inyang M, Gao B, Zimmerman A, Zhou Y, Cao X (2015) Sorption and cosorption of lead and sulfapyridine on carbon nanotube-modified biochars. Environ Sci Pollut Res 22(3):1868–1876

Jung KW, Hwang MJ, Jeong TU, Ahn KH (2015) A novel approach for preparation of modified-biochar derived from marine macroalgae: dual purpose electro-modification for improvement of surface area and metal impregnation. Bioresour Technol 191:342–345

Kim S, Park CM, Jang A, Jang M, Hernández-Maldonado AJ, Yu M, Heo J, Yoon Y (2019) Removal of selected pharmaceuticals in an ultrafiltration-activated biochar hybrid system. J Membr Sci 570–571:77–84

Krishna Murthy TP, Gowrishankar BS, Krishna RH, Chandraprabha MN, Sreenivasa Rao R (2019) Influence of fuel nature on dye adsorption efficiency of solution combustion derived zinc oxide nanoparticles: a comparative study. Mater Res Express 6(5):55512. https://doi.org/10.1088/2053-1591/aafe3f

Kyzas G, Kostoglou M, Kyzas GZ, Kostoglou M (2014) Green adsorbents for wastewaters: a critical review. Materials 7(1):333–364. https://doi.org/10.3390/ma7010333

Lai WY, Lai CM, Ke GR, Chung RS, Chen CT, Cheng CH, Pai CW, Chen SY, Chen CC (2013) The effects of woodchip biochar application on crop yield, carbon sequestration and greenhouse gas emissions from soils planted with rice or leaf beet. J Taiwan Inst Chem Eng 44(6):1039–1044

Li L, Rowbotham JS, Greenwell CH, Dyer PW (2013) An introduction to pyrolysis and catalytic pyrolysis: versatile techniques for biomass conversion. Elsevier, pp 173–208

Li Q, Tang L, Hu J, Jiang M, Shi X, Zhang T, Li Y, Pan X (2018) Removal of toxic metals from aqueous solution by biocharsderived from long-rootEichhornia crassipes. R Soc Open Sci 5:180966

Liu C, Wang W, Wu R, Liu Y, Lin X, Kan H, Yunwu Zheng (2020) Preparation of acid-and alkali-modified biochar for removal of methylene blue pigment. ACS Omega 5:30906–30922

Lyu H, Tang J, Cui M, Gao B, Shen B (2020) Biochar/iron (BC/Fe) composites for soil and groundwater remediation: synthesis, applications, and mechanisms. Chemosphere 246:125609

Ma L, Wang Q, Islam SM, Liu Y, Ma S, Kanatzidis MG (2016) Highly selective and efficient removal of heavy metals by layered double hydroxide intercalated with the MoS42–ion. J Am Chem Soc 138(8):2858–2866

Manikandan SK, Giannakoudakis DA, Prekodravac JR, Nair V, Colmenares JC (2023) Role of catalyst supports in biocatalysis. J Chem Technol Biotechnol 98:7–21

Manikandan SK, Pallavi P, Sujatha SKB, Bhattacharjee D, Giannakoudakis DA, Katsoyiannis IA, Nair V (2023) Effective usage of biochar and microorganism for removal of heavy metalions and pesticides. Molecules 28:719

Mao Q, Zhou Y, Yang Y, Zhang J, Liang L, Wang H, Luo S, Luo L, Jeyakumar P, Ok YS, Rizwan M (2019) Experimental and theoretical aspects of biochar-supported nanoscale zero-valentiron activating H2O2 for ciprofloxacin removal from aqueous solution. J Hazard Mater 380:120848

Medha I, Chandra S, Vanapalli KR, Samal B, Bhattacharya J, Das BK (2021) (3-Aminopropyl)triethoxysilane and iron rice straw biochar composites for the sorption of Cr (VI) and Zn (II) using the extract of heavy metals contaminated soil. Sci Total Environ 771:144764

Murtaza G, Ditta A, Ullah N, Usman M, Ahmed Z (2021) Biochar for the management of nutrient impoverished and metal contaminated soils: preparation, applications, and prospects. *J Soil Sci Plant Nutr*:1–23

Ndoun MC, Elliott HA, Preisendanz HE, Williams CF, Knopf A, Watson JE (2021) Adsorption of pharmaceuticals from aqueous solutions using biochar derived from cotton gin waste and guayule bagasse. Biochar 3:89–104

Nicolaou E, Philippou K, Anastopoulos I, Pashalidis I (2019) Copper adsorption by magnetized pine-needle biochar. Processes 7:903

Panahi HKS, Dehhaghi M, Ok YS, Nizami AS, Khoshnevisan B, Mussatto SI, Aghbashlo M, Tabatabaei M, Lam SS (2020) A comprehensive review of engineered biochar: production, characteristics, and environmental applications. J Clean Prod 270:122462

Panwar NL, Pawar A (2020) Influence of activation conditions on the physicochemical properties of activated biochar: a review. *Biomass Conver Biorefinery*:pp 1–23

Park J, Wang JJ, Meng Y, Wei Z, DeLaune RD, Seo D (2019) Adsorption/desorption behavior of cationic and anionic dyes by biochars prepared at normal and high pyrolysis temperatures. Colloids Surf A: Physicochem Eng Asp 572:274–282.

Paunovic O, Pap S, Maletic S, Taggart MA, Boskovic N, Turk Sekulic M (2019) Ionisable emerging pharmaceutical adsorption onto microwave functionalised biochar derived from novel lignocellulosic waste biomass. J Colloid Interface Sci 547:350–360.

Pentananunt R, Rahman AM, Bhattacharya SC (1990) Upgrading of biomass by means of torrefaction. Energy 15(12):1175–1179

Pradhan S, Abdelaal AH, Mroue K, Al-Ansari T, Mackey HR, McKay G (2020) Biochar from vegetable wastes: agro-environmental characterization. Biochar 2(4):439–453

Sadegh H, Ali GAM, Gupta VK, Makhlouf ASH, Shahryari-ghoshekandi R, Nadagouda MN, Sillanpää M, Megiel E (2017) The role of nanomaterials as effective adsorbents and their applications in wastewater treatment. J Nanostr Chem 7(1):1–14. https://doi.org/10.1007/s40097-017-0219-4

Shah Maulin P (2020) Microbial bioremediation & biodegradation. Springer

Shah Maulin P (2021a) Removal of refractory pollutants from wastewater treatment plants. CRC Press

Shah Maulin P (2021b) Removal of emerging contaminants through microbial processes. Springer

Shen Z, Zhang J, Hou D, Tsang DC, Ok YS, Alessi DS (2019) Synthesis of MgO-coated corncob biochar and its application in lead stabilization in a soil washing residue. Environ Int 122:357–362

Singh AK, Giannakoudakis DA, Arkas M, Triantafyllidis KS, Nair V (2023) Composites of Lignin-based Biochar with BiOCl for photocatalytic water treatment: RSM studies for process optimization. Nanomater 13:735.

Singh NB, Nagpal G, Agrawal S, Rachna (2018) Water purification by using adsorbents: a review. Environ Technol Innov 11:187–240. https://doi.org/10.1016/J.ETI.2018.05.006

Thines KR, Abdullah EC, Mubarak NM, Ruthiraan M (2017) Synthesis of magnetic biochar from agricultural waste biomass to enhancing route for waste water and polymer application: a review. Renewable Sustain Energy Rev 67:257–276

Tran HN, Tomul F, Ha NTH, Nguyen DT, Lima EC, Le GT, Chang C, Masindi V, Woo SH (2020) Innovative spherical biochar for pharmaceutical removal from water: insight into adsorption mechanism. J Hazard Mater 394:122255

Tripathi M, Sahu JN, Ganesan P, Monash P, Dey TK (2015) Effect of microwave frequency on dielectric properties of oil palm shell (OPS) and OPS char synthesized by microwave pyrolysis of OPS. J Anal Appl Pyrolysis 112:306–312

Wang S, Gao B, Li Y, Zimmerman AR, Cao X (2016) Sorption of arsenic onto Ni/Fe layered double hydroxide (LDH)-biochar composites. RSC Adv 6(22):17792–17799

Wang RZ, Huang DL, Liu YG, Zhang C, Lai C, Wang X, Zeng GM, Gong XM, Duan A, Zhang Q, Xu P (2019a) Recent advances in biochar-based catalysts: properties, applications and mechanisms for pollution remediation. Chemical Engineering Journal, 371:380–403. Elsevier B.V. https://doi.org/10.1016/j.cej.2019.04.071

Wang X, Gu Y, Tan X, Liu Y, Zhou Y, Hu X, Cai X, Xu W, Zhang C, Liu S (2019b) Functionalized biochar/clay composites for reducing the bioavailable fraction of arsenic and cadmium in river sediment. Environ Toxicol Chem 38(10):2337–2347

Wang L, Li X, Tsang DC, Jin F, Hou D (2020) Green remediation of Cd and Hg contaminated soil using humic acid modified montmorillonite: immobilization performance under accelerated ageing conditions. J Hazard Mater 387:122005

Wang L, Ok YS, Tsang DC, Alessi DS, Rinklebe J, Mašek O, Bolan NS, Hou D (2021) Biochar composites: emerging trends, field successes, and sustainability implications. Soil Use Manag.

Wei M, Liu X, He Y, Xu X, Wu Z, Yu K, Zheng X (2020) Biochar inoculated with Pseudomonas putida improves grape (Vitis vinifera L.) fruit quality and alters bacterial diversity. Rhizosphere, 16:100261

Wu P, Wang Z, Wang H, Bolan NS, Wang Y, Chen W (2020) Visualizing the emerging trends of biochar research and applications in 2019: a scientometric analysis and review. Biochar 2(2):135–150. https://doi.org/10.1007/s42773-020-00055-1

Xiao R, Wang JJ, Li R, Park J, Meng Y, Zhou B, Pensky S, Zhang Z (2018) Enhanced sorption of hexavalent chromium [Cr(VI)] from aqueous solutions by diluted sulfuric acid-assisted MgO-coated biochar composite. Chemosphere 208:408–416

Xiong B, Zhang Y, Hou Y, Arp HPH, Reid BJ, Cai C (2017) Enhanced biodegradation of PAHs in historically contaminated soil by M. gilvum inoculated biochar. Chemosphere 182:316–324

Xue L, Gao B, Wan Y, Fang J, Wang S, Li Y, Muñoz-Carpena R, Yang L (2016) High efficiency and selectivity of MgFe-LDH modified wheat-straw biochar in the removal of nitrate from aqueous solutions. J Taiwan Inst Chem Eng 63:312–317

Yang F, Zhang S, Sun Y, Tsang DC, Cheng K, Ok YS (2019) Assembling biochar with various layered double hydroxides for enhancement of phosphorus recovery. J Hazard Mater 365:665–673

Yang Z, Xing R, Zhou W, Zhu L (2020) Adsorption characteristics of ciprofloxacin onto g-MoS 2 coated biochar nanocomposites. Front Environ Sci Eng 14(3):1–10

Yu Z, Qiu W, Wang F, Lei M, Wang D, Song Z (2017) Effects of manganese oxide-modified biochar composites on arsenic speciation and accumulation in an indica rice (Oryza sativa L.) cultivar. Chemosphere 168:341–349

Zhang P, O'Connor D, Wang Y, Jiang L, Xia T, Wang L, Tsang DC, Ok YS, Hou D (2020) A green biochar/iron oxide composite for methylene blue removal. J Hazard Mater 384:121286

Zhang S, Zhang H, Liu F, Yang F, Zhou S, Zheng K, Chu C, Liu L, Ju M (2019) Effective removal of Cr(VI) from aqueous solution by biochar supported manganese sulfide. RSC Adv 9:31333–31342

Zhou M, Zhang C, Yuan Y, Mao X, Li Y, Wang N, Wang S, Wang X (2020) Pinewood outperformed bamboo as feedstock to prepare biochar-supported zero-valent iron for Cr6+reduction. Environ Res 187:109695

Zubair M, Manzar MS, Mu'azu ND, Anil I, Blaisi NI, Al-Harthi MA (2020) Functionalized MgAl-layered hydroxide intercalated date-palm biochar for Enhanced Uptake of Cationic dye: Kinetics, isotherm and thermodynamic studies. Appl Clay Sci 190:105587

Emerging Methods Used in Bioremediation and Nano Techniques for the Removal of Heavy Metals in Contaminated Soil and Industrial Effluents

Anisha Susan Johnson, T. Franklin Rupa, and K. Veena Gayathri

1 Introduction

There has been a significant rise in the discharge of heavy metals into nature as a result of increased urbanization, industrialization and agricultural production. They are naturally released as a result of volcanic eruptions and metal corrosion, causing minimal environmental damage. On the other hand, anthropogenic activities like tannery, mining, smelting and the disposal of untreated industrial effluents pose an alarming threat to the entire ecosystem (Zhang et al. 2020). Heavy metals, unlike other pollutants, are extremely difficult to degrade and eliminate from contaminated areas. These metals display toxicity at concentrations as low as 1.0–10 mg/L (Ahemad 2019). Direct contact, ingestion and inhalation can all result in major health implications, gene mutations, neurological disorders and cancer (Yaghmaeian and Jaafari 2018). A variety of remediation methods are available owing to the pressing concerns of this issue. Earlier, conventional methods such as membrane filtration, electrodeposition, floatation, ion exchange and electro-coagulation were employed for heavy metal remediation. However, these methods are very expensive and can moreover result in the production of secondary toxic sludge (Yaghmaeian and Jaafari 2018). Thus, bioremediation provides the best solution. It is an environmentally friendly approach of removing heavy metal ions from contaminated soil and industrial wastewater by employing the natural abilities of microorganisms, fungi, algae, and plants. It is broadly classified into in-situ and ex-situ mode of bioremediation (Chibueze et al. 2016). In-situ bioremediation methods treat heavy metal contaminants directly on site, whereas ex-situ bioremediation methods collect contaminants from the source and treat them somewhere else. The main objectives of this review is

A. S. Johnson · T. F. Rupa · K. V. Gayathri (✉)
Department of Biotechnology, Stella Maris College (Autonomous), Chennai, India
e-mail: veenagayathri@stellamariscollege.edu.in

M. P. Shah (ed.), *Advanced and Innovative Approaches of Environmental Biotechnology in Industrial Wastewater Treatment*,
https://doi.org/10.1007/978-981-99-2598-8_12

to (1) discuss the sources and factors involved in heavy metal pollution (2) outline the different methods of bio-remediation using bacteria, fungi, algae, plant and nanoparticles (3) highlight the role of bioreactors in heavy metal remediation. This review specifically considers an attempt to summarize the possible applications along with the advantages and limitations of various bioremediation strategies.

2 Heavy Metal Pollution

Heavy metals, also referred to as trace elements or metallic elements, are elements with atomic densities greater than 5 g/cm^3 (Akpor 2014). Heavy metals are difficult to degrade in the environment and as a result, they contaminate the ecosystem and pose health risks to both animals and plants. These metals include Arsenic (As), Cadmium (Cd), Chromium (Cr), Copper (Cu), Lead (Pb), Mercury (Hg), Nickel (Ni), Selenium (Se), Silver (Ag), Zinc (Zn), Gold (Au), Manganese (Mn) (Das and Osborne 2018; Briffa et al. 2020). Its growing popularity has created significant concern in both terrestrial and aquatic environments. They are also known for their persistence since they do not dissipate quickly. They are also less mobile in water, which leads to silt accumulation beneath the water body (Dash et al. 2021). They accumulate in living beings, including humans, as they enter the food chain through the process of bioaccumulation. Furthermore, they are difficult to digest and might cause serious health issues. They are known to cause organ damage and are also classified as human carcinogens (Tchounwou et al. 2012). In nature, they are extremely poisonous and non-biodegradable. Heavy metals, such as copper, produce reactive oxygen species which can act as a soluble transporter of electrons, resulting in significant cellular and nerve damage (Kaur and Roy 2020). As a result of these consequences, they must be treated and removed from the natural environment (Barakat 2011).

3 Sources of Heavy Metal Pollution

Heavy metal pollution is caused from a variety of sources including water pollution, road construction, the automobile industry, agriculture (pesticides, insecticides and fertilizers) and natural causes such as volcanic activity, metal corrosion, metal evaporation, soil erosion and geological weathering. Anthropogenic activities like mining and the disposal of untreated industrial effluent, as well as the use of herbicides, insecticides, and fertilizers in agricultural operations, all contribute to heavy release of untreated heavy metals. A graphical representation of the different sources of heavy metals is shown in Fig. 1. Use of pesticides, insecticides and fertilizers in agriculture can lead to heavy metal leaching into nearby soil and water bodies. Other major sources of heavy metal pollution includes domestic and industrial effluents (Zhang et al. 2020). Water bodies naturally contain small concentrations of heavy metal that are required for ecological balance. Unfortunately, these extensive pollution-causing

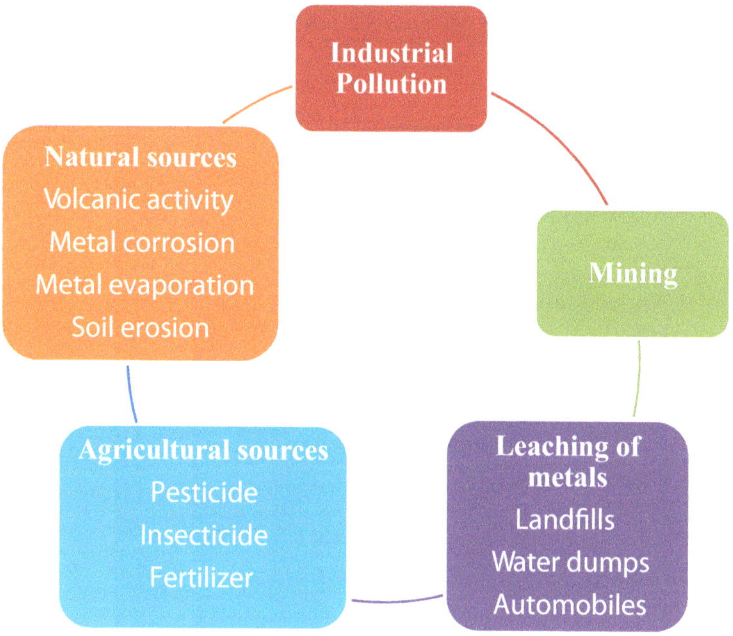

Fig. 1 Representation of source of heavy metals

activities disrupt this ecological equilibrium to a great extent (Xu et al. 2021). The hazardous effects of these heavy metals on microorganisms, plants, animals and humans are listed in Tables 1 and 2, respectively. According to recent studies, the long-term dumping of untreated wastewater from industrial sources can degrade water quality, rendering it unsafe for human consumption (Kapahi and Sachdeva 2019).

4 Bioindicators of Heavy Metals

Bioindicators are living organisms that are used to measure or track the changes in the biological and non-biological components of an environment (Han et al. 2015). The significance of using bioindicators is to assess pollution in a specific area, to see the state of the environment and to modify the contaminated component in a short period of time (Table 3).

Table 1 Types of heavy metals and their toxic effect on animals, plants and microorganisms

Heavy metal	Effect on animals	Effect on plants	Effect on microbes	Reference
Cadmium	Affects calcium & phosphorus metabolism in bone	Decreased seed germination & lipid content	Denaturation of protein & damage nucleic acid, nitrogen mineralization	Verma et al. (2018)
Chromium	Affects reproduction system	Decreased enzyme activity & plant growth	Slow growth, slow uptake of oxygen	Verma et al. (2018)
Mercury	Alters the enzyme activity	Inhibition of photosynthesis	Inhibition of enzyme activity, denaturation of nucleic acid	Azizi et al. (2016)
Lead	Affects reproductive system	Reduces plant growth and photosynthesis	Inhibition of enzyme activity	Akpor (2014)

Table 2 Types of heavy metals, their sources and toxic effects on human

Heavy metals	Sources	Effect on humans	Reference
Arsenic	Pesticide, metal smeller	Bronchitis, dermatitis, gastrointestinal tract infection, carcinogen	Grace et al. (2020)
Cadmium	Ceramic, metal finishing industries, tobacco smoking	Gastrointestinal tract infection, kidney damage, renal disorder, carcinogen	Yadav et al (2017b)
Copper	Water, food, mining, plating, copper polish	Abdominal pain, vomiting, gastrointestinal side effect, liver and kidney damage	Gupta et al. (n.d.)
Zinc	Metal plating, plastic items, pumps, water	Kidney failure, stomach ache, liver failure, nervous system damage	Azizi et al. (2016)

Table 3 Examples of few bioindicators used for the indication of heavy metals

Bioindicator species	Indication of heavy metals
Daphnia Magna	Cu, Cd, Zn, Se
Fishes	Mn, Co, Fe, Cu, Zn
Klebsormidium	Fe
Caesalpinia pulcherrima and grass	Pb, Cu, Cd, Mn, Ni
Macro invertibrates	Chlorine
Litchen	Sulfur dioxide, Nitrogen

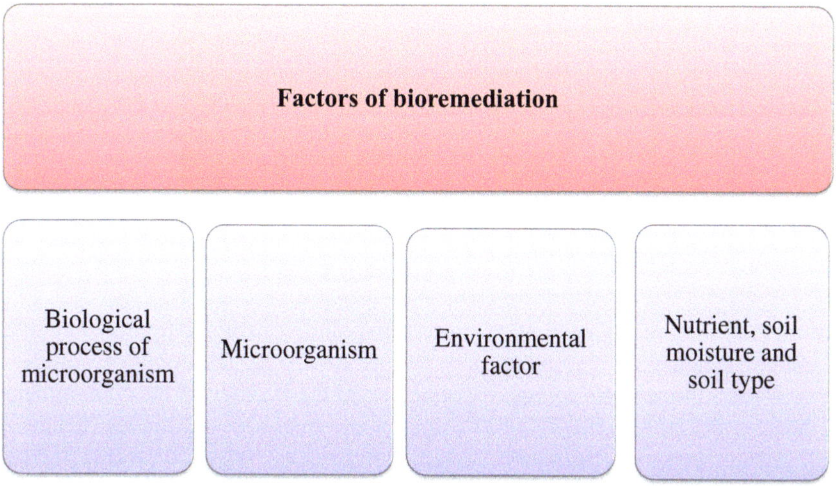

Fig. 2 Representation of factors affecting the bioremediation of heavy metals (Science and Sharma 2012)

5 Bioremediation of Heavy Metals

Bioremediation is the technique of eliminating hazardous pollution and chemicals using microbes to transform them into less hazardous pollutants or chemicals. More than fungi, algae or protozoans, bacteria are commonly used in the bioremediation process. It is a green method that involves the use of biological agents to break down heavy metals. Bioremediation systems usually work in an aerobic environment (Arora et al. 2017). Different enzyme activities are engaged in the potential to breakdown the pollutants (Kumar and Gunasundari, n.d.). However, traditional approaches are not as effective as bioremediation methods (Hou 2014). There are different factors that influence the rate of bioremediation like the biological process of the microorganism, environmental factors, concentration of contaminants, etc. A graphical representation of the factors affecting the bioremediation of heavy metals is shown in Fig. 2.

6 Types of Bioremediation Methods

6.1 In-situ Bioremediation

This method entails treating polluted substances at the source of the contamination. Since it does not necessitate any excavation, it causes little to no soil disturbance. As there is no additional cost for the excavation procedure, these techniques are less expensive than ex-situ bioremediation. In-situ bioremediation approaches have successfully treated chlorinated solvents, dyes, heavy metals and hydrocarbons. Important environmental variables such as electron acceptor status, moisture content, nutrient availability, pH and temperature must be maintained for a successful in-situ bioremediation strategy (Chibueze et al. 2016).

6.2 *Ex-situ Bioremediation*

This method entails digging contaminants from polluted areas and transporting them to another location for treatment. The cost of treatment, the depth of pollution, the types of pollutant, the degree of pollutant and the geology of the contaminated site are the factors to consider when using ex-situ treatment techniques (Chibueze et al. 2016). The most effective bioremediation strategy focuses on increasing soil quality so that uniform oxygenation, nutrient distribution and moisture control may be achieved (Kumar and Gunasundari, n.d.). A graphical representation of the types of bioremediations is given in Fig. 3 (Tables 4 and 5).

7 Microbial Remediation of Heavy Metals

In the past two decades, researchers have been working on advanced bioremediation methods that are environmentally friendly and cost-effective (Kit and Chang 2020). Microorganisms are versatile, thriving even under the harshest conditions on the planet. The prime bioremediators of heavy metals are bacteria, fungi, algae and plants. They can be used as an effective alternative to physio-chemical remediation methods as they have the capacity to biodegrade and detoxify harmful heavy

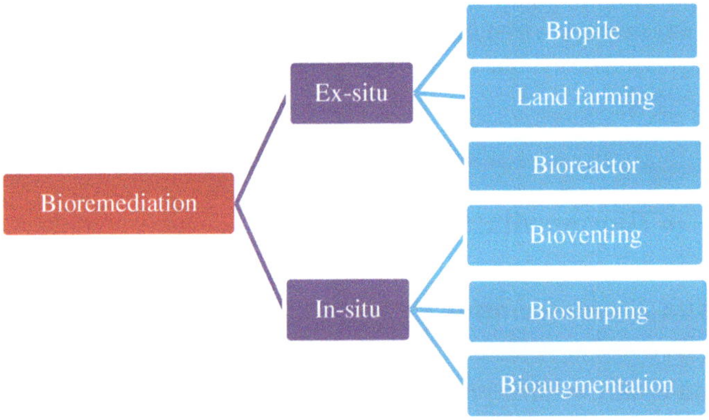

Fig. 3 Representation of types of bioremediation (Azubuike et al. 2016)

Table 4 Types of in-situ and ex-situ bioremediation and their advantages

Bioremediation	Types of in-situ	Advantages	Reference
In situ			
Contaminant is removed at the site	Bioventing	• Quickly transforms the pollutant into a harmless state • Less time consuming	Azubuike et al. (2016)
	Bioslurping	• Increases groundwater oxygen concentration	Azubuike et al. (2016)
	Bioaugmentation	• Completely degrades the contaminant to ethylene & chloride	Science and Sharma (2012)
Ex situ			
Contaminant is taken away from the site and treated to remove the harmful chemical and pollutant	Biopile	• It helps in the increase of microbial population • Cost effective • It is very effective in the removal of pollutants in cold regions	Science and Sharma (2012)
	Bioreactor	• Parameters of bioprocess can be easily maintained and controlled • Wide range of pollutants is removed with the help of various types of bioreactor	Gahlawat and Choudhury (2019)
	Land farming	• Less equipment is required at lower cost • It can be used in hydrocarbon pollutant sites • Minimal energy requirement	Gahlawat and Choudhury (2019)

Table 5 An overview of the advantages and disadvantages of in-situ and ex-situ method of bioremediation

Advantages of in-situ bioremediation	**Disadvantages of in-situ bioremediation**
Removes the contaminants present on site	Time-consuming process
Large amount of soil can be treated	Difficult to monitor
Less expensive when compared to ex-situ	Does not completely remove the contaminants
Reduces handling exposure	Lower treatment efficiency
Advantage of ex-situ	**Disadvantage of ex-situ**
Faster biodegradation process	Expensive & potentially dangerous
Easy to monitor the environment	Increased exposure to toxic metals

metal components into less toxic forms sustainably (Arora et al. 2017). Some of the most prominent microbial species employed for heavy metal bioremediation include *Flavobacterium, Pseudomonas, Bacillus, Arthrobacter, Corynebacterium, Methosinus, Rhodococcus, Mycobacterium, Stereum hirsutum, Nocardia, Methanogens, Aspergillus niger, Pleurotus ostreatus, Rhizopus arrhizus, Azotobacter, Alcaligenes* and *Phormidium valderium, Ganoderma applanatum* (S. Verma and Kuila 2019). However, anaerobic bacteria are not widely used in the bioremediation process (Tarekegn et al. 2020).

8 Metal-Microbe Interactions

Metal-microbe interactions such as biotransformation, bio-mineralization, bioleaching, biodegradation, bioaccumulation, intracellular chelation, intracellular precipitation and adsorption takes place when microorganisms are constantly exposed to metals toxins either in water or in contaminated soil (Fig. 4) (Arora et al. 2017). Biosorption is a rapid and reversible process in which metal ions interact with functional groups (carboxyl, phosphate, amine, hydroxyl) on the microbial cell surface through physico-chemical interactions such as ion exchange, crystallization, and precipitation (Kumar & Gunasundari, n.d.). Some heavy metals can also be remediated by microorganisms through metabolic and enzymatic reduction (Kumar and Gunasundari, n.d.).

Specific biotechnological methods are employed to develop microorganisms that produce degradative enzymes for bioremediation (Dangi et al. 2019). However, prolonged exposure of these microorganisms to heavy metals can cause harmful effects like DNA damage, disruption of the microbial cell membrane, hindrance in protein synthesis and cell division (Volarić et al. 2021). To overcome this and to resist heavy metals like arsenic and cadmium, they evolve many defense mechanisms such as solvent efflux pumps, chemi-osmotic pumps and energy-dependent efflux systems (Dixit et al. 2015).

9 Bioremediation Using Bacteria

Bacteria are employed to eliminate heavy metals from the environment as they are inexpensive and environmentally acceptable. They can either be used in the form of extracts or as a whole to remove the metal contaminants (Bertolino et al. 2014). *Bacillus subtilis, Pseudomonas aeruginosa, Pseudomonas fluorescens,* and *Escherichia coli* are used for reducing toxic hexavalent chromium to a less toxic trivalent form (Volarić et al. 2021). Studies have also shown that the use of bacterial consortium is more efficient in the bioremediation process than the use of a

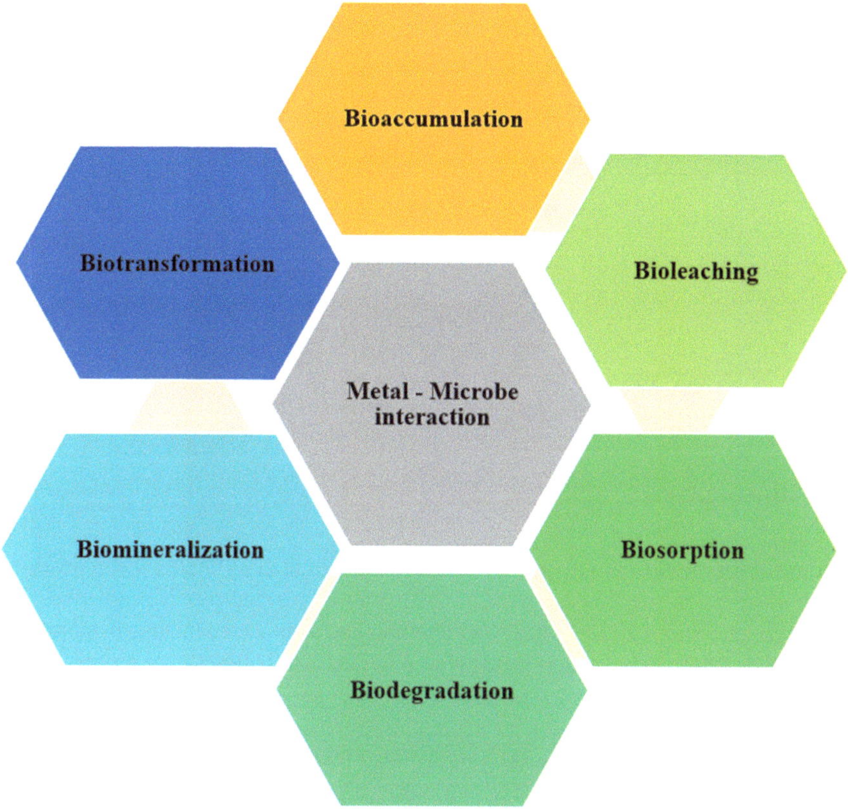

Fig. 4 Representation of metal-microbe interactions in bioremediation

single bacterial strain. Heavy metals such as Cu, Zn, Pb, Cd, and Cr are successfully removed from industrial wastewater and contaminated soil by using bacterial consortiums (Volarić et al. 2021). However, the efficiency of the procedure is highly dependent on the type of bacteria and the concentration of metal pollutants.

10 Mechanism of Bacterial Remediation of Heavy Metals

Bacterial heavy metal removal is accomplished through a variety of mechanisms. The metal ions are accumulated or entrapped onto the cell surface by the process of biosorption followed by its bioaccumulation in the intracellular region of bacteria. Metals containing organic compounds are degraded by enzymatic reduction during the biosynthesis and/or biodegradation processes. Few microbes such as sulfate-reducing bacteria solubilize and precipitate the metal complex into simpler forms (Ahemad 2019). A graphical representation of the above-mentioned mechanisms is shown in Fig. 5.

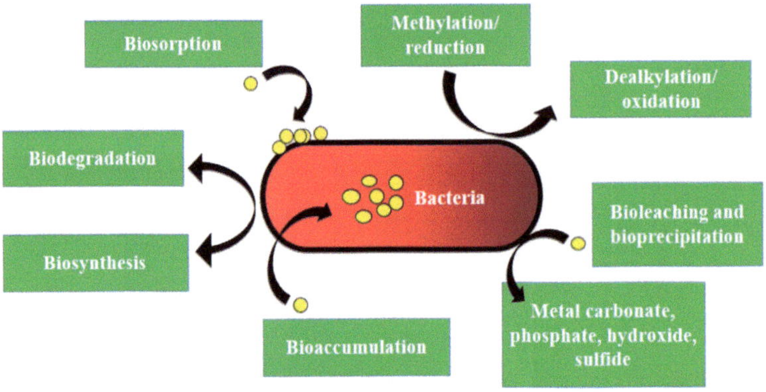

Fig. 5 Representation of bacterial remediation of heavy metals (Modified from Ahemad 2019)

11 Biosorption

Binding of the metal ions onto the bacterial cell wall is the main phenomenon involved in the biosorption mechanism. Simple diffusion, active and passive transport, ion exchange, chelation, and physical and chemical adsorption occur in this approach (Yang et al. 2015). The bacterial cell wall contains active functional groups which act as an excellent binding site for metal ions. Since phosphates and carboxyl groups are negatively charged, they attract metal cations easily. This process also involves hydrogen bonding and electrostatic interactions. Reactive compounds on the bacterial cell wall, such as Extracellular polymeric substance (EPS), have a high metal absorption capacity. Numerous studies have reported its significance in heavy metal bioremediation (Dixit et al. 2015). Heavy metals absorb faster than organic contaminants because metal ions can form stronger bonds with the functional groups present on the bacterial cell wall (Zhang et al. 2020 Shah Maulin 2020, 2021a, b).

12 Bioaccumulation

After the absorption of heavy metals on the cell surface, they are transported into the cytoplasm and cellular components. Microorganisms can normally withstand chemical concentrations up to a certain level, after which the chemicals become toxic and potentially harm the organism. Prolonged concentration of these contaminants can cause detrimental effects on bacterial growth and metabolism. The sensitivity of organisms to chemicals varies greatly depending on the organisms and chemicals used (Dangi et al. 2019). The degree of bioaccumulation of metals is also determined by the presence of nearby organic pollutants which decreases the rate of heavy metal transportation (Zhang et al. 2020).

13 Biotransformation

In most cases, bacteria cannot absorb the heavy metals directly. They must be transformed into a low-toxic state. Heavy metals are transformed by microbial oxidation, reduction, methylation and demethylation. These reactions are accomplished through the use of enzymes and biosurfactants, which alter the properties of the metal (Dixit et al. 2015). Studies have shown a wide range of bacteria that are capable of converting Cr (IV) into Cr (III) which is a less toxic and soluble form (Tarekegn et al. 2020).

14 Biodegradation

Many soil bacteria encode genes that produce degradative enzymes such as hydrolase, dehalogenase, peroxidase, hydrolytic enzymes and laccases (Zhang et al. 2020). These metabolites or enzymes help in the breakdown of metals. For example, *Pseudomonas putida* has the ability to degrade organophosphorus pesticides into carbonates and phosphates. This in turn helps in the biomineralization of heavy metals such as cadmium (Li et al. 2016) (Table 6).

Toxic Cr (VI) is converted into less toxic Cr (III) by bacterial species like *Bacillus subtilis*, *Enterobacter cloacae* and *Pseudomonas putida* (Kumar and Gunasundari, n.d.). Some bacteria convert heavy metals into metal carbonates and sulfides. *Escherichia coli* and *Staphylococcus sp.* are used to reduce the metals to their simplest forms, thus making the removal process much easier (Yin et al. 2019). Actinobacteria consortium is used for the conversion of Cr (IV) to Cr (III) from contaminated soil (Juan et al. 2017). Pesticides and chemical residues in agricultural run-off water eventually penetrate deep into the soil or neighbouring water bodies. Pesticide-reducing bacteria such as *Bacillus sp.* AKD1 and *Cupriavidus sp.* significantly acts on these chemicals and reduces the dissolved heavy metals such as arsenic and cadmium (Liu et al. 2017).

15 Bioremediation of Heavy Metals by Fungi (Mycoremediation)

Mycoremediation is a type of bioremediation that employs fungal species. Fungi can grow well in different habitats and few of them are even capable of proliferation in contaminated soils and in areas that are in constant exposure to industrial toxins and effluents. They absorb and accumulate the heavy metals in their fruiting bodies. Such bioremediators are called hyperaccumulators (Fayyad et al. 2020). White-rot fungi contains degradative enzymes such as lignin peroxidase, laccase, hydrogen peroxide and manganese peroxide that act on heavy metals (Hanif and Bhatti 2015). Mycelium

Table 6 List of few bacterial species that are used in heavy metal bioremediation

Bacterial species	Heavy metal	References
Sporosarcina saromensis (M52)	Cr	Zhao et al. (2016)
Bacillus sp. SFC	Cr	Ontañon et al. (2018)
Bacillus subtilis	Cr	Min et al. (2015)
Cellulosimicrobium sp.	Pb	Bharagava and Mishra (2018)
Gemella sp.	Pb	Marzan et al. (2017)
Micrococcus sp.	Pb	Marzan et al. (2017)
Vibrio fluvialis	Hg	Saranya et al. (2017)
Geobacter sulfurreducens	Cr	He et al. (2019)
Pseudomonas aeruginosa	Cd, As	Tariq et al. (2018)
Rhodopseudomonas sp.	Co	Gao et al. (2017)
Microbacterium profundi	Fe	Wu et al. (2015)
Pseudomonas fluorescens	Cr	Dragana et al. (2017)
Lysinibacillus sphaericus CBAM5	Co, Cu, Cr and Pb	Peña-montenegro et al. (2015)

Table 7 Factors determining the efficiency of mycoremediation

Factors	Effect on fungal degradation
Temperature	Higher temperature is preferred for faster rate of degradation
Nature of contaminants	Absorption rate of smaller molecules is greater than complex form of metal
pH	Optimal degradation rate is between pH 4–5

acts like a large underground web absorbing nutrients and water. The mycelium-secreted extracellular enzymes aid in the removal of metal contaminants from the soil. Several factors such as pH, temperature, toxin concentration and agitators determine the efficiency of mycoremediation (Table 7) (Fayyad et al. 2020). *Trametes versicolor, Bjerkandera adusta, Lentinula edodes, Irpex lacteus, Agaricus bisporus, Pleurotus tuberregium* and *Pleurotus pulmonarius* are some examples of white-rot fungi used in the bioremediation of heavy metals (Rhodes and Rhodes 2015).

Among the different fungal species, white-rot fungi is most preferred as it can degrade metal toxins and insoluble molecules such as pesticides, herbicides, chlorophenols, dyes, dioxins and heavy metals effectively (Fayyad et al. 2020). They are also rich in extracellular ligninolytic enzymes that are capable of absorbing harmful dyes and heavy metal from the industrial effluents. The mechanisms involved in mycoremediation include binding of the metal ions onto the cell wall, bioaccumulation of toxins by mycelium, volatilization, extracellular precipitation and intracellular chelation (Das and Osborne 2018). *Aspergillus niger* is an example of white-rot fungus that assimilates dyes and metal toxins with the help of its mycelium (Fayyad et al. 2020). According to a research study, an integrated approach involving a combination of bacteria and fungi can be effectively used to increase the rate of heavy metal removal. When *Agrocybe aegerita* (fungus) and *Serratia sp.* (bacteria) were used in combination, they were able to degrade higher concentrations of Ni and Cd more effectively X (Li et al. 2016).

16 Mechanism of Mycoremediation

Different types of fungi carry out different mechanisms based on its characteristics and environmental conditions. Biosorption is seen in fungal species with dense mycelium structures. They absorb the heavy metal contaminants from the soil and concentrate them into their fruiting bodies. Biodegradation is the process in which the fungi release degradative enzymes that break down the complex metal component into a soluble and less toxic form (Mendy et al. 2021). Another mechanism involved includes biotransformation of heavy metals where the fungi transforms and converts the metals into its respective hydroxides, carbonates, oxalates and phosphates. Based on the research work done by Benjuwan et al., fungal isolates such as *Formitopsis cf. meliae* and *Ganoderma aff. steyaert anum* have the ability to transform zinc sulfate into zinc oxalate hydrate, cadmium sulfate into its oxalates and lead nitrate into lead oxalates (Kaewdoung et al. 2016). A graphical representation of mycoremediation of heavy metals by biosorption and bioaccumulation through the mycelium is shown in Fig. 6.

Heavy metal resistance in fungi is carried out by extracellular and intracellular sequestration. In extracellular sequestration, metal ions are chelated and bind to the cell wall with the help of extracellular polymeric substances (EPS) (Arora et al. 2017). The fungal cell wall is made up of chitin, chitosan, glucans and glycoproteins which act as the binding site for free carboxyl and hydroxyl groups of heavy metals. In intracellular sequestration, the metals bind to cellular proteins which prevent cytotoxicity of the fungi. These proteins either expel the metal ions from the cytosol or transport it into vacuoles (Kaewdoung et al. 2016).

Fig. 6 Mycoremediation of heavy metals by biosorption and bioaccumulation through the mycelium

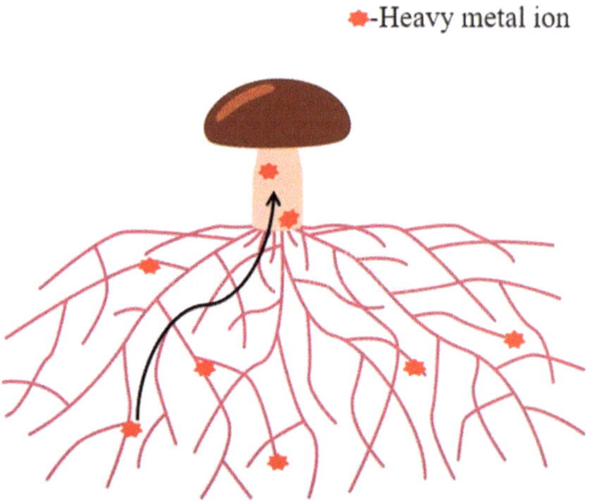

17 Bioremediation of Heavy Metals by Algae (Phycoremediation)

Compared to bacteria and fungi, algal species have a higher absorption rate and tolerance to heavy metals. Larger biomass production can be achieved with less nutrition supply. Algae can also grow well in heavy metal-contaminated soil and other unfavorable conditions (Cameron et al. 2018). Apart from being highly effective, phycoremediation is a simple process with value-added end product formation such as biofertilizers and biofuels (Abinandan et al. 2019). Phaeophyta, Rhodophyta and Chlorophyta are the three main classifications of algae out of which phaeophyta (brown algae) has a better absorption rate of metal ions (Kapahi and Sachdeva 2019). Both living as well as dead biomass can be used in phycoremediation. In recent years, cyanobacteria are grown near metal industries to remove the contaminants from the nearby industrial wastewater and soil (Chabukdhara et al. 2017). Researchers are focusing more on the utilization of microalgae for the bioremediation process because of its high metal binding affinity. The cell wall of microalgae is made up of lipids, organic proteins and polysaccharides such as cellulose and alginate along with functional groups such as phosphate, amide, carboxyl, hydroxyl, sulphides and thiol, all of which act as excellent heavy metal binding sites (Adnan and Hasan 2018). They can effectively assimilate heavy metals with lower toxicity, such as copper, zinc and manganese.

18 Mechanism of Phycoremediation

Metallic ion sequestration occurs when metal ions interact with functional groups present on the algal cell surface. The metal ions, which are cationic in aqueous media, are easily absorbed at these binding sites because the functional groups impart an overall negative charge on the surface (Kaplan 2013). Figure 7 shows a graphical illustration of phycoremediation of heavy metals in which the metal ions bind to the functional groups present on the algal cell surface. They are first absorbed by passive physical absorption which occurs rapidly. This is completely independent of cellular metabolism. They are then absorbed or diffused into the cytoplasm at a much slower rate by an active metabolism-dependent chemi-sorption process (Senthil Kumar and Gunasundari 2018). Different mechanisms are developed by algal species for the bioaccumulation of heavy metals. This includes compartmentalization, complex formation with carboxylic groups, translocation, covalent binding, electrostatic interactions, detoxification and exclusion (Das and Osborne 2018). Phytochelatins (PCs) are binding proteins produced by algae to translocate the heavy metal ions to cell vacuoles (Kumar and Gunasundari, n.d.). They also form organometallic compounds that are translocated from the cytoplasm into the vacuoles in order to reduce cell toxicity. It was also studied that heavy metals at a low concentration can actually enhance the growth rate of many microalgae (Balaji et al. 2015). Few examples of phycoremediators include *Chlorophyta, Spirogyra, Sargassum natans, Cylindracea, Cyanophyta, Fucus vesiculosus and Phormidium.* Phaeophyta has a high amount of alginate and sulfated polysaccharides which makes them act like hyperaccumulators and hyper absorbents of metal ions (Liu et al. 2017). *Anabaena* and *Spirogyra* can grow naturally in contaminated water due to their high metal resistance capacity (Balaji et al. 2015). Microalgae are effective in the removal of zinc, copper and cobalt from industrial wastewater. *Caulerpa racemosa* is used for the remediation of boron. *Phormidium* is a blue green algae that hyperaccumulates copper, zinc, lead and cadmium (Kumar and Gunasundari, n.d.) (Table 8).

Fig. 7 A graphical illustration of phycoremediation of heavy metals (Biosorption mechanism)

Table 8 List of algal species for bioremediation of heavy metals

Phylum	Algal species	Heavy metal	References
Bacillariophyta	*Chaetoceros sp.*	Cd	Gao et al. (2017)
	Nitzschia sp.	Cd	Gao et al. (2017)
	Thalassiosira sp.	Cd	Gao et al. (2017)
Chlorophyta	*Chlamydomonas microsphaera*	Cd	Sun et al. (2019)
	Scenedesmus quadricauda	Cd	Sun et al. (2019)
	Chlorella vulgaris	Cd, Cu, Pb	Ali et al. (2016)
	Merismopedia tenuissima	Cd, Cu	Fawzy (2016)
Cyanophyta	*Arthrospira platensis*	Cd	Sun et al. (2019)
	Microcystis aeruginosa	Cd	Sun et al. (2019)
	Sargassum myriocystum	Cr	Jayakumar et al. (2015)
Phaeophyta	*Sargassum sp.*	Co	Soleymani et al. (2015)
Rhodophyta	*Galaxaura oblongata*	Pb, Cu, Co, Cd	Ibrahim (2011)
	Jania rubens	Pb, Cu, Co, Cd	Ibrahim (2011)
	Pterocladia capillacea	Pb, Cu, Co, Cd	Ibrahim (2011)
	Hypnea valentiae	Cd	Aravindhan et al. (2010)

19 Phytoremediation

Phytoremediation is a remediation process that entails the use of several plant species to remove and degrade heavy metals from contaminated soil and groundwater. The word 'phyto' means plants and 'remediation' means to remove. Phytoremediation can be accomplished through a variety of methods such as phytoextraction, phytovolatilization, phytostabilization, translocation, rhizoremediation and plant absorption (Hou 2014). Different methods of phytoremediation are illustrated in Fig. 8.

Out of these techniques, phytoextraction is the most commonly studied (Yin et al. 2019). Both genetically modified plants and wild plants can be used for phytoremediation (Dixit et al. 2015). *Alfalfa, legumes, Populus, Eucalyptus, Salix are few plant species that* have a high phytoremediation capacity (Sivarajasekar et al. 2018).

20 Plant—Metal Interaction

The plant root systems are semi- permeable in nature. Not all heavy metals can be absorbed by the plants. Essential minerals like calcium and zinc are required for plant metabolism. Only those heavy metals with similar properties to these minerals are easily absorbed by the plants (Khalid et al. 2017). After the absorption of metals via the root system, the metalloids are translocated into shoots, leaves, seeds and plant

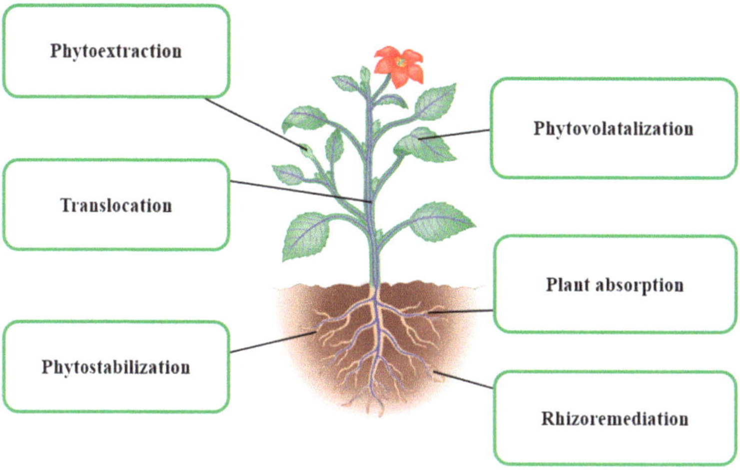

Fig. 8 Different methods of phytoremediation (Modified from Hou 2014)

vacuoles (Khalid et al. 2017). There are many factors that influence the plant–metal interaction such as the properties of soil, plant rooting system, plant density, plant size, mineral uptake capacity, concentration of heavy metals, presence of other soil components, pH and temperature of the environment in which the plant grows and the presence of microorganisms and fungi (Zhang et al. 2020).

21 Phytoextraction

Phytoextraction, also known as phytoaccumulation, is the process of concentrating heavy metal ions from the groundwater or soil into the root and shoot systems of plants. According to research done by (Das and Osborne 2018), it was discussed that plants with a well-developed root system like *Pennisetum purpureum* (Napier grass) were able to absorb, translocate and store lead into their tissues efficiently. Figure 9 depicts a graphical representation of phytoextraction of heavy metals from a contaminated site. One of the major limitations of phytoextraction is that many plant species begin to show a very slow growth rate and produce considerably less biomass due to toxin bioaccumulation. According to (Yang et al. 2015), plants that bioaccumulate metal toxins in non-edible parts are preferred because they are safe for animal consumption and do not introduce toxins into the food chain.

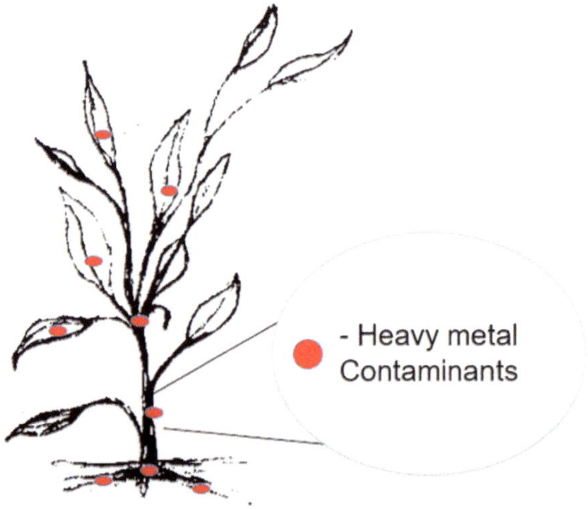

Fig. 9 Phytoextraction of heavy metals (indicated in red circle) from contaminated site Modified from Sivarajasekar et al. (2018)

22 Phytostabilization

Phytostabilization is a method of immobilizing metal ions rather than degrading them. This helps to prevent the risk of heavy metal leaching into the soil and water bodies quickly (Chibuike and Obiora 2014). It is termed so because metal pollutants are immobilized by the plant root system and the rate at which toxins leach into the soil is reduced, hence limiting the contamination spread. The main mechanism involved in phytostabilization is the attachment of metals to the active binding sites such as phytochelatins and metallothioneins on the plant cell surface (Shackira and Puthur 2019). There are many factors that have to be considered for the selection of plants. These factors include dense rooting system, ability to translocate low concentration of metal ions, faster growth rate for better underground root coverage and the ability to self-propagate (Chibuike and Obiora 2014). Adding soil stimulants and organic amendments that increase the availability of heavy metals to the plant roots can improve the efficiency of phytostabilization.

23 Phytovolatilization

Metal ions such as selenium, mercury and arsenic are absorbed by plants and converted into components that are easily volatilized from the leaves' stomata into the environment (Gupta et al., n.d.). Although this method does not completely remove heavy metals, it does aid in the conversion of toxic metals into their less toxic forms. The main disadvantage of this method is that it can only be used with volatile heavy metals like mercury and selenium. Furthermore, there is a high probability of metal ions from the atmosphere re-entering the soil or water (Das and Osborne 2018).

24 Rhizosphere Biodegradation (Rhizoremediation)

Rhizoremediation is a method that involves the interaction of soil microorganisms, metal ions and plant root system. It is mainly suitable for remediation of metals such as Cu, Zn, Ni, Cr and Cd. According to (Kumar and Gunasundari, n.d.), tobacco, corn, spinach, sunflower and Indian mustard are effective in removal of lead from contaminated soil and groundwater. This method is more effective for plants that have expanded rooting system because the roots secrete many extracellular enzymes, sugars, organic substrates and amino acids that are utilized by the soil microorganisms for the bioremediation process (Jeevanantham et al. 2019). Plant growth-promoting rhizobacteria (PGPR) and arbuscular mycorrhizal fungus (AMF) can be found in close proximity to plant roots. These microbes interact with heavy metals in the soil, converting them into more soluble forms that can be easily absorbed by the plants (Praveen et al. 2019) (Table 9).

Table 9 Plant species with different types of bioremediation of heavy metals

Plant species	Type of phytoremediation	Heavy metal	Mechanism	References
Sedum alfredii	Phytostabilization	Pb, Cd	Glutathione biosynthesis that bind metals to the roots	Anjum et al. (2012)
Nicotiana tabacum, Zea mays, Brassica juncea	Rhizoremediation	Pb	Microorganisms enhance the biodegradation process	Jeevanantham et al. (2019)
Corrigiola telephiifolia	Phytoaccumulation	As	Hyperaccumulation in root and shoot	García-Salgado et al. (2012)
Crotalaria juncea	Rhizoremediation	Cd	Heavy metals remediation in association with *Achromobacter sp.* AO22	Stanbrough et al. (2013)
Triticum aestivum	Rhizoremediation	Zn	Heavy metal remediation in association with *Pseudomonas flurescens*	Sirohi et al. (2015)
Brassica nigra	Rhizoremediation	Cu	Heavy metals remediation in association with *Kocuria sp.* CRB15	Hansda et al. (2017)
Suaeda nudiflora	Rhizoremediation	Zn, Pb	Heavy metals remediation in association with *Bacillus megaterium*	Jha et al. (2017)
Helianthus annuus	Rhizoremediation	Cr	Heavy metal remediation in association with *Pseudomonas sp.* CPSB21	Gupta et al. (n.d.)

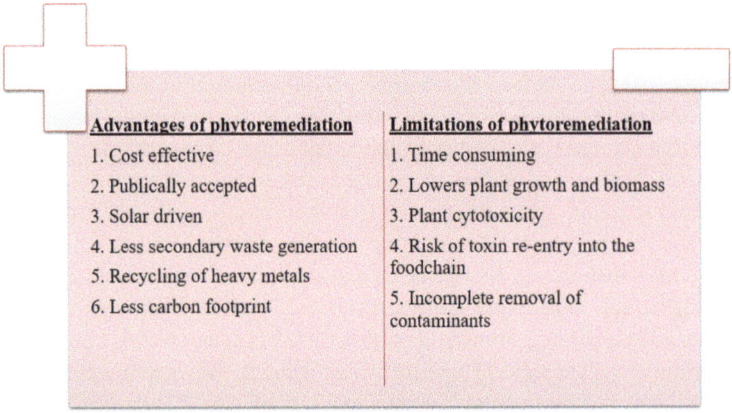

Fig. 10 Advantages and limitations of phytoremediation of heavy metals

25 Advantages and Limitations of Phytoremediation

Phytoremediation is publicly accepted, relatively less expensive and is an environmentally friendly approach. Despite the fact that it offers a lot of advantages, there are many drawbacks which must be considered. When plants absorb metal toxins, it can cause an imbalance in the ecosystem by making the plants unfit for consumption. Phytoremediation also requires proper maintenance and the entire clean-up process may take many years to complete. High metal concentration in plants can also result in cytotoxicity and low plant growth and biomass (Gupta et al., n.d.). A combination method of phytoremediation of biofortification can be employed to elevate the efficiency of bioremediation (Kaur and Roy 2020). An overview of advantages and limitations of phytoremediation is shown in Fig. 10.

26 Nanoparticles in Bioremediation of Heavy Metals

Nanoparticles are nano-sized materials of dimensions in the range of 1–100 nm. Nanoremediation is the process of using nanoparticles to remove heavy metals from polluted sites. Since nanoparticles have a high ratio of surface area to volume, they have greater absorption capacity, reactivity, catalyzing properties and efficiency (Abdallah et al. 2019). The types of nanoparticles used in the removal of heavy metals are shown in Fig. 11. This includes carbon nanotubes (CNTs), carbon-based nanoparticles (NPs), graphene oxide (GO), zero-valent metal (ZVM), zeolites, nanospheres, nanofibers, nano composites and nanowires (Kumar et al. 2019).

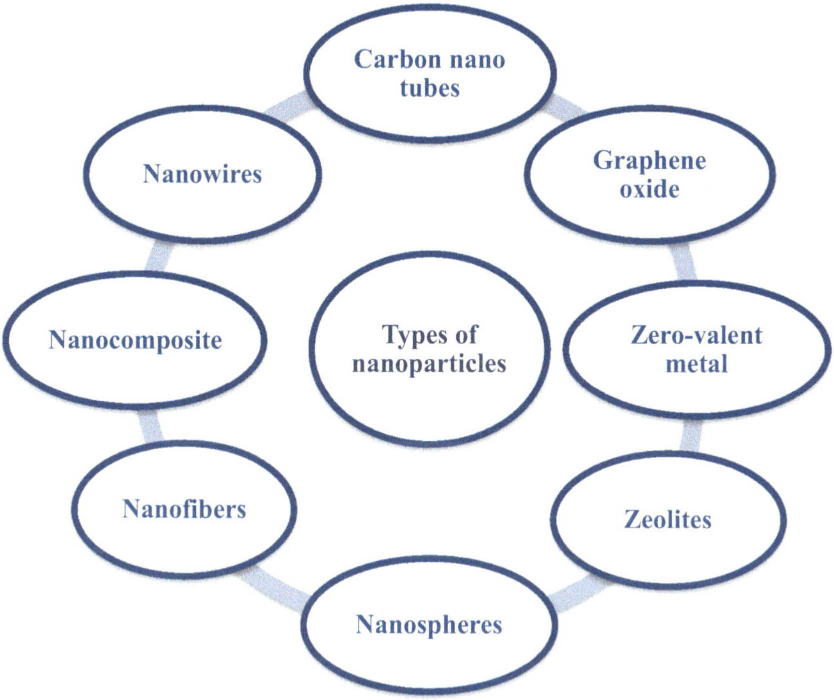

Fig. 11 Types of nanoparticles used in the removal of heavy metals

27 Carbon Nanotubes (CNTs)

Carbon nanotubes are a form of carbon-based nanomaterial with a diameter ranging from 1 to 3 nm. They are classified into two major categories which are single-walled carbon nanotubes (SWCNT) and multi-walled carbon nanotubes (MWCNT) (Cai et al. 2019). Representation of SWCNT and MWCNT is shown in Fig. 12. These nanotubes are made using graphene sheets. Based on the synthesis and purification process, the composition of the nanoparticles varies greatly. MWCNTs are preferred over SWCNTs because they have more surface area for adsorption due to their multilayered structure (Arsenov 2020). According to (Kumar et al. 2019), CNTs were found to have a high absorption capacity for heavy metals like copper, lead, and manganese. In order to improve its absorption capacity, they are incorporated or doped with cations, functional groups like carboxylic acid and various conducting polymers. High positive charge, surface area and conducting polymers on the nanomaterial enhance the adsorption of metals like chromium and copper (S. Kumar et al. 2019). Though CNTs have many desirable properties for bioremediation, they are significantly expensive which adds up to their high treatment costs. Moreover, CNTs can be toxic in nature. The toxicities are usually minimized by adding biosurfactants to the reactive surfaces (Yoo and Pak 2013). As a result, these nanoparticles have been rendered safe for environmental applications.

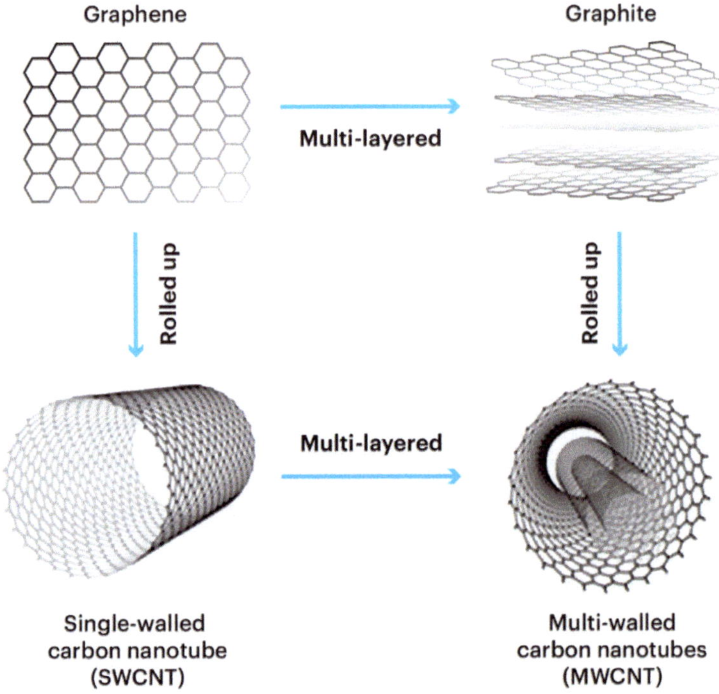

Fig. 12 Representation of SWCNT and MWCNT (*Image source* https://tuball.com/articles/single-walled-carbon-nanotubes)

28 Synthesis of CNTs and Graphene Oxide (GO)

Arc discharge, laser ablation, electrochemical, hydrothermal and chemical vapour deposition (CVD) are some of the main techniques used for the synthesis of carbon nanotubes (Yin et al. 2019). Hydrothermal method is used for the removal of chromium from industrial wastewater by doping the CNTs with molybdenum sulphide and tin sulphide (Li et al. 2016).

An oxidation reaction transforms graphite into graphene oxide. Sulphuric acid is used to integrate graphite with chemicals such as potassium permanganate and potassium ferrite at 5 °C, resulting in monolayers (Abdul et al. 2021). As a result, graphite intercalation compounds (GIC) and pure graphite oxide (PGO) are produced. They are subsequently oxidized by oxidizing agents such as KMnO4 at 35 °C, converting GIC to PGO. Water treatment is used to exfoliate PGO into GO sheets (Yumei et al. 2017). Alkali treatment can also be used to decrease GO sheets. They are ultrasonically de-mineralized and subsequently combined with amino polymers containing a very small amount of KOH. To extract the resulting complex from the solution, centrifugation is often used. The functionalized GO complex is then dispersed in ethanol using sonication. At pH 3–4, reducing dopants like zinc and iron are stirred into the solution (Abdul et al. 2021). The GO complex is then activated by raising the pH to 11 with NaOH and hydrothermally treating it at 150 °C. After 10 h of treatment, the solution is centrifuged to get doped graphene oxide (Baruah et al. 2019).

29 Adsorption Mechanism of CNTs and GOs

The adsorption process of CNTs and GOs is determined by the physico-chemical characteristics of the functional groups doped on their walls. The major mechanisms are carboxylated CNTs being amidated or esterified, and functional groups being covalently attached to the walls of carbon nanotubes (Gong et al. 2021). There are two types of pores in carbon nanotubes: one on the outermost wall and the other on the interior layers of multi-walled CNTs. When functional groups are placed onto conducting polymers, the adsorption capacity of heavy metals like Cr (VI) is considerably increased (Jeevanantham et al. 2019). The rate of adsorption is influenced by a variety of extrinsic factors. A positive entropy is created by a high temperature, allowing for a quicker adsorption rate of heavy metals from contaminated locations (Jeevanantham et al. 2019). Increased temperature also causes more Brownian movement, which improves the chances of making contact with the metal ions.

30 Zero-Valent Metal (ZYM)-Based Nanoparticles in Heavy Metal Remediation

30.1 Silver-Based Nanoparticles

Silver nanoparticles interact efficiently with heavy metal contaminants such as mercury, chromium and cadmium. Ag-based nanomaterials can effectively remove Hg^{2+} because of the reduction potential of silver (Vélez et al. 2018). There are many studies conducted that report the efficiency of silver nanoparticles in the uptake of mercury ions. Many researchers synthesize silver nanoparticles from plant extracts. Ag nanoparticles were produced from leaf extract of *Piliostigma thonningii* for the removal of toxic metal ions from laboratory-prepared effluent (Shittu and Ihebunna 2017) The gums of *Azadirachta indica, Araucaria heterophylla* and *Prosopis chilensis* were used to synthesize Ag nanoparticles for effective removal of chromium from wastewater (Samrot et al. 2019). Recently, zeolite-derived coal fly ash (CFA) doped with Ag nanomaterial has been employed for the optimized removal of Hg^{2+} ions from an aqueous medium, with a removal rate of 99% (Tauanov et al. 2019).

30.2 Gold-Based Nanoparticles

Gold-based nanoparticles (AuNPs) have a strong affinity for Hg ions, creating complexes like AuHg, Au_3Hg, and $AuHg_3$ (Yao et al. 2016). AuNPs have excellent catalytic properties, electrical conductivity, and biocompatibility with enzymes. In a model aqueous solution, For spectrophotometric detection of selected heavy metal ions (Cd^{2+}, Cu^{2+}, Co^{2+}, Pb^{2+}, and Ni^{2+}), AgNPs and AuNPs were employed. The researchers have found that silver and gold nanoparticles can be used as chemical probes to detect heavy metals in contaminated areas (Modrzejewska-sikorska and Konowa 2020).

30.3 Zero-Valent Iron (ZVI)

ZVI is a zero-valent iron that consists of a combination of ferrous and ferric oxides. FeO is a good reducing agent and Fe_2O_3 provides active sites which improve the interactions with heavy metals (Hashim et al. 2011). ZVI is mainly focused in the bioremediation of copper, nickel, mercury, chromium and cadmium. When compared to the other nano-adsorbents, iron-based nanoparticles adsorb 46 times more (Parvin et al. 2019). ZVI has high specificity and reduction ability on metal ions, making it a suitable option for bioremediation of heavy metals (Li et al. 2017). The ZVI method of nanoremediation is ideal because of the simple synthesis process and feasibility (Abdul et al. 2021).

31 Microbial Synthesis of Nanoparticles

Microorganisms, particularly bacteria, act as nanofactories because they are extensively used for the synthesis of nanoparticles (Yadav et al. 2017a). Scientists are focusing on this method of synthesis because they are more sustainable, cost-effective and easier maintenance of growth conditions such as oxygenation, temperature, pH and incubation time on a large-scale basis. Various bacterial strains such as *Escherichia coli, Bacillus subtilis, Bacillus megaterium, Pseudomonas aeruginosa, Klebsiella pneumoniae, Bacillus cereus, Alteromonas* and *Ochrobactrum* have been used for the synthesis of nanoparticles (Gahlawat and Choudhury 2019). A variety of factors influence the reduction of metal ions into nanoparticles. The presence of functional groups on the microbial cell surface, pH, temperature, metal salt concentration, and the medium used are all important aspects to consider (Yumei et al. 2017).

32 Bacterial Synthesis of Silver Nanoparticles

Bacteria are present in abundance and are mostly preferred for biosynthesis of nanoparticles over other microbial groups. The two main mechanisms involved in the biosynthesis of nanoparticles include intracellular and extracellular synthesis (Gahlawat and Choudhury 2019). A graphical representation of bacterial synthesis of nanoparticles is shown in Fig. 13

This biosynthesis depends greatly on the electron transport cofactors and NADH released by the bacteria along with other enzymatic reduction factors such as nitrate reductase (Zhu et al. 2019). Nitrate reductase plays a major role in AgNPs synthesis when accompanied by the coenzyme nicotinamide adenine dinucleotide phosphate (NADPH).

In the process of extracellular synthesis, nanoparticles are synthesized outside bacterial cells and the reductase enzyme is released outside (Ovais et al. 2018). In the case of intracellular synthesis, cell lysis is performed in order to recover the nanoparticles. This is achieved by sonication, autoclaving and the use of salts and detergents (Ovais et al. 2018). The overview of bacterial production of nanoparticles includes inoculation and harvesting of bacterial cells, production of NPs by adding substrates, separation of cells, homogenization, isolation of nanoparticles, purification and product formulation ((Iravani and Varma 2020). Biosynthesis of AgNPs can

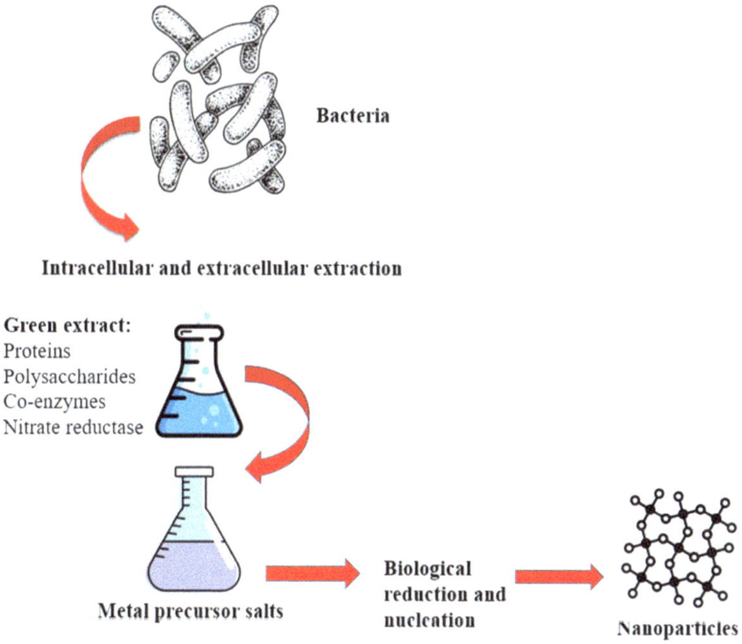

Fig. 13 Representation of bacterial synthesis of nanoparticles (Modified from Gahlawat and Choudhury 2019)

also be attained by using bacterial exo-polysaccharides (Jeevanantham et al. 2019). Nowadays, researchers are exploring marine bacterial species as a potential source for nanoparticle synthesis for heavy metal remediation.

33 Bioreactors in Heavy Metal Remediation

A bioreactor is a system that uses microorganisms to digest pollutants found in groundwater and soil. This process might be aerobic or anaerobic in nature. Different criterias are used to create these bioreactors for cellular production (Ramesh et al. 2019). Microorganisms and other types of biomass are used in some bioreactors to boost their effectiveness in heavy metal removal (Jeevanantham et al. 2019). Since bioreactors are made up of multiple processes and remedial applications, they are employed in the removal of heavy metals. They also have other applications like wastewater treatment in the environmental sector, cell culture and tissue engineering and production of chemicals in industrial biotechnology (Wang et al. 2020). A representation of types of bioreactors used in heavy metal remediation is shown in Fig. 14.

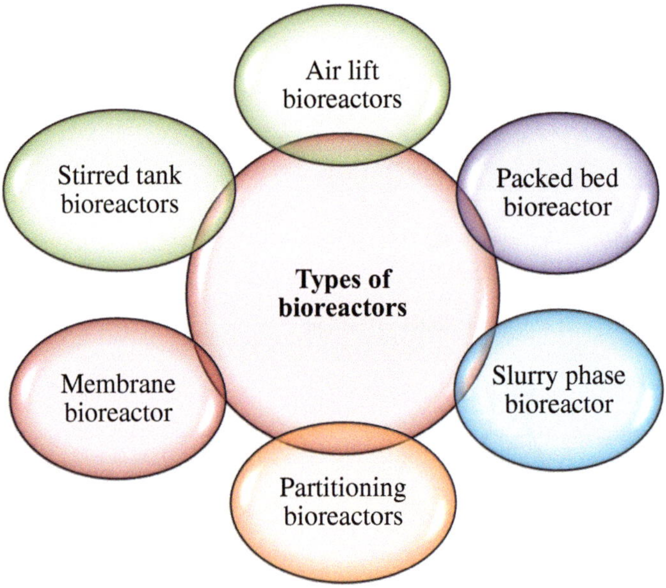

Fig. 14 Representation of types of bioreactors used in heavy metal remediation

34 Slurry Phase Bioreactor

Slurry phase bioreactors are used to treat polluted mediums in the slurry phase. The bio-slurry reactor is the other name for this bioreactor. A slurry phase bioreactor is mainly used for the treatment of polycyclic aromatic hydrocarbon-polluted loamy soil. Microbial transformation and degradation can remove polycyclic aromatic hydrocarbons from environmental compartments such as soils (Forján et al. 2020). Ex-situ remediation of most soils and sediments from petrochemicals, herbicides, pesticides and heavy metal contaminants is possible with a slurry phase bioreactor.

35 Stirred Tank Bioreactor

Bioreactors must be built in order to achieve the most efficient manifestation of the biological characteristics of living cells. In the stirred tank bioreactor, the stirrer is included. This system can operate in either batch or continuous mode. A continuous stirred tank bioreactor is made up of a cylindrical vessel with a motor in the middle. State mixing is halted at the bioreactor for phase sedimentation during sedimentation. The stirred tank bioreactor's bioremediation technique is quite straightforward when compared to other types of bioreactors. Due to its lower operating costs, the stirred tank bioreactor is the most essential bioreactor for industrial applications (Spier et al. 2011).

36 Fluidized Bed Bioreactor

A fluidized bed reactor is a chemical reactor in which catalytic particles interact with a gas stream fed from the bottom, resulting in a mixture that acts like a fluid in the emulsion phase (Guda et al. 2015). A microbial biofilm develops solid particles in a fluidized bed bioreactor, which encourages bacterial growth (Kumar and Gunasundari, n.d.). Fluidized bed processes have a larger number of variables to handle due to gas–solid interactions and chemical reactions, making the manufacturing process more complex. The fluidized bed consists of particulate solid fuel combustion in an inert material bed, commonly sand, that is fluidized by the movement of gas (Philippsen et al. 2015).

37 Membrane Bioreactor

Membrane bed reactors combine filtration and biological processes through the use of a membrane. In a membrane bioreactor, the activated sludge process is combined with a membrane separation process (Melin et al. 2006). A membrane bioreactor is usually used in the removal of micropollutants (Goswami et al. 2018). The bioreactor process can be set up depending on the nutrient removal needs of the project. They can also be used to treat home and industrial waste water biologically. To meet the increasing demands, existing technology must be improved to fulfil the needs of a wastewater treatment plant (Radjenović et al. 2008).

38 Airlift Bioreactor

For the treatment of gaseous or volatile air pollutants, the Airlift bioreactor may be a viable option. The medium of the vessels in these reactors is separated into two interconnected zones by a draught or a baffle tube (Kumar and Gunasundari, n.d.). When they operate at high cell densities, they are sensitive to being limited by gas–liquid mass transfer and poor liquid phase mixing (AL-Mashhadani et al. 2015). The two main configurations of the airlift bioreactor are external loop reactors and internal loop reactors. The creation of a dead zone due to poor mixing and non-uniform nutrition input produced by high biomass density is a key disadvantage of this reactor (Rawat et al. 2019).

39 Packed Bed Bioreactor

Microbial growth on a fixed film substrate is possible with a packed bed reactor system. Fixed film reactors are used to create small reactors. This reactor is best used in big substrates with a lot of flow. It is critical to distinguish the system's tolerance for heavy metal. A variety of fungi and bacteria are used in the cleanup process. Organic and inert solid materials were used as packing materials (Azizi et al. 2016) (Table 10).

Table 10 Types of bioreactors and their applications

Types of bioreactor	Applications	Reference
Packed bed bioreactor	Essential to distinguish the system's tolerance for heavy metal	Azizi et al. (2016)
Airlift bioreactor	It's employed in filamentous fermentation, biological wastewater treatment, animal and plant cell culture, and other applications	Rawat et al. (2019)
Membrane bioreactor	It is used to remove the micro pollutants and heavy metals in wastewater	Zhang et al. (2018)
Stirred tank bioreactor	They can treat hydrocarbon rich industrial wastewater	Spier et al. (2011)
Slurry phase bioreactor	They increase the mass transfer rates and increase the contact of microorganisms, pollutants or nutrients. They are effective in the use of biostimulation and bioaugmentation	Robles-González et al. (2008)

40 Recent Advancements

40.1 Rhizosphere Engineering

The rhizosphere is a zone dominated by the plant root system at the root-soil interface. The roots release metabolites into the soil and are a home to a range of physically, chemically, and biologically interconnected processes. Microbes and metal ions occupy the space between cells in the endorhizosphere, which comprises the cortex and endodermis (Seshadri et al. 2015). Figure 15 depicts a section showing bulk soil and rhizosphere soil with three rhizosphere sections i.e. ectorhizosphere, rhizoplane, and endorhizosphere—as well as metalloid and microorganism. For example, Cd accumulation occurs in most plant roots but it is resistant to other parts of the plants. As a result, the physiological responses of plants are critical in heavy metal resistance (Nocito et al. 2011). Engineered bioremediation methods include addition of growth stimulators to the rhizosphere, supplying nutrients to the polluted soil to increase the bioremediation properties of microorganisms and genetically modifying the microorganisms. Several genetically modified plants and organisms are known for the formation of a special compound that improves the rhizospheric transformation of heavy metals. The rhizosphere ecosystem has thus been used to remove heavy metals from the soil in this way (Kumar and Gunasundari, n.d.).

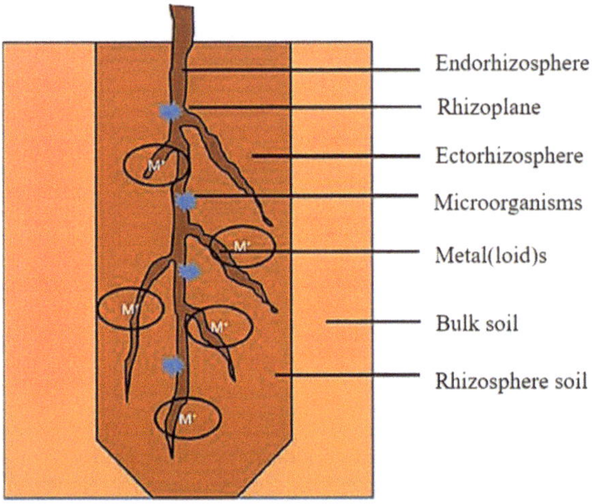

Fig. 15 A section of soil showing bulk soil and rhizosphere soil with three regions of rhizosphere– Ectorhizosphere, rhizoplane and Endorhizosphere with metalloid and microorganisms (Seshadri et al. 2015)

40.2 Genetic Modification of Plants for Phytoremediation of Metals

Plants should have a large root system and grow quickly in order to attain improved phytoremediation efficiency. These plants must be able to withstand soil pollution and generate more biomass in a contaminated environment. Biotechnological precision, particularly in genetic engineering, contributes to a rapid and significant change in the crop by enhancing novel genes and features that can be efficiently put into plants to boost their phytoremediation capability for metal removal (Jagtap and Bapat 2015). The capacity to build an effective and cost-effective plant for soil metals is based on the absorption and translocation mechanisms and the ability to hyperaccumulate the heavy metals. *Populus angustifolia, Nicotiana tabacum,* and *Silene cucubalis* have all been studied and modified genetically to accumulate more heavy metals than their wild counterparts (Fulekar et al. 2009).

41 Conclusion

Bioremediation is an environmentally friendly approach of removing heavy metals from the contaminated sites. It is a promising and cost-effective approach. Advances in bioremediation appear to be very appealing, but for these solutions to attain market value, field trials are required. Bioremediation techniques have proven to be an ideal technique for remediating heavy metals from the environment while simultaneously replenishing the site and maintaining the ecological balance. The integrated strategy involving a combination of microorganisms has recently grown in interest as it helps to tackle the constraints of a single microbial strain or biosystem. Environmental standards have gotten more stringent, demanding that the treated effluent is of higher quality. Although a variety of approaches can be employed to treat heavy metal-contaminated wastewater, it's crucial to remember that the optimum treatment depends on a number of parameters, including pH, initial metal concentration, environmental impact and cost-effectiveness.

References

Abdallah BB, Zhang X, Andreu I, Gates BD, El Mokni R, Rubino S, Landoulsi A, Chatti A (2019) Differentiation of nanoparticles isolated from distinct plant species naturally growing in a heavy metal polluted site. J Hazard Mater 386:121644. https://doi.org/10.1016/j.jhazmat.2019.121644

Abdul N, Sucharita S, Naidu S, Sudarshana V (2021) Chemosphere nano-remediation of toxic heavy metal contamination: Hexavalent chromium [Cr (VI)]. Chemosphere 266:129204. https://doi.org/10.1016/j.chemosphere.2020.129204

Abinandan S, Subashchandrabose SR, Pannerselvan L (2019) Bioresource technology potential of acid-tolerant microalgae, Desmodesmus sp. MAS1 and Heterochlorella sp. MAS3, in heavy metal removal and biodiesel production at acidic pH. Bioresour Technol 278(January):9–16. https://doi.org/10.1016/j.biortech.2019.01.053

Adnan SNAM, Hasan S (2018) Phycobiliproteins production and heavy metals reduction ability of Porphyridium sp Phycobiliproteins production and heavy metals reduction ability of Porphyridium sp.

Ahemad M (2019) Remediation of metalliferous soils through the heavy metal resistant plant growth promoting bacteria : Paradigms and prospects. Arab J Chem 12(7):1365–1377. https://doi.org/10.1016/j.arabjc.2014.11.020

Akpor OB (2014) Heavy metal pollutants in wastewater effluents: sources, effects and remediation. Adv Biosci Bioeng 2(4):37. https://doi.org/10.11648/j.abb.20140204.11

Al-Mashhadani MKH, Wilkinson SJ, Zimmerman WB (2015) Airlift bioreactor for biological applications with microbubble mediated transport processes. Chem Eng Sci 137:243–253. https://doi.org/10.1016/j.ces.2015.06.032

Ali MH, Hussian AEM, Abdel-Satar AM, Goher ME, Napiórkowska-Krzebietke A, Abd El-Monem AM (2016) The isotherm and kinetic studies of the biosorption of heavy metals by non-living cells of chlorella vulgaris. J Elementology 21(4):1263–1276. https://doi.org/10.5601/jelem.2016.21.1.1040

Anjum NA, Ahmad I, Mohmood I, Pacheco M, Duarte AC, Pereira E, Umar S, Ahmad A, Khan NA, Iqbal M, Prasad MNV (2012) Modulation of glutathione and its related enzymes in plants' responses to toxic metals and metalloids—a review. Environ Exp Bot 75:307–324. https://doi.org/10.1016/j.envexpbot.2011.07.002

Aravindhan R, Maharshi B, Janardhanan SK, Jonnalagadda RR, Nair BU (2010) Biosorption of cadmium metal ion from simulated wastewaters using Hypnea valentiae biomass: A kinetic and thermodynamic study. Biores Technol 101(5):1466–1470. https://doi.org/10.1016/j.biortech.2009.08.008

Arora S, Singh AK, Singh YP (2017) Bioremediation of salt affected soils: an Indian perspective, pp 1–313. Springer. https://doi.org/10.1007/978-3-319-48257-6

Arsenov D (2020) Nanotechnology and remediation of agrochemicals. https://doi.org/10.1016/B978-0-08-103017-2.00019-2

Azizi S, Kamika I, Tekere M (2016) Evaluation of heavy metal removal from wastewater in a modified packed bed biofilm reactor. PLoS ONE 11(5):1–13. https://doi.org/10.1371/journal.pone.0155462

Azubuike CC, Chikere CB, Okpokwasili GC (2016) Bioremediation techniques–classification based on site of application: principles, advantages, limitations and prospects. World J Microbiol Biotechnol 32(11):1–18. https://doi.org/10.1007/s11274-016-2137-x

Balaji S, Kalaivani T, Sushma B, Pillai CV, Shalini M (2015) Characterization of sorption sites and differential stress response of microalgae isolates against tannery effluents from Ranipet industrial area—an application towards phycoremediation: 6514(January 2016). https://doi.org/10.1080/15226514.2015.1115960

Barakat MA (2011) New trends in removing heavy metals from industrial wastewater. Arab J Chem 4(4):361–377. https://doi.org/10.1016/j.arabjc.2010.07.019

Barbosa LDP, Costa PF, Bertolino SM, Silva JCCD, Guerra-Sá R, Leão VA, Teixeira MC (2014) Nickel, manganese and copper removal by a mixed consortium of sulfate reducing bacteria at a high COD/sulfate ratio. World J Microbiol Biotechnol 30:2171–2180. https://doi.org/10.1007/s11274-013-1592-x

Baruah A, Mondal S, Sahoo L, Gautam UK (2019) Ni-Fe-layered double hydroxide/N-doped graphene oxide nanocomposite for the highly efficient removal of Pb(II) and Cd(II) ions from water. J Solid State Chem 280(II):120963. https://doi.org/10.1016/j.jssc.2019.120963

Bharagava RN, Mishra S (2018) Ecotoxicology and environmental safety hexavalent chromium reduction potential of Cellulosimicrobium sp. isolated from common effluent treatment plant of tannery industries. Ecotoxicol Environ Saf 147(May 2017):102–109. https://doi.org/10.1016/j.ecoenv.2017.08.040

Briffa J, Sinagra E, Blundell R (2020) Heavy metal pollution in the environment and their toxicological effects on humans. Heliyon 6(9):e04691. https://doi.org/10.1016/j.heliyon.2020.e04691

Cai C, Zhao M, Yu Z, Rong H, Zhang C (2019) Science of the Total Environment Utilization of nanomaterials for in-situ remediation of heavy metal (loid) contaminated sediments: a review. Sci Total Environ 662:205–217. https://doi.org/10.1016/j.scitotenv.2019.01.180

Cameron H, Mata MT, Riquelme C (2018) The effect of heavy metals on the viability of Tetraselmis marina AC16- MESO and an evaluation of the potential use of this microalga in bioremediation. https://doi.org/10.7717/peerj.5295

Chabukdhara M, Gupta SK, Gogoi M (2017) Phycoremediation of heavy metals coupled with generation of bioenergy, pp 163–188. https://doi.org/10.1007/978-3-319-51010-1

Chibueze C, Chioma A, Chikere B (2016) Bioremediation techniques—classification based on site of application: principles, advantages, limitations and prospects. World J Microbiol Biotechnol. https://doi.org/10.1007/s11274-016-2137-x

Chibuike GU, Obiora SC (2014) Heavy metal polluted soils: effect on plants and bioremediation methods

Dangi AK, Sharma B, Hill RT, Shukla P (2019) Critical Reviews in Biotechnology Bioremediation through microbes: systems biology and metabolic engineering approach. Crit Rev Biotechnol 39(1):79–98. https://doi.org/10.1080/07388551.2018.1500997

Das A, Osborne JW (2018) Bioremediation of heavy metals. Springer, Cham

Dash S, Borah SS, Kalamdhad AS (2021) Heavy metal pollution and potential ecological risk assessment for surficial sediments of Deepor Beel, India. Ecological Indicators 122(May 2020):107265. https://doi.org/10.1016/j.ecolind.2020.107265

Dixit R, Malaviya D, Pandiyan K, Singh UB, Sahu A, Shukla R, Singh BP, Rai JP, Sharma PK, Lade H, Paul D (2015) Bioremediation of heavy metals from soil and aquatic environment: an overview of principles and criteria of fundamental processes. Sustainability 7(2):2189–2212. https://doi.org/10.3390/su7022189

Dragana IČ, Spasojevi JM, Babi OB, Maleti P, Simeunovi JB, Ron D, Dalmacija D, Tama I (2017) RESEARCH ARTICLE A chemical and microbiological characterization and toxicity assessment of the Pan č evo industrial complex wastewater canal sediments. Serbia. https://doi.org/10.1007/s11356-017-8513-8

Fawzy MA (2016) Phycoremediation and adsorption isotherms of cadmium and copper ions by Merismopedia tenuissima and their effect on growth and metabolism. Environ Toxicol Pharmacol 46:116–121. https://doi.org/10.1016/j.etap.2016.07.008

Fayyad RJ, Muslim SN, Naseer A, Ali M (2020) Application strategies for using fungi and algae as bioremediators: a review. Plant Archives 20:788–792

Forján R, Lores I, Sierra C, Baragaño D, Gallego JLR, Peláez AI (2020) Bioaugmentation treatment of a PAH-polluted soil in a slurry bioreactor. Appl Sci (Switzerland) 10(8):2837. https://doi.org/10.3390/APP10082837

Fulekar MH, Singh A, Bhaduri AM (2009) Genetic engineering strategies for enhancing phytoremediation of heavy metals. Afr J Biotech 8(4):529–535. https://doi.org/10.4314/ajb.v8i4.59858

Gahlawat G, Choudhury AR (2019) A review on the biosynthesis of metal and metal salt nanoparticles by microbes. RSC Adv 23:12944–12967. https://doi.org/10.1039/c8ra10483b

Gao R, Wang Y, Zhang Y, Tong J, Dai W (2017) Cobalt (II) bioaccumulation and distribution in Rhodopseudomonas palustris. 2818(March). https://doi.org/10.1080/13102818.2017.1292148

García-Salgado S, García-Casillas D, Quijano-Nieto MA, Bonilla-Simón MM (2012) Arsenic and heavy metal uptake and accumulation in native plant species from soils polluted by mining activities. Water Air Soil Pollut 223(2):559–572. https://doi.org/10.1007/s11270-011-0882-x

Gong Z, Chan HT, Chen Q (2021) Application of nanotechnology in analysis and removal of heavy metals in food and water resources. Nanomaterials 11:1792

Goswami L, Vinoth Kumar R, Borah SN, Arul Manikandan N, Pakshirajan K, Pugazhenthi G (2018) Membrane bioreactor and integrated membrane bioreactor systems for micropollutant removal from wastewater: A review. Journal of Water Process Engineering 26(September):314–328. https://doi.org/10.1016/j.jwpe.2018.10.024

Grace K, Senthil PP, Kilaru VJ, Vardhan H (2020) Microalgae for biofuel production and removal of heavy metals: a review. Environ Chem Lett 0123456789. https://doi.org/10.1007/s10311-020-01046-1

Guda VK, Steele PH, Penmetsa VK, Li Q (2015) Fast pyrolysis of biomass: recent advances in fast pyrolysis technology. In: Recent advances in thermochemical conversion of biomass (Issue 1). Elsevier. https://doi.org/10.1016/B978-0-444-63289-0.00007-7

Han Y, Kwon O, Cho Y (2015) A study of bioindicator selection for long-term ecological monitoring. J Ecol Environ 38(January):119–122

Hanif MA, Bhatti HN (2015) Remediation of heavy metals using easily cultivable, fast growing, and highly accumulating white rot fungi from hazardous aqueous streams. Desalin Water Treat 53(1):238–248. https://doi.org/10.1080/19443994.2013.848413

Hansda A, Kumar V, Anshumali (2017) Cu-resistant Kocuria sp. CRB15: a potential PGPR isolated from the dry tailing of Rakha copper mine. 3 Biotech 7(2). https://doi.org/10.1007/s13205-017-0757-y

Hashim MA, Mukhopadhyay S, Narayan J, Sengupta B (2011) Remediation technologies for heavy metal contaminated groundwater. J Environ Manage 92(10):2355–2388. https://doi.org/10.1016/j.jenvman.2011.06.009

He Y, Gong Y, Su Y, Zhang Y, Zhou X (2019) Chemosphere Bioremediation of Cr (VI) contaminated groundwater by Geobacter sulfurreducens: environmental factors and electron transfer flow studies. Chemosphere 221:793–801. https://doi.org/10.1016/j.chemosphere.2019.01.039

Hou D (2014) Metal contamination and bioremediation of agricultural soils for food safety and sustainability. Nat Rev Earth Environ 1:366–381. https://doi.org/10.1038/s43017-020-0061-y

Ibrahim WM (2011) Biosorption of heavy metal ions from aqueous solution by red macroalgae. J Hazard Mater 192(3):1827–1835. https://doi.org/10.1016/j.jhazmat.2011.07.019

Iravani S, Varma RS (2020) Bacteria in heavy metal remediation and nanoparticle biosynthesis. https://doi.org/10.1021/acssuschemeng.0c00292

Jagtap UB, Bapat VA (2015) Genetic engineering of plants for heavy metal removal from soil. https://doi.org/10.1007/978-3-319-14526-6_22

Jayakumar R, Rajasimman M, Karthikeyan C (2015) Sorption and desorption of hexavalent chromium using a novel brown marine algae Sargassum myriocystum. Korean J Chem Eng 32(10):2031–2046. https://doi.org/10.1007/s11814-015-0036-8

Jeevanantham S, Saravanan A, Hemavathy RV, Kumar PS (2019) Environmental Technology & Innovation Removal of toxic pollutants from water environment by phytoremediation: A survey on application and future prospects. Environ Technol Innov 13:264–276. https://doi.org/10.1016/j.eti.2018.12.007

Jha Y, Subrmanian RB, Mishra KK (2017) Role of plant growth promoting rhizobacteria in accumulation of heavy metal in metal contaminated soil. Emergent Life Sci Res. https://doi.org/10.7324/elsr.2017.314856

Juan A, Aparicio D, Raimondo EE, Benimeli CS, Alejandra M (2017) Actinobacteria consortium as an efficient biotechnological tool for mixed polluted soil reclamation: experimental factorial design for bioremediation process optimization. J Hazard Mater. https://doi.org/10.1016/j.jhazmat.2017.08.041

Kaewdoung B, Sutjaritvorakul T, Gadd GM, Whalley AJS, Sihanonth P (2016) Heavy metal tolerance and biotransformation of toxic metal compounds by new isolates of wood-rotting fungi from Thailand. Heavy Metal Tolerance and Biotransformation of Toxic Metal Compounds by New 0451(March). https://doi.org/10.1080/01490451.2015.1048394

Kapahi M, Sachdeva S (2019) Bioremediation options for heavy metal pollution. J Health Pollut 9(24):191203

Kaplan D (2013) Absorption and adsorption of heavy metals by microalgae, pp 602–611. Wiley

Kaur S, Roy A (2020) Bioremediation of heavy metals from wastewater using nanomaterials. Environ Dev Sustain 23:9617. https://doi.org/10.1007/s10668-020-01078-1

Khalid S, Shahid M, Khan N, Murtaza B, Bibi I, Dumat C (2017) A comparison of technologies for remediation of heavy metal contaminated soils. J Geochem Explor 182:247–268. https://doi.org/10.1016/j.gexplo.2016.11.021

Kit Y, Chang J (2020) Bioresource Technology Bioremediation of heavy metals using microalgae: Recent advances and mechanisms. Biores Technol 303(January):122886. https://doi.org/10.1016/j.biortech.2020.122886

Kumar PS, Gunasundari E (n.d.) Bioremediation of heavy metals

Kumar S, Prasad S, Yadav KK, Shrivastava M, Gupta N, Nagar S, Bach Q, Kamyab H, Khan SA, Yadav S, Malav LC (2019) Hazardous heavy metals contamination of vegetables and food chain: Role of sustainable remediation approaches—a review. Environ Res 179:108792. https://doi.org/10.1016/j.envres.2019.108792

Li S, Wang W, Liang F, Zhang WX (2017) Heavy metal removal using nanoscale zero-valent iron (nZVI): Theory and application. J Hazard Mater 322:163–171. https://doi.org/10.1016/j.jhazmat.2016.01.032

Li X, Dong S, Yao Y, Shi W, Wu M, Xu H (2016) RSC Advances Inoculation of bacteria for the bioremediation of heavy metals contaminated soil by Agrocybe. RSC Adv 6:65816–65824. https://doi.org/10.1039/C6RA11767H

Liu S, Zeng G, Niu Q, Liu Y, Zhou L, Jiang L, Tan X, Xu P, Zhang C, Cheng M (2017) Bioresource technology bioremediation mechanisms of combined pollution of PAHs and heavy metals by bacteria and fungi : A mini review. Biores Technol 224:25–33. https://doi.org/10.1016/j.biortech.2016.11.095

Marzan LW, Hossain M, Mina SA, Akter Y, Chowdhury AMMA (2017) Isolation and biochemical characterization of heavy-metal resistant bacteria from tannery effluent in Chittagong city, Bangladesh: Bioremediation viewpoint. Egypt J Aquat Res. https://doi.org/10.1016/j.ejar.2016.11.002

Melin T, Jefferson B, Bixio D, Thoeye C, De Wilde W, De Koning J, van der Graaf J, Wintgens T (2006) Membrane bioreactor technology for wastewater treatment and reuse. Desalination 187(1–3):271–282. https://doi.org/10.1016/j.desal.2005.04.086

Mendy PA, Kargbo A, Entonu ME (2021) Bioremediation of heavy metal ions from contaminated soil and water by microbes: A review. Afr J Biol Sci 3(2):1–8. https://doi.org/10.33472/AFJBS.3.2.2021.1-8

Min K, Gong G, Woo HM, Kim Y, Um Y (2015) A dye-decolorizing peroxidase from Bacillus subtilis exhibiting substrate-dependent optimum temperature for dyes and β-ether lignin dimer. Sci Rep 5(1):8245. https://doi.org/10.1038/srep08245

Modrzejewska-sikorska A, Konowa E (2020) Silver and gold nanoparticles as chemical probes of the presence of heavy metal ions. J Mol Liq 302:112559. https://doi.org/10.1016/j.molliq.2020.112559

Nocito FF, Lancilli C, Dendena B, Lucchini G (2011) Cadmium retention in rice roots is influenced by cadmium availability, chelation and translocation. Plant Cell Environ 34(6):994–1008. https://doi.org/10.1111/j.1365-3040.2011.02299.x

Ontañon OM, Fernandez M, Agostini E, González PS (2018) Identification of the main mechanisms involved in the tolerance and bioremediation of Cr (VI) by Bacillus sp. SFC 500-1E. Vi. Environ Sci Pollut Res 25(16):16111

Ovais M, Khalil AT, Ayaz M, Ahmad I, Nethi SK, Mukherjee S (2018) Biosynthesis of metal nanoparticles via microbial enzymes: a mechanistic approach 1–20. https://doi.org/10.3390/ijms19124100

Parvin F, Rikta SY, Tareq SM (2019) Application of nanomaterials from wastewater. In: Nanotechnology in water and wastewater treatment. Elsevier Inc. https://doi.org/10.1016/B978-0-12-813902-8.00008-3

Peña-montenegro TD, Lozano L, Dussán J (2015) Genome sequence and description of the mosquitocidal and heavy metal tolerant strain Lysinibacillus sphaericus CBAM5. Stand in Genomic Sci 10(1):1–10

Philippsen CG, Vilela ACF, Zen LD (2015) Fluidized bed modeling applied to the analysis of processes: review and state of the art. J Market Res 4(2):208–216. https://doi.org/10.1016/j.jmrt.2014.10.018

Praveen A, Pandey VC, Marwa N, Singh DP (2019) Harnessing plant à microbe interactions. In: Phytomanagement of polluted sites. Elsevier Inc. https://doi.org/10.1016/B978-0-12-813912-7.00015-6

Radjenović J, Matošić M, Mijatović I, Petrović M, Barceló D (2008) Membrane bioreactor (MBR) as an advanced wastewater treatment technology. In: Handbook of environmental chemistry, volume 5: water pollution, 5 S2(May 2014), pp 37–101. https://doi.org/10.1007/698_5_093

Ramesh D, Muniraj IK, Thangavelu K (2019) Knowledge update on bioreactor technology for cellulase production. In: From cellulose to cellulase: strategies to improve biofuel production. Elsevier B.V. https://doi.org/10.1016/B978-0-444-64223-3.00011-4

Rawat JM, Bhandari A, Raturi M, Rawat B (2019) Agrobacterium rhizogenes mediated hairy root cultures: a promising approach for production of useful metabolites. In: New and future developments in microbial biotechnology and bioengineering: microbial secondary metabolites biochemistry and applications. Elsevier B.V. https://doi.org/10.1016/B978-0-444-63504-4.00008-6

Rhodes CJ, Rhodes CJ (2015) Mycoremediation (bioremediation with fungi)—growing mushrooms to clean the earth. Chem Speciat Bioavailab 26(3):196. https://doi.org/10.3184/095422914X14047407349335

Robles-González IV, Fava F, Poggi-Varaldo HM (2008) A review on slurry bioreactors for bioremediation of soils and sediments. Microbial Cell Factories 7(February 2014). https://doi.org/10.1186/1475-2859-7-5

Samrot AV, Agnes JL, Roshini ASM, Stefi PRSM, Jenifer RPA, Madankumar SA (2019) Bioactivity and heavy metal removal using plant gum mediated green synthesized silver nanoparticles. J Cluster Sci 30:1599–1610. https://doi.org/10.1007/s10876-019-01602-y

Saranya K, Sundaramanickam A, Shekhar S, Swaminathan S, Balasubramanian T (2017) Bioremediation of mercury by vibrio fluvialis screened from industrial effluents. BioMed Res Int 2017:1–6

Science L, Sharma S (2012) Bioremediation: features, strategies and applications. Asian J Pharm Life Sci 2(2):202–213

Seshadri B, Bolan NS, Naidu R (2015) Rhizosphere-induced heavy metal(Loid) transformation in relation to bioavailability and remediation. J Soil Sci Plant Nutr 15(2):524–548. https://doi.org/10.4067/s0718-95162015005000043

Senthil Kumar P, Gunasundari ES (2018) Bioremediation: applications for environmental protection and management (Energy, Environment, and Sustainability). Springer Nature Singapore Pte Ltd. https://doi.org/10.1007/978-981-10-7485-1_9

Shackira AM, Puthur JT (2019) Phytostabilization of heavy metals: understanding of principles and practices. https://doi.org/10.1007/978-3-030-20732-8

Shah Maulin P (2020) Microbial bioremediation & biodegradation. Springer

Shah Maulin P (2021a) Removal of refractory pollutants from wastewater treatment plants. CRC Press

Shah Maulin P (2021b) Removal of emerging contaminants through microbial processes. Springer

Shittu KO, Ihebunna O (2017) Purification of simulated waste water using green synthesized silver nanoparticles of Piliostigma thonningii aqueous leaves extract. Adv Nat Sci Nanosci Nanotechnol 8(4):045003. https://doi.org/10.1088/2043-6254/aa8536

Sirohi G, Upadhyay A, Shankar Srivastava P, Srivastava S (2015) PGPR mediated zinc biofertilization of soil and its impact on growth and productivity of wheat. J Soil Sci Plant Nutr 15(1):202–216. https://doi.org/10.4067/S0718-95162015005000017

Sivarajasekar SMN, Paramasivan JSVT, Naushad M (2018) Phytoremediation of heavy metals: Mechanisms, methods and enhancements. Environ Chem Lett. https://doi.org/10.1007/s10311-018-0762-3

Soleymani F, Khani MH, Pahlavanzadeh H, Manteghian M (2015) Study of cobalt (II) biosorption on Sargassum sp. by experimental design methodology. Int J Environ Sci Technol 12(6):1907–1922. https://doi.org/10.1007/s13762-014-0739-0

Spier MR, Vandenberghe LPDS, Medeiros ABP, Soccol CR (2011) Application of different types of bioreactors in bioprocesses. In: Bioreactors: design, properties and applications (Issue January)

Stanbrough R, Chuaboonmee S, Palombo EA, Malherbe F, Bhave M (2013) Heavy metal phytoremediation potential of a heavy metal resistant soil bacterial isolate, Achromobacter sp. Strain AO22. APCBEE Proc 5:502–507. https://doi.org/10.1016/j.apcbee.2013.05.085

Sun X, Liu Z, Jiang Q, Yang Y (2019) Concentrations of various elements in seaweed and seawater from Shen'ao Bay, Nan'ao Island, Guangdong coast, China: environmental monitoring and the bioremediation potential of the seaweed. Sci Total Environ 659:632–639. https://doi.org/10.1016/j.scitotenv.2018.12.364

Tarekegn MM, Salilih FZ, Ishetu AI, Tarekegn MM, Salilih FZ, Ishetu AI (2020) Cogent food & agriculture microbes used as a tool for bioremediation of heavy metal from the environment. Food Science & Technology | Review Article Microbes used as a tool for bioremediation of heavy metal from the environment. Cogent Food Agric 6(1). https://doi.org/10.1080/23311932.2020.1783174

Tariq A, Ullah U, Asif M, Sadiq I (2018) Biosorption of arsenic through bacteria isolated from Pakistan. Int Microbiol 22(1):59–68

Tauanov Z, Tsakiridis PE, Shah D, Inglezakis VJ (2019) Toxic/Hazardous Substances and Environmental Engineering Synthetic sodalite doped with silver nanoparticles: Characterization and mercury (II) removal from aqueous solutions and mercury (II) removal from aqueous solutions. J Environ Sci Health, Part A 54(9):951–959. https://doi.org/10.1080/10934529.2019.1611129

Tchounwou PB, Yedjou CG, Patlolla AK, Sutton DJ (2012) Molecular, clinical and environmental toxicicology Volume 3: environmental toxicology. Mol Clin Environ Toxicol 101:133–164. https://doi.org/10.1007/978-3-7643-8340-4

Vélez E, Campillo G, Morales G, Hincapié C, Osorio J, Arnache O (2018) Silver nanoparticles obtained by aqueous or ethanolic aloe vera extracts: an assessment of the antibacterial activity and mercury removal capability. J Nanomater 2018: 1–7

Verma R, Vijayalakshmy K, Chaudhiry V (2018) Detrimental impacts of heavy metals on animal reproduction: A review. J Entomol Zool Stud 6(6):27–30
Verma S, Kuila A (2019) Bioremediation of heavy metals by microbial process. Environ Technol Innov 14:100369. https://doi.org/10.1016/j.eti.2019.100369
Volarić A, Svirčev Z, Tamindžija D, Radnović D (2021) Microbial bioremediation of heavy metals. Hemijska Industrija 75(2):103–115
Wang B, Wang Z, Chen T, Zhao X (2020) Development of novel bioreactor control systems based on smart sensors and actuators. Front Bioeng Biotechnol 8(February):1–15. https://doi.org/10.3389/fbioe.2020.00007
Wu Y, Zhou P, Cheng H, Wang C, Wu M, Xu X (2015) Draft genome sequence of microbacterium profundi Shh49 T, an Actinobacterium isolated from deep-sea sediment of a Polymetallic nodule environment. Genome Announcements 3(3):4–5. https://doi.org/10.1128/genomeA.00642-15. Copyright
Xu Y, Shi H, Fei Y, Wang C, Mo L, Shu M (2021) Article identification of soil heavy metal sources in a large-scale area affected by industry. Sustainability (Switzerland) 13(2):1–18. https://doi.org/10.3390/su13020511
Yadav KK, Singh JK, Gupta N, Kumar V (2017a) *A review of nanobioremediation technologies for environmental cleanup: a novel biological approach.* J Mater Environ Sci 8(2):740–757
Yadav, Krishna & Gupta, Neha & Kumar, Vinit & Singh, Jitendra. (2017b). Bioremediation of heavy metals from contaminated sites using potential species: A review. Indian J Environ Prot 37:65–84
Yaghmaeian K, Jaafari J (2018) Optimization of heavy metal biosorption onto freshwater algae (*Chlorella coloniales)* algae cells using response surface methodology (RSM). Chemosphere. https://doi.org/10.1016/j.chemosphere.2018.10.205
Yang T, Chen M, Wang J (2015) Trends in Analytical Chemistry Genetic and chemical modification of cells for selective separation and analysis of heavy metals of biological or environmental significance. Trends Anal Chem 66:90–102. https://doi.org/10.1016/j.trac.2014.11.016
Yao L, Teng J, Zhu M, Zheng L, Zhong Y, Liu G, Xue F, Chen W (2016) MWCNTs based high sensitive lateral flow strip biosensor for rapid determination of aqueous mercury ions. Biosens Bioelectron 85:331–336. https://doi.org/10.1016/j.bios.2016.05.031
Yin K, Wang Q, Lv M, Chen L (2019) Microorganism remediation strategies towards heavy metals. Chem Eng J 360(August 2018):1553–1563. https://doi.org/10.1016/j.cej.2018.10.226
Yoo H, Pak J (2013) Synthesis of highly fluorescent silica nanoparticles in a reverse microemulsion through double-layered doping of organic fluorophores. https://doi.org/10.1007/s11051-013-1609-2
Yumei L, Yamei L, Qiang L, Jie B (2017) Rapid biosynthesis of silver nanoparticles based on flocculation and reduction of an exopolysaccharide from Arthrobacter sp. B4: its antimicrobial activity and phytotoxicity
Zhang H, Yuan X, Xiong T, Wang H, Jiang L (2020) Bioremediation of co-contaminated soil with heavy metals and pesticides: influence factors, mechanisms and evaluation methods. Chem Eng J 398(October 2019):125657. https://doi.org/10.1016/j.cej.2020.125657
Zhang X, Ma J, Chen M, Wu Z, Wang Z (2018) Microbial responses to transient shock loads of quaternary ammonium compounds with different length of alkyl chain in a membrane bioreactor. AMB Express 8(1). https://doi.org/10.1186/s13568-018-0649-5
Zhao R, Wang B, Cai QT, Li XX, Liu M, Hu D, Guo DB, Wang J, Fan C (2016) Bioremediation of Hexavalent Chromium Pollution by Sporosarcina saromensis M52 Isolated from Offshore Sediments in Xiamen, China. Biomed Environ Sci 29(2):127–136. https://doi.org/10.3967/bes2016.014
Zhu Y, Xu F, Liu Q, Chen M, Liu X, Wang Y, Sun Y, Zhang L (2019) Science of the Total Environment Nanomaterials and plants : Positive effects, toxicity and the remediation of metal and metalloid pollution in soil. Sci Total Environ 662:414–421. https://doi.org/10.1016/j.scitotenv.2019.01.234

Therapeutic and Diagnostic Potential of Nanomaterials for Enhanced Biomedical Applications

Nick Vordos, Despina A. Gkika, Nikolaos Pradakis, Athanasios C. Mitropoulos, and George Z. Kyzas

1 Introduction

Nanotechnology has affected daily life by addressing an array of concerns and has contributed to the treatment of a wide range of illnesses (Barzinjy et al. 2020). The collaboration of nanotechnology, material sciences and biotechnology, has quickly become popular in the field of therapeutic medicine (Eftekhari et al. 2021). Nanoparticles can go past biological boundaries and get to their destination due to their specific physicochemical attributes. Nanomedicine takes advantage of materials with nano-dimensions, commonly at the 1–100 nm range. Such materials are part of the delivered drugs or included in the related devices and production thereof. When a material reaches nano-dimensions, its properties may change, thus exhibiting different optical, magnetic and electrical behaviours compared to their larger versions. Therapeutic nanomaterials may act as a crucial drug delivery mechanism. Despite the observed growth and variety of users, however, their particular architecture is still a topic of constant interest, mainly because of the multitude of aspects to consider, such as their unique material attributes, therapeutic agents to be used or activation approaches. Theranostics couples the diagnosis with the therapeutic treatment, which can be enabled through the formulation of nanomaterials with complex structures (Cheng et al. 2021). This chapter covers the diagnostic and therapeutic abilities of nanomaterials in detail, featuring the recent advancements in therapeutic applications, that are dependent on the use of nanomaterials. Further nanomedicine research and

N. Vordos (✉) · N. Pradakis
Department of Physics, International Hellenic University, Kavala, Greece
e-mail: nvordos@physics.ihu.gr

D. A. Gkika · A. C. Mitropoulos · G. Z. Kyzas
Department of Chemistry, International Hellenic University, Kavala, Greece

M. P. Shah (ed.), *Advanced and Innovative Approaches of Environmental Biotechnology in Industrial Wastewater Treatment*,
https://doi.org/10.1007/978-981-99-2598-8_13

better healthcare services are expected to be the product of the development of nanotherapeutics, as scientists exert better control over the attributes of manufactured nanomaterials.

2 Therapeutic Nanomaterials for Advanced Biomedical Applications

2.1 General

Nanomaterials are capable to enter cancer cells and possibly result in successful tumour destruction with minimal impact on other functions (Harvey and Plé 2021). The suitability of nanomaterials as drug carriers has been thoroughly explored, as depicted by a large number of clinical research works. Nanoparticles are thought to be some of the most hopeful drug delivery mechanisms (Singh et al. 2019). To further explore them depending on their type, nanomaterials can be classified as polymer- or lipid-based and non-polymeric (Fig. 1).

Table 1 presents the group of non-polymeric materials, along with their potential applications and related findings.

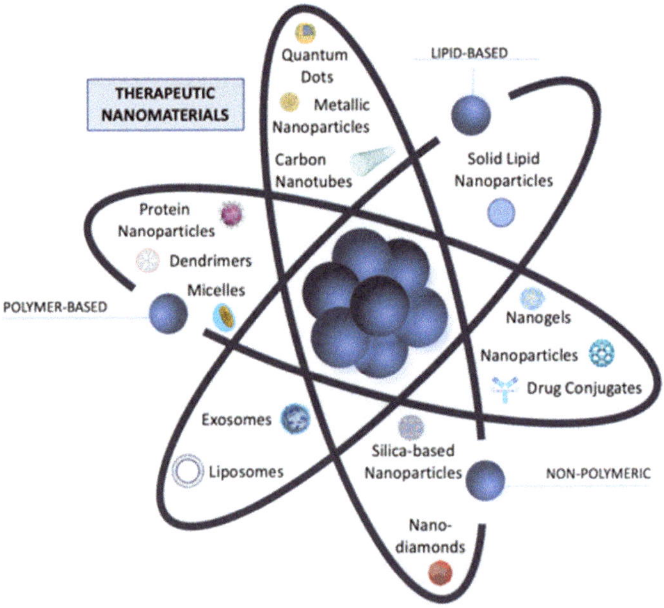

Fig. 1 Illustration of the types of therapeutic nanoparticles

Table 1 Group of non-polymeric materials

Type	Nanoparticle	Cancer applications	Findings	Other applications	Reference
Non-polymeric nano-materials	Metal Nanoparticles (Gold)	Various types of cancers	Displays photothermal conversion; the high temperature in the targeted cancer cells results in their death	Glioma therapy (Dhar et al. 2011) Antimicrobial coatings (Noimark et al. 2014) Diabetes (Cho et al. 2014)	(Tian et al. 2016)
	Metal Nanoparticles (Platinum)	Lung, head and neck, ovarian and testicular cancers	Has the ability to absorb X-rays and create free radicals	Drug delivery (Dhar et al. 2009)	(Devasena 2017)
	Metal Nanoparticles (Gallium)	Lung cancer, Prostate cancer and breast cancer	It might prolong a patient's life	Alzheimer's Disease (Wang and Luo 2020)	(Niesvizky et al. 2002)
	Metal Nanoparticles (silver)	Hepatocellular carcinoma	It has singular physicochemical attributes such as great conductivity and chemical stability, while it also displays antiviral, anticancer and other properties	Antimicrobial therapy (Kanmani and Lim 2013)	(Sánchez-López et al. 2020)
	Carbon Nanotubes	Various types of cancers	It is stable in nano-form, offers s a high aspect ratio and possesses a large surface area	Parkinson's disease (Marei et al. 2017)	(Novoselov 2004)

(continued)

Table 1 (continued)

Type	Nanoparticle	Cancer applications	Findings	Other applications	Reference
	Nanodiamonds	Hepatic cancer stem cells	It has great potential in drug delivery and is extremely biocompatible	Neurodegenerative diseases (Saraf et al. 2019)	(Zhang et al. 2016)
	Silica Nanoparticles (MCM, SBA, HMS, MSU)	Various types of cancers	Its porosity is useful containing the drug within the pore structure thus resulting in a high drug load	Parkinson's Disease (Swar et al. 2019)	(Mehmood et al. 2017)
	Quantum dots	Drug delivery	It is photostable and fluorescent	Rheumatism (Hanada et al. 2013) Stem cell therapy (Yano et al. 2015)	(Bilan et al. 2017)

Metal compounds have been proven to be effective against tumours. Metal nanoparticles of various shapes and sizes can be used to transport drugs, proteins and other substances, as well as to detect, diagnose and treat patients (Bhattacharya and Mukherjee 2008). More specifically, metal oxide nanoparticles have been found to be very stable and have simple preparation process; They can be easily manufactured with the desirable size and other attributes and are easily integrated with either hydrophobic or hydrophilic systems; They can be used in conjunction with other molecules, because of their negative surface charge, thus are considered very promising for future uses in biomedicine (Gobi et al. 2021).

Gold nanoparticles (GNPs) can be defined as colloidal suspensions of gold particles at nano-dimensions. They are typically created through a reduction of chloroauric acid in an environment containing a variety of capping and/or stabilizing substances (Ganesh Kumar et al. 2017). Gold nanoparticles have been used effectively for diagnosing and treating cancer, due to its singular optical and localized surface plasmon resonance and comparatively low levels of cytotoxicity. When gold nanoparticles are exposed to light of specific wavelengths, there is photothermal conversion and the resulting high temperatures kill cancer cells of the targeted tissue. Furthermore, gold nanoparticles can be used as drug carriers, in cases where the use of light can cause the release of the drug (Tian et al. 2016).

Platinum compounds are often used for treating lung and other cancers (Rademaker-Lakhai et al. 2004). There are multiple anti-cancer drugs containing platinum that have been clinically used to address more than 50% of cancers (Bruno et al. 2017). In addition, platinum can absorb X-rays, enhance doses and create free radicals, which may cause cancer cells to die (Devasena 2017).

Gallium is a common alternative to platinum when it comes to tumour treatment (Foster et al. 1986; Kandil et al. 2018). Gallium may act similarly to Iron and thus disturb iron-related processes in cancer cells (Chitambar 2012). Gallium Nitrate is able to address elevated bone turnover in cases where bone metastasis has occurred, while there are various works establishing the advantages for people who suffer from breast cancer or myeloma (Warrell et al. 1987). Furthermore, it is believed that low doses of gallium nitrate might prolong the life of people suffering from late-stage multiple myeloma (Niesvizky et al. 2002; Niesvizky 2003). Hart and Adamson (1971) initially reported the anti-cancer abilities of gallium nitrate around 1971; it has been constantly studied ever since and used to treat lung, breast and prostate cancer (Webster et al. 2000). More specifically, the clinical trials consisted of the use of gallium nitrate intravenously, exhibiting great results versus lymphomas and bladder cancer and milder results versus prostate, cervix and ovarian tumours (Collery et al. 2002). Gallium Phosphide (GaP) nanowires have been found to display improved results in terms of cell adhesion and survivability (Vidu et al. 2014). Past research suggested that gallium may predict AD due to its similar binding tendency with transferrin as iron (Wang and Luo 2020).

Silver nanoparticles (SNPs) have also been put under the spotlight because of their desirable physicochemical attributes such as being conductive and chemically stable, as well as possessing antiviral, anticancer and multiple other useful properties (Sharma et al. 2009). Silver nanoparticles have been proven effecting in delivering drugs due to these physicochemical properties. They have exhibited biocompatibility, stability and improved solubility; they can be manufactured in a high yield without needing to aggregate (Baig et al. 2019). After conducting a research in vitro with HepG2 cells, Ahmadian et al. (2018) reported that SNPs have cytotoxicity potential and could serve as a possible contender for treating Hepatocellular carcinoma. In general, noble metal nanoparticles, have been researched in depth for their possible applications in nanomedicine; however, their practical use is still currently limited because of the increased related costs (Sánchez-López et al. 2020; Gkika et al. 2017, 2021).

Carbon-based nanomaterials have also been deemed as significant due to their potential biomedical applications. **Carbon nanomaterials** are a type of nanomaterial with multiple subcategories that are carbon-based. Their biocompatibility and safety are considered to be better than metal-based nanomaterials (Saleem et al. 2018). Carbon nanotubes can be utilized in applications such as cancer diagnosis and treatment due to their exhibited stability in the nano-scale dimensions, their high aspect ratio and large surface area. Furthermore, their ability to be chemically functionalized makes them ideal for active targeting applications (Baig et al. 2019). Fullerenes have been found to be effective in delivering antibiotics as well as antiviral and anti-cancer drugs (Mroz et al. 2007). Graphene possesses outstanding mechanical attributes and thus has been extensively researched for use in biomedicine, cancer treatment included (Novoselov 2004). The works of Bhirde et al. (2009) reported that modified carbon nanotubes have been effecting in destroying head and neck squamous cancer cells, while Dhar et al. (2008) demonstrated that carbon nanotubes loaded with cisplatin were successful in targeting human choriocarcinoma and nasopharyngeal carcinoma cells. Further research on multiwalled carbon nanotubes (MWCNTs) and their drug delivery abilities has shown that they have promising potential in the treatment of tumors that previously resisted multi-drugs (Kumar et al. 2018). Diamond nanoparticles are also significant since they can be used as biomarkers in fluorescence imaging. **Nanodiamonds** are very effective when delivering genes or hydrophobic treatments and chemotherapy with continuous release. Furthermore, the high refractive index and displayed Raman optical activity (ROA) suggest that nanodiamonds possess scattering attributes, rendering them ideal for cellular imaging (Badea and Kaur 2013). Nanodiamonds have further been explored for their potential uses in targeted treatment and cancer imaging (Cheng et al. 2021). Nanodiamonds (NDs) have been found to be very biocompatible and generally exhibit exemplary delivery capabilities compared to other nano-carriers (Zhang et al. 2016). Zhao et al. have proven that detonation nanodiamonds coated with hyperbranched polyglycerol-carrying cancer treatment drugs can be extremely effective in causing selective toxicity to the targeted cancer cells, requiring minimal uptake and causing minimal toxicity in macrophages (Zhao et al. 2014).

Silica-based nanoparticles possess singular surface attributes, including porosity and functionalization abilities. Nanosilica appears to have promising results in effective drug delivery. MCM-41 and various other mesoporous nanomaterials that possess larger pore sizes, such as SBA-15 and 16, MSU and more have been effectively used for delivering drugs. Their porosity is useful in trapping the drug within the pore structure, thus accomplishing high drug load (Mehmood et al. 2017). As a result, these nanomaterials have proven their effective use as drug carriers, able to locally deliver antibiotics battling biofilm-related infections, that has been challenging from a biomedical point of view (Ganesh Kumar et al. 2017). A sono-sensitive silica nanoparticle is able to destroy cancer cells when using ultrasound radiation (Osminkina et al. 2011).

Quantum dots (QD) can be useful as an innovative type of fluorescent probe for biomolecular and cellular imaging in vivo due to their photostability. Their quantum yield reaches almost 100%, and they exhibit wide absorption and narrow emission ranges (Bilan et al. 2017). In addition, quantum dots are widely accepted as effective nanotheranostic platforms (Matea et al. 2017). Olerile et al. (2017) have created a theranostics system using a combination of quantum dots and anti-cancer treatments in lipid carriers with nano-structure. The nanoparticles had a spherical shape, displaying an improved encapsulation effectiveness of paclitaxel (around 81%) and a reduction of the rate of tumor growth in the range of about 78%. The researchers also reported that this system had the capacity to expressly aim for and detect H22 cancer cells (Olerile et al. 2017). The group of the polymer-based materials, shown in Table 2.

Synthetic and natural **polymer-based nanomaterials** (PBNMs) have also captured the attention of researchers boasting desirable properties, such as being biocompatible, non-toxic and biodegradable (Crucho and Barros 2017). PBNMs can be categorized as nanocapsules or as nanospheres according to their architecture. In the case of capsules, a polymeric membrane wraps around therapeutic agents, while these agents are distributed around or within the polymer structure in nanospheres (Letchford et al. 2009). Figure 2 depicts a nanosystem with specific architecture and physicochemical attributes that enables getting past the barrier between blood and brain, improving the delivery of dopamine to the brain (Monge-Fuentes et al. 2021).

Nanoparticles have been used as innovative instruments of renal disease theranostics. They are also able to **target the kidney** because of their special properties via size, shape, charge and other property adjustments. Furthermore, there is promising research on their use during the creation of implantable artificial kidneys (Eftekhari et al. 2021). Recent works have proven the positive impact of inorganic nanoparticles in oxidative-stress diseases, such as acute kidney injury (AKI) (Soh et al. 2017). Molybdenum-based polyoxometalate (POM) nanoclusters have been reported to have remarkable renal uptake and corresponding antioxidant impact on the kidney during AKI (Ni et al. 2018).

Protein nanoparticles initially utilized the inherent protein attributes when flowing in serum, allowing the drug composites to be dissolved and transferred during circulation. Later on, natural proteins were mixed with various drugs in an attempt to address the toxic effects (Paredes et al. 2019). Protein nanoparticles have multiple benefits as a drug delivery method, due to their enhanced stability (Hong et al. 2020). The interactions between the proteins and the nanoparticles define the behaviours of nanomaterials in biological systems (Yang et al. 2013). Protein materials, such as particles that act like viruses, silk proteins and various polypeptides

Table 2 Group of polymer-based materials

Type	Nanoparticles	Application	Findings	Other applications	Reference
Polymeric nanomaterials	Polymer-based Nanoparticles	Renal transplant	Biocompatible, non-immunogenic, non-toxic and biodegradable	Parkinson's Disease (Pahuja et al. 2019)	(Crucho and Barros 2017)
	Protein Nanoparticles	Drug delivery	Very stable	Alzheimer's disease (Hong et al. 2020)	(Hong et al. 2020)
	Drug-Conjugates	Control delivery	Can sustainably release drugs, very stable	Alzheimer's disease (Kuo and Lee 2016)	(Ekladious et al. 2019)
	Dendrimers	Drug delivery	Molecules are very uniform, the distribution of weight is narrow in range, adjustable size and shape attributes	Neurodegeneration (Igartúa et al. 2018)	(Wu et al. 2015)
	Nanogels	Cancer	Very biocompatible and biodegradable, exhibits quick swelling/contracting attributes, able to achieve high drug loads	Alzheimer's disease (Paul et al. 2017)	(Paul et al. 2017)
	Micelles	Drug delivery	Enhances solubility in the aqueous means	Alzheimer's disease (Mirzaie et al. 2019)	(Ray Banerjee 2020)

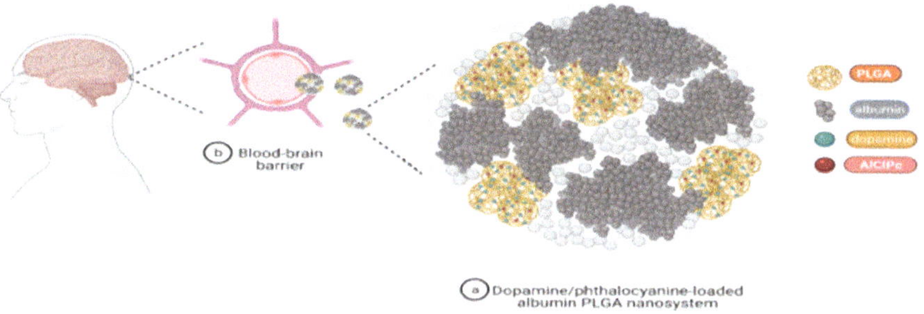

Fig. 2 Illustration of a dopamine/phthalocyanine-loaded nanosystem's architecture (**a**) and its interaction with the barrier between brain and blood (**b**) to before getting past it and releasing its load. *AlClPc* aluminum chloride phthalocyanine, *PLGA* poly (lactic-*co*-glycolic acid). Reproduced with permission from Monge-Fuentes et al. (2021), Copyright Nature, Ltd

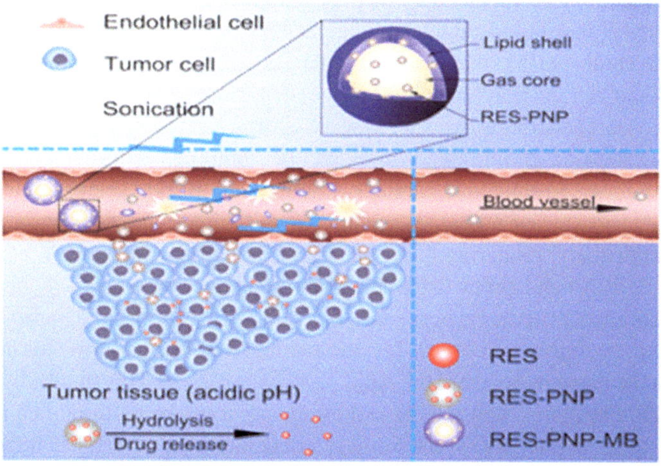

Fig. 3 Illustration of RES-PNP-MB as an anti-tumor drug carrier. Reproduced with permission from Lv et al. (2016), Copyright Nature, Ltd

have been broadly used in cell recognition and tissue engineering applications (Ren et al. 2012). Manufactured protein nanomaterials have further been considered for use in scaffolding platforms (Khmelinskaia et al. 2021). Natural nanoparticles, such as lipoproteins, have been effective in the treatment of Alzheimer's disease (Hong et al. 2020).

Figure 3 depicts a microbubble carrying resveratrol nanoparticles, which is susceptible to PH changes (RES-PNP-MB). It is an innovative multi-purpose drug carrier, created to improve the drug effect, while minimizing potential side effects (Lv et al. 2016).

Polymer–drug conjugates are pharmacologically active composites consisting of one or more therapeutic agents (Ekladious et al. 2019). Polymer drug conjugates exhibit high levels of drug loading, and allow for the sustained delivery of drugs. They are very stable compared to other nano-sized carriers that encompass drugs (Kumar et al. 2017), and were some of the first formulations created to only dispense the drug inside the target site (Zhang et al. 2012). The composites have increased total weight, which encourages the pharmacokinetic inclination within the cells (Yetisgin et al. 2020). Polymer drug conjugates usually display longer half-life (most often more than 1 h) compared to free drugs (b5 min) during blood circulation, resulting in increased drug presence in tumors (Singer et al. 2005). They have been proven to be more effective and dependable for the sustained delivery of drugs (Yang et al. 2017). Some conjugates are sensitive to PH changes; it is thus possible to take advantage of the pH-sensitive chemical bonds connecting the polymer and drug to reach a selective release of the drug in the acidic tumour setting (Pang et al. 2016). Conjugates reportedly also enhance the drug's bioavailability (Chen et al. 2014).

Dendrimers are a type of molecules that have a heavily branched structure and easy-to-modify surfaces. They display a significant level of molecular uniformity, narrowly distributed molecular weight and have adjustable dimensions and shape (Wu et al. 2015). The most popular dendrimers in biomedicine include poly(amidoamine) and poly(propylene imine) dendrimers (Thakore et al. 2019). They have been researched for the capacity to serve as drug carriers against various diseases (Patel et al. 2021). Cationic dendrimers can be combined with nucleic acids, thus acting as effective nucleic acid carriers (Cheng et al. 2021). The exhibited polyvalency and distribution of potential capabilities on their surface are important advantages of dendrimers, rendering them ideal for combating cancer, HIV, etc., as well as delivering drugs and genes (Mignani et al. 2018).

Nanogels are nanoscale carriers based on colloidal gel formation (Sardoiwala et al. 2018). They are highly biocompatible and biodegradable, and allow for larger drug loads, with the capacity to avoid the rapid renal exclusion and the uptake by the reticuloendothelial system. They also have the capacity to make hydrophobic molecules soluble in their gel networks, have lower concentrations of critical micelles, and display slower dissociation rates (Paul et al. 2017). Nanogels have various uses in cancer treatment. Dickerson et al. suggested that siRNAs delivered by nanogels may enhance the efficiency of chemotherapy treatments (Dickerson et al. 2010). A different study showed that nanogels loaded with paclitaxel resulted in improved cytotoxic effects in HEPG-2 cells compared to free drugs (Li et al. 2011). Nanogels are also highly effective versus autoimmune diseases, neurodegenerative and inflammatory disorders and diabetes. They are also great options for intracellular delivery (Paul et al. 2017). Nanogel-PEG carriers have been proven to be effective in delivering inhibitors for HIV-1 into brain cells (Vinogradov et al. 2010).

Polymer micelle nanomaterials are typically utilized for the delivery of very hydrophobic treatments, peptides or contrast imaging molecules (Wakaskar 2018). Polymeric micelles (Ahmad et al. 2014) enhance the solubility in aqueous means by trapping the less soluble drugs and releasing the drug at the targeted site (Ray Banerjee 2020). Peng et al. created a polymer micelle by mixing temozolomide and siRNA with a folic acid polymer to get past the barrier of acquired resistance to drugs during the delivery of the treatment (Peng et al. 2018). Table 3 presents a more detailed description of the lipid-based materials.

Table 3 Group of lipid-based materials

Type	Nanoparticle	Application	Findings	Other applications	Reference
Lipids	Solid Lipid Nanomaterials	Drug delivery	Biocompatible, enhanced drug stability and capacity for carrying huge loads	Solid tumors (Rizwanullah et al. 2020)	(Ahmad et al. 2014)
	Liposomes	Drug Delivery	Imitates biological membranes	Parkinson Disease (Yue et al. 2018)	(Navya and Daima 2016)
	Exosomes	Drug delivery	Biocompatible, chemically stable, manages intercellular communications	Alzheimer Disease (Yin et al. 2020)	(Wei et al. 2021)

Solid lipid nanomaterials exhibit properties such as biocompatibility, enhanced drug stability, better load capacity and offer controlled drug delivery (Wang et al. 2020). Typical uses of solid lipid nanomaterials include the effective delivery of vaccines, genes and various drugs. They have the capacity to enter the brain area and thus can be used for the treatment of neurodegenerative diseases, and are crucial for ultrasonic drug delivery (Devasena 2017). Doxil is an example of a liposome drug that gained an FDA approval for treating Kaposi's sarcoma. It encapsulates the drug in a liposome carrier to notably prolong its half-life and enhance its concentration in the tumor site (Duncan 2006).

Liposomes contain a lipid membrane with two layers and an aqueous interior imitating biological membranes for the improvement of the efficiency and safe release of anticancer, anesthetics and various other drugs, with the delivery of gene drugs (Navya and Daima 2016). The results revealed that (Dex)-associated liposomes possess promising pharmacokinetic and tumor-regressing abilities that could target aggressive and drug-resistant cancer cells (Ahmad et al. 2016).

Exosomes are cell lipid bilayer vesicles that work like membranes and consist of various substances, including but not limited to proteins, DNA and RNA. They are often found in the central nervous system and may bear a small amount of molecular genetic material and proteins that are crucial for communication in the cells (Yin et al. 2020). Exosomes are naturally biocompatible, chemically stable and allow for intercellular communication at longer distances; Allogenic exosomes are superior over the immune system in that they may prevent the load from releasing too early and enhance the delivery of drugs to specific locations (Wei et al. 2021). Researchers (Yetisgin et al. 2020) have used **exosomes** to deliver nucleic acids, molecules and proteins (Hadla et al. 2016). They can be manufactured to target cancer treatment specifically. Jeong et al. successfully delivered miRNA-497 into A549 cells, suppressing tumour growth and the expression of the related genes. Exosomes may also be used to combat Alzheimer's Disease (Yin et al. 2020).

3 Diagnostic Abilities of Nanomaterials

3.1 Bioimaging

Bioimaging is a very significant tool for the detection of a wide range of diseases (Wang et al. 2019). Nanomaterials addressing bioimaging issues are summarized in Table 4.

Fluorescent nanomaterials have been in the spotlight for the past decade due to their remarkable optical attributes (Oliveira et al. 2018). Fluorescent nanomaterials could be categorized as inorganic (carbon, silica & semiconductor quantum dots, metal nanoparticle quantum dots) and organic (polymers conjugates, organic dyes, etc.) (Zhao et al. 2018). The organic ones are preferred for use in bioimaging, because they are more biodegradable, biocompatible and less toxic (Svechkarev and Mohs 2019). Nevertheless, they are unstable in aqueous means and hydrophobic, thus constricting their fluorescent quantum yield.

Zhang et al., described the composition of innovative **polydopamine-based fluorescent organic nanoparticles** through the oxidation of polydopamine nanoparticles in a hydrogen peroxide environment. They exhibit high photoluminescence and are well-dispersible; they are biocompatible and water-soluble, which renders them very suitable for bioimaging.

Organic dyes offer simplicity, sensitivity and responsiveness, thus have often been utilized in bioimaging (Rizwanullah et al. 2020). They are able to absorb visible light, acting as chromophores. The most popular dyes include fluorescein, which transmits bright signals and is not toxic (Vinogradov et al. 2010). Furthermore, cyanine dyes are commonly utilized due to being bright, biocompatible and offering better molar absorptivity (An and Wang 2018). Coumarins demonstrated high fluorescence and offer good quantum yields (Ayare et al. 2019).

Table 4 Nanomaterials addressing bioimaging issues

Application	Type	Findings	Reference
Bioimaging	Fluorescent Nanomaterials	More biodegradable, biocompatible and less toxic	(Svechkarev and Mohs 2019)
	Polydopamine-based fluorescent organic Nanoparticles	Well-dispersed through the cytoplasm of biological cells, biocompatible and water soluble	(Zhang et al. 2012)
	Organic Dyes	Exhibit brightness, are biocompatible and have higher molar ability to absorb	(Ayare et al. 2019)
	Aggregation-induced emission	Remarkable dispersity, fluorescence, highly compatible and less toxic	(Cao et al. 2017a)
	Magnetic resonance imaging (Iron Oxide)	Cost-effective, better magnetization and the capacity to create stable suspensions	(Arbab et al. 2005)

The **aggregation-induced emission** process has been taken into consideration for various biomedical applications (Huang et al. 2017), since a**ggregation-induced emission** materials possess a variety of attractive features, including colour change, which makes them useful as entities emitting light (Jiang et al. 2017). Such nanoparticles have been deemed as the most appropriate agents for bioimaging due to their excellent dispersity, strong fluorescence, biocompatibility and low levels of toxicity (Cao et al. 2017b). Typical examples include acetylene polymers, fluorescent silica nanoparticles and fluorescent nanodiamond-based polymer compounds (Huang et al. 2019).

Magnetic resonance imaging (MRI) has various advantages over fluorescence imaging, due to the quick capture of images, high resolutions and the use of radiation. Its use is, however, hindered in biological environments due to its reduced sensitivity (Sun et al. 2008). Superparamagnetic iron oxide nanoparticles have captured the attention of researchers in the past few years because of their broad biomedical applications (Gao et al. 2009). They are cost-effective, and offer improved magnetization and the capacity to formulate stable suspensions that are imperative for bio-applications. Figure 4 depicts in *vivo* dual T1-T2 contrast and biodistribution of nCP:Fe.

3.2 Biosensors

Biosensors are described by their organic receptor units, such as nucleic acids or amino-acid chains, produced by a living organism (Wu et al. 2012). There are various nanomaterials used in biosensor applications in order to enhance the sensitivity of diagnostic tests and improve upon their detection limitations. Table 5 summarizes these materials and key findings about them.

Nanoparticles are widely used due to the variety of available attributes that can be adjusted by altering factors such as shape, dimensions and composition. Nanoparticles offer better sensitivity and improved detection limits in electrochemical diagnostic sensors. Liu et al. (2020) described the creation of a nanosensor for detecting hydrogen sulphide (H_2S) through ratiometric luminescence in aqueous means and live cells, using mesoporous silica nanoparticles. The difference in the quantity of H_2S in the body is attributed to various diseases, such as Alzheimer's disease and cancers (Lin et al. 2015). Gold nanoparticles have been frequently used in biosensors because of exhibiting great biocompatibility, optical and electronic attributes, comparatively simple production and modifications processes (Liu and Lu 2004).

Figure 5a depicts the steps required for the construction of a GraFET biosensor, as well as the activation process, the antibody binding and the identification of antigen through tracking of the electrical attributes while Fig. 5b illustrates the steps required for the binding of the graphene-Ab conjugate (Roberts et al. 2020).

Nanowires provide a high surface area to mass ratio, and based on the components they consist of, they may allow for adjustments in conductivity levels. Current processes for diagnosing prostate cancer cannot detect below certain serum PSA

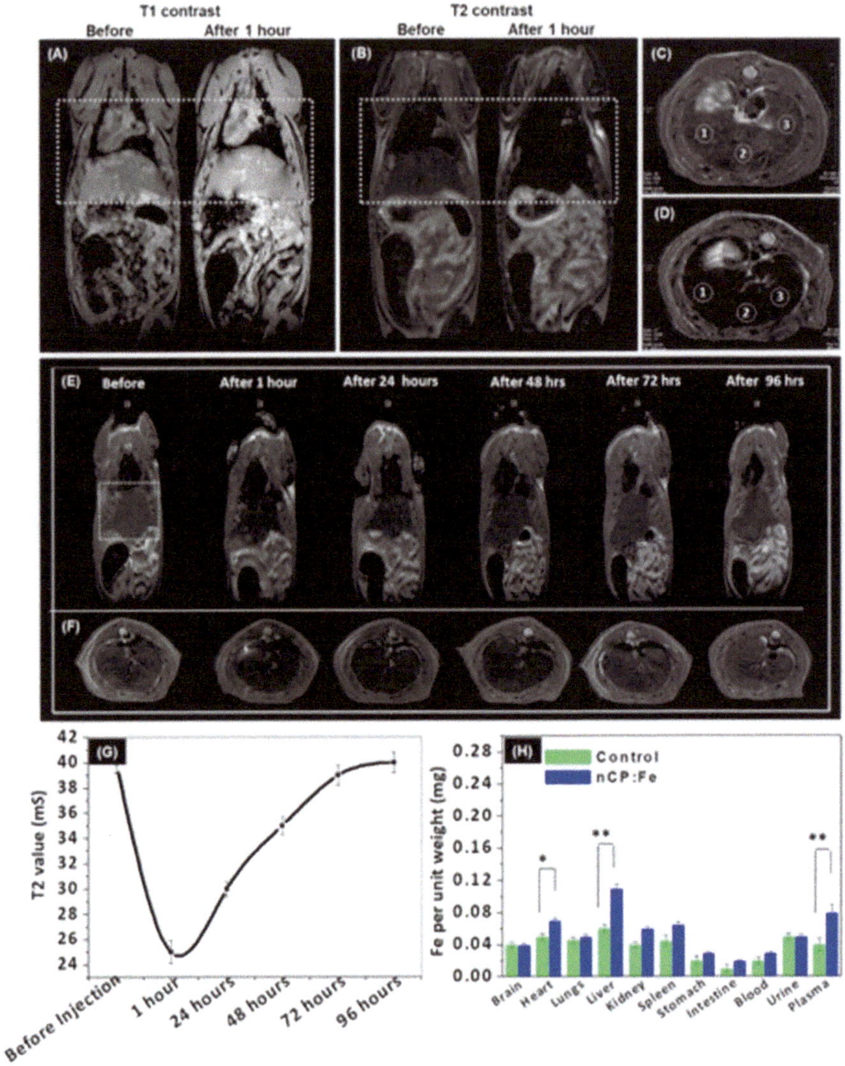

Fig. 4 In *vivo* dual T1–T2 contrast and biodistribution of nCP:Fe. In vivo (**A**) T1 weighted and (**B**) T2 weighted MRI (coronal section) of Wistar rat before and after being injected with nCP:Fe. Improvement of both T1 and T2 contrast can be observed after the injection of the sample in the subject's liver and heart (depicted as a white dotted box) (**C**) Axial liver section pre- injection (**D**) Axial liver section 30 min post-injection (**E**) T2 weighted MRI of wistar rat through the next 96 h following the injection of nCP:Fe (**F**) Liver sections through the 96-h-period (**G**) Variation in T2 time of liver in the post-injection 96-h-period (**H**) Iron concentration in various organs, one hour post-injection; comparison to PBS control, appraised through ICP analysis. Reproduced with permission from Ashokan et al. (2017), Copyright Nature, Ltd

Table 5 Nanomaterials address the various diagnostic challenges

Application	Type	Findings	Reference
Biosensors	Nanoparticles	Highly sensitive, enhance the detection limits	(Barbosa et al. 2019)
	Nanowires	Improve conductivity	(Puppo et al. 2016)
	Carbon Nanotubes	May serve as scaffolds for immobilizing biomolecules	(Tîlmaciu and Morris 2015)
	Graphene	Improved the loading amount, orientation and ability to bind to antibodies	(Zhang et al. 2013)
	Quantum dots	Improved the electrochemical activity of the electrode	(Gu et al. 2019)
	Chitosan	Remarkably biocompatible, biodegradable and non-toxic	(Cheung et al. 2015)
	Dendrimers	Improved stability, sensitivity and reproducibility while limiting non-specific interactions	(Jain et al. 2010)

levels, however solutions taking advantage of nanowires may assist in the detection of protein cancer markers (Puppo et al. 2016).

Carbon nanotubes (CNTs), which possess a large surface area may assist in immobilizing biological molecules and exhibit remarkable physical, electrical and optical attributes (Tîlmaciu and Morris 2015). The use of CNTs in biosensors has allowed for the early discovery of biomarkers for many diseases. Specific types of biosensors using carbon nanotubes have been tested for the identification of malignant biomarkers by combining them with nucleic acids, proteins and antibodies (Wang and Dai 2015; Shah 2021b). Multi-walled carbon nanotubes (MWCNTs) carrying the P450 biomolecule were successfully employed to detect drugs for the electrochemical treatment of cancers (Baj-Rossi et al. 2012; Ray Banerjee 2020). Despite their potential, further research is required to improve the conjugation of CNTs customized receptors. Improving the morphology, appropriateness to biological environments, reducing toxicity and enhancing cell interaction should be taken into account.

Graphene (Feng et al. 2012) has outstanding attributes, hence is one of the most frequently used two-dimensional nanomaterials (Morales-Narváez and Merkoçi 2018). When using graphene sheets, the connection between graphene and the antibodies was exploited for the immobilization of antibodies. The use of graphene nanomaterials increased the loads and improved the orientation and ability to bind with antibodies (Zhang et al. 2013).

Quantum dots have multiple applications, including luminescence sensors, as well as electrochemical, optical and photoelectrochemical biosensors. They are particularly useful for the detection of enzymes, proteins and neurotransmitters and provide a great alternative to typical fluorophores. Such attributes could eventually

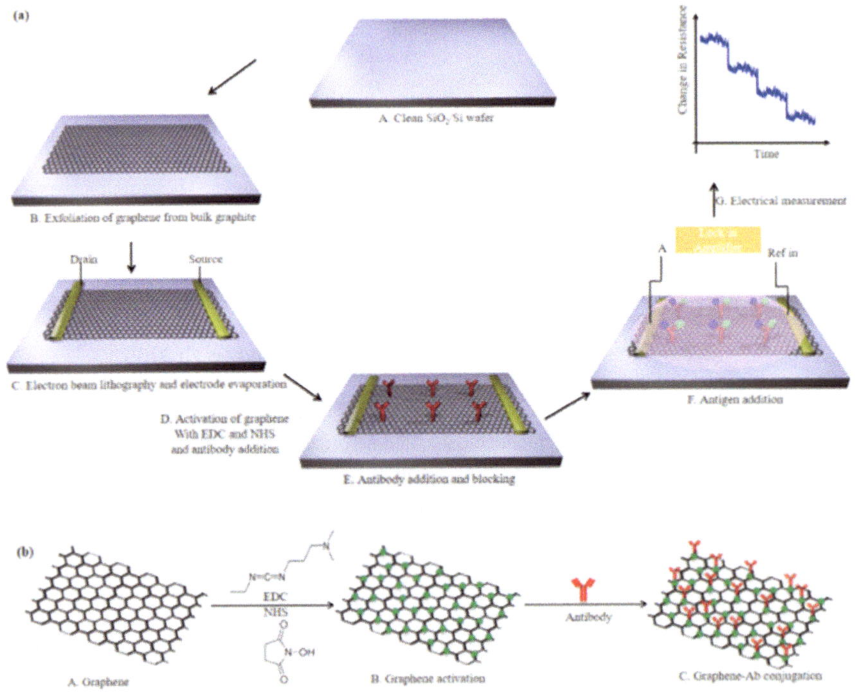

Fig. 5 **(a)** Illustration of the steps required for the construction of a GraFET biosensor: **(A, B)** Graphene exfoliation on pre-cleaned Silicon/Silica substrate using 3 M tape. **(C, D)** E-beam lithography was employed to determine the source and drain pads; thermal evaporation of 5/50 nm Chromium/Gold was used to electrically contact graphene **(E, F)** Graphene activation through EDC-NHS carbomiide reaction and functionalization. BSA as the blocker for the rest of the non-specific binding locations. **(G, H)** Biosensing through the addition of silver to the antibody in multiple concentrations, connected to the functionalized graphene. **(I)** The sensing abilities were tracked through constant measurement of graphene channel's resistance to the various silver concentrations **(b)** Illustration of the binding steps **(A)** exfoliated graphene before being activated. **(B)** EDC-NHS activation of carboxylic groups on graphene. **(C)** Binding on amine group of Ab with the activated carboxylic group of graphene. Reproduced with permission from Roberts et al. (2020), Copyright Nature, Ltd.

make bioconjugate quantum dots applicable for use in cancer diagnosis. The inclusion of carbon dots can improve the binding capacity further improve the performance of the electrode (Gu et al. 2019).

Chitosan is highly biocompatible and biodegradable and does not exhibit toxicity, thus has great potential as a nanostructured material for use in medicine (Cheung et al. 2015). Chemical alterations could be conducted to improve its attributes, like solubility. Chitosan has, as a result, been often used in biosensors and bioanalysis. Other studies suggested that chitosan's levels of toxicity rely on its concentration proportionately (Shah 2020, 2021a).

Dendrimers may assist in improving stability, sensitivity and reproducibility while limiting non-specific interactions (Shah 2021a). They are applicable among others in drug delivery and biocatalysis. The associated risks include hemolytic, hematological and cytotoxicity (Jain et al. 2010).

4 Conclusions

In this chapter, we explored various subjects of great interest in the field of theranostics. Instead of trying to offer a simple historical run-down, the focus lies on determining the developing, promising areas of current and future research. These include the design of therapeutic nanomaterials for improved permeability and selective treatment and the exploration of innovative nanomaterials for comparatively less explored purposes. Despite the enormous amount of research work in the field, our focus lies on nanomaterials that might be externally stimulated, such as metal nanoparticles, quantum dots and polymer nanoparticles. The use of nanomaterials in bioimaging and biosensors for diagnosis has great research potential. A wide range of nanomaterial-boosted sensors has been reviewed and analysed in the point of view of disease detection. Despite the fact that most will not be easily applied practically in clinical environments due to their own toxicity risks, we mainly aim to shed some light on the capacity to use basic chemical principles to produce nanomaterials exhibiting exciting therapeutic abilities. Testing nanomaterial behaviour in vivo is also a significant matter to consider.

References

Ahmad Z, Shah A, Siddiq M, Kraatz H-B (2014) Polymeric micelles as drug delivery vehicles. RSC Adv 4(33):17028–17038. https://doi.org/10.1039/C3RA47370H

Ahmad A, Mondal SK, Mukhopadhyay D, Banerjee R, Alkharfy KM (2016) Development of liposomal formulation for delivering anticancer drug to breast cancer stem-cell-like cells and its pharmacokinetics in an animal model. Mol Pharm 13(3):1081–1088. https://doi.org/10.1021/acs.molpharmaceut.5b00900

Ahmadian E et al (2018) Effect of silver nanoparticles in the induction of apoptosis on human hepatocellular carcinoma (HepG2) cell line. Mater Sci Eng C 93:465–471. https://doi.org/10.1016/j.msec.2018.08.027

An HW, Wang MD (2018) The self-assembly of cyanine dyes for biomedical application in vivo. In: Wang H, Li LL (eds), In vivo self-assembly nanotechnology for biomedical applications, Singapore: Springer, pp 31–55. https://doi.org/10.1007/978-981-10-6913-0_2

Arbab AS, Wilson LB, Ashari P, Jordan EK, Lewis BK, Frank JA (2005) A model of lysosomal metabolism of dextran coated superparamagnetic iron oxide (SPIO) nanoparticles: implications for cellular magnetic resonance imaging. NMR Biomed 18(6):383–389. https://doi.org/10.1002/nbm.970

Ashokan A et al (2017) Biomineral nano-theranostic agent for magnetic resonance image guided, augmented radiofrequency ablation of liver tumor. Sci Rep 7(1):14481. https://doi.org/10.1038/s41598-017-14976-8

Ayare NN, Ramugade SH, Sekar N (2019) Photostable coumarin containing azo dyes with multifunctional property. Dyes Pigm 163:692–699. https://doi.org/10.1016/j.dyepig.2018.12.050

Badea I, Kaur R (2013) Nanodiamonds as novel nanomaterials for biomedical applications: drug delivery and imaging systems. IJN 203. https://doi.org/10.2147/IJN.S37348
Baig B, Halim SA, Farrukh A, Greish Y, Amin A (2019) Current status of nanomaterial-based treatment for hepatocellular carcinoma. Biomed Pharmacother 116:108852. https://doi.org/10.1016/j.biopha.2019.108852
Baj-Rossi C, Micheli GD, Carrara S (2012) Electrochemical detection of anti-breast-cancer agents in human serum by cytochrome P450-coated carbon nanotubes. Sensors 12(5):6520–6537. https://doi.org/10.3390/s120506520
Barbosa AI et al (2019) Development of label-free plasmonic Au-TiO2 thin film immunosensor devices. Mater Sci Eng C 100:424–432. https://doi.org/10.1016/j.msec.2019.03.029
Barzinjy AA, Hamad SM, Aydın S, Ahmed MH, Hussain FHS (2020) Green and eco-friendly synthesis of nickel oxide nanoparticles and its photocatalytic activity for methyl orange degradation. J Mater Sci Mater Electron 31(14):11303–11316. https://doi.org/10.1007/s10854-020-03679-y
Bhattacharya R, Mukherjee P (2008) Biological properties of 'naked' metal nanoparticles. Adv Drug Deliv Rev 60(11):1289–1306. https://doi.org/10.1016/j.addr.2008.03.013
Bhirde AA et al (2009) Targeted killing of cancer cells in vivo and in vitro with EGF-directed carbon nanotube-based drug delivery. ACS Nano 3(2):307–316. https://doi.org/10.1021/nn800551s
Bilan RS et al (2017) Engineering of optically encoded microbeads with FRET-free spatially separated quantum-dot layers for multiplexed assays. ChemPhysChem 18(8):970–979. https://doi.org/10.1002/cphc.201601274
Bruno PM et al (2017) A subset of platinum-containing chemotherapeutic agents kills cells by inducing ribosome biogenesis stress. Nat Med 23(4):461–471. https://doi.org/10.1038/nm.4291
Cao Q et al (2017a) Microwave-assisted multicomponent reactions for rapid synthesis of AIE-active fluorescent polymeric nanoparticles by post-polymerization method. Mater Sci Eng C 80:578–583. https://doi.org/10.1016/j.msec.2017.07.006
Cao Q et al (2017b) Preparation of AIE-active fluorescent polymeric nanoparticles through a catalyst-free thiol-yne click reaction for bioimaging applications. Mater Sci Eng C 80:411–416. https://doi.org/10.1016/j.msec.2017.06.008
Chen Z et al (2014) Controlled release of free doxorubicin from peptide–drug conjugates by drug loading. J Control Release 191:123–130. https://doi.org/10.1016/j.jconrel.2014.05.051
Cheng Z, Li M, Dey R, Chen Y (2021) Nanomaterials for cancer therapy: current progress and perspectives. J Hematol Oncol 14(1):85. https://doi.org/10.1186/s13045-021-01096-0
Cheung R, Ng T, Wong J, Chan W (2015) Chitosan: an update on potential biomedical and pharmaceutical applications. Mar Drugs 13(8):5156–5186. https://doi.org/10.3390/md13085156
Chitambar CR (2021) Gallium-containing anticancer compounds. Fut Med Chem 4(10):Art. no. 10. https://doi.org/10.4155/fmc.12.69
Cho H-J, Oh J, Choo M-K, Ha J-I, Park Y, Maeng H-J (2014) Chondroitin sulfate-capped gold nanoparticles for the oral delivery of insulin. Int J Biol Macromol 63:15–20. https://doi.org/10.1016/j.ijbiomac.2013.10.026
Collery P, Keppler B, Madoulet C, Desoize B (2002) Gallium in cancer treatment. Crit Rev Oncol Hematol 42(3):Art. no. 3
Crucho CIC, Barros MT (2017) Polymeric nanoparticles: a study on the preparation variables and characterization methods. Mater Sci Eng C 80:771–784. https://doi.org/10.1016/j.msec.2017.06.004
Devasena T (2017) Therapeutic and diagnostic nanomaterials. Singapore, Springer. https://doi.org/10.1007/978-981-10-0923-5
Dhar S, Liu Z, Thomale J, Dai H, Lippard SJ (2008) Targeted single-wall carbon nanotube-mediated Pt(IV) prodrug delivery using folate as a homing device. J Am Chem Soc 130(34):11467–11476. https://doi.org/10.1021/ja803036e

Dhar S, Daniel WL, Giljohann DA, Mirkin CA, Lippard SJ (2009) Polyvalent oligonucleotide gold nanoparticle conjugates as delivery vehicles for platinum(IV) warheads. J Am Chem Soc 131(41):14652–14653. https://doi.org/10.1021/ja9071282

Dhar S, Mali V, Bodhankar S, Shiras A, Prasad BLV, Pokharkar V (2011) Biocompatible gellan gum-reduced gold nanoparticles: cellular uptake and subacute oral toxicity studies. J Appl Toxicol 31(5):411–420. https://doi.org/10.1002/jat.1595

Dickerson EB, Blackburn WH, Smith MH, Kapa LB, Lyon LA, McDonald JF (2010) Chemosensitization of cancer cells by siRNA using targeted nanogel delivery. BMC Cancer 10(1):10. https://doi.org/10.1186/1471-2407-10-10

Duncan R (2006) Polymer conjugates as anticancer nanomedicines. Nat Rev Cancer 6(9):688–701. https://doi.org/10.1038/nrc1958

Eftekhari A et al (2021) Application of advanced nanomaterials for kidney failure treatment and regeneration. Materials 14(11):2939. https://doi.org/10.3390/ma14112939

Ekladious I, Colson YL, Grinstaff MW (2019) Polymer–drug conjugate therapeutics: advances, insights and prospects. Nat Rev Drug Discov 18(4):273–294. https://doi.org/10.1038/s41573-018-0005-0

Feng T, Feng D, Shi W, Li X, Ma H (2012) A graphene oxide-peptide fluorescence sensor for proteolytically active prostate-specific antigen. Mol BioSyst 8(5):1441. https://doi.org/10.1039/c2mb05379a

Foster BJ, Clagett-Carr K, Hoth D, Leyland-Jones B (1986) Gallium nitrate: the second metal with clinical activity. Cancer Treat Rep 70(11):1311–1319

Ganesh Kumar C, Poornachandra Y, Pombala S (2017) Therapeutic nanomaterials: from a drug delivery perspective. In: Nanostructures for drug delivery, Elsevier, pp 1–61. https://doi.org/10.1016/B978-0-323-46143-6.00001-4

Gao Q et al (2009) The study of novel Fe3O4@γ-Fe2O3 core/shell nanomaterials with improved properties. J Magn Magn Mater 321(8):1052–1057. https://doi.org/10.1016/j.jmmm.2008.10.022

Gkika DA et al (2017) Price tag in nanomaterials? J Nanopart Res 19(5):Art. no. 5. https://doi.org/10.1007/s11051-017-3875-x.

Gkika DA, Vordos N, Magafas L, Mitropoulos AC, Kyzas GZ (2021) Risk return profile of nanomaterials. J Mol Struct 1228:129740. https://doi.org/10.1016/j.molstruc.2020.129740

Gobi R, Ravichandiran P, Babu RS, Yoo DJ (2021) Biopolymer and synthetic polymer-based nanocomposites in wound dressing applications: a review. Polymers 13(12):1962. https://doi.org/10.3390/polym13121962

Gu C et al (2019) Bimetallic ZrHf-based metal-organic framework embedded with carbon dots: ultra-sensitive platform for early diagnosis of HER2 and HER2-overexpressed living cancer cells. Biosens Bioelectron 134:8–15. https://doi.org/10.1016/j.bios.2019.03.043

Hadla M et al (2016) Exosomes increase the therapeutic index of doxorubicin in breast and ovarian cancer mouse models. Nanomedicine 11(18):2431–2441. https://doi.org/10.2217/nnm-2016-0154

Hanada S, Fujioka K, Futamura Y, Manabe N, Hoshino A, Yamamoto K (2013) Evaluation of anti-inflammatory drug-conjugated silicon quantum dots: their cytotoxicity and biological effect. IJMS 14(1):1323–1334. https://doi.org/10.3390/ijms14011323

Hart MM, Adamson RH (1971) Antitumor activity and toxicity of salts of inorganic group IIIa metals: aluminum, gallium, indium, and thallium. Proc Natl Acad Sci 68(7):1623–1626. https://doi.org/10.1073/pnas.68.7.1623

Harvey PD, Plé J (2021) Recent advances in nanoscale metal-organic frameworks towards cancer cell cytotoxicity: an overview. J Inorg Organomet Polym 31(7):2715–2756. https://doi.org/10.1007/s10904-021-02011-3

Hong S, Choi DW, Kim HN, Park CG, Lee W, Park HH (2020) Protein based nanoparticles as drug delivery systems. Pharmaceutics 12(7):604. https://doi.org/10.3390/pharmaceutics12070604

Hu YL, Qi W, Feng H, Shao J-Z, Gao J-Q (2011) Toxicity evaluation of biodegradable chitosan nanoparticles using a zebrafish embryo model. IJN 3351 https://doi.org/10.2147/IJN.S25853

Huang H et al (2017) Direct encapsulation of AIE-active dye with β cyclodextrin terminated polymers: self-assembly and biological imaging. Mater Sci Eng C 78:862–867. https://doi.org/10.1016/j.msec.2017.04.080

Huang L et al (2019) A facile surface modification strategy for fabrication of fluorescent silica nanoparticles with the aggregation-induced emission dye through surface-initiated cationic ring opening polymerization. Mater Sci Eng C 94:270–278. https://doi.org/10.1016/j.msec.2018.09.042

Igartúa DE, Martinez CS, Temprana CF, del V Alonso S, Prieto MJ (2018) PAMAM dendrimers as a carbamazepine delivery system for neurodegenerative diseases: a biophysical and nanotoxicological characterization. Int J Pharm 544(1):191–202. https://doi.org/10.1016/j.ijpharm.2018.04.032

Jain K, Kesharwani P, Gupta U, Jain NK (2010) Dendrimer toxicity: let's meet the challenge. Int J Pharm 394(1–2):122–142. https://doi.org/10.1016/j.ijpharm.2010.04.027

Jiang R et al (2017) A facile one-pot Mannich reaction for the construction of fluorescent polymeric nanoparticles with aggregation-induced emission feature and their biological imaging. Mater Sci Eng C 81:416–421. https://doi.org/10.1016/j.msec.2017.08.048

Kandil EI, El-sonbaty SM, Moawed FS, Khedr OM (2018) Anticancer redox activity of gallium nanoparticles accompanied with low dose of gamma radiation in female mice. Tumour Biol 40(3):101042831774967. https://doi.org/10.1177/1010428317749676

Kanmani P, Lim ST (2013) Synthesis and structural characterization of silver nanoparticles using bacterial exopolysaccharide and its antimicrobial activity against food and multidrug resistant pathogens. Process Biochem 48(7):1099–1106. https://doi.org/10.1016/j.procbio.2013.05.011

Khmelinskaia A, Wargacki A, King NP (2021) Structure-based design of novel polyhedral protein nanomaterials. Curr Opin Microbiol 61:51–57. https://doi.org/10.1016/j.mib.2021.03.003

Kumar L, Verma S, Vaidya B, Mehra NK (2017) Nanocarrier-assisted antimicrobial therapy against intracellular pathogens. In: Nanostructures for antimicrobial therapy, Elsevier, pp 293–324. https://doi.org/10.1016/B978-0-323-46152-8.00013-5

Kumar M et al (2018) Desmethyl tamoxifen and quercetin-loaded multiwalled CNTs: a synergistic approach to overcome MDR in cancer cells. Mater Sci Eng C 89:274–282. https://doi.org/10.1016/j.msec.2018.03.033

Kuo Y-C, Lee Y-J (2016) Rescuing cholinergic neurons from apoptotic degeneration by targeting of serotonin modulator- and apolipoprotein E-conjugated liposomes to the hippocampus. IJN 11:6809–6824. https://doi.org/10.2147/IJN.S123442

Letchford K, Liggins R, Wasan KM, Burt H (2009) In vitro human plasma distribution of nanoparticulate paclitaxel is dependent on the physicochemical properties of poly(ethylene glycol)-block-poly(caprolactone) nanoparticles. Eur J Pharm Biopharm 71(2):196–206. https://doi.org/10.1016/j.ejpb.2008.08.003

Li N, Wang J, Yang X, Li L (2011) Novel nanogels as drug delivery systems for poorly soluble anticancer drugs. Colloids Surf B 83(2):237–244. https://doi.org/10.1016/j.colsurfb.2010.11.027

Lin VS, Chen W, Xian M, Chang CJ (2015) Chemical probes for molecular imaging and detection of hydrogen sulfide and reactive sulfur species in biological systems. Chem Soc Rev 44(14):4596–4618. https://doi.org/10.1039/C4CS00298A

Liu J, Lu Y (2004) Colorimetric biosensors based on DNAzyme-assembled gold nanoparticles. J Fluoresc 14(4):343–354. https://doi.org/10.1023/B:JOFL.0000031816.06134.d3

Liu J et al (2020) Responsive nanosensor for ratiometric luminescence detection of hydrogen sulfide in inflammatory cancer cells. Anal Chim Acta 1103:156–163. https://doi.org/10.1016/j.aca.2019.12.056

Lv Y, Hao L, Hu W, Ran Y, Bai Y, Zhang L (2016) Novel multifunctional pH-sensitive nanoparticles loaded into microbubbles as drug delivery vehicles for enhanced tumor targeting. Sci Rep 6(1):29321. https://doi.org/10.1038/srep29321

Marei HE et al (2017) Nanotubes impregnated human olfactory bulb neural stem cells promote neuronal differentiation in Trimethyltin-induced neurodegeneration rat model. J Cell Physiol 232(12):3586–3597. https://doi.org/10.1002/jcp.25826
Matea C et al (2017) Quantum dots in imaging, drug delivery and sensor applications. IJN 12:5421–5431. https://doi.org/10.2147/IJN.S138624
Mehmood A, Ghafar H, Yaqoob S, Gohar UF, Ahmad B (2017) Mesoporous silica nanoparticles: a review. J Dev Drug 6(2). https://doi.org/10.4172/2329-6631.1000174
Mignani S et al (2018) Recent therapeutic applications of the theranostic principle with dendrimers in oncology. Sci China Mater 61(11):1367–1386. https://doi.org/10.1007/s40843-018-9244-5
Mirzaie Z, Ansari M, Kordestani SS, Rezaei MH, Mozafari M (2019) Preparation and characterization of curcumin-loaded polymeric nanomicelles to interference with amyloidogenesis through glycation method. Biotechnol Appl Biochem 66(4):537–544. https://doi.org/10.1002/bab.1751
Monge-Fuentes V et al (2021) Dopamine-loaded nanoparticle systems circumvent the blood–brain barrier restoring motor function in mouse model for Parkinson's disease. Sci Rep 11(1):15185. https://doi.org/10.1038/s41598-021-94175-8
Morales-Narváez E, Merkoçi A (2018) Graphene oxide as an optical biosensing platform: a progress report. Adv Mater 1805043. https://doi.org/10.1002/adma.201805043
Mroz P et al (2007) Functionalized fullerenes mediate photodynamic killing of cancer cells: type I versus type II photochemical mechanism. Free Radical Biol Med 43(5):711–719. https://doi.org/10.1016/j.freeradbiomed.2007.05.005
Navya PN, Daima HK (2016) Rational engineering of physicochemical properties of nanomaterials for biomedical applications with nanotoxicological perspectives. Nano Converg 3(1):Art. no. 1. https://doi.org/10.1186/s40580-016-0064-z
Ni D et al (2018) Molybdenum-based nanoclusters act as antioxidants and ameliorate acute kidney injury in mice. Nat Commun 9(1):5421. https://doi.org/10.1038/s41467-018-07890-8
Niesvizky R (2003) Gallium nitrate in multiple myeloma: prolonged survival in a cohort of patients with advanced-stage disease. Semin Oncol 30(2 Suppl 5):20–24. https://doi.org/10.1016/S0093-7754(03)00172-6
Niesvizky R, Choy CG, Siegel D, Lyons L, Michaeli J (2002) Extended survival in advanced-stage multiple myeloma patients treated with gallium nitrate. Leuk Lymphoma 43(3):603–605. https://doi.org/10.1080/10428190210321
Noimark S et al (2014) Functionalised gold and titania nanoparticles and surfaces for use as antimicrobial coatings. Faraday Discuss 175:273–287. https://doi.org/10.1039/C4FD00113C
Novoselov KS (2004) Electric field effect in atomically thin carbon films. Science 306(5696):666–669. https://doi.org/10.1126/science.1102896
Olerile LD et al (2017) Near-infrared mediated quantum dots and paclitaxel co-loaded nanostructured lipid carriers for cancer theragnostic. Colloids Surf B 150:121–130. https://doi.org/10.1016/j.colsurfb.2016.11.032
Oliveira E et al (2018) Green and red fluorescent dyes for translational applications in imaging and sensing analytes: a dual-color flag. ChemistryOpen 7(1):3–3. https://doi.org/10.1002/open.201700177
Osminkina LA, Gongalsky MB, Motuzuk AV, Timoshenko VY, Kudryavtsev AA (2011) Silicon nanocrystals as photo- and sono-sensitizers for biomedical applications. Appl Phys B 105(3):665–668. https://doi.org/10.1007/s00340-011-4562-8
Pahuja R et al (2019) Correction to trans-blood brain barrier delivery of dopamine-loaded nanoparticles reverses functional deficits in Parkinsonian rats. ACS Nano 13(7):8490–8490. https://doi.org/10.1021/acsnano.9b04446
Pang X, Jiang Y, Xiao Q, Leung AW, Hua H, Xu C (2016) pH-responsive polymer–drug conjugates: design and progress. J Control Release 222:116–129. https://doi.org/10.1016/j.jconrel.2015.12.024
Paredes KO, Ruiz-Cabello J, Alarcón DI, Filice M (2019) The state of the art of investigational and approved nanomedicine products for nucleic acid delivery. In Nucleic acid nanotheranostics, Elsevier, pp 421–456. https://doi.org/10.1016/B978-0-12-814470-1.00015-0

Patel D, Talele D, Pathak D, Misra A (2021) Polymers used in Peyer's patch targeting. In Applications of polymers in drug delivery, Elsevier, pp 159–185. https://doi.org/10.1016/B978-0-12-819659-5.00006-9

Paul SD, Sharma H, Jeswani G, Jha AK (2017) Novel gels: implications for drug delivery. In Nanostructures for drug delivery, Elsevier, pp 379–412. https://doi.org/10.1016/B978-0-323-46143-6.00012-9

Peng Y, Huang J, Xiao H, Wu T, Shuai X (2018) Codelivery of temozolomide and siRNA with polymeric nanocarrier for effective glioma treatment. IJN 13:3467–3480. https://doi.org/10.2147/IJN.S164611

Puppo F et al (2016) SiNW-FET in-air biosensors for high sensitive and specific detection in breast tumor extract. IEEE Sensors J 16(10):3374–3381. https://doi.org/10.1109/JSEN.2015.2433836

Rademaker-Lakhai JM et al (2004) A phase I and pharmacological study of the platinum polymer AP5280 given as an intravenous infusion once every 3 weeks in patients with solid tumors. Clin Cancer Res 10(10):3386–3395. https://doi.org/10.1158/1078-0432.CCR-03-0315

Ray Banerjee, E (ed) (2020) Nanomaterials and biomedicine: therapeutic and diagnostic approach. Singapore, Springer, https://doi.org/10.1007/978-981-15-5274-8

Ren D, Dalmau M, Randall A, Shindel MM, Baldi P, Wang S-W (2012) Biomimetic design of protein nanomaterials for hydrophobic molecular transport. Adv Funct Mater 22(15):3170–3180. https://doi.org/10.1002/adfm.201200052

Rizwanullah M, Alam M, Harshita SR, Rizvi MMA, Amin S (2020) Polymer-lipid hybrid nanoparticles: a next-generation nanocarrier for targeted treatment of solid tumors. CPD 26(11):1206–1215. https://doi.org/10.2174/1381612826666200116150426

Roberts A et al (2020) Graphene functionalized field-effect transistors for ultrasensitive detection of Japanese encephalitis and Avian influenza virus. Sci Rep 10(1):14546. https://doi.org/10.1038/s41598-020-71591-w

Saleem J, Wang L, Chen C (2018) Carbon-based nanomaterials for cancer therapy via targeting tumor microenvironment. Adv Healthcare Mater 7(20):1800525. https://doi.org/10.1002/adhm.201800525

Sánchez-López E et al (2020) Metal-based nanoparticles as antimicrobial agents: an overview. Nanomaterials 10(2):292. https://doi.org/10.3390/nano10020292

Saraf J, Kalia K, Bhattacharya P, Tekade RK (2019) Growing synergy of nanodiamonds in neurodegenerative interventions. Drug Discov Today 24(2):584–594. https://doi.org/10.1016/j.drudis.2018.10.012

Sardoiwala MN, Kaundal B, Roy Choudhury S (2018) Development of engineered nanoparticles expediting diagnostic and therapeutic applications across blood–brain barrier. In: Handbook of nanomaterials for industrial applications, Elsevier, pp 696–709. https://doi.org/10.1016/B978-0-12-813351-4.00038-9

Shah MP (2020) Microbial Bioremediation & Biodegradation. Springer

Shah MP (2021a) Removal of refractory pollutants from wastewater treatment plants. CRC Press

Shah MP (2021b) Removal of emerging contaminants through microbial processes. Springer

Sharma VK, Yngard RA, Lin Y (2009) Silver nanoparticles: green synthesis and their antimicrobial activities. Adv Coll Interface Sci 145(1–2):83–96. https://doi.org/10.1016/j.cis.2008.09.002

Singer JW et al (2005) Paclitaxel poliglumex (XYOTAX; CT-2103): an intracellularly targeted taxane. Anticancer Drug 16(3):243–254. https://doi.org/10.1097/00001813-200503000-00003

Singh AP, Biswas A, Shukla A, Maiti P (2019) Targeted therapy in chronic diseases using nanomaterial-based drug delivery vehicles. Sig Transduct Target Ther 4(1):33. https://doi.org/10.1038/s41392-019-0068-3

Soh M et al (2017) Ceria-zirconia nanoparticles as an enhanced multi-antioxidant for sepsis treatment. Angew Chem Int Ed 56(38):11399–11403. https://doi.org/10.1002/anie.201704904

Sun C, Lee JSH, Zhang M (2008) Magnetic nanoparticles in MR imaging and drug delivery. Adv Drug Deliv Rev 60(11):1252–1265. https://doi.org/10.1016/j.addr.2008.03.018

Svechkarev D, Mohs AM (2019) Organic fluorescent dye-based nanomaterials: advances in the rational design for imaging and sensing applications. CMC 26(21):4042–4064. https://doi.org/10.2174/0929867325666180226111716

Swar S, Máková V, Stibor I (2019) Effectiveness of diverse mesoporous silica nanoparticles as potent vehicles for the drug L-DOPA. Materials 12(19):3202. https://doi.org/10.3390/ma12193202

Thakore SI, Solanki A, Das M (2019) Exploring potential of polymers in cancer management. In: Materials for biomedical engineering, Elsevier, pp 113–133. https://doi.org/10.1016/B978-0-12-818433-2.00004-2

Tian L, Lu L, Qiao Y, Ravi S, Salatan F, Melancon M (2016) Stimuli-responsive gold nanoparticles for cancer diagnosis and therapy. JFB 7(3):19. https://doi.org/10.3390/jfb7030019

Tîlmaciu CM, Morris MC (2015) Carbon nanotube biosensors. Front Chem 3. https://doi.org/10.3389/fchem.2015.00059

Vidu R, Rahman M, Mahmoudi M, Enachescu M, Poteca TD, Opris I (2014) Nanostructures: a platform for brain repair and augmentation. Front Syst Neurosci 8. https://doi.org/10.3389/fnsys.2014.00091

Vinogradov SV, Poluektova LY, Makarov E, Gerson T, Senanayake MT (2010) Nano-NRTIs: efficient Inhibitors of HIV Type-1 in macrophages with a reduced mitochondrial toxicity. Antivir Chem Chemother 21(1):1–14. https://doi.org/10.3851/IMP1680

Wakaskar RR (2018) General overview of lipid–polymer hybrid nanoparticles, dendrimers, micelles, liposomes, spongosomes and cubosomes. J Drug Target 26(4):311–318. https://doi.org/10.1080/1061186X.2017.1367006

Wang Z, Dai Z (2015) Carbon nanomaterial-based electrochemical biosensors: an overview. Nanoscale 7(15):6420–6431. https://doi.org/10.1039/C5NR00585J

Wang B, Luo X (2020) A first-principles study on potential chelation agents and indicators of Alzheimer's disease. RSC Adv 10(58):35574–35581. https://doi.org/10.1039/D0RA06855A

Wang Z et al (2019) Three novel camphor-based fluorescence probes for ratiometric detection of hypochlorite and bio-imaging in living cells. Sens Actuators B Chem 284:148–158. https://doi.org/10.1016/j.snb.2018.12.104

Wang K et al (2020) Therapeutic nanomaterials for neurological diseases and cancer therapy. J Nanomater 2020:1–18. https://doi.org/10.1155/2020/2047379

Warrell RP, Alcock NW, Bockman RS (1987) Gallium nitrate inhibits accelerated bone turnover in patients with bone metastases. JCO 5(2):292–298. https://doi.org/10.1200/JCO.1987.5.2.292

Webster LK, Olver IN, Stokes KH, Sephton RG, Hillcoat BL, Bishop JF (2000) A pharmacokinetic and phase II study of gallium nitrate in patients with non-small cell lung cancer. Cancer Chemother Pharmacol 45(1):55–58. https://doi.org/10.1007/PL00006743

Wei W et al (2021) Mesenchymal stem cell-derived exosomes: a promising biological tool in nanomedicine. Front Pharmacol 11:590470. https://doi.org/10.3389/fphar.2020.590470

Wu J et al (2012) A molecular peptide beacon for the ratiometric sensing of nucleic acids. J Am Chem Soc 134(4):1958–1961. https://doi.org/10.1021/ja2103845

Wu L, Ficker M, Christensen JB, Trohopoulos PN, Moghimi SM (2015) Dendrimers in medicine: therapeutic concepts and pharmaceutical challenges. Bioconjugate Chem 26(7):1198–1211. https://doi.org/10.1021/acs.bioconjchem.5b00031

Yang S-T, Liu Y, Wang Y-W, Cao A (2013) Biosafety and bioapplication of nanomaterials by designing protein-nanoparticle interactions. Small 9(9–10):1635–1653. https://doi.org/10.1002/smll.201201492

Yang R, Mondal G, Wen D, Mahato RI (2017) Combination therapy of paclitaxel and cyclopamine polymer-drug conjugates to treat advanced prostate cancer. Nanomed Nanotechnol Biol Med 13(2):391–401. https://doi.org/10.1016/j.nano.2016.07.017

Yano S et al (2015) Pharmacokinetics for once-daily modified release formulation of tacrolimus hydrate in unrelated hematopoietic stem cell transplantation. Ann Hematol 94(3):491–496. https://doi.org/10.1007/s00277-014-2233-7

Yetisgin AA, Cetinel S, Zuvin M, Kosar A, Kutlu O (2020) Therapeutic nanoparticles and their targeted delivery applications. Molecules 25(9):2193. https://doi.org/10.3390/molecules25092193

Yin Q et al (2020) Targetting exosomes as a new biomarker and therapeutic approach for Alzheimer's disease. CIA 15:195–205. https://doi.org/10.2147/CIA.S240400

Yue P, Gao L, Wang X, Ding X, Teng J (2018) Ultrasound-triggered effects of the microbubbles coupled to GDNF- and Nurr1-loaded PEGylated liposomes in a rat model of Parkinson's disease. J Cell Biochem 119(6):4581–4591. https://doi.org/10.1002/jcb.26608

Zhang X et al (2012) Biocompatible polydopamine fluorescent organic nanoparticles: facile preparation and cell imaging. Nanoscale 4(18):5581. https://doi.org/10.1039/c2nr31281f

Zhang X-Q, Xu X, Bertrand N, Pridgen E, Swami A, Farokhzad OC (2012) Interactions of nanomaterials and biological systems: implications to personalized nanomedicine. Adv Drug Deliv Rev 64(13):1363–1384. https://doi.org/10.1016/j.addr.2012.08.005

Zhang J et al (2013) A novel surface plasmon resonance biosensor based on graphene oxide decorated with gold nanorod–antibody conjugates for determination of transferrin. Biosens Bioelectron 45:230–236. https://doi.org/10.1016/j.bios.2013.02.008

Zhang Y et al (2016) One-shot immunomodulatory nanodiamond agents for cancer immunotherapy. Adv Mater 28(14):2699–2708. https://doi.org/10.1002/adma.201506232

Zhao L et al (2014) Polyglycerol-coated nanodiamond as a macrophage-evading platform for selective drug delivery in cancer cells. Biomaterials 35(20):5393–5406. https://doi.org/10.1016/j.biomaterials.2014.03.041

Zhao Y, Chang C, Gai P, Han L, Li F, Li B (2018) One-step synthesis of fluorescent organic nanoparticles: the application to label-free ratiometric fluorescent pH sensor. Sens Actuators B Chem 273:1479–1486. https://doi.org/10.1016/j.snb.2018.07.047

Nanomaterials and Their Properties: Thermal Analysis, Physical, Mechanical and Chemical Properties

Despina A. Gkika, Nick Vordos, Athanasios C. Mitropoulos, Dimitra A. Lambropoulou, and George Z. Kyzas

1 Introduction

Nanotechnology is a fast-developing research field, led by the continuous improvements of the past few years, particularly because of the increased attention on nanoparticles. Nanomaterials have dimensions at the nanoscale, i.e., at the range 10–6 m (Gkika et al. 2020). Nanomaterial production is a collaborative process covering multiple disciplines, including physics, chemistry, biology, economics, materials science and engineering. The synergy between researchers from diverse disciplines will assuredly result in the construction of innovative materials with tunable properties (Charitidis et al. 2014).

Their singular, remarkable attributes render such materials superior and essential in multiple facets of human activities. They can possess excellent magnetic, mechanical and other attributes that considerably deviate from those of their larger-sized versions. Due to these properties, nanomaterials can be made use of in a vast range of applications (Physical Fundamentals of Nanomaterials 2018). When reaching closer to nano-dimensions the ratio between fraction surface and volume increases accordingly, thus the amount of atoms found at the material surface is rendered more important (Koçak and Karasu 2018; Guisbiers et al. 2012).

This chapter aims to cover the physical, mechanical and chemical attributes of nanomaterials and the manner in which they allowed nanomaterials to be exploited in various fields. The potential hazardous attributes of nanomaterials are not fully

D. A. Gkika (✉) · A. C. Mitropoulos · G. Z. Kyzas
Department of Chemistry, International Hellenic University, Kavala, Greece
e-mail: dgkika@chem.ihu.gr

N. Vordos
Department of Physics, International Hellenic University, Kavala, Greece

D. A. Lambropoulou
Department of Chemistry, Aristotle University of Thessaloniki, Thessaloiki, Greece

M. P. Shah (ed.), *Advanced and Innovative Approaches of Environmental Biotechnology in Industrial Wastewater Treatment*,
https://doi.org/10.1007/978-981-99-2598-8_14

explored, however even the risks can be transformed into opportunities if any desirable returns can be identified (Gkika et al. 2021).

Nanomaterials' thermal features rely on various aspects that are largely unimportant in the larger versions of the materials. Such features include, among others, the thermal conductivity, melting point and glass-transition temperature, which all depend on the material's particle size (Seifi et al. 2020). Although there is a wide range of publications on the subject of thermal analysis methods and nanomaterial properties, they are often not easy to follow in order to get a better understanding. As a result, this chapter will attempt to serve as an alternative, concise and easier to comprehend point of reference.

2 Thermal Analysis of Nanomaterials

Nanomaterials and their characterization are a research topic of great interest. Researchers suggest that the outcomes of these analyses and the insurability of nanomaterial production could potentially hinder commercialization and further research (Borca-Tasciuc et al. 2005; Qi, 2005; Singh et al. 2016). A variety of thermal analysis (TA) methods may be employed to identify and measure the properties of nanomaterials. They comprise a subsection of Materials Science that examines the attributes of the nanomaterial in relation to a variety of temperature levels (Corcione and Frigione 2012; Jain and Sharma 2011). These techniques all analyse the effect on specified physical attributes, while the temperature and potentially the atmospheric pressure are controlled.

Figure 1 depicts the most common techniques, including information on the measured attribute, as the most frequently used thermal analysis methods that track the nanomaterial properties based on temperature or according to time in a variety of temperatures:

- Evolved Gas Analysis, measuring gaseous decomposition.
- Thermogravimetric Analysis, mainly measuring mass changes.
- Differential Thermal Analysis, mainly measuring temperature difference and the glass transition temperature.
- Thermomechanical Analysis, mainly measuring changes in dimensions (expansion or reduction).
- Differential Scanning Calorimetry, mainly measuring heat difference.

Nowadays, some of these methods have been combined with various techniques in order to achieve a more accurate characterization of nanomaterials (Seifi et al. 2020). TA can be utilized to get indirect proof of the nano-dispersion, which holds great potential.

The most frequent combinations of methodologies share the same sample and thermal environment. Furthermore, any gas analysis method can be used for EGA. The most commonly employed techniques are mass- and Fourier transform infrared

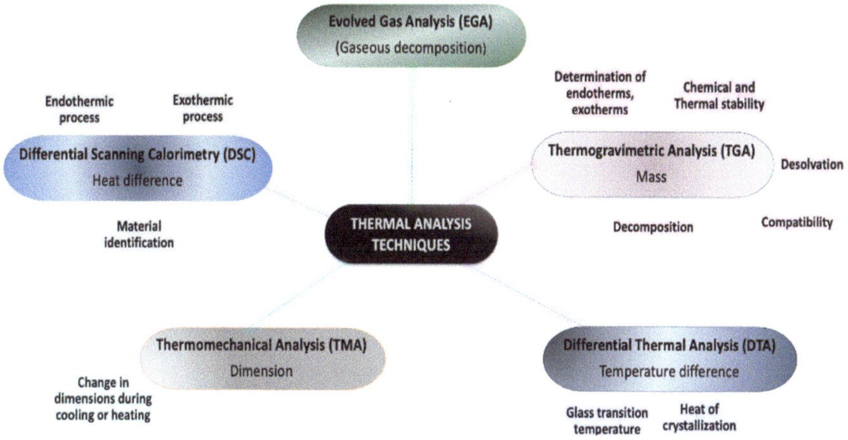

Fig. 1 Schematic illustration of thermal analysis techniques

spectroscopy. Most companies now offer the option to connect the thermogravimetric analysis with mass- and Fourier transform infrared spectroscopy. Thermogravimetric Analysis-Mass Spectroscopy (TGA/MS) and Thermogravimetric Analysis—Fourier transform infrared Spectroscopy (TGA/FTIR) can be crucial in the identification of intricate processes related to mass loss.

The difference between true concurrent methods such as Thermogravimetric Analysis—Differential Thermal Analysis (TGA/DTA) and Thermogravimetric Analysis—Differential Scanning Calorimetry (TGA/DSC), lies in the lack of time delay between the measurements, as opposed to the almost concurrent calculations of TGA/MS and TGA/FTIR. Concurrent TGA/DTA or TGA/DSC has multiple benefits. Besides the time and cost savings, the main benefit is the removal of ambiguity in the interpretation of the outcomes that might occur due to different samples, conditions, or other parameters (Thermal Analysis of Polymers 2009).

2.1 Differential Scanning Calorimetry (DSC)

Calorimetry refers to measuring of thermodynamic changes of materials as a result of chemical reactions, phase and other physical changes. Carious calorimetry types are applicable to nanoparticles. Differential Scanning Calorimetry (DSC) can be used on both thermodynamic data -such as heat capacity, enthalpy and entropy- as well as kinetic data (reaction rates, energy of activation) (Tewary 2015). In regards to the explanation of DSC graphs (Fig. 2), the most frequently encountered events are:

(i) Melting, appearing as a peak on the endotherm, because energy needs to be consumed to achieve melting.
(ii) Fusion heat, that can be identified through the integration of the peak area.

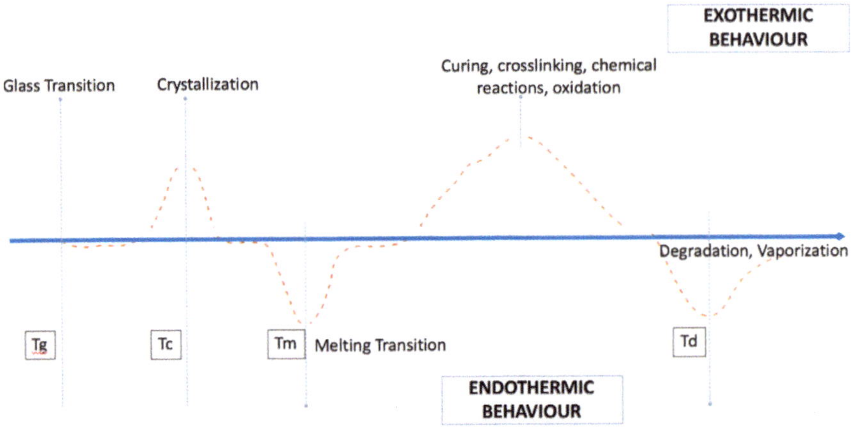

Fig. 2 The major events that DSC can identify (Tanzi 2017)

(iii) Loss of moisture, depicted as relatively broad and shallow endotherms.
(iv) Crystallization, illustrated as an exothermic mechanism.
(v) Cure reactions, illustrated, similarly to loss of moisture, as broad and shallow exotherms.
(vi) Glass transition (Tg) is the transition that is usually the most frequently measured. Glass transition is depicted as a specific point in the curve.

DSC can track every event that might cause differences in energy levels; DSC is, thus, the main thermal analysis process used by the industry. DSC can measure transitions like thermal stability, melting, crystallization and glass transition (Tg), and might determine the different temperature levels among the samples. Chemical reactions, such as heat capacity, purity assessment and thermal curing can also be measured (Millot et al. 2015; Wang and Lu 2013).

The nano-products industry is quite invested in researching topics such as thermal stability or thermal degradation behaviour. DSC is already frequently used in nano-research and development for their assessment. DSC is considered to be a simple, unambiguous and fairly quick characterization technique, although it has certain limitations (K et al. 2019; Ren et al. 2018; Sahraee et al., 2017; Shivam et al., 2019; Sumesh et al., 2019; Tang et al., 2016).

The melting point is identified using a melting point device; however, the outcome is not always accurate and easy to reproduce. Using DSC means that the Tm is obtained from a very accurate and well-calibrated instrument. Furthermore, this method provides a lot more information about the examined nanomaterial. For example, Tm changes may assist in understanding more about the amorphous content. As a result, the melting endotherm may determine the purity of the sample. The impact of modifications on the melting point of nanocomposites has been analysed using the DSC in various research works (Krishnaswamy et al., 2012; Nourani et al. 2016; Panaitescu et al., 2013; Wang et al. 2019).

The crystallization behaviour of nanomaterials is a compelling research field. There are various elements that could affect a material's structure, including temperature and processing conditions. It would be very difficult to reach appropriate, predictive structure and property correlations without an in-depth comprehension of the crystallization behaviour. Results from past studies suggest that the nanocomposites' structure and crystallization behaviours are dissimilar to those of neat materials (Dadbin & Kheirkhah, 2014; Dudhipala & Veerabrahma, 2016; Kumar et al., 2013; Madhukar et al., 2014; Morais et al., 2014; Nguyen et al., 2016; N. Wang et al., 2017).

The glass transition is mainly evaluated via DSC and DMA. DSC is the most frequently used technique for the measurement of Tg. This factor may be assessed through thermomechanical analysis as well, however this technique is less sensitive (Mach et al. 2019).

The Tg a significant attribute of amorphous and partially crystalline nanomaterials. It helps outline the temperature range within which the nanomaterials' mechanical attributes transition from hard to soft. The effect of nanomaterial composition and modification Tg has been examined through DSC analysis in multiple prior research works (Agrawal et al., 2013; Fotiadou et al., 2013; L. Gao et al., 2016; Khoramishad et al., 2018; H. Liu et al., 2011; Muhannad Mahdi Abd, 2013; Patidar et al., 2011; G. Wu et al., 2019). Being aware of the Tg allows for better comprehension of the subject, clarifying whether using nanoparticles results in better assembly, production and utilization of conductive adhesives (Mach et al. 2019).

2.2 *Thermogravimetric Analysis (TGA)*

TGA can provide information about mass changes in a specified temperature radius that enables specific measurements. By tracking the mass of material in a furnace, then any thermal events that result in a mass difference may be identified and assessed. This is ideal for the evaluation of a nanomaterial's decomposition temperature, as well as other temperature milestones, such as adsorption and/or desorption. This technique tracks the rise in temperature and heating, which was caused by the formation of degraded volatile components, bringing about a change of weight (Chowdhury et al. 2010; Fitaroni et al. 2015).

Nano-additives have been in the spotlight due to their unique attributes compared to their larger versions. The impact of modifying additive nanocomposites can be analysed using thermogravimetric analysis in plain air or, in some cases, nitrogen atmosphere. So far, the outcomes suggest that the inclusion of the nano-additive caused a notable change in the thermal stability, the ability to act as a flame retardant and the resulting nanomaterial char yields as opposed to the plain polymers (Hashemifard et al., 2011; Olad & Hayasi, 2011; Shabanian et al., 2014; Song et al., 2010).

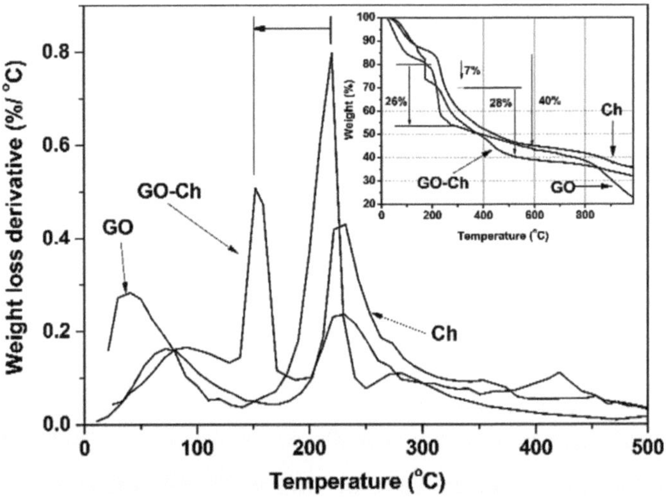

Fig. 3 Differential thermogravimetry (DTG) curves in nitrogen (Inset: Thermogravimetric analysis (TGA) of Ch, GO and GO–Ch in nitrogen) Reproduced with permission from Ref (Travlou et al. 2013), Copyright Elsevier 2013, Ltd

The nanomaterial composition and purity are two of the most crucial parameters to consider, and TGA is a fast method for characterizing them, through decomposition, with a minimum requirement for preparation. An important aspect of assessing the components comprising the nanomaterials and their purity is the characterization process, especially during the synthesis and formation stages (Mohandes and Salavati-Niasari 2013; Mohandes and Salavati-Niasari 2013; Salavati-Niasari et al., 2010).

The surfaces appear to have chemical differences, as observed on the DTG curves evaluated in nitrogen (Fig. 3). Differences in mass at the particular range are depicted through the peaks, while the area below them corresponds to the scope of said mass difference.

The thermogravimetric analysis of the studied samples was performed in a nitrogen environment and the corresponding graphs are presented in Fig. 3. GO displays about 20% difference in mass at temperatures under 200 °C, due to the water that was absorbed evaporating and a swift 26% mass drop between 200 °C and 250 °C as a result of the removal of the functional groups containing oxygen. The ability to create combination techniques, pairing the qualitative and quantitative assessment, further enhances the TGA technique. GC is a compound separation method, ideal for volatile compounds. The separation of gases occurs based on the differences in the component distribution between phases (Seifi et al. 2020). MS can identify even the smallest amount of materials, and differentiate elements of 1 ppm size or more, which occurs in the duration of the actual TGA scan (Lv and Wu 2012; Ding et al. 2017).

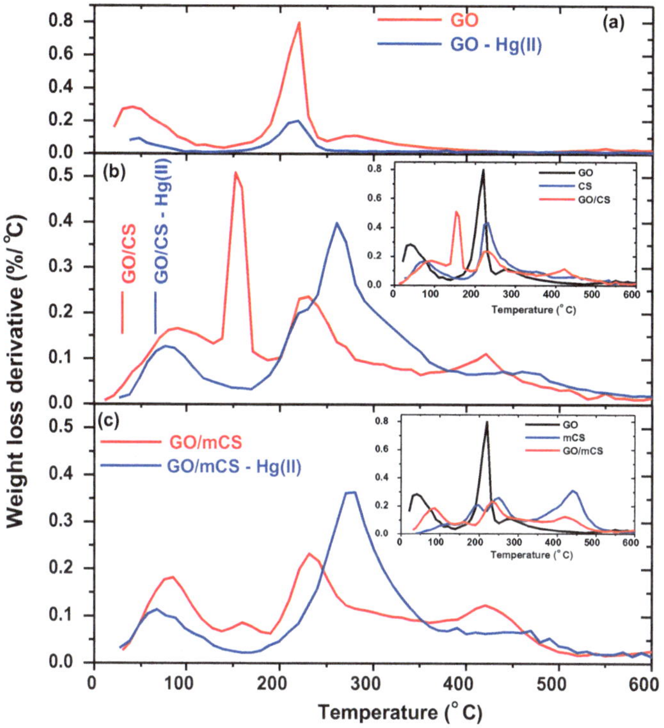

Fig. 4 Differential thermogravimetry (DTG) curves in nitrogen before and after Hg(II) adsorption onto: (a) GO; (b) GO/CS; (c) GO/mCS Reproduced with permission from Ref (Kyzas et al. 2014), Copyright Elsevier 2014, Ltd

The DTG curves (Fig. 4) describe the chemical characteristics of the surface of the studied nanomaterials. For all cases, the highest point happened between 80 °C and 100 °C, which could potentially be ascribed to the mass change owing to the evaporated water.

TGA is a frequently employed technique for the analysis of the weight changes of nanomaterials depending on either time or a change in temperature in a specific environment. FT-IR has been successfully utilized for gas identification. The combination of techniques allows for an extensive characterization in regard to decomposition and thermal stability (Wang et al. 2011; Si et al. 2014).

2.3 *Differential Thermal Analysis (DTA)*

DTA allows for a quantitative assessment and identification of the nanomaterial components, by analysing their thermal performance while they are heated. The method relies on the basic principle that when a material is heated, it will be subject

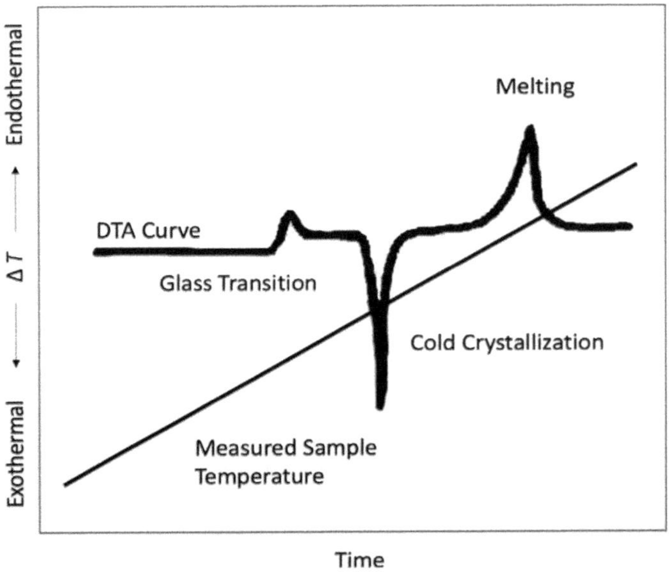

Fig. 5 The DTA curve over time (Hatakeyama and Quinn 1999)

to phase transitions and reactions, including heat release or absorption (Gao et al. 2005; Alkan et al. 2011). DTA is often used for the identification of melting and decomposition points. There are multiple works researching the melting and decomposition of nanomaterials through DTA (Si et al. 2014; Khan et al. 2016; Soltanzadeh and Morsali 2010). DTA has also been used for verifying the Tg of nanomaterials (Suzuki et al. 2018).

DTA has a similar pattern to DSC. If the studied material goes through a phase transition after being heated, energy is either released or consumed and a difference in temperature is noted. The most notable difference is the variable used on the vertical axis, DTA refers to a change in temperature (Fig. 5), whereas DSC refers to the flow of heat.

2.4 *Dynamic Mechanical Analysis (DMA)*

DMA is a useful characterization method, with the ability to expose the complex elastic modulus of solids, thus becoming essential for any materials science laboratory working on correlating the structure and properties of solid materials (Patra et al. 2020).

DMA is usually a more sensitive technique, as opposed to DTA and DSC. These techniques are more useful after the transition from crystalline to amorphous form has been completed. In these cases, a notable change has happened in the nanomaterial's mechanical attributes. As a result, DMA is preferrable for the identification of the

Tg as well as other minor structural or phase alterations of the nanomaterial (Saba et al. 2016).

2.5 Thermomechanical Analysis (TMA)

TMA can quickly and effortlessly measure the extent of the nanomaterial's deformation under stress conditions depending on either temperature changes or time differences, while a specified temperature is set. TMA is commonly used to identify the Tg, the potential softening points, as well as the material's linear expansion, through the application of a constant force on a nanomaterial at various temperature levels. TMA results can help comprehend the structure, components and potential applications of nanomaterials (Park et al. 2011; Zha and Fang 2010).

TMA is a very sensitive technique that can be used to measure the enlargement and reduction of materials that are cross-linked or filled. TMA outcomes are able to indirectly help perceive the dimensional orientation of nanocomposites.

TMA can be further utilized to determine the Tg, as the material transitions from glass state to rubber, dramatically changing its free molecular volume. The thermal expansion curve and its resulting reduction can be used to identify the Tg. On the other hand, the Tg remained almost identical to the Tg of pure epoxy, suggesting that the pure multi-walled nanotubes do not take part in the curing of the epoxy and are not completely included in the epoxy network (Liang and Tjong 2006).

2.6 Evolved Gas Analysis

Evolved gas analysis (EGA) (Gkika et al. 2020) is a method allowing for the identification of the nature and quantity of volatile product(s) created while the materials degrade due to thermal effects. This method analyses the gaseous components that evolved during the process, during which a sequence of reactions take place depending on the temperature levels, and are examined using various other techniques. EGA is generally used to assess reactions related to the degradation of materials by identifying the components of the decomposition products (Xie and Pan 2001). The use of MS and FTIR in EGA is already well-researched. Characterizing the composition of the evolved gases is significant in thermal analyses, especially when reviewing the reactions between gas–solid or the decomposition taking place in multicomponent systems (Cho et al. 2016; Martínez-Esaín et al. 2019).

3 Properties of Nanomaterials

This work conceptualized the nanomaterial properties as a network connecting the three categories of properties to the nanomaterial profile (Fig. 6). The selection of these connections stemmed from in-depth literature research. The bigger nodes constitute the major categories, that are surrounded by the more detailed subcategories of properties. Once the nodes were identified, they were labelled and colored accordingly in order to differentiate the groups of properties that are part of the nanomaterial profile network.

The depicted network highlights the most prominent clusters of profile drivers. The 3 main clusters consist of the mechanical, the chemical and the physical properties of the nanomaterial.

The nanomaterials' properties might be classified as mechanical, physical or chemical (Abd Elkodous et al. 2019).

3.1 Physical Properties of Nanomaterials

Physical properties (Table 1) include porosity, viscosity, melting point size, shape, specific surface area, agglomeration or aggregation, surface morphology, crystallinity, state of dispersion, density, temperature solubility and purity (Bierkandt et al., 2018; Flowers et al. 2018; Liu et al. 2009; Sharma et al., 2019).

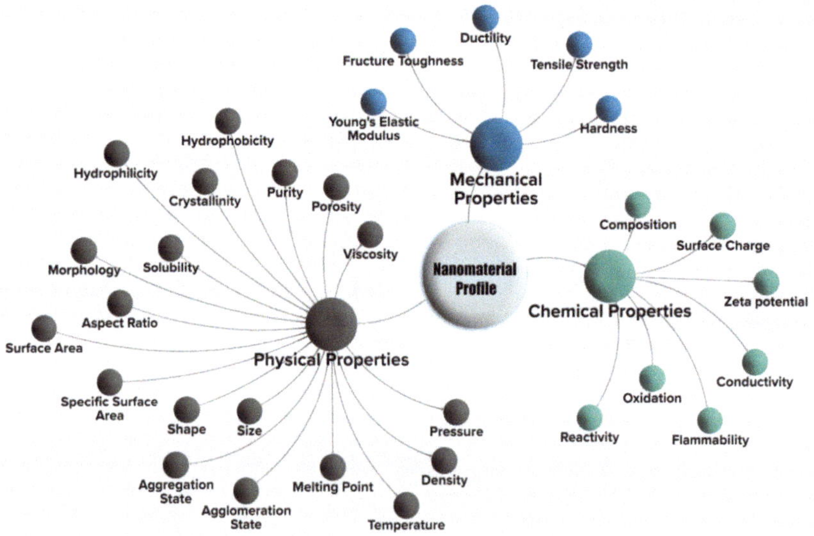

Fig. 6 Physical, chemical and mechanical properties of Nanomaterials

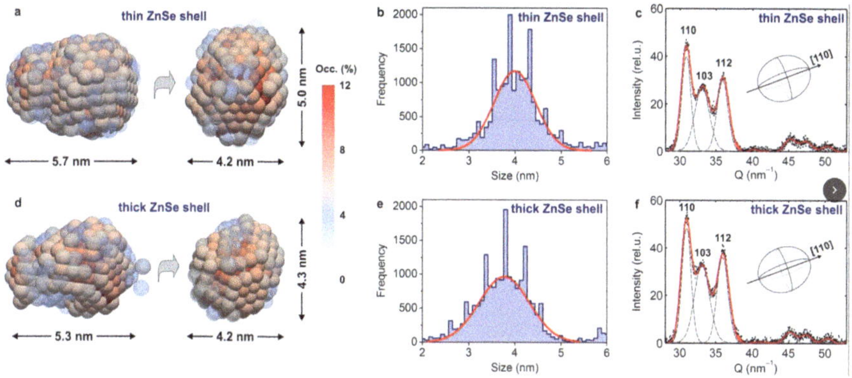

Fig. 7 Depiction of the shape and crystalline structure of nanocrystal scatterer. (a, d) Average shapes and (b, e) distribution of the diameter of Ag-In-Se/ZnSe core/shell nanocrystals, as acquired from SAXS using the ab initio shape-retrieval method. (c, f) Wide angle X-ray scattering spectra of Ag-In-Se/ZnSe core/shell nanocrystals, plotted along with Gaussian fits of their crystalline Bragg reflections. Information about the thin ZnSe shell sample are displayed in the upper panels while the ones for the thick shells appear in the panels below Reproduced with permission from Ref, Copyright Nature 2017 (Yarema et al. 2017), Ltd

Knowledge about a material's state of agglomeration or aggregation and the relevant responses is crucial for understanding its distribution and fate. Per the IUPAC's definition, an agglomerate comprises of dispersed particles that are linked through weak physical interactions, which might result in a physical separation due to the formation of precipitates that are bigger than a colloidal. This process can be reversed. An aggregate, on the other hand, comprises of strongly linked colloidal particles, clustering cannot be reversed (Sokolov et al. 2015). The stability of the manufactured nanomaterials relies on the pH value of the dispersion medium and the concentration of the solvent's electrolytes (Campbell 2002).

Depending on size and dimensions, the nanomaterials can be categorized into the following groups: (a) Nanoparticles, nanoclusters (0D), (b) Nanotubes, Nanowires (1D), (c) Nanoplates, nanolayers, (2D), and (d) Batches of multi-nanolayers, nanowires, nanotubes (3D) (Asha and Narain 2020). The particle size is important for regulatory reasons since it dictates if a substance is classified as a nanomaterial. Determining the size of a nanoparticle is usually the first phase in the physicochemical characterization process. There is no single technique to evaluate the size and dimensions of all known materials (Rasmussen et al. 2018). The material's size significantly impacts the interphase properties; the aggregation/agglomeration mainly lowers the interphase concentration, resulting in poor nanocomposites modulus and strength (Ashraf et al. 2018).

The impact of shape on material properties has been heavily researched (Guisbiers et al. 2012). Shape is one of the significant factors with an effect on the "nanoform" definition (Agency and How to prepare registration dossiers that cover nanoforms:

Table 1 Nanomaterial physical attributes' impact

Nanomaterial attributes	Attribute impact	Reference
Aggregation/ agglomeration state	Affects exposure evaluation	Sokolov et al. (2015)
Particle Size	Utilized for the nanomaterial determination and is important in regulatory context	Rasmussen et al. (2018)
Shape	One of the most significant properties for a nanomaterial	Temmerman et al. (2012)
Surface area	The surface area determines how a system responds to, disseminates or disposes of the nanomaterial	Powers et al. (2007)
Specific Surface area	Indicative of nanomaterial nature	Bushell et al. (2020)
Aspect ratio	Affects the behaviour in vivo	Shukla et al. (2015)
Solubility	Has an effect on stability in biological media	Landsiedel et al. (2017)
Morphology	Performance indicator	Shen et al. (2014)
Melting Point	Impacts solubility, which affects toxicity	Habibi-Yangjeh et al. (2008)
Temperature	May affect the efficiency of the delivery	Lee et al. (2014)
Particle density	Provides information about the composition	Minelli et al. (2018)
Pressure	Has an effect on dimensions, shape and dislocation	Abhilash and Pandey (2012)
Crystallinity	Crystallinity directly affects the chemistry of solid nanomaterials' surface	Nanomaterials: risks and benefits (2009)
Purity	Affects the accurate evaluation of the optical and electronic attributes	Wu et al. (2014)
Porosity	Can deeply impact a system's electrochemical properties by affecting the amount and type of the catalytic sites	Batchelor-McAuley and Compton (2020)
Viscosity	Affects the flow resistance of fluids, significant for transfer of heat	Mishra et al. (2014)
Hydrophobicity	May have an effect on dispersibility in water, toxicity and ability to connect to surfaces in the immediate environment	Gao and Lowry (2018)
Hydrophilicity	Impacts the adsorption of proteins and adhesion to other cells	Kuo et al. (2018)

best practices 2017). It might vary to a considerable degree, which is why it is considered a challenging process (Rasmussen et al. 2018). This has led to the definition of a wide range of quantifiable shape descriptors (Temmerman et al. 2012), including concepts such as sphericity and aspect ratios (Rasmussen et al. 2018) (Fig. 7)

The surface area may help identify nanomaterials (Wohlleben 2015), and affect the interaction with biological systems. Reducing the size seems to cause an exponential increase in surface area vs volume, resulting in a more reactive nanomaterial surface (Powers et al. 2007).

Volume-specific surface area (VSSA) may—under certain circumstances—clarify if a material has nanoscale dimensions. Specific surface area (SSA) has been proposed as a suitable alternate choice to that of mass or number-based dose metrics (Bushell et al. 2020). The SSA is commonly determined through a gas absorption technique referred to as the BET-method that can measure porosity and/or surface area down to 1 nm level (Doren et al. 2011). The high specific surface area renders ultrathin 2D nanomaterials into promising building blocks for the creation of functional nanocomposites (Zhang 2015). Because of their small size, most nanomaterials possess less volume, but higher surface. The surface energy is thus increased, and they become very reactive (Soares et al. 2018).

Aspect ratio can be described as the ratio of the length of the two axes (Baig et al. 2021). There are low-aspect-ratio nanoparticles (LARN) such as spherical, cubic, etc., nanoparticles, as opposed to the nanomaterials with high aspect ratio (HARN), e.g. nano-wires/tubes (Fadeel et al. 2012). The term is commonly used for structures with an aspect ratio equal to or greater than 10:1 (Higgins et al. 2020). The aspect ratio significantly affects the in vivo behaviour of nanoparticles, with an impact on biodistribution, longevity and tumor penetration (Shukla et al. 2015). Zhao et al. examined the way low aspect ratio quartz nanopillars and nanobars may provoke protein recruitment during endocytosis mediated by clathrin (Zhao et al. 2017). High-aspect-ratio nanostructures affect the cellular and intracellular environment; there is a wide variety of nanomaterials that can incite fundamental cell behaviour, such as mechanotransduction. The safety of using such nanostructures in vivo is yet another important challenge to overcome (Higgins et al. 2020). High aspect ratio nanomaterials (HARNs) possess a physical similarity to asbestos (Fubini et al. 2011). The toxicity of fibers with high aspect ratio is influenced by their biodurability, that is impacted by their mechanical properties and dissolution process (Lippmann 1990).

Solubility is a key factor with an impact on biopersistence and biokinetics of nanomaterials affecting their toxicity (Sohal et al. 2018). There have been efforts to group nanomaterials with the purpose of risk assessment, and solubility has been defined as a significant factor to be considered (Landsiedel et al. 2017). Solubility also impacts stability in biological media, with an effect on the formation of agglomerates/aggregates with different sedimentation rates, interactions with cells and biological responses (Avramescu et al. 2020). Water-soluble proteins have promising applications in nanomedicine. (Baig et al. 2021).

Controlled morphology is important for the efficiency of nanomaterials. For example, Laifa Shen et al. created an electrode structure consisting of mesoporous nanowire arrays used on cloths, thus enhancing the effectiveness of the electrode (Shen et al. 2014). Nanoparticles in powder format displayed an improved tap density as opposed to nanoparticles in non-spherical form, and the material significantly enhanced the secondary lithium's power density. Using the same pair of materials with alternative morphologies will lead to variant results. The catalyst selection also affects the morphology and type of nanomaterial produced (Baig et al. 2021). Morphologies that are core–shell based may result in the reduced requirement of critical elements for a multitude of applications as well (Ying et al. 2001). Special attention is being paid to manufacturing nanomaterials with specific morphologies and nano-dimensions, in order to reach the preferred well-organized nanostructures (Baig et al. 2021).

Figure 8 depicts the morphological analysis conducted via scanning electron microscopy (SEM). The WO3 nanoparticles manufactured with Na2WO4?2H2O exhibit diverse morphologies: (i) nanocubes assigned to m-WO3, (ii) nanosheets assigned to o-WO3?0.33H2O, (iii) nanowires assigned to h-WO3 and (iv) bundle structures (Marques et al. 2015). Figure 9 illustrates the structural morphology of graphene, h-BN and MoS2 at $\gamma = 0$. It can be observed that there are random variations in their height (z-direction). After the application of the strain to the internal boundary of the structures, regular types of wrinkles start to appear (Arabha and Rajabpour 2020).

The suppression of the melting point when dealing with nanomaterials has been a long-established concern (Schmidt et al. 1998). The melting point impacts solubility, which affects toxicity; For example, if a material is not easily soluble, its concentration may be too minimal to induce a toxic result, and as such, it would be meaningful to evaluate the melting point of a material based on its chemical properties (Habibi-Yangjeh et al. 2008). The intricacies of the melting mechanism, are still not known. Adjustments in the melting point happen because nanomaterials possess a considerably bigger surface-to-volume ratio, significantly changing their attributes (Antoniammal and Arivuoli 2012). It has been verified both in practice and theory, that a nanomaterial's melting temperature (T) relies on its size (Guisbiers et al. 2012; Sar and Nanda 2010). The shape however also affects the nanoparticle's melting point. Changes in morphology all affect the melting point and the extent of that effect (Nanda et al. 2002).

The behaviour of nanoparticles at higher temperatures is interesting by itself (Foster et al. 2019), however it was discovered that higher levels might result in a larger degree of nanoparticle coalescence, with an impact on its atomic diffusion rate and the extent of plastic deformation (Hu et al. 2020). The temperature affects the reaction rate as well as the morphologies (Lee et al. 2014). The evaluation of the self-diffusivity coefficients implies that temperature has a significant impact on the reorientation stage (Foroutan et al. 2017). In other research work, where endocytosis is used as the delivery mechanism, the temperature is likely to affect the efficiency of the delivery (Higgins et al. 2020). The temperature has an impact on both viscosity and interfacial tension (Ting et al. 2006).

Color changes between phases indicate the presence and formation of nanoparticles (Vanaja et al. 2013; Chauhan and Upadhyay 2018). Simulations uncovered that the color of nanoparticles is significantly impacted by the size when density is constant. Furthermore, the visual observation of chemically-induced color changes could reveal information on the energy changes, when the Pr oxidation state is changed (Mourdikoudis et al. 2018).

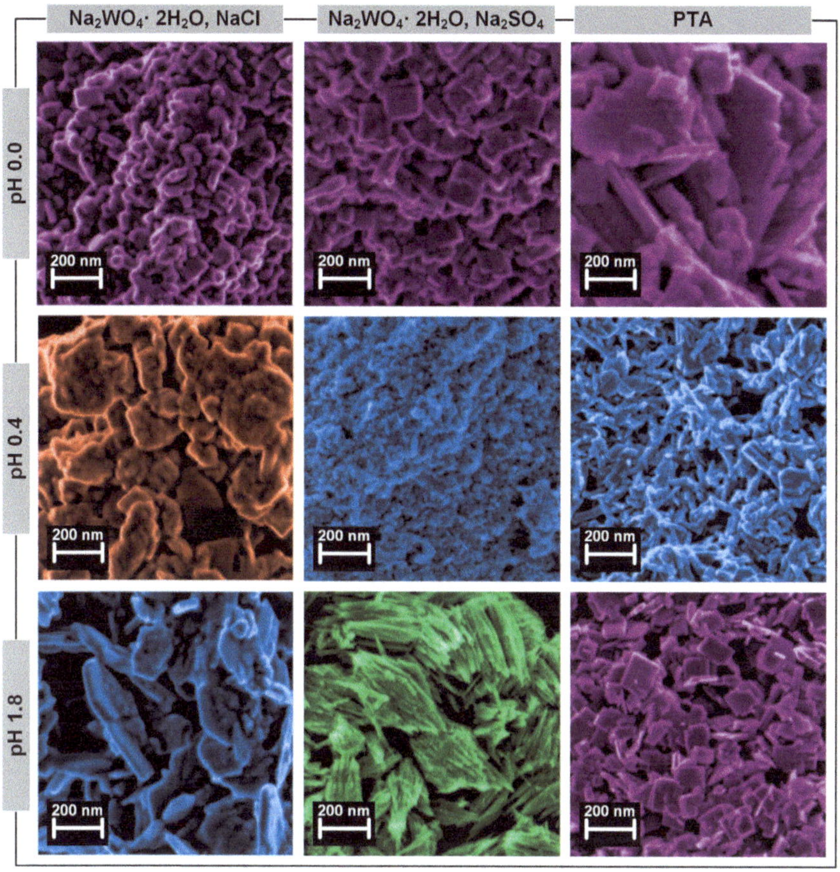

Fig. 8 SEM images of the manufactured WO3 nanoparticles. The images have been colored using the GIMP software for better comprehension. The various colors correspond to the XRD diffractograms for each crystalline structure Reproduced with permission from Ref (Marques et al. 2015), Copyright Nature 2015, Ltd

The density attribute is useful to understand the material's composition. If the particle density is known, it may help access additional details, such as the chemical composition and drug loading (Minelli et al. 2018).

Pressure is significant for the manufacturing of nanomaterials (Patra and Baek 2014). Adjusting the pressure impacts the shape as well as the size of the produced nanomaterials (Abhilash and Pandey 2012). Researchers are still investigating exactly how pressure has an impact on surface and bulk transport (Hu et al. 2020).

Crystallinity is an attribute that describes the pattern in which the elements consisting of a nanomaterial are arranged, and is significant because if affects the chemistry of solid nanomaterials' surface (Nanomaterials: risks and benefits. 2009). It also affects tensile strength, modulus and flow resistance, thus has an impact on their

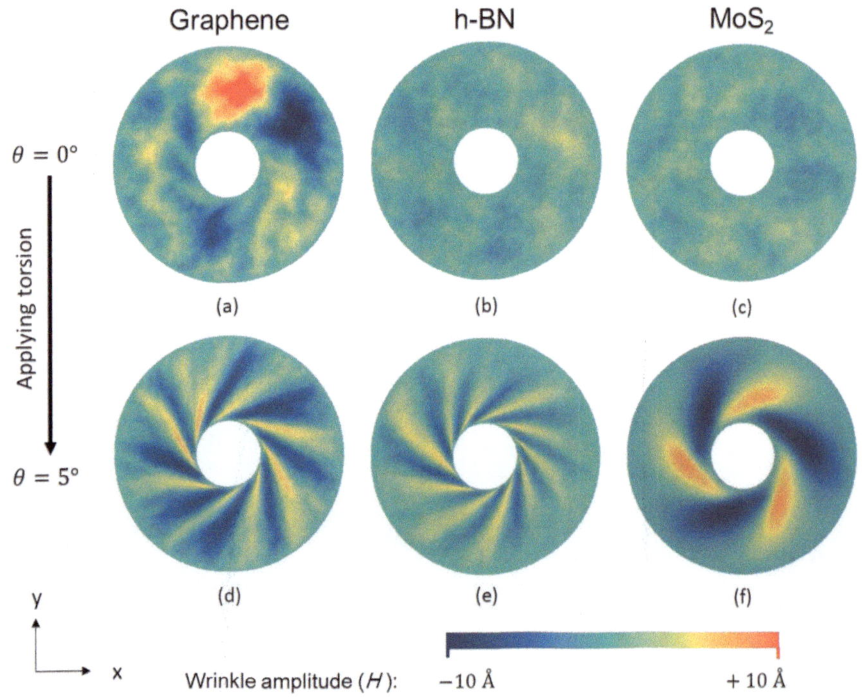

Fig. 9 Contours of wrinkles amplitude in two-dimensional nanostructures (graphene, h BN, MoS2) in pristine forms (a, b, c) and under torsional deformations θ = 5Åã (d, e, f) Reproduced with permission from Ref (Arabha and Rajabpour 2020), Copyright Elsevier 2020, Ltd

reinforcing capabilities. Moreover, crystalline materials have less of a tendency to sustain physical or chemical changes compared to amorphous materials, i.e. they have better stability. Differences in crystallinity may affect purity, such as in cellulose nanocrystals (Barrera and Cornish 2019). Temperature changes might affect the nanomaterial crystallinity, trigger phase transformation and impact photocatalysis (Muthee and Dejene 2021).

A significant part of evaluating the purity of a nanomaterial revolves around identifying the chemical elements of a nanoparticle, specifically during the composition and synthesis process (Mansfield 2015). Purity is crucial for the assessment of potential optical and electronic attributes, as well as for the evaluation of the relationship between structure and functions when conducting a toxicology analysis. It is further important for identifying the role of the ligand on the surface of the nanopmaterial (Wu et al. 2016). Purity may also impact conductivity (Sun and Wang 2013).

Porosity is significant for the identification of the microstructures of a material. The pore system, described as a series of small pores and channels that might communicate or not, can be broken down into different porosity categories (Issaadi et al. 2018). The quality and potential uses of a nanoparticle are majorly affected by the porosity of the material (Patra and Baek 2014). Porosity may significantly impact the electrochemical behaviour by altering the number as well as the type of the catalytic sites that can be used, and affecting the transport of mass in the area around the electrochemical interface (Batchelor-McAuley and Compton 2020). The size of the pores and the porosity may also have an effect on the hardness and potential osteogenic attributes of biomaterials (Cheng et al. 2021).

Viscosity determines a fluid material's flow resistance and has a significant impact on heat transfer ability (Mishra et al. 2014). Furthermore, intraluminal viscosity has an effect on oral drug absorption through various methods. Radwan et al. (2013) suggested that the rate of disintegration of trospium chloride was slowed down when used in a viscous solution, which delayed the delivery of the drug (Tanaka et al. 2019).

Researchers agree that a surface is considered as hydrophobic in cases where its static water contact angle θ is greater than 90° while it is considered as hydrophilic when said angle is less than that. There is generally very little justification about why a surface may switch from being hydrophilic to hydrophobic (Law 2014). Hydrophilic materials have an affinity to fluids. Hydrophilicity affects blood compatibility and cellular compatibility (Kuo et al. 2018). A surface's hydrophobicity or hydrophilicity may be the product of a differentiation in molecular chemical properties or a change in polarity resulting in a modified surface chemistry. Such a modification would have an impact on the adsorption of proteins and the way in which the material interacts with the target cell (Banik and Brown 2015). Nanowires may imitate various tissues due to their shape and surface attributes (Ghezzi et al. 2021). Despite the importance of hydrophobicity of materials, little research has been conducted regarding the measurement of this property reliably. It is of note that hydrophobicity might be affected by interacting with other elements in water, such as ions, pH, etc. Hydrophobicity can thus be regarded as system dependent and to an extent extrinsic (Gao and Lowry 2018).

3.2 Mechanical Properties of Nanomaterials

Mechanical properties indicate the characteristic under various circumstances. Different materials possess different characteristics (Wu et al. 2020). There are multiple factors affecting them, namely the nanoparticle options, the manufacturing process, the grain size and boundary structure, which interact and have an effect on each other. The use of diverse materials and processing methods will result in nanomaterials with diverse structures and properties. Based on prior research, the importance of each factor has been examined and summarized below (Wu et al. 2020). The mechanical properties (Table 2) include hardness, Young's (elastic) modulus, ultimate tensile strength, ductility, as well as fracture toughness (Abd Elkodous et al. 2019; Domun et al. 2015).

The nanoparticle's hardness is often different from their larger versions and some exhibit apparent size-dependent behaviors. For spherical polymer nanoparticles, there are currently no observations of similar behaviours depending on size (Guo

Table 2 Nanomaterial mechanical attributes' importance

Nanomaterial attributes	Attribute importance	Reference
Hardness	Size-dependent behaviour	Guo et al. (2014)
Young's elastic modulus	Determines the strain behaviour	Zhang et al. (2018)
Strength	Affects the lifetime of the material	Domun et al. (2015)
Toughness	Affects the suitability of the material	Day et al. (2001)
Ductility	The capacity to be drawn or deformed without breaking	Sun and Wang (2013)

et al. 2014). Paik et al. 2007) suggested that polypropylene (PP) nanoparticles have higher elastic modulus as compared to their larger counterpart. A different work, however, about polystyrene nanoparticles has estimated them to be marginally lower than the larger materials, because of the presence of hydrated ionic functional groups inside them (Tan et al. 2004).

Young's modulus is considered one of the most significant mechanical attributes, characterizing the materials' strain behaviour (Zhang et al. 2018). For crystalline metal nanoparticles, dislocations have been proven to have an impact on the mechanical behavior of the nanoparticles. Ramos et al. (2013) observed that the hardness and elastic modulus of six-fold symmetry gold nanoparticles reached higher levels than the larger material. Gerberich et al. (2003) noted that the hardness at 40 nm diameter was 4 times higher than larger particles of silicon. They suggested that the occasional dislocations or line defects in the interior of the particle are the major drivers against high pressures. Jing et al. (2006), Cuenot et al. (2004) have observed that the elastic moduli of nanowires decreased as the radial diameter increased for the cases of lead and silver. Yu et al. (2000) mentioned that the Young's modulus was significantly better after adding CNT in the epoxy, same as the tensile strength. Furthermore, Liang and Pearson examined (2009) two different types of silicon nanoparticles with different dimensions. The Young's moduli increased after adding the nanoparticle moderately.

The strength and toughness of materials characterize the suitability and determine the lifetime of the materials Domun et al. (2015). The use of nanomaterials does not necessarily enhance the mechanical attributes of the resulting product (AlKahtani 2018). The inclusion of TO2 nanoparticles in the thermo-polymerized acrylic resins has been proven to negatively impact the flexural strength, as the nanoparticle concentration rises (Nazirkar et al. 2014). It is notable that in similar works, and in different ones conducted by Balos et al. (2014), and Nazirkar et al. (2014), the best outcomes in terms of flexural strength have been achieved when of nanoparticle loads were used in low levels. Toughness is considered an important factor in deciding if the final product is suitable for a specified application without breaking (Day et al. 2001; Zhang et al. 2008) examined the impact of improving DGEBF with silicon nanoparticles at a variety of temperatures. The observed enhancement in fracture toughness was considerably better compared to the stiffness. Other works have concentrated on using rubber particles to improve the fracture toughness (Domun et al. 2015).

The ductility of nano-structured metals has been a popular research subject for more than a decade (Zhu and Wu Jun 2018). Nevertheless, nanostructured materials are characterized by low ductility which slowed down their development beyond laboratory-scale (Lu et al. 2011). Low ductility prevents the use of metallic nanomaterials (Ovid'ko et al. 2018). The mixture of rubber with other nanofillers in epoxy resins has thus been a subject of discussion for many scientists. It has been noted that in such cases there is a significant improvement in fracture toughness and ductility, however there is also a concurrent decrease in the stiffness of the cured polymers (Domun et al. 2015).

It is notable that the measurement of mechanical properties of individual nanoparticles is very complicated since there are various factors affecting the results, including the requirement for a consistent nanoparticle dispersion on a hard substrate, the accurate particle location and the correct addition of loads, along with the appraisal of the minimum level of nanoparticle deformation (Guo et al. 2014).

3.3 Chemical Properties of Nanomaterials

The chemical properties of nanomaterials (Table 3) include the chemical composition, reactivity, oxidation, flammability, thermal conductivity, zeta potential and surface charge (Abd Elkodous et al. 2019; Liu et al. 2013).

Based on external circumstances, atoms may organize themselves in diverse crystal structures compared to materials with variant proportions of the same components. According to these external conditions, a nanomaterial may have distinct crystal structures, and corresponding physiochemical attributes (Asha and Narain

Table 3 Nanomaterial chemical attributes' importance

Nanomaterial attributes	Attribute importance	Reference
Chemical Composition	It helps describe its attributes and behaviour	Ambrosi et al. (2012)
Reactivity	Is important to sorting the environmental fate	Leeuwen et al. (2013)
Oxidation	Produces nanoscale heterostructures	Sutter et al. (2013)
Flammability	Important when using bio-based composites	Sienkiewicz and Czub (2020)
Thermal conductivity	Impacts the charging and discharging rate	Khodadadi and Hosseinizadeh (2007)
Zeta potential	Has impact on the stability of a dispersion	Rasmussen et al. (2018)
Surface charge	Affects functionality and attributes	Asha and Narain (2020)

2020). The components on a nanomaterial's surface chemical composition is especially important, along with determining any potential impurities on either the interior or the surface or the nanoparticle, which could alter the material's attributes and not permit accurate assessments versus other materials that have the same components (Ambrosi et al. 2012).

Nanomaterial reactivity is currently a popular subject (Magro et al. 2018) since nnanomaterials with high surface energy commonly are very reactive and unstable. They are easy to degrade or oxidize when exposed to the environment (Asha and Narain 2020). Information on the reactivity of natural and synthetic nanoparticles is crucial to understanding their effect on ecotoxicology and health (Leeuwen et al. 2013). Furthermore, it is required in the creation of nanoparticles with specified functionality that can have multiple applications (Dong et al. 2007). Surface atoms display react differently compared to those in the larger versions (Magro et al. 2018).

Oxidation has different types of effects on nanoparticles (Xanthopoulou et al. 2019; Yin et al. 2012) as well as other researchers, suggested that the chemical structure of the interior and exterior of nanomaterials is the deciding factor about whether oxidation should occur (Fetisov et al. 2015). The use of metal nanoparticles for many important applications has brought the spotlight on comprehending and controlling their oxidation. Oxidation of metal nanostructures is a size/shape-dependent process that has been extensively researched (Leitner et al. 2020); it takes place often during long-term storage (Xanthopoulou et al. 2019). Oxidation is an impressive mechanism that can create nanoscale heterostructures (Sutter et al. 2013).

Flammability might often be the critical factor in regard to the possibility of using bio-based composites (Sienkiewicz and Czub 2020). Flame retardants have been accepted as safety measures that are able to reduce the occurrences of accidents and deaths related to fire. This term is used for a group of materials that can be added during the synthesis to avert or weaken a potential explosion (Vahidi et al. 2021). A flammability analysis suggests that an appropriate quantity of HARN with exfoliation attributes successfully assists in reducing the combustion (Subasinghe et al. 2016).

The thermal conductivity (TC) of a material affects the melting-solidification rate, frequently mentioned as the charging and discharging rate (Khodadadi and Hosseinizadeh 2007; Navya and Daima 2016). The addition of nanoparticles may result in increased or reduced thermal conductivity by suppressing natural convection (Esapour et al. 2018). It has been suggested that special attention should be paid when incorporating thermally conductive nanoparticles into hydrogels, because besides enhancing the phase change rate, it also affects the temperature levels at which a phase change would happen. For high performance thermoelectric materials, the effectiveness also relies on a material's capacity to exhibit a high electrical conductivity. Single crystals of nanoparticles display the best electrical conductivity due to the lack of grain boundaries scattering charge carriers (Coetzee et al. 2020; Barisik et al. 2014).

The zeta may be defined as the difference between the dispersion means and a fluid attached to the dispersed material (Rasmussen et al. 2018). The surface of charged colloids can be described by the zeta potential; Further results can be extracted from the analysis of features such as distribution and adsorption (Shah 2020; Jastrzębska and Olszyna 2015). Even minor adjustments in the parameters, like a small pH change, could considerably impact the aggregation rate of dispersed nanoparticles, and thus the assessment of the zeta potential according to the pH levels is quite common, beyond estimating the isoelectric point (Rasmussen et al. 2018).

Surface charge helps understand a nanomaterial's attributes and potential functionality (Asha and Narain 2020) and determines the colloidal behaviour, thus affecting the response to nanomaterial exposure via changes in shape and size (Shah 2021a). It also impacts the starting electrostatic interaction; For example toxicity versus living beings is linked with positive surface charges (Shah 2021b).

4 Conclusions

It is evident that nanotechnology has been consequential in the biomedicine, resulting in the outstanding improvement of a series of useful nanomaterials, with multiple applications in medicine. There are various aspects that impact the quality and quantity of manufactured nanomaterials as well as their possible uses in areas of interest. This chapter summarized the current status of our knowledge about the attributes of nanomaterials. Certain properties of nanomaterials have been identified as key to controlling the abilities of the synthesized nanoparticles, and thus should be further researched in the future.

This chapter further described the various techniques for the thermal analysis of nanomaterials. Throughout this analysis, the uses of each technique were presented, focusing on their advantages, as well as on clarifying how they can be efficiently combined by complementing each other. Presenting each technique individually as well as in comparison to its alternatives, renders this chapter a guide, assisting the research community to better comprehend the subject. As a result, scientists will be able to make informed choices about the appropriate techniques for their needs and requirements.

Acknowledgements The financial support received for this study from the Greek Ministry of Development and Investments (General Secretariat for Research and Technology) through the research project "Intergovernmental International Scientific and Technological Innovation-Cooperation. Joint declaration of Science and Technology Cooperation between China and Greece" with the topic "Development of monitoring and removal strategies of emerging micro-pollutants in wastewaters" (Grant no: T7ΔKI-00220) and it is gratefully acknowledged.

References

Abhilash Pandey BD (2012) Synthesis of zinc-based nanomaterials: a biological perspective. IET Nanobiotechnology 6(4):144–148. https://doi.org/10.1049/iet-nbt.2011.0051.

Agrawal SL, Rai N, Chand N (Jan 2013) Dynamic mechanical, DSC, and electrical investigations on nano Al_2O_3 filled PVA:NH_4SCN:DMSO polymer composite dried gel electrolytes. Int J Polym Mater Polym Biomater 62(2):61–67. https://doi.org/10.1080/00914037.2011.617341

Alkan C, Sarı A, Karaipekli A (Jan 2011) Preparation, thermal properties and thermal reliability of microencapsulated n-eicosane as novel phase change material for thermal energy storage. Energy Convers Manage 52(1):687–692. https://doi.org/10.1016/j.enconman.2010.07.047

AlKahtani RN (2018) The implications and applications of nanotechnology in dentistry: a review. Saudi Dent J 30(2):107–116. https://doi.org/10.1016/j.sdentj.2018.01.002

Ambrosi A, Chee SY, Khezri B, Webster RD, Sofer Z, Pumera M (2012) Metallic impurities in graphenes prepared from graphite can dramatically influence their properties. Angew Chem Int Ed 51(2):500–503. https://doi.org/10.1002/anie.201106917

Antoniammal P, Arivuoli D (2012) Size and shape dependence on melting temperature of gallium nitride nanoparticles. J Nanomater 2012:1–11. https://doi.org/10.1155/2012/415797

Arabha S, Rajabpour A (2020) Effect of planar torsional deformation on the thermal conductivity of 2D nanomaterials: a molecular dynamics study. Mater Today Commun 22:100706. https://doi.org/10.1016/j.mtcomm.2019.100706

Asha AB, Narain R (2020) Nanomaterials properties. In: polymer Science and Nanotechnology, Elsevier, pp 343–359. https://doi.org/10.1016/B978-0-12-816806-6.00015-7

Ashraf MA, Peng W, Zare Y, Rhee KY (2018) Effects of size and aggregation/agglomeration of nanoparticles on the interfacial/interphase properties and tensile strength of polymer nanocomposites. Nanoscale Res Lett 13(1):214. https://doi.org/10.1186/s11671-018-2624-0

Avramescu M-L, Chénier M, Palaniyandi S, Rasmussen PE (2020) Dissolution behavior of metal oxide nanomaterials in cell culture medium versus distilled water. J Nanopart Res 22(8):222. https://doi.org/10.1007/s11051-020-04949-w

Baig N, Kammakakam I, Falath W (2021) Nanomaterials: a review of synthesis methods, properties, recent progress, and challenges. Mater. Adv. 2(6):1821–1871. https://doi.org/10.1039/D0MA00807A

Balos S, Pilic B, Markovic D, Pavlicevic J, Luzanin O (2014) Poly(methyl-methacrylate) nanocomposites with low silica addition. J Prosthet Dent 111(4):327–334. https://doi.org/10.1016/j.prosdent.2013.06.021

Banik BL, Brown JL (2015) Interaction of responsive/switchable surfaces with cells. In: Switchable and responsive surfaces and materials for biomedical applications, Elsevier, pp 189–201. doi: https://doi.org/10.1016/B978-0-85709-713-2.00008-0

Barrera CS, Cornish K (2019) Characterization of agricultural and food processing residues for potential rubber filler applications. J. Compos. Sci. 3(4):102. https://doi.org/10.3390/jcs3040102

Barisik M, Atalay S, Beskok A, Qian S (2014) Size dependent surface charge properties of silica nanoparticles. J Phys Chem C 118(4):1836–1842. https://doi.org/10.1021/jp410536n

Batchelor-McAuley C, Compton RG (2020) Characterising and evidencing the effects of porosity in nano-electrochemistry. Curr Opin Electrochem 22:35–43. https://doi.org/10.1016/j.coelec.2020.03.012

Bierkandt FS, Leibrock L, Wagener S, Laux P, Luch A (2018) The impact of nanomaterial characteristics on inhalation toxicity. Toxicol Res 7(3):321–346. https://doi.org/10.1039/c7tx00242d

Borca-Tasciuc T, Vafaei S, Borca-Tasciuc D-A, Wei BQ, Vajtai R, Ajayan PM (Sep 2005) Anisotropic thermal diffusivity of aligned multiwall carbon nanotube arrays. J Appl Phys 98(5):054309. https://doi.org/10.1063/1.2034079

Bushell M et al (2020) Characterization of commercial metal oxide nanomaterials: crystalline phase, particle size and specific surface area. Nanomater 10(9):1812. https://doi.org/10.3390/nano10091812

Campbell H (2020)"A non-inferiority test for R-squared with random regressors. arXiv:2002.8476 null. Accessed 31Jul 2020. [Online]. http://arxiv.org/abs/2002.08476

Charitidis CA, Georgiou P, Koklioti MA, Trompeta A-F, Markakis V (2014) Manufacturing nanomaterials: from research to industry. Manufacturing Rev 1:11. https://doi.org/10.1051/mfreview/2014009
Chauhan S, Upadhyay LSB (2018) An efficient protocol to use iron oxide nanoparticles in microfluidic paper device for arsenic detection. MethodsX 5:1528–1533. https://doi.org/10.1016/j.mex.2018.10.017
Cheng L et al (2021) The preparation and application of calcium phosphate biomedical composites in filling of weight-bearing bone defects. Sci Rep 11(1):4283. https://doi.org/10.1038/s41598-021-83941-3
Cho S-P et al (2016) One step synthesis of Au nanoparticle-cyclized polyacrylonitrile composite films and their use in organic nano-floating gate memory applications. J. Mater. Chem. C 4(7):1511–1516. https://doi.org/10.1039/C5TC04166J
Chowdhury A, Bould J, Zhang Y, James C, Milne SJ (2010) Nano-powders of Na0.5K0.5NbO3 made by a sol–gel method. J Nanopart Res 12(1):209–215. https://doi.org/10.1007/s11051-009-9595-0
Coetzee D, Venkataraman M, Militky J, Petru M (2020) Influence of nanoparticles on thermal and electrical conductivity of composites. Polymers 12(4):742. https://doi.org/10.3390/polym12040742
Corcione C, Frigione M (Dec 2012) Characterization of nanocomposites by thermal analysis. Mater 5(12):2960–2980. https://doi.org/10.3390/ma5122960
Cuenot S, Frétigny C, Demoustier-Champagne S, Nysten B (2004) Surface tension effect on the mechanical properties of nanomaterials measured by atomic force microscopy. Phys Rev B 69(16):165410. https://doi.org/10.1103/PhysRevB.69.165410
Dadbin S, Kheirkhah Y (2014) Gamma irradiation of melt processed biomedical PDLLA/HAP nanocomposites. Radiat Phys Chem 97:270–274. https://doi.org/10.1016/j.radphyschem.2013.12.001
Day RJ, Lovell PA, Wazzan AA (2001) Toughened carbon/epoxy composites made by using core/shell particles. Compos Sci Technol 61(1):41–56. https://doi.org/10.1016/S0266-3538(00)00169-X
De Temmerman P-J, Van Doren E, Verleysen E, Van der Stede Y, Francisco MA, Mast J (2012) Quantitative characterization of agglomerates and aggregates of pyrogenic and precipitated amorphous silica nanomaterials by transmission electron microscopy. J Nanobiotechnol 10(1):24. https://doi.org/10.1186/1477-3155-10-24
Ding H, Huang K, Li S, Xu L, Xia J, Li M (Nov.2017) Synthesis of a novel phosphorus and nitrogen-containing bio-based polyol and its application in flame retardant polyurethane foam. J Anal Appl Pyrol 128:102–113. https://doi.org/10.1016/j.jaap.2017.10.020
Domun N, Hadavinia H, Zhang T, Sainsbury T, Liaghat GH, Vahid S (2015) Improving the fracture toughness and the strength of epoxy using nanomaterials – a review of the current status. Nanoscale 7(23):10294–10329. https://doi.org/10.1039/C5NR01354B
Dong Y-P, Cui H, Xu Y (2007) Comparative studies on electrogenerated chemiluminescence of luminol on gold nanoparticle modified electrodes. Langmuir 23(2):523–529. https://doi.org/10.1021/la0617107
Dudhipala N, Veerabrahma K (Feb 2016) Candesartan cilexetil loaded solid lipid nanoparticles for oral delivery: characterization, pharmacokinetic and pharmacodynamic evaluation. Drug Deliv 23(2):395–404. https://doi.org/10.3109/10717544.2014.914986
Elkodous MAbd et al. (2019) Therapeutic and diagnostic potential of nanomaterials for enhanced biomedical applications. Colloids SurfS B: Biointerfaces 180:411–428. https://doi.org/10.1016/j.colsurfb.2019.05.008
Esapour, M, Hamzehnezhad A, Rabienataj Darzi AA, Jourabian M (2018) Melting and solidification of PCM embedded in porous metal foam in horizontal multi-tube heat storage system. Energy Convers Manag 171:398–410. https://doi.org/10.1016/j.enconman.2018.05.086
European Chemicals Agency (2017) How to prepare registration dossiers that cover nanoforms: best practices. Accessed 31 Jul 2020. [Online]. http://dx.publications.europa.eu/

Fadeel B, Pietroiusti A, Shvedova AA (eds) (2012) Adverse effects of engineered nanomaterials: exposure, toxicology, and impact on human health (1st edn). London ; Waltham, MA: Elsevier/ Academic Press.

Fetisov, AV Kozhina GA, Estemirova, SKh, Mitrofanov VYa (2015) On the room-temperature aging effects in YBa 2 Cu 3 O 6+δ. Phys C: Supercond Its Appl 515:54–61. https://doi.org/10.1016/ j.physc.2015.05.008.

Fitaroni LB, de Lima JA, Cruz SA, Waldman WR (Jan.2015) Thermal stability of polypropylene–montmorillonite clay nanocomposites: limitation of the thermogravimetric analysis. Polym Degrad Stab 111:102–108. https://doi.org/10.1016/j.polymdegradstab.2014.10.016

Flowers P, Theopold K, Langley R, Robinson WR, OpenStax College, Chemistry

Foroutan M, Zahedi H, Esmaeilian F (2017) Temperature effects on spreading of water nano-droplet on poly(methyl methacrylate): a molecular dynamics simulation study. J Polym Sci Part b: Polym Phys 55(20):1532–1541. https://doi.org/10.1002/polb.24409

Foster DM, Pavloudis Th, Kioseoglou J Palmer RE Atomic-resolution imaging of surface and core melting in individual size-selected Au nanoclusters on carbon. Nat Commun 10(1):2583. https:/ /doi.org/10.1038/s41467-019-10713-z

Fotiadou S et al (Apr 2013) Structure and dynamics of hyperbranched polymer/layered silicate nanocomposites. Macromol 46(7):2842–2855. https://doi.org/10.1021/ma302405q

Fubini B, Fenoglio I, Tomatis M, Turci F (2011) Effect of chemical composition and state of the surface on the toxic response to high aspect ratio nanomaterials. Nanomedicine 6(5):899–920. https://doi.org/10.2217/nnm.11.80

Gao M, Ling B, Yang S, Zhao M (Mar 2005) Flame retardance of wood treated with guanidine compounds characterized by thermal degradation behavior. J Anal Appl Pyrol 73(1):151–156. https://doi.org/10.1016/j.jaap.2005.01.006

Gao L et al (Sep 2016) Effects of the amine/epoxy stoichiometry on the curing behavior and glass transition temperature of MWCNTs-NH 2 /epoxy nanocomposites. Thermochim Acta 639:98–107. https://doi.org/10.1016/j.tca.2016.07.017

Gao X, Lowry GV (Jan 2018) Progress towards standardized and validated characterizations for measuring physicochemical properties of manufactured nanomaterials relevant to nano health and safety risks. NanoImpact 9:14–30. https://doi.org/10.1016/j.impact.2017.09.002

Ghezzi B et al (2021) Hydrogen plasma treatment confers enhanced bioactivity to silicon carbide-based nanowires promoting osteoblast adhesion. Mater Sci Eng, C 121:111772. https://doi.org/ 10.1016/j.msec.2020.111772

Gerberich WW et al (2003) Superhard silicon nanospheres. J Mech Phys Solids 51(6):979–992. https://doi.org/10.1016/S0022-5096(03)00018-8

Gkika DA et al (2020) Patents of nanomaterials related with cancer treatment applications. J Nanopart Res 22(10):335. https://doi.org/10.1007/s11051-020-05052-w

Gkika DA, Vordos N, Magafas L, Mitropoulos AC, Kyzas GZ (Mar 2021) Risk return profile of nanomaterials. J Mol Struct 1228:129740. https://doi.org/10.1016/j.molstruc.2020.129740

Guisbiers G, Mejía-Rosales S, Leonard Deepak F (2012) "Nanomaterial properties: size and shape dependencies." J Nanomater: 1–2. doi: https://doi.org/10.1155/2012/180976.

Guo D, Xie G, Luo J (2014) Mechanical properties of nanoparticles: basics and applications. J Phys d: Appl Phys 47(1):013001. https://doi.org/10.1088/0022-3727/47/1/013001

Habibi-Yangjeh A, Pourbasheer E, Danandeh-Jenagharad M (Apr 2008) Prediction of melting point for drug-like compounds using principal component-genetic algorithm-artificial neural network. Bull Korean Chem Soc 29(4):833–841. https://doi.org/10.5012/BKCS.2008.29.4.833

Hashemifard SA, Ismail AF, Matsuura T (2011) Effects of montmorillonite nano-clay fillers on PEI mixed matrix membrane for CO2 removal. Chem Eng J 170(1):316–325. https://doi.org/ 10.1016/j.cej.2011.03.063

Hatakeyama T, Quinn FX (1999) Thermal analysis: fundamentals and applications to polymer science, 2nd edn. Wiley, Chichester; New York

Higgins SG, Becce M, Belessiotis-Richards A, Seong H, Sero JE, Stevens MM (2020) High-aspect-ratio nanostructured surfaces as biological metamaterials. Adv Mater 32(9):1903862. https://doi.org/10.1002/adma.201903862

Hu D, Cui Z, Fan J, Fan X, Zhang G (2020) Thermal kinetic and mechanical behaviors of pressure-assisted Cu nanoparticles sintering: a molecular dynamics study. Results in Physics 19:103486. https://doi.org/10.1016/j.rinp.2020.103486

Issaadi N, Aït-Mokhtar A, Belarbi R, Hamami A (2018) Effect of variability of porous media properties on drying kinetics: application to cement-based materials. In: *Advances in Multi-Physics and Multi-Scale Couplings in Geo-Environmental Mechanics*, Elsevier, pp 243–289. doi: https://doi.org/10.1016/B978-1-78548-278-6.50008-6

Jain S, Sharma MP (Jan 2011) Thermal stability of biodiesel and its blends: a review. Renew Sustain Energy Rev 15(1):438–448. https://doi.org/10.1016/j.rser.2010.08.022

Jastrzębska AM, Olszyna AR (2015) The ecotoxicity of graphene family materials: current status, knowledge gaps and future needs. J Nanoparticle Res 17(1) Art(1). https://doi.org/10.1007/s11051-014-2817-0.

Jing GY et al (2006) Surface effects on elastic properties of silver nanowires: contact atomic-force microscopy. Phys Rev B 73(23):235409. https://doi.org/10.1103/PhysRevB.73.235409

K SS, I MP R GR (2019) "Mahua oil-based polyurethane/chitosan/nano ZnO composite films for biodegradable food packaging applications." Int J Biol Macromol 124:163–174. https://doi.org/10.1016/j.ijbiomac.2018.11.195

Khan MF et al (Jun.2016) Sol-gel synthesis of thorn-like ZnO nanoparticles endorsing mechanical stirring effect and their antimicrobial activities: potential role as nano-antibiotics. Sci Rep 6(1):27689. https://doi.org/10.1038/srep27689

Khodadadi JM, Hosseinizadeh SF (2007) Nanoparticle-enhanced phase change materials (NEPCM) with great potential for improved thermal energy storage. Int Commun Heat Mass Transfer 34(5):534–543. https://doi.org/10.1016/j.icheatmasstransfer.2007.02.005

Khoramishad H, Ashofteh RS, Pourang H, Berto F (2018) Experimental investigation of the influence of temperature on the reinforcing effect of graphene oxide nano-platelet on nanocomposite adhesively bonded joints. Theoret Appl Fract Mech 94:95–100. https://doi.org/10.1016/j.tafmec.2018.01.010

Koçak A, Karasu B (Jan.2018) General evaluations of nanoparticles. El-Cezeri Fen Ve Mühendislik Dergisi 5(1):191–236. https://doi.org/10.31202/ecjse.361663

Krishnaswamy K, Orsat V, Thangavel K (2012) Synthesis and characterization of nano-encapsulated catechin by molecular inclusion with beta-cyclodextrin. J Food Eng 111(2):255–264. https://doi.org/10.1016/j.jfoodeng.2012.02.024

Kumar NN, Rama Sreekanth PS, Kanagaraj S (2013) Effect of γ-irradiation on thermal properties of MWCNTs reinforced HDPE. In: Giri PK, Goswami DK, Perumal A (ed) Advanced nanomaterials and nanotechnology, vol 143Berlin, Heidelberg: Springer Berlin Heidelberg. pp 409–420. https://doi.org/10.1007/978-3-642-34216-5_40.

Kuo Z-K et al (2018) Hydrophilic films: How hydrophilicity affects blood compatibility and cellular compatibility. Adv Polym Technol 37(6):1635–1642. https://doi.org/10.1002/adv.21820

Kyzas GZ, Travlou NA, Deliyanni EA (2014) The role of chitosan as nanofiller of graphite oxide for the removal of toxic mercury ions. Colloids Surf, B 113:467–476. https://doi.org/10.1016/j.colsurfb.2013.07.055

Landsiedel R, Ma-Hock L, Wiench K, Wohlleben W, Sauer UG (2017) Safety assessment of nanomaterials using an advanced decision-making framework, the DF4nanoGrouping. J Nanopart Res 19(5):171. https://doi.org/10.1007/s11051-017-3850-6

Law K-Y (Feb.2014) Definitions for hydrophilicity, hydrophobicity, and superhydrophobicity: getting the basics Right. J Phys Chem Lett 5(4):686–688. https://doi.org/10.1021/jz402762h

Liang GD, Tjong SC (2006) Electrical properties of low-density polyethylene/multiwalled carbon nanotube nanocomposites. Mater Chem Phys 100(1):132–137. https://doi.org/10.1016/j.matchemphys.2005.12.021

Linkov I, Steevens J (eds) Nanomaterials: risks and benefits. Dordrecht: Springer : In cooperation with NATO Public Diplomacy Division.

Lippmann M (1990) Effects of fiber characteristics on lung deposition, retention, and disease. Environ Health Perspect 88:311–317. https://doi.org/10.1289/ehp.9088311

Lee S-W et al (2014) Effect of temperature on the growth of silver nanoparticles using plasmon-mediated method under the irradiation of green LEDs. Materials 7(12):7781–7798. https://doi.org/10.3390/ma7127781

Leitner J, Sedmidubský D, Lojka M, Jankovský O (2020) The effect of nanosizing on the oxidation of partially oxidized copper nanoparticles. Materials 13(12):2878. https://doi.org/10.3390/ma13122878

Lv G, Wu S (Sep.2012) Analytical pyrolysis studies of corn stalk and its three main components by TG-MS and Py-GC/MS. J Anal Appl Pyrol 97:11–18. https://doi.org/10.1016/j.jaap.2012.04.010

Liu K, Cao X, Bai Q, Wen H, Gu Z (2009) Relationships between physical properties of brown rice and degree of milling and loss of selenium. J Food Eng 94(1):69–74. https://doi.org/10.1016/j.jfoodeng.2009.03.001

Liu H, Chaudhary D, Yusa S, Tadé MO (Feb 2011) Glycerol/starch/Na+-montmorillonite nanocomposites: a XRD, FTIR, DSC and 1H NMR study. Carbohyd Polym 83(4):1591–1597. https://doi.org/10.1016/j.carbpol.2010.10.018

Liu, X, Harper Tang, Steevens J, Xu R (2013) NEIMiner: nanomaterial environmental impact data miner. *IJN*: 15. https://doi.org/10.2147/IJN.S40974

Liang YL, Pearson RA (2009) Toughening mechanisms in epoxy–silica nanocomposites (ESNs). Polymer 50(20):4895–4905. https://doi.org/10.1016/j.polymer.2009.08.014

Lu J, Chan HL, Chen AY, Kou HN (2011) Mechanics of high strength and high ductility materials. Procedia Engg 10:2202–2207. https://doi.org/10.1016/j.proeng.2011.04.364

Mach P, Geczy A, Polanský R, Bušek D (Mar 2019) Glass transition temperature of nanoparticle-enhanced and environmentally stressed conductive adhesive materials for electronics assembly. J Mater Sci: Mater Electron 30(5):4895–4907. https://doi.org/10.1007/s10854-019-00784-5

Madhukar K et al (2014) Thermal properties of single walled carbon nanotubes composites of polyamide 6/poly(methyl methacrylate) blend system. J Therm Anal Calorim 115(1):345–354. https://doi.org/10.1007/s10973-013-3320-z

Mansfield E (2015) Recent advances in thermal analysis of nanoparticles. In: Modeling, characterization, and production of nanomaterials, Elsevier, pp 167–178. doi: https://doi.org/10.1016/B978-1-78242-228-0.00006-5

Magro M, De Liguoro M, Franzago E, Baratella D, Vianello F (2018) The surface reactivity of iron oxide nanoparticles as a potential hazard for aquatic environments: a study on Daphnia magna adults and embryos. Sci Rep 8(1):13017. https://doi.org/10.1038/s41598-018-31483-6

Martínez-Esaín J et al (2019) Using evolved gas analysis – mass spectrometry to characterize adsorption on a nanoparticle surface. Nanoscale Adv. 1(7):2740–2747. https://doi.org/10.1039/C9NA00098D

Marques AC et al (2015) Office paper platform for bioelectrochromic detection of electrochemically active bacteria using tungsten trioxide nanoprobes. Sci Rep 5(1):9910. https://doi.org/10.1038/srep09910

Menczel, JD Prime, RB (eds) (2009) Thermal analysis of polymers. Hoboken, NJ, USA: John Wiley & Sons, Inc. https://doi.org/10.1002/9780470423837

Millot C, Fillot L-A, Lame O, Sotta P, Seguela R (Oct.2015) Assessment of polyamide-6 crystallinity by DSC: temperature dependence of the melting enthalpy. J Therm Anal Calorim 122(1):307–314. https://doi.org/10.1007/s10973-015-4670-5

Minelli C et al (2018) Measuring the size and density of nanoparticles by centrifugal sedimentation and flotation. Anal Methods 10(15):1725–1732. https://doi.org/10.1039/C8AY00237A

Mishra PC, Mukherjee S, Nayak SK, Panda A (2014) A brief review on viscosity of nanofluids. Int Nano Lett 4(4):109–120. https://doi.org/10.1007/s40089-014-0126-3

Mohandes F, Salavati-Niasari M (2013) Application of a new coordination compound for the preparation of AgI nanoparticles. Mater Res Bull 48(10):3773–3782. https://doi.org/10.1016/j.materresbull.2013.05.094

Mohandes F, Salavati-Niasari M (Jan.2013) Sonochemical synthesis of silver vanadium oxide micro/nanorods: solvent and surfactant effects. Ultrason Sonochem 20(1):354–365. https://doi.org/10.1016/j.ultsonch.2012.05.002

Morais DDS, Barbosa R, Medeiros KM, Araújo EM, de Mélo TJA (2014) Preparation of poly(lactic acid)/bentonite clay bio-nanocomposite. MSF 775–776:233–237. https://doi.org/10.4028/www.scientific.net/MSF.775-776.233

Mourdikoudis S, Pallares RM, Thanh NTK (2018) Characterization techniques for nanoparticles: comparison and complementarity upon studying nanoparticle properties. Nanoscale 10(27):12871–12934. https://doi.org/10.1039/C8NR02278J

Muhannad Mahdi Abd (2013) "Mechanical properties of micro and nano TiO2/epoxy composites. https://doi.org/10.13140/RG.2.2.30316.31360

Muthee DK, Dejene BF (2021) Effect of annealing temperature on structural, optical, and photocatalytic properties of titanium dioxide nanoparticles. Heliyon 7(6):e07269. https://doi.org/10.1016/j.heliyon.2021.e07269

Nanda KK, Sahu SN, Behera SN (2002) Liquid-drop model for the size-dependent melting of low-dimensional systems. Phys Rev A 66(1):013208. https://doi.org/10.1103/PhysRevA.66.013208

Navya PN, Daima HK (2016) Rational engineering of physicochemical properties of nanomaterials for biomedical applications with nanotoxicological perspectives. Nano Converg 3(1) Art(1). https://doi.org/10.1186/s40580-016-0064-z

Nazirkar G, Bhanushali S, Singh S, Pattanaik B, Raj N (2014) Effect of anatase titanium dioxide nanoparticles on the flexural strength of heat cured poly methyl methacrylate resins: an in-vitro study. J Indian Prosthodont Soc 14(S1):144–149. https://doi.org/10.1007/s13191-014-0385-8

Nguyen TA, Nguyen TH, Nguyen TV, Thai H, Shi X (2016) Effect of nanoparticles on the thermal and mechanical properties of epoxy coatings." j nanosci nanotechnol 16(9):9874–9881. https://doi.org/10.1166/jnn.2016.12162.

Nourani M, Hamdami N, Keramat J, Moheb A, Shahedi M (Apr 2016) Thermal behavior of paraffin-nano-Al2O3 stabilized by sodium stearoyl lactylate as a stable phase change material with high thermal conductivity. Renewable Energy 88:474–482. https://doi.org/10.1016/j.renene.2015.11.043

Ovid'ko IA, Valiev RZ, Zhu YT (2018) Review on superior strength and enhanced ductility of metallic nanomaterials. Prog Mater Sci 94:462–540. https://doi.org/10.1016/j.pmatsci.2018.02.002.

Olad A, Hayasi M (Oct.2011) A comparative study of polystyrene/layered silicate nanocomposites, synthesized by emulsion and bulk polymerization methods. Polym-Plast Technol Eng 50(14):1487–1495. https://doi.org/10.1080/03602559.2011.593083

Paik P, Kar KK, Deva D, Sharma A (2007) Measurement of mechanical properties of polymer nanospheres by atomic force microscopy: effects of particle size. Micro Nano Lett. 2(3):72. https://doi.org/10.1049/mnl:20070030

Panaitescu DM, Frone AN, Nicolae C (Dec 2013) Micro- and nano-mechanical characterization of polyamide 11 and its composites containing cellulose nanofibers. Eur Polymer J 49(12):3857–3866. https://doi.org/10.1016/j.eurpolymj.2013.09.031

Park J-J, Lee C-H, Lee J-Y, Kim H-D (Jun.2011) Preparation of epoxy/micro- and nano- composites by electric field dispersion process and its mechanical and electrical properties. IEEE Trans. Dielect. Electr. Insul. 18(3):667–674. https://doi.org/10.1109/TDEI.2011.5931051

Patidar D, Agrawal S, Saxena NS (Dec 2011) Glass transition activation energy of CdS/PMMA nano-composite and its dependence on composition of CdS nano-particles. J Therm Anal Calorim 106(3):921–925. https://doi.org/10.1007/s10973-010-1150-9

Patra JK, Baek K-H (2014) Green Nanobiotechnology: factors affecting synthesis and characterization techniques. J Nanomater 2014:1–12. https://doi.org/10.1155/2014/417305

Patra S, Ajayan PM, Narayanan TN (2020) Dynamic mechanical analysis in materials science: the Novice's Tale. Oxf Open Mater Sci 1(1):itaa001. https://doi.org/10.1093/oxfmat/itaa001.

Physical fundamentals of nanomaterials. Elsevier 2018. https://doi.org/10.1016/C2012-0-02139-4

Powers KW, Palazuelos M, Moudgil BM, Roberts SM (2007) Characterization of the size, shape, and state of dispersion of nanoparticles for toxicological studies. Nanotoxicology 1(1):42–51. https://doi.org/10.1080/17435390701314902

Qi WH (Nov.2005) Size effect on melting temperature of nanosolids. Physica B 368(1–4):46–50. https://doi.org/10.1016/j.physb.2005.06.035

Radwan A et al (2013) Mechanistic understanding of food effects: water diffusivity in gastrointestinal tract is an important parameter for the prediction of disintegration of solid oral dosage forms. Mol Pharmaceutics 10(6):2283–2290. https://doi.org/10.1021/mp3006209

Ramos M et al (2013) Hardness and elastic modulus on six-fold symmetry gold nanoparticles. Materials 6(1):198–205. https://doi.org/10.3390/ma6010198

Rasmussen K et al (2018) Physico-chemical properties of manufactured nanomaterials - Characterisation and relevant methods. An outlook based on the OECD testing programme. Regul Toxicol Pharmacol 92:8–28. https://doi.org/10.1016/j.yrtph.2017.10.019

Ren XD et al (2018) Thermal stability of surface nano-crystallization layer in AZ91D magnesium alloy induced by laser shock peening. Surf Coat Technol 334:182–188. https://doi.org/10.1016/j.surfcoat.2017.09.037

Saba N, Paridah MT, Abdan K, Ibrahim NA (2016) Dynamic mechanical properties of oil palm nano filler/kenaf/epoxy hybrid nanocomposites. Constr Build Mater 124:133–138. https://doi.org/10.1016/j.conbuildmat.2016.07.059

Sahraee S, Milani JM, Ghanbarzadeh B, Hamishehkar H (Mar 2017) Effect of corn oil on physical, thermal, and antifungal properties of gelatin-based nanocomposite films containing nano chitin. LWT Food Sci Technol 76:33–39. https://doi.org/10.1016/j.lwt.2016.10.028

Salavati-Niasari M, Mir N, Davar F (Apr.2010) A novel precursor for synthesis of metallic copper nanocrystals by thermal decomposition approach. Appl Surf Sci 256(12):4003–4008. https://doi.org/10.1016/j.apsusc.2010.01.067

Sar DK, Nanda KK (2010) Melting and superheating of nanowires—A nanotube approach. Nanotechnol 21(20):205701. https://doi.org/10.1088/0957-4484/21/20/205701

Schmidt M, Kusche R, von Issendorff B, Haberland H (1998) Irregular variations in the melting point of size-selected atomic clusters. Nat 393(6682):238–240. https://doi.org/10.1038/30415

Seifi H, Gholami T, Seifi S, Ghoreishi SM, Salavati-Niasari M (Aug.2020) A review on current trends in thermal analysis and hyphenated techniques in the investigation of physical, mechanical and chemical properties of nanomaterials. J Anal Appl Pyrol 149:104840. https://doi.org/10.1016/j.jaap.2020.104840

Shabanian M, Faghihi K, Raeisi A, Varvanifarahani M, Khonakdar HA, Wagenknecht U (Jul.2014) New poly(ether-imide)/MWCNT nanocomposite: flammability, thermal and electrical properties. J Therm Anal Calorim 117(1):293–299. https://doi.org/10.1007/s10973-014-3682-x

Shah MP (2020) Microbial bioremediation & biodegradation. Springer

Shah MP (2021a) Removal of refractory pollutants from wastewater treatment plants. CRC Press

Shah Maulin P (2021b) Removal of emerging contaminants through microbial processes. Springer

Sharma S, Jaiswal S, Duffy B, Jaiswal A (Mar 2019) Nanostructured materials for food applications: spectroscopy, microscopy and physical properties. Bioengineering 6(1):26. https://doi.org/10.3390/bioengineering6010026

Shukla S et al (Apr.2015) The impact of aspect ratio on the biodistribution and tumor homing of rigid soft-matter nanorods. Adv Healthcare Mater 4(6):874–882. https://doi.org/10.1002/adhm.201400641

Shen L, Che Q, Li H, Zhang X (2014) Mesoporous $NiCo_2O_4$ nanowire arrays grown on carbon textiles as binder-free flexible electrodes for energy storage. Adv Funct Mater 24(18):2630–2637. https://doi.org/10.1002/adfm.201303138

Si M, Feng J, Hao J, Xu L, Du J (Feb.2014) Synergistic flame retardant effects and mechanisms of nano-Sb2O3 in combination with aluminum phosphinate in poly(ethylene terephthalate). Polym Degrad Stab 100:70–78. https://doi.org/10.1016/j.polymdegradstab.2013.12.023

Singh M, Hlabana KK, Singhal S, Devlal K (Jul.2016) Grain-size effects on the thermal conductivity of nanosolids. Journal of Taibah University for Science 10(3):375–380. https://doi.org/10.1016/j.jtusci.2015.04.006

Shivam V, Shadangi Y, Basu J, Mukhopadhyay NK (2019) Alloying behavior and thermal stability of mechanically alloyed nano AlCoCrFeNiTi high-entropy alloy. J Mater Res 34(5):787–795. https://doi.org/10.1557/jmr.2019.5

Sienkiewicz A, Czub P (2020) Flame retardancy of biobased composites—Research development. Materials 13(22):5253. https://doi.org/10.3390/ma13225253

Soares S, Sousa J, Pais A, Vitorino C (2018) Nanomedicine: principles, properties, and regulatory issues. Front Chem 6:360. https://doi.org/10.3389/fchem.2018.00360

Sohal IS, Cho YK, O'Fallon KS, Gaines P, Demokritou P, Bello D (2018) Dissolution behavior and biodurability of ingested engineered nanomaterials in the gastrointestinal environment. ACS Nano 12(8):8115–8128. https://doi.org/10.1021/acsnano.8b02978

Soltanzadeh N, Morsali A (Jan.2010) Sonochemical synthesis of a new nano-structures bismuth(III) supramolecular compound: New precursor for the preparation of bismuth(III) oxide nano-rods and bismuth(III) iodide nano-wires. Ultrason Sonochem 17(1):139–144. https://doi.org/10.1016/j.ultsonch.2009.05.003

Song G, Ma S, Tang G, Yin Z, Wang X (May2010) Preparation and characterization of flame retardant form-stable phase change materials composed by EPDM, paraffin and nano magnesium hydroxide. Energy 35(5):2179–2183. https://doi.org/10.1016/j.energy.2010.02.002

Sokolov SV, Tschulik K, Batchelor-McAuley C, Jurkschat K, Compton RG (2015) Reversible or not? distinguishing agglomeration and aggregation at the nanoscale. Anal Chem 87(19):10033–10039. https://doi.org/10.1021/acs.analchem.5b02639

Subasinghe A, Das R, Bhattacharyya D (2016) Study of thermal, flammability and mechanical properties of intumescent flame retardant PP/kenaf nanocomposites. Int J Smart Nano Mater 7(3):202–220. https://doi.org/10.1080/19475411.2016.1239315

Sumesh KR, Kanthavel K, Ajithram A, Nandhini P (Sep 2019) Bioalumina nano powder extraction and its applications for sisal, coir and banana hybrid fiber composites: mechanical and thermal properties. J Polym Environ 27(9):2068–2077. https://doi.org/10.1007/s10924-019-01496-x

Sutter EA, Tong X, Jungjohann K, Sutter PW (2013) Oxidation of nanoscale Au-In alloy particles as a possible route toward stable Au-based catalysts. Proc Natl Acad Sci 110(26):10519–10524. https://doi.org/10.1073/pnas.1305388110

Sun J, Wang R (2013) Carbon nanotube transparent electrode. In: Suzuki S (ed) Syntheses and applications of carbon nanotubes and their composites. InTech. doi: https://doi.org/10.5772/51783

Suzuki K et al (2018) Nano-crystallization of amorphous alloys by ultra-rapid annealing: an effective approach to magnetic softening. J Alloy Compd 735:613–618. https://doi.org/10.1016/j.jallcom.2017.11.110

Tang Y, Su D, Huang X, Alva G, Liu L, Fang G (Oct 2016) Synthesis and thermal properties of the MA/HDPE composites with nano-additives as form-stable PCM with improved thermal conductivity. Appl Energy 180:116–129. https://doi.org/10.1016/j.apenergy.2016.07.106

Tan S, Sherman RL, Ford WT (2004) Nanoscale compression of polymer microspheres by atomic force microscopy. Langmuir 20(17):7015–7020. https://doi.org/10.1021/la049597c

Tanaka Y, Matsubara R, Furukawa K, Satonaka S, Kasaoka S (2019) The influence of viscosity-enhancing agents on oral absorption of drugs. Pharmazie 11:661–664. https://doi.org/10.1691/ph.2019.9097

Tanzi, MC (2017) "Characterization of thermal properties and crystallinity of polymer biomaterials. In: Characterization of Polymeric Biomaterials, Elsevier, pp 123–146. https://doi.org/10.1016/B978-0-08-100737-2.00006-6.

Tewary VK (ed) (2015) Modeling, characterization, and production of nanomaterials. Elsevier, Amsterdam

Ting TH, Yap YF, Nguyen N-T, Wong TN, Kiong Chai JC Yobas L (2006) Active control for droplet-based microfluidics. Adelaide, Australia, Dec. 2006, p 64160E. https://doi.org/10.1117/12.696496

Travlou NA, Kyzas GZ, Lazaridis NK, Deliyanni EA (2013) Graphite oxide/chitosan composite for reactive dye removal. Chem Eng J 217:256–265. https://doi.org/10.1016/j.cej.2012.12.008

Vahidi G, Bajwa DS, Shojaeiarani J, Stark N, Darabi A (2021) Advancements in traditional and nanosized flame retardants for polymers—A review. J Appl Polym Sci 138(12):50050. https://doi.org/10.1002/app.50050

Vanaja M, Gnanajobitha G, Paulkumar K, Rajeshkumar S, Malarkodi C, Annadurai G (2013) Phytosynthesis of silver nanoparticles by Cissus quadrangularis: influence of physicochemical factors. J Nanostruct Chem 3(1):17. https://doi.org/10.1186/2193-8865-3-17

Van Doren EA, De Temmerman P-JR, Francisco M, Mast J (2011) Determination of the volume-specific surface area by using transmission electron tomography for characterization and definition of nanomaterials. J Nanobiotechnol 9(1):17. https://doi.org/10.1186/1477-3155-9-17

van Leeuwen HP, Buffle J, Duval JFL, Town RM (2013) Understanding the extraordinary ionic reactivity of aqueous nanoparticles. Langmuir 29(33):10297–10302. https://doi.org/10.1021/la401955x

Wang H-Y, Lu S-S (2013) Study on thermal properties of phase change material by an optical DSC system. Appl Therm Eng 60(1–2):132–136. https://doi.org/10.1016/j.applthermaleng.2013.06.055

Wang C, Cai L, Shi SQ, Wang G, Cheng H, Zhang S (2019) Thermal and flammable properties of bamboo pulp fiber/high-density polyethylene composites: influence of preparation technology, nano calcium carbonate and fiber content. Renewable Energy 134:436–445. https://doi.org/10.1016/j.renene.2018.09.051

Wang N, Si Y, Yu J, Fong H, Ding B (Apr 2017) Nano-fiber/net structured PVA membrane: effects of formic acid as solvent and crosslinking agent on solution properties and membrane morphological structures. Mater Des 120:135–143. https://doi.org/10.1016/j.matdes.2017.02.007

Wu G, Liu D, Chen J, Liu G, Kong Z (Feb 2019) Preparation and properties of super hydrophobic films from siloxane-modified two-component waterborne polyurethane and hydrophobic nano SiO2. Prog Org Coat 127:80–87. https://doi.org/10.1016/j.porgcoat.2018.06.016

Wang X, Rathore R, Songtipya P, Jimenez-Gasco MdelM, Manias E, Wilkie CA (2011) EVA-layered double hydroxide (nano)composites: Mechanism of fire retardancy. Polym Degrad Stab 96(3):301–313. https://doi.org/10.1016/j.polymdegradstab.2010.03.014

Wohlleben W (ed) (2015) Safety of nanomaterials along their lifecycle: release, exposure, and human hazards. CRC Press/Taylor & Francis Group, Boca Raton

Wu R, Nguyen T, Marquart G, Miesen T, Mau T, Mackiewicz M (2014) A facile route to tailoring peptide-stabilized gold nanoparticles using glutathione as a synthon. Molecules 19(5):6754–6775. https://doi.org/10.3390/molecules19056754

Wu W, Wan Z, Zhu M, Zhang D (2016) A facile route to aqueous phase synthesis of mesoporous alumina with controllable structural properties. Microporous Mesoporous Mater 223:203–212. https://doi.org/10.1016/j.micromeso.2015.11.004

Wu Q, Miao W, Zhang Y, Gao H, Hui D (2020) Mechanical properties of nanomaterials: a review. Nanotechnol Rev 9(1):259–273. https://doi.org/10.1515/ntrev-2020-0021

Xanthopoulou et al (2019) Effects of precursor concentration in solvent and nanomaterials room temperature aging on the growth morphology and surface characteristics of Ni–NiO nanocatalysts produced by dendrites combustion during SCS. Appl Sci 9(22):4925. https://doi.org/10.3390/app9224925

Xie W, Pan W-P (2001) No title found. J Therm Anal Calorim 65(3):669–685. https://doi.org/10.1023/A:1011946707342

Yarema M et al (2017) Mapping the atomistic structure of graded core/shell colloidal nanocrystals. Sci Rep 7(1):11718. https://doi.org/10.1038/s41598-017-11996-2

Yin F, Wang ZW, Palmer RE (2012) Ageing of mass-selected Cu/Au and Au/Cu core/shell clusters probed with atomic resolution. J Exp Nanosci 7(6):703–710. https://doi.org/10.1080/17458080.2012.710856

Ying J, Wan C, Jiang C, Li Y (2001) Preparation and characterization of high-density spherical LiNi0.8Co0.2O2 cathode material for lithium secondary batteries. J Power Sources 99(1–2):78–84. https://doi.org/10.1016/S0378-7753(01)00477-3

Yu M (2000) Strength and breaking mechanism of multiwalled carbon nanotubes under tensile load. Sci 287(5453):637–640. https://doi.org/10.1126/science.287.5453.637

Zha L, Fang Z (Jul.2010) Polystyrene/$CaCO_3$ composites with different $CaCO_3$ radius and different nano-$CaCO_3$ content-structure and properties. Polym Compos 31(7):1258–1264. https://doi.org/10.1002/pc.20915

Zhang H (2015) Ultrathin two-dimensional nanomaterials. ACS Nano 9(10):9451–9469. https://doi.org/10.1021/acsnano.5b05040

Zhang J, Panwar A, Bello D, Isaacs JA, Jozokos T, Mead J (2018) The effects of recycling on the structure and properties of carbon nanotube-filled polycarbonate. Polym Eng Sci 58(8):1278–1284. https://doi.org/10.1002/pen.24692

Zhao W et al (2017) Nanoscale manipulation of membrane curvature for probing endocytosis in live cells. Nature Nanotech 12(8):750–756. https://doi.org/10.1038/nnano.2017.98

Zhang H, Tang L-C, Zhang Z, Friedrich K, Sprenger S (2008) Fracture behaviours of in situ silica nanoparticle-filled epoxy at different temperatures. Polymer 49(17):3816–3825. https://doi.org/10.1016/j.polymer.2008.06.040

Zhu YT, Wu XL (2018) Ductility and plasticity of nanostructured metals: differences and issues. Mater Today Nano 2:15–20. https://doi.org/10.1016/j.mtnano.2018.09.004

Bioremediation of Industrial Wastewater: An Overview with Recent Developments

Pranjali Mahamuni-Badiger, Pratikshkumar R. Patel, Pooja M. Patil, Ranjit Gurav, Sangchul Hwang, and Maruti J. Dhanavade

1 Introduction

Water is the most important yet scarce resource throughout the world. About 97% of Earth is covered with seawater which is unsuitable for human use, whereas only 3% of the fresh water is suitable for consumption. As a result, about 1.6 billion people worldwide are not getting safe water for their daily use (Sharma et al. 2020). In addition, wastewater from leather, textile, pharmaceutical, seafood processing, cosmetics, paint, plastic, and paper industries are polluting natural water reservoirs like rivers, lakes, etc. (Choi et al. 2020; Gurav et al. 2021a; Cho et al. 2021). This process resulted in polluting freshwater sources with varying amounts of dyes, metals/metalloids,

The original version of this chapter was revised: Author's proof corrections have been updated. The correction to this chapter is available at
https://doi.org/10.1007/978-981-99-2598-8_21

P. Mahamuni-Badiger
Department of Microbiology, Rayat Shikshan Sanstha's, S.M. Joshi College, Pune, India

P. R. Patel
Council of Scientific and Industrial Research-National Chemical Laboratory, Pune, India

P. M. Patil
Department of Environment Management, Chhatrapati Shahu Institute of Business Education and Research, Kolhapur, India

R. Gurav (✉) · S. Hwang
Ingram School of Engineering, Texas State University, San Marcos, TX 78666, USA
e-mail: ranjit.g@txstate.edu

S. Hwang
e-mail: ranjit.g@txstate.edu

M. J. Dhanavade (✉)
Department of Microbiology, Bharati Vidyapeeth's Dr. Patangrao Kadam Mahavidyalaya, Sangli, India
e-mail: maruti.dhanavade@bharatividyapeeth.edu.in

M. P. Shah (ed.), *Advanced and Innovative Approaches of Environmental Biotechnology in Industrial Wastewater Treatment*,
https://doi.org/10.1007/978-981-99-2598-8_15

salts, and organic pollutants. Similarly, tons of synthetic dyes are released in natural water bodies containing about 60% of azo dyes having a negative impact on the human and environment (Gurav et al. 2021b). This also significantly compromises the aesthetic quality of water bodies, increases the biochemical and chemical oxygen demand (BOD and COD), impairs photosynthesis, inhibits plant growth; and enters into the food chain causing a potential threat to the ecosystem and human health. The presence of dyes in water bodies not only add dangerous chemicals such as aromatic amines but also creates a toxic, allergic, carcinogenic, and mutagenic effect on living organisms (Suryawanshi et al. 2023; Vyavahare et al. 2019).

A variety of approaches are available for wastewater treatment including physical, chemical, and biological or a combination of them, out of them the biological treatment is the most eco-friendly and cost-effective method (Gurav et al. 2021b, 2022). Other physical–chemical methods generates sludge and are associated with high costs together with inputs or operations. Bioremediation process for pollutant removal is a collective phenomenon involving use of biological systems to either restore or clean-up contaminated sites by reducing the pollutant levels up to undetectable, nontoxic, or acceptable ranges (Gurav et al. 2016, 2017). Moreover, bioremediation can be used along with physical and chemical methods to handle different groups of environmental pollutants (Liu et al. 2015). Although the process of bioremediation was invented by George M. Robinson, the concept of bioremediation was first applied on a big scale for the cleaning of the Sun Oil pipeline spill at Ambler, Pennsylvania in 1972 (Mahamuni et al. 2019). The following are the protocols developed by Environmental Protection Agency (EPA) for bioremediation.

The different microbial species like bacteria, fungi, and algae can adsorb and/ or degrade contaminants using the extracellular enzymes such as azoreductases, laccases, and peroxidases via oxidization, immobilization, or transformation (Hlihor et al. 2017). Although monoculture strains are efficient to degrade organic pollutants, microbial consortia, and combining bacteria with fungi also efficiently degrade contaminants (Mahamuni-Badiger et al. 2020a). Various mechanisms including bioaccumulation, biodegradation pathways, and different modes for biosorption have also been investigated for the removal of pollutants. The modern bioremediation approaches include the constant exploration of novel microorganisms having a strong potential to remediate pollutants. In addition, studies that divulge the basis of microbial metabolism, understanding what types of organisms play role in bioremediation, engineering techniques for supplying stimulative supplementary material for the growth of microorganisms, and research advances to increase the availability of contaminants to microbes together would increase the efficiency of bioremediation. The use of genetically modified strains addressing these issues will in turn push the boundaries of bioremediation.

In this chapter, we highlighted and discussed the bioremediation process, basic strategies for bioremediation, potential microbes used in the processes, recent advances in the bioremediation technology along future goals to achieve maximum degradation. Lastly, the chapter summarizes new research advances for expanding the future capabilities of bioremediation.

2 Industrial Wastewater Pollution

Industrial wastes disposal pose major challenges to human health and the environment. Pollution from industrial wastes includes a variety of persistent compounds, which have been linked to toxic, allergenic, carcinogenic, and mutagenic effects on living organisms including species extinction (Medfu Tarekegn et al. 2020). There are commonly five different kinds of contaminants such as munitions wastes, organic solvents, halogenated aromatic hydrocarbon, pesticides, and polycyclic aromatic hydrocarbons (PAHs). At present, over half of all municipal solid waste contaminates drinking water and disrupts natural ecosystems.

To treat industrial effluents, traditional physico-chemical methods including advanced oxidation processes, adsorption, ozonation, membrane filtration, photocatalytic degradation, coagulation and flocculation, electrocoagulation, photoelectrocatalysis, and electrochemical oxidation have been used (Gurav et al. 2019a, b, 2020; Gurav et al. 2021c; Kim et al. 2020). But these methods require complex infrastructures and high economic and energy costs. Also, there are some environmental effects associated with the use of these methods like the production of secondary pollutants or the generation of large amounts of contaminated sludge and toxic by-products. Hence, it is necessary to find more sustainable alternatives for the treatment of industrial waste. Developing economically viable approaches for sustainable waste management and treating pollutants in situ will reduce the costs of achieving sustainability targets which is also important for ensuring a high quality of life (Srivastava et al. 2014).

In recent years, wastewater treatment systems focusing on minimizing the impact on human health and the environment and allowing the reuse of wastewater in response to the issue of the growing demand for water and the depletion of natural water sources have pushed technological development toward the use of microorganisms. Microorganisms can degrade industrial effluents utilizing biosorption mechanisms and/or enzymatic degradation and their effectiveness depends on various factors such as survival, adaptability, activity of the microorganism, and chemical structure of the pollutant. The microbial community including the bacterial, fungal, and algal species degrade contaminants using the extracellular enzymes such as azoreductases, laccases, and peroxidases via oxidization, immobilization, or transformation. Hence, bioremediation is the most effective, economically feasible, and eco-friendly way to treat contaminated environment.

3 Basic Concept of Bioremediation

Utilization of microorganisms like bacteria, fungi, and algae are employed to clean environmental pollutants by the different metabolic processes is termed as bioremediation. The principle of bioremediation is the degradation and conversion of toxic pollutants to less hazardous forms. Microorganisms can decrade and/or biotransform certain residual effluents, pesticides, and hydrocarbons. The chemicals in such compounds supply the required amount of carbon along with some other essential factors required to promote the growth of these microbes. The enzymatic metabolic pathways of microorganisms can increase the rate of biochemical reactions to enhance the degradation of the pollutant. Hence, bioremediation involves the microbial degradation, immobilization, or detoxification of different types of pollutants present in the wastes either by aerobic or anaerobic microorganisms (Azubuike et al. 2016a, b).

Remediation of contaminated areas with the use of microbes (bioremediation) is a powerful and reliable method with eco-friendly features. Recent advancements in bioremediation techniques are reaching the goal to keep a pollution-free environment in the past two decades. Researchers have developed different bioremediation techniques that restore cleaner and pollution-free environments wherein, the ability of microorganisms could be enhanced by providing optimum conditions for their growth or by employing genetically modified microorganisms. This technology has been implemented to eradicate environmentally hazardous chemicals and detoxify them into nontoxic forms.

4 General Strategies Used in Bioremediation

The bioremediation process can be classified based on the organisms and strategies applied.

4.1 *Classification of Bioremediation Based on Organisms Used*

Based on the organism used, bioremediation is classified mainly into three types:

a. *Microbial Remediation*
b. *Phytoremediation*
c. *Mycoremediation*

4.1.1 Microbial Remediation

The use of microorganisms in bioremediation to convert the contaminants into less toxic forms is referred to as microbial remediation. In the bioremediation process, breakdown of compounds is carried out through microbial metabolism. The microbial ability to grow at extreme temperatures, salinity, pH, different hazardous chemicals or effluent concentration are the important factors responsible for effective microbial remediation. The microorganisms can degrade effluent by employing several types of mechanisms like biosorption and enzymatic degradation. Biosorption is dependent on the attraction between the toxic compound and the components of microbial cell wall or exopolysaccharides. However, the factors such as pH, temperature, ionic strength, contact time, adsorbent concentrations, chemical structure, and type of microorganism are mainly affecting the biosorption process. However, in enzymatic degradation, the reduction reaction was mediated by azoreductase enzymes. Whereas oxidative degradation was catalyzed by enzymes from peroxidizes and phenoloxidases classes like laccase, tyrosinase, N-demethylase manganese peroxidase, lignin peroxidase (Solano et al. 2015).

In azo dye degradation,the azoreductase and laccase enzymes transfer electrons to the azo bond present on dye-producing aromatic amines. Laccases are nothing but copper oxidases that can degrade dyes aerobically by the process of oxidation using redox mediators to increase the reaction rate in which hydrogen atom is removed from the hydroxyl group and replaced with phenolic substrates and aromatic amines (Tišma et al. 2020). Whereas the bacterial peroxidase enzymes are involved in the process of dye degradation using H_2O_2 as a terminal electron. The mode of action of peroxidases is similar to laccase and can degrade dye without producing toxic aromatic amines.

The effectiveness of microorganisms for dye or any other compound degradation depends on various factors such as pH, pollutant concentration, temperature, growth, adaptability, activity of the microorganism, and pollutant chemical structure. Any microorganism used for bioremediation purposes may contain the resistant genotype for the pollutant. In uptake and reflux of metals, the biosorption, intracellular assimilation, immobilization, complexation, precipitation, and their release are included (Stelting et al. 2012). Two groups of microbes are employed in bioremediation are aerobic, and anaerobic. Aerobic bacteria need an oxygen source, and the by-products at the end of the process are typically water, salts, and carbon dioxide; Eg. *Pseudomonas, Sphingomonas, Flavobacterium, Nocardia, Acinetobacter, Mycobacterium,* and *Rhodococcus.* These microbes have been reported to degrade pesticides, hydrocarbons, alkanes, and PAHs compounds. The anaerobic bioremediation processes are carried out in the absence of oxygen, and by-products of this process are methane, sulfides, hydrogen gas, and elemental sulfur. These reports and studies suggested that microorganisms are playing a crucial role in the degradation of toxic compounds.

4.1.2 Phytoremediation

Phytoremediation includes the application of green plants and their related microbial communities to decrease the number of contaminants present in soils, sediments, and surface water. The plants used in the process disseminate the toxic material from the soil and hold within their plant tissues and constrain them until they are broken down at the roots. Plants sorb the contaminants with their roots and accumulate in the stems of the plant. Pollutants that can be removed using the plants are metals, pesticides, chlorinated solvents, dyes, polychlorinated biphenyls, and petroleum hydrocarbons.

Some plants that can be used for phytoremediation are Indian Mustard, Indian Grass, Brown mustard, Sunflower plants, Barley Grass, Pumpkin, Poplar trees, Pine trees, White Willows, etc. The use of plants that can accumulate metals are employed to remove toxic heavy metals from soil and water by using advanced technologies.

4.1.3 Mycoremediation

Fungi are well-known as nature's best decomposers. They break down most of the plants and hard wood material, resulting in the regeneration of the soil. Fungi use their metabolic enzymes to decompose chemicals like metals and varied types of pesticides acting as a catalyst for microorganisms and plants by breaking down the larger hydrocarbon chains into smaller pieces, thereby making their process easy. Different carbon sources can be added to polluted sites to initiate and speed up the degradation processes. The ability of fungi to adapt their metabolism for exploitation of various sources of carbon and nitrogen makes them a viable option for the degradation of various dyes. *Trametes Versicolor* which degrades red dye 27 through lignin peroxidases (Rekik et al. 2019), *Aspergillus niger* and *Aspergillus terreus* degrade and absorb the red azo dye MX-5 by reducing its toxicity (Janusz et al. 2017).

4.2 Based on the Strategies Applied

Based on the application, there are two major types of bioremediation techniques, i.e. in-situ, and ex-situ that are depending on the contaminated site or fraction to be handled (Concetta Tomei and Daugulis 2013).

4.2.1 In-situ Bioremediation

In-situ bioremediation involve leaving the soil in its original place and bringing the biological mechanisms to the soil to perform on-site bioremediation. According to the saturation and aeration levels of a contaminated area (soil or water), different in-situ and ex-situ bioremediation strategies can be applied. In-situ techniques include bioventing, biosparging, biostimulation, and bioaugmentation, whereas ex-situ ones involve biopiles, bioreactors, biofilters, landfarming, and composting (Concetta Tomei and Daugulis 2013).

There are two categories of in-situ bioremediation:

i. *Intrinsic in-situ bioremediation*
ii. *Engineered in-situ bioremediation.*

i. Intrinsic In-situ Bioremediation

The process of converting harmful pollutants into non-toxic forms through the inherent abilities of the naturally occurring microbial population is referred to as intrinsic in-situ bioremediation. Here, the natural ability of microorganisms to degrade the contaminants is examined and used for the bioremediation process (Azubuike et al. 2016b). The sewage water can be treated by the processes of intrinsic bioremediation with the help of aerobic microbes provided with aeration for the bacteria to thrive and grow. The by-product of this process is nitrogen gas, which is later released into the atmosphere.

ii. Engineered/Accelerated In-situ Bioremediation

As intrinsic bioremediation may not be suitable when site conditions are not matching with the microbial growth requirement due to the slow process as the growth and availability of microorganisms are not adequate, there is the limited capacity of electron acceptor and nutrients, cold temperature, and high concentration of contaminants, hence in these conditions, engineered in-situ bioremediation is employed. Here, substrates, nutrients, or specific microbial species are added to the sub-surface to enhance the bioremediation efficiency (Janusz et al. 2017). Engineered in-situ bioremediation enhances the required biodegradation reactions by enhancing the growth of more microorganisms under optimum physicochemical growth conditions. Oxygen, electron acceptors and nutrients increase the microbial growth at the contaminated surface. Hence this method would be useful to carry out the degradation of various pollutants located at different sites. Various methods can be employed to carry out engineered in-situ bioremediation.

Types of Engineered In-Situ Bioremediation

a. Biosparging:
In biosparging, to increase the groundwater O_2 concentration, the air is injected under pressure. It also enhances the biodegradation rate of contaminants using bacteria. The efficiency of biosparging relies on two factors: soil permeability, which decides pollutant bioavailability to microorganisms, and pollutant biodegradability (Atlas 2009). Biosparging works due to the high rate of airflow to get pollutant volatilization, whereas biosparging enhances biodegradation (Azubuike et al. 2016b; Atlas 2009).

b. Bioventing:
Bioventing is a combination of soil venting and bioremediation. Bioventing provides oxygen to the existing soil microorganisms and increases the rate of in-situ biodegradation of compounds. Oxygen is provided through direct air injection into contaminated soil utilizing wells. This technique utilizes only the required amount of air that is necessary for degradation. It also minimizes the volatilization and discharge of contaminants into the environment (Vaccari et al. 2020).

c. Bioaugmentation:
Bioaugmentation is the process of introducing a specific combination of naturally occurring or genetically engineered microbial strains having enhanced degradation capabilities in contaminated sites for augmenting the natural degradation process (Garbisu et al. 2017).

d. Bioslurping:
Bioslurping combines the efficiency of vacuum-enhanced pumping, and soil vapor extraction with bioventing for soil treatment, and groundwater contaminated with volatile and semi-volatile organic compounds (Gidarakos and Aivalioti 2007).

e. Biostimulation:
Biostimulation involves the addition of various aqueous solutions, having nutrients or other amendments to the contaminated site for increasing the degradation abilities and the growth of indigenous microbial populations (Sarkar et al. 2016).

f. Natural Bioattenuation:
Bioattenuation includes passive remediation of polluted sites. Here the existing indigenous microorganisms capable of degrading the contaminants are used.

In-situ bioremediation techniques are employed to remove chlorinated solvents, dyes, heavy metals, and hydrocarbons from polluted sites. In-situ bioremediation is tedious as compared to other remedial methods (Urionabarrenetxea et al. 2021; Concetta Tomei and Daugulis 2013).

4.2.2 Ex-Situ Bioremediation

Ex-situ bioremediation techniques involve excavating pollutants from contaminated areas and subsequently carrying them to another site for treatment (Urionabarrenetxea et al. 2021). Ex-situ bioremediation technologies have high predictability and high efficiency. It can effectively reduce environmental and human health risks. The environmental risk is reduced based on the expense of treatment, intensity of pollution, type of pollutant, degree of pollution, location, and geology of the polluted site. Ex-situ bioremediation is classified into two types based on the status of the contaminated material under treatment.

i. *Solid-phase system*
ii. *Slurry-phase systems*

i. Solid-Phase Treatment

A Solid-phase system includes organic wastes present in solid form (e.g., leaves, animal manures, and agricultural wastes), and other problematic wastes (e.g., domestic and industrial wastes, sewage sludge, and municipal solid wastes). A Solid-phase system requires a large amount of area and degradation requires more time to complete as compared to slurry-phase processes. The Solid-phase treatment include biopiles or windrows, land farming, composting, etc. (Kumar et al. 2018).

a. Composting:
Composting involves the biological decomposition of wastes under specific conditions to a state in which they will be easy to handle, store, and/or employed on the land without any harm to the environment (Kumar et al. 2018). Composting can be

done in an open or closed system. The addition of biodegradable substrates (either solid or liquid substrates) (e.g., molasses, compost) for microbial proliferation has been used to enhance biodegradation efficiencies.

b. Land Farming:
Land farming includes the spread of excavated contaminated soil on a ground surface to increase aerobic microbial activity by the addition of nutrients and water. This is very simple to design and apply and needs low capital. By using this technique very large area of contaminated soil can be treated. In this technique, autochthonous microorganisms are used that will biodegrade pollutants aerobically (Rao et al. 2010).

c. Biopiling/Windrowing:
Biopiling involves above-ground piling of excavated polluted soil, with the addition of nutrients, and maintaining aeration that will enhance bioremediation by enhancing microbial activities. This system consists of a treatment bed, an aeration system, an irrigation/nutrient system, and a leachate collection system. Other environmental parameters are also controlled. The nutrient system is buried under the soil which allows the aeration and passing of nutrients either by vacuum or pressure. This technique is economically feasible, and in this method, effective biodegradation can take place.

d. Biofilter:
In microbial filters or biofilters, microbes grow to degrade volatile compounds. The bed is mainly composed of materials like peat, composted yard, bark, soil, and plastic shapes. The best example of a biofilter is the trickling filter which has a wide range of applications in the treatment of different liquid effluents (Chaudhary et al. 2003).

ii. Slurry-Phase Treatment

Slurry bioreactors are highly engineered treatment systems for soil bioremediation. It is mainly composed of a vessel or container in which the biodegradation of contaminants in soil and water occurs with the help of bacteria. A rotating drum bioreactor is the most common type of slurry system (Wang et al. 2007). Besides this, there are different types of bioreactors, which include: batch, fed-batch, sequencing batch, continuous, and multistage. The preference of the type of operating mode mainly depends on the economical aspect.

5 Different Types of Microorganisms Used in Bioremediation

The varieties of microorganisms used in bioremediation can be isolated from indigenous contaminated sites or elsewhere and then applied to the targeted sites. In the bioremediation process, microorganisms must act enzymatically on the pollutant and convert them into the less toxic products. This process is favorable only when environmental conditions allow microbial growth and activity. Different types of microorganisms like bacteria, yeast, and fungi have been used to clean-up contaminated environments. Bacterial species like *Flavobacterium, Pseudomonas, Enterobacter, Bacillus*, and *Micrococcus* possess high surface-to-volume ratios and active chemisorption sites on the cell surface that help them for the biosorption process. Mixed cultures of bacteria are more metabolically active that enhances the biosorption process of the metals. Microorganisms uptake heavy metals either by bioaccumulation (actively) or adsorption (passively).

6 Role of Microrganisms in Bioremediation

Bacteria have a unique ability to withstand and adapt to variations in COD and BOD, high salinity, variable pH, dissolved oxygen, and pollutant concentration, hence they are the most relevant species to be used in the bioremediation process. The major advantage of bacteria over other species is their short growth cycle, ability to adapt, and diverse metabolic activities. The monoculture or consortium of bacteria can remove individual pollutant, and mixture of pollutant. When more than one bacteria are used in a consortium there are few known syngenetic metabolisms, which improve the efficiency of degrading hydrocarbons, and other chemicals (d'Errico et al. 2020).

Bacterial species use different resistance mechanisms including the capture of metal ions by extracellular barriers like the capsule, cell wall, plasma membrane, intracellular sequestration of metal ions, the extrusion of metal ions through efflux or diffusion pumps, biotransformation of toxic metal ions, and decreased sensitivity of cellular targets to metal ions (Sarkar et al. 2016). Hence, the metal detoxification process can be attributed to different pathways like biosorption, bioaccumulation, oxidation, reduction, biomethylation intracellular/extracellular sequestration, intracellular/extracellular precipitation, volatilization, and biosurfactant production (Table 1). The thermophilic and hyperthermophilic bacterial species use an alternative mechanism for enzymatic production to resist metals and transfer ions at the active site. Highly metal-resistant bacterial strains have been isolated and characterized from metal-contaminated sites and explored for their ability to detoxify toxic heavy metal ions. Numerous microbial strains, like *Bacillus, Pseudomonas, Alcaligenes, Lysinibacillus, Enterobacter, Aspergillus,* and *Penicillium* have been reported for the efficient transformation of toxic heavy metals into less toxic forms. Also, the consortiums or synergistic associations act as biological inducers to enhance the azo dye degradation ability of bacterial species.

Especially in a large-scale operation, it is reported that microorganisms due to their union of catabolic function become more useful to degrade dyes because they possess greater resistance to abiotic conditions and have low rates of enzyme inactivation. During the interaction of bacteria and metal, biofilm formation plays an important role in the survival of bacterial species in presence of high metal concentration, and efflux systems allow bacteria to interact with different amino acids as a mechanism

Table 1 Different microorganisms used in the bioremediation process

Microorganisms	Compound
Saccharomyces cerevisiae	Pb, and Ni (Dangi et al. 2019)
Halomonas sp. strain MCTG39a	Hexadecane (Wang et al. 2019a, b)
Bacillus firmus, Pseudomonas sp.	Zn (Singh et al. 2017)
Agaricus bisporus	Cd, Zn (Bilal and Iqbal 2020)
Aspergillus niger	Ni, Co (Bilal and Iqbal 2020)
Aspergillus, Mucor, Penicillium, and Rhizopus	Cd, Cu, Fe (Medfu Tarekegn et al. 2020)
Spirogyra sp. and Cladophora sp.	Pb (Wang et al. 2020)
Candida tropicalis	Cd, Cr, Ni (Yuan et al. 2021)
Pseudomonas veronii	Cd, Zn, Cu (Diaconu et al. 2020)
Phormidium valderianum	Ni, Cd (Ojuederie and Babalola 2017)

of adaptation to the environment. In *Pseudomonas aeruginosa* biofilm, the cells demonstrated considerably higher resistance to ions of Pb, Zn, and Cu than free cells, whereas cells located at the surface of the biofilm were killed. In the previous report, it was found that extracellular polymers of the biofilm accumulated metal ions protecting bacterial cells inside the biofilm (Mahamuni-Badiger et al. 2020a). However, in the case of mixed contaminants, finding a suitable consortium is a major challenge (Singh et al. 2011).

7 Mechanism of Bioremediation

The interaction between extracellular polymeric substances and heavy metals takes place by different mechanisms like precipitation or proton exchange. The cell wall surface possesses a negative charge because of the presence of carboxyl, amino, phosphoryl, and sulfo groups that act as ion-exchange sites for metal ions (Comte et al. 2008). Bioremediation occurs via different mechanisms like redox reactions, adsorption, ion exchange, precipitation, and electrostatic interaction. The redox reactions in microorganisms mobilize or immobilize metal ions which is important for bioremediation.

During the bioremediation process, pollutant get converted from their insoluble and stationary form into their mobile and soluble state. For example, some bacteria reduce Hg (II) to the elemental and volatile form of Hg (0) (Ojuederie and Babalola 2017). Reduction reactions in microbes facilitate the solubility of ions like Fe (III) and As (V) by converting them to Fe (II) and As (III). Biomethylation of heavy metals in soil and water can alter their toxicity, volatility, and mobility. Mobilization of heavy metal ions also takes place through the microbial decomposition of organic matter, where the carboxylic acid and amino acids generated by microbes are involved in the chelating of metal ions. Immobilization of metal ions takes place through different mechanisms like biosorption, bioaccumulation, bioconversion, and precipitation. The *Methanothermobacter thermautotrophicus* converts Cr (VI) to Cr (III) by reduction reaction and immobilizes it in hydroxide-oxide forms (Fernández et al. 2018).

Many heavy metals like Cu and Zn are crucial for the regulation of biological processes of the microbes while some others generate oxidative stress, denature macromolecules, and reduce the bioremediation ability of microbes to heavy metals like Cr and Cd (Coelho et al. 2015). The excess concentration of heavy metals destroys microbial cells by altering their physiological and biochemical properties. The Cr (III) alters the structure and activity of enzymes by reacting with their carbonyl and thiol groups. Oxidative stress causes the toxicity of Cr (VI) to the cells (Nagda et al. 2021). In *Alishewanella* sp. WH16-1, H_2O_2 content increases remarkably under Cr (VI) pressure (Cheung and Gu 2007). The electrostatic interaction between cationic Cr (III) complexes and negatively charged phosphate groups of DNA affects transcription, replication, and cause mutagenesis. In the case of *P.aeruginosa*, the immobilization of cells in the agarose-alginate gel increases the reduction of Cr (VI). According to studies it is observed that the treatment with Fe^{+2} decreases extractable As while increasing the concentration of Cu, Mn, and Zn in soil. Heavy metals like Cu (I) and Cu (II) generate ROS through Fenton and Haber–Weiss reactions that act as electron acceptors or donors and rupture cytoplasmic molecules, DNA, lipids, and proteins (Amezcua-Allieri et al. 2005). Heavy metals hamper enzymatic functions by competitive or non-competitive interaction with the substrate that changes the structure of an enzyme (Fig. 1).

Bioremediation can be categorized into biosorption and bioaccumulation. The bioaccumulation process is dependent upon cellular metabolism in which metal uptake takes place by living cells. This method will not be convenient as heavy toxic metals can deposit in the cell and hamper metabolic activities that can lead to cell death. On the other hand, in biosorption, dead biomass remains unaffected and doesn't require any growth or nutritional medium. This is a flexible and fast mechanism that involves the retention of metals utilizing physicochemical interactions like ion exchange, adsorption, precipitation, and crystallization. This process can be influenced by different factors like pH, ionic strength, biomass concentration, temperature, particle size, and ions present in the solution. Biosorption is independent of cellular metabolism so, both living and non-living biomass can be used. The adsorption of heavy metals takes place on the surface in passive mode without any energy expenditure until equilibrium is reached. Therefore, biosorption is more reliable than bioaccumulation as it is independent of cellular metabolism and relies on the type of biomass and contaminant involved. These methods can be also modified by using metallic nanoparticles along with bacteria and genetically engineered microorganisms (Tanvi et al. 2020).

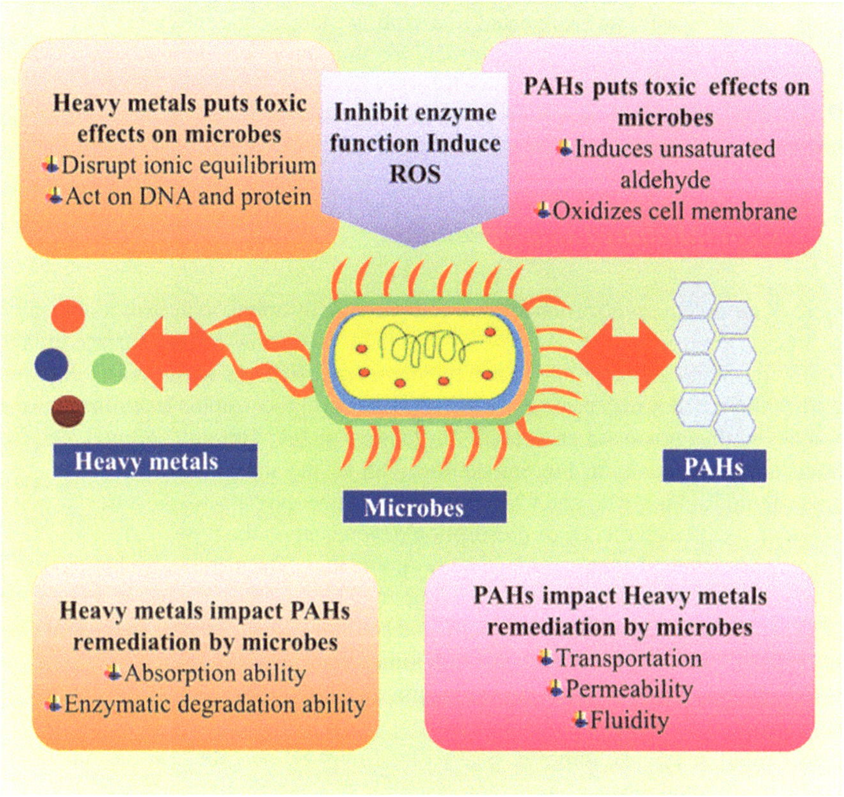

Fig. 1 Different mechanisms of microbes-based bioremediation (Malla et al. 2018)

7.1 Bacteria-Mediated Bioremediation

Biosorption is a cost-effective and convenient technique to remove heavy metals from the environment. Bacteria show a unique mechanism for the bioremediation of heavy metals, through efflux pumps, intra-, and extra-sequestration, conversion to less toxic chemical species by redox reactions, alkylation or dealkylation, and reduction. Bacterial biomass removes metal ions like Cu, Co, Fe, Pb, and Cd. The rate of bioremediation depends upon the bacterial cell as Gram-positive and negative bacteria possess different cell wall compositions in terms of N-acetylmuramic acid and N-acetylglucosamine. Bacterial cells possess an overall negative charge due to the presence of anionic functional groups like sulfate, amine, hydroxyl, and carboxyl groups that act as metal-binding sites. Bacteria can act as biosurfactants in the bioremediation of metal-contaminated soil. This technology can be very effective and reliable in the case of Cd and Pb-polluted soil (Amezcua-Allieri et al. 2005; Comte et al. 2008). The metal-resistant strains of *P. aeruginosa* for the bioremediation of metal-polluted wastewater subjected to mutation increase the inhibitory concentration of Cd for their growth. It was concluded that the *P. aeruginosa* strain can be used for the bioremediation of Cd-polluted wastewater. Different bacteria used in bioremediation of heavy metal are given in Table 1.

7.2 Algae-Mediated Bioremediation

Microalgae are eukaryotic, unicellular, and photoautotrophic organisms abundant in the aquatic environment. Due to their small size, they show a large surface area to volume ratio and are easily available for interaction with the surrounding environment. Although they may pass heavy metal cations to higher trophic levels through the food chain, they possess a great ability to remove metals from the polluted medium which is advantageous in bioremediation. Out of the three classes of algae, i.e., Phaeophyta, Rhodophyta, and Chlorophyta, the Phaeophyta possess better biosorption capacity. The efficiency of biosorption depends upon the type of algal biomass, charge, and chemical composition of heavy metal ions. The negatively charged functional groups on the cell surface serve as potential binding sites for heavy metal ions. Different ions like calcium, magnesium, and sodium that are present in the cell wall get replaced by heavy metal ions through ion exchange.

The accumulation of heavy metal by microalgae takes place in two steps, i.e., rapid removal by the cell surface and slow removal which occur inside the cell. During rapid or passive removal, heavy metal ions get adsorbed on the functional groups on the cell surface. This step is rapid, non-metabolic, and reversible that takes place in living and nonliving cells. The second step that occurs inside the cell is metabolism dependent that involves the transfer of ions across the cell membrane and accumulation inside the cell. This is a slow and irreversible step and occurs only in living cells (Abdel-Razek et al. 2019). Microalgae serve as good biosorbents because of their abundant availability in seas and oceans.

7.3 Fungal-Mediated Bioremediation/Mycoremediation

Bioremediation by fungi takes place through extracellular and intracellular mechanisms. Extracellular mechanisms occur by chelation, precipitation, and cell wall interaction, while intracellular mechanism occurs by binding to organic acids, peptides, sulfate, polyphosphates, and organic acids. Fungi can withstand harsh stress conditions like water availability, nutrients, pH, temperature, etc. This bioremediation is economically feasible and doesn't leave any harmful waste product after detoxification. Fungi deposit heavy metals in their bodies so become unavailable in the media. The *Aspergillus* species have been studied for the removal of Cr from tannery wastewater. The growth and yield of the plants get affected by the presence of heavy metals (Zhan et al. 2018). However arbuscular mycorrhizal fungi (AM) develop a symbiotic relationship with the plant and result in the establishment of tolerance toward heavy metals. AM fungi detoxify heavy metals by different mechanisms like chelation of heavy metals, adsorption by cell chitin, and immobilization of metals. AM fungi inhibit the intake of Cd and Pb into plant roots.

Generally, mushrooms are used not only in the human diet but also for mycoremediation due to their efficient ability of heavy metal uptake. This uptake gets influenced by contact time, age of mycelia, and fructification. Heavy metals in the underground parts of the plants get retained by *Diversispora sputum* and *Funneliformis mosseae* which reduce Zn, Pb, and Cd content in the shoot of maize plants (Zhan et al. 2018). According to some studies, AM fungi reduce heavy metal stress by preventing its uptake by the host plant. For example, AM fungi decreased the heavy metal impact on *Calendula officinalis* development by decreasing the heavy metal (Cd, Pb) uptake and promoting the secondary metabolism (Hristozkova et al. 2016) (Table 2).

Table 2 Detoxification mechanism by different microbial species

Detoxification mechanism	Microbial species
Biomethylation	*Pseudomonas, Alcaligenes, Acinetobacter, Flavobacterium, Aeromonas* (Yin et al. 2017)
Biosorption	*Alcaligenes* sp., *Bacillus cereus, Bacillus thuringiensis, Candida tropicalis, Cyberlindnera Fabiani, Enterobacter* sp., *Kocuria rosea, Lysinibacillus* sp., *Microbacterium oxydans, Ochrobactrum* sp., *Penicillium chrysogenum, Serratia marcescens, Wickerhamomyces anomalus* (El-Naggar et al. 2018)
Oxidation	*Microbacterium lacticum, Brevibacillus brevis, Pseudomonas stutzeri* (Wang et al. 2019a, b)
Reduction	*Aeromonas* sp., *Alcaligenes faecalis, Aspergillus niger, Bacillus* sp., *Cellulosimicrobium* sp., *Exiguobacterium* sp., *Klebsiella pneumonia Pseudochrobactrum saccharolyticum, Pseudomonas stutzeri, Streptomyces* sp. (Kim et al. 2013)
Bioaccumulation	*Yarrowia* sp. (Kim et al. 2013)
Intracellular sequestration	*Alcaligenes* sp., *Bacillus* sp., *Lysinibacillus fusiformis* (Kumar et al. 2017)
Extracellular sequestration	*Alcaligenes faecalis, Arthrobacter, Bacillus cereus, Kocuria rosea, Lysinibacillus* sp., *Microbacterium oxydans, Ochrobactrum* sp. *Pseudomonas* sp., *Serratia marcescens, Yarrowia* sp. (Gonzalez-Muñoz et al. 2012)
Intracellular precipitation	*Citrobacter freundii, Staphylococcus aureus* (Zhang and Klapper 2010)
Extracellular precipitation	*Pseudomonas aeruginosa* (Amezcua-Allieri et al. 2005)
Biosurfactant production	*Pseudomonas aeruginosa* (Varjani and Upasani 2017)
Volatilization	*Alcaligenes faecalis, Klebsiella pneumonia, Lysinibacillus fusiformis, Pseudomonas aeruginosa, Yarrowia* sp. (Mello et al. 2020)

8 Factors Affecting Bioremediation

Microorganisms can degrade various heavy metals due to their metabolic system and ability to adapt to unfavorable environments. Microorganisms degrade heavy metals by facilitating biochemical reactions and by acting as a biocatalyst. Their ability to bioremediation is dependent upon varieties of factors like nature, stability, the concentration of pollutants, availability to microorganisms, and the physico-chemical nature. These factors affect the degradation rates of heavy metals by microorganisms that depend on the nature of microorganisms and their nutritional requirements (biotic factors) or are related to the environment (abiotic factors).

8.1 Biotic Factors

The biotic factor affects the bioremediation of heavy metals is their metabolic ability to degrade heavy metals. Biotic factors influence the degradation of pollutants by competition between microorganisms, antagonistic interaction between microorganisms, or predation by protozoa or bacteriophages. The rate of degradation mainly relies on the number of contaminants and catalysts present. Therefore, the quantity of catalyst expresses the number of organisms that can metabolize the contaminant and the quantity of the enzyme. Biotic factors that affect the rate of degradation are mutation, gene transfer, interaction population size, and compositions (Srivastava et al. 2014).

8.2 Abiotic Factors

Different abiotic factors affect the bioremediation are pH, temperature, moisture, redox potential, water-solubility, nutrient availability, oxygen content, physico-chemical properties of metals, concentration, and chemical structure of contaminant. For the bioremediation process, microorganisms need several nutrients like carbon, nitrogen, and phosphorous and addition of nutrients is an effective way to enhance the biodegradation rate (Akintelu et al. 2021).

8.2.1 Temperature

Temperature is one of the most important factors that affect bioremediation. It has a direct effect on the physicochemical properties of contaminants, and the rate of microbial metabolism. In a cold environment, the rate of bioremediation is very slow as the rate of metabolic activities of microbes becomes slow. Temperature higher than required affects ribosomal confirmations and decreases the rate of protein synthesis, while temperature less than required negatively affects the membrane permeability and hampers the function of a transport system. These changes affect the transport of substrate through cells and affect the growth rate. The required temperature for the bioremediation process is about 20–40 °C (Akintelu et al. 2021). According to many studies, the rate of bioremediation is higher in mesophilic conditions. Microorganisms like *Aspergillus niger*, *Bacillus cereus*, *Enterobacter*, *Rhizopus oryzae*, *Rhizobium species*, and *Botrytis cinerea* showed maximum bioremediation under mesophilic conditions. Thermophilic microorganisms which show a higher rate of bioremediation at the temperature range of above 45 °C to 80 °C can be beneficial in the remediation of compost.

8.2.2 Effect of pH

The pH is another important factor in bioremediation. pH has a great impact on the microbial metabolic rate that affects the bioremediation process. Rate of metabolic activity is highly susceptible to minute changes in pH. Different microorganisms like *Actinobacter junii, E. coli, Micrococcus leutus, Pleurotus platypus, Pseudomonas aeruginosa*, and *Trichoderma* showed a higher potential to degrade heavy metals at neutral pH conditions. At low values of pH, functional groups on the cell surface may change their properties and unable to interact with metal ions. In the same manner, higher pH values also affect the ability of microorganisms to degrade contaminants. Alkaline pH increases the logarithmic phase of microorganisms which negatively affects the bioremediation rate.

8.2.3 Effect of Metal Ions Concentration

The presence of heavy metals affects the cellular functions of microorganisms. Low concentration of heavy metals favors the biodegradation process, while the higher level negatively affects microbial growth by slowing down metabolic activities. Higher concentrations of heavy metals affect the growth of microorganisms by interacting with cell membranes that damage membrane structure, metals in cytoplasm inactivate enzymes, and genetic material gets damaged by metal ion removal efficiency depending upon the ion concentration and type of microorganism present. For example, the *Aspergillus niger* strain removed 15.6 mg/g of Cu after 1 h incubation (Kumar and Dwivedi 2021).

8.2.4 Effect of Other Factors

Some other factors can also affect the bioremediation rates. Nutrient availability is one of the most important factors that have a great influence on biodegrading microorganisms. Nutrients that are involved in microbial growth are used by microorganisms for metabolic reactions and the synthesis of essential enzymes. A sufficient concentration of nutrients is very essential, where nutrients like carbon, nitrogen, calcium, magnesium, sulfur, and phosphorous are very important for different cellular functions. These functions include cellular protein, cell wall synthesis, and nucleic acid synthesis. Climate change is also one of the important factors that affect microbial growth and rate of metabolism. An elevated level of CO_2 concentration is related to an increase in the bacterial population and a decrease in the population of fungi (Wu et al. 2018). Hence these are some other factors that proved to be very crucial for the microbial bioremediation process.

9 Recent Advances in Bioremediation

Bioremediation is a sustainable approach to industrial and environmental pollution management and can be simultaneously used with other physical and chemical treatments for the proper management of different groups of environmental pollutants (Singh et al. 2011). Although the technology requires a prior thorough understanding and exposure to microbial processes, it is the most promising technology for water reuse to meet the current demand for clean water for the ever-increasing population. The natural ability of microorganisms and plants shows the bioremediation of complex hazardous molecules/toxic metals into a simple, non-toxic form with very low efficacy. Therefore, the engineering of microorganisms and plants provides a strong potential for powerful degradation and reduces the accumulation of pollutants at a much higher rate.

Keeping in view the above fact, bioremediation can be tailored by using the various genetic approaches for optimizing enzyme production, metabolic pathways, and growth conditions along with the needs of the polluted site and the specific microbes require for degrading pollutants. This can be further improved by using synthetic biology tools to adapt microbes to the pollution in the environment to which they are to be added.

10 Genetic Manipulation for Enhanced Bioremediation

To enhance the bioremediation potential of any bacteria, modern tools of genomics, transcriptomics, proteomics, metabolomics, phenomics, and lipidomics can be applied. These enhanced use of integrated research approaches were found useful, to investigate interactions and networks at the molecular, cellular, community, and ecosystem levels. One such approach to increase the bioremediation potential is the insertion of certain functional genes into the bacterial genome for the change in metabolic pathways like transport and chemotaxis, and the adaption of features toward extreme environmental conditions. There are several reports for the development of engineered bacteria for enhanced bioremediation where functional genes are inserted into the bacterial genomes through molecular biology techniques. The engineered microorganisms produce trehalose and reduces 1 mM Cr (VI) to Cr (III) (Sharma et al. 2018). Insertion of a small gene in *Synechococcus* sp. accounted for enhanced heavy metal tolerance. Engineered *Chlamydomonas reinhardtii* generated a significant increase in the tolerance to Cd toxicity and accumulation (Janssen and Stucki 2020). The *Corynebacterium glutamicum* was genetically modified using overexpression of ars operons (*ars*1 and *ars*2) to decontaminate As contaminated sites. *Deinococcus radiodurans*, the most radio-resistant organism, has been modified genetically to consume and degrade toluene and the ionic form of Hg from nuclear wastes (Brim et al. 2020). Similarly, genetically engineered microbes for heavy metal, hydrocarbons, and industrial effluent remediation involve the use of *Escherichia coli, Staphylococcus aureus, Thalassospira lucentensis*, and *Saccharomyces cerevisiae.*

Genetically modified organisms can also degrade azo dyes through mechanisms involving genetic modification or gene transfer that encode enzymes with different characteristics or biochemical pathway variants in a microorganism. Insertion of the bmtA gene coding for metallothionein into marine bacteria has been done and successfully employed in highly metal-contaminated environments (Arashiro 2018).

Recently developed CRISPR-Cas9 system has also been used for improving the microbial and plant for rapid and efficient bioremediation of pollutants. Omics approaches such as genomics, transcriptomics, metabolomics, and proteomics aid the systems biology studies of microbes for analyzing the genetic level regulation for bioremediation. Advancement in sequencing and next-generation sequencing resolves the novel genes involved in the biodegradation pathways of various persistent pollutants. Metagenomics is the direct analysis of the genome and can be applied to oil, xenobiotics, and heavy metals remediation using *Marinobacterium, Marinobacter, Cycloclasticus, Sphingomonas, Candidatus Solibacter*. Similarly, metaproteomics is the protein study derived from environmental samples. Recent reports highlighted the metaproteomics approach to study the bacterial adaptation tactics in different contaminated sites, *i.e.*, heavy metals, oil, xenobiotics, and other pollutants (Sharma et al. 2018). Remediation of organophosphorus insecticides using *Aspergillus, Pseudomonas, Chlorella, and Arthrobacter* has also been reported. There are several bacterial strains known for pesticide bioremediation, and the whole genome has been sequenced, *i.e., Pseudomonas putida and Rhodococcus sp.* (Frederick et al. 2013). These engineered microbes have arisen possibilities to use the bacterial species with bioremediation potential, incorporated for achieving greater bioremediation rates at different pH, temperature, and even at different ecosystems.

In addition to this, the application of nanotechnology for the remediation of contaminants is another promising approach that enhances the bioremediation potential where nanoremediation methods have the application for the transformation and detoxification of pollutants. These nanomaterials possess both chemical reduction and catalysis to degrade the pollutants of concern (Mahamuni-Badiger et al. 2020b; Mahamuni et al. 2019).

10.1 Genetically Engineered Microorganisms Used in Bioremediation

The use of microbes present in the environment, *i.e.*, soil, water, etc., for bioremediation has become reliable, but there is a need for further improvements in this technology. For this purpose, the use of genetically engineered microorganisms has become significant which can enhance the degradation rates of environmental toxins and waste (Fig. 2).

Desired genetic manipulation in the microorganisms can develop the ability to remove different kinds of hydrocarbons. The *Pseudomonas* strain containing multiple plasmids can oxidize PAH, terpenic aromatic, and aliphatic hydrocarbons. Recombinant DNA technology is a boon that alters the genetic characteristics of microorganisms. Heavy metals are the major parts of contaminants including radionuclides, and bioremediation of such waste may produce toxicity to most bacteria of radiation from these radionuclides. The *Deinococcus radiodurans* are highly resistant to exposure to ionizing radiation. So, this organism can be the perfect host for the genetic engineering purpose that can be useful for the bioremediation of such mixed waste. The genetic engineering in microorganisms used in bioremediation involves 4 important approaches that are (i) bioaffinity bioreporter sensor application, and toxicity reduction (ii) bioprocess development, observation, and control, (iii) increasing enzyme specificity, and (iv) designing pathways (Pande et al. 2020).

The genetically modified *E. coli* encodes the Hg^{2+} transport system and metal-binding protein, *i.e.*, metallothionein. This organism can accumulate 8 μmole Hg^{2+}/g of cell dry weight (Raghu et al. 2008). Hg is a very toxic heavy metal occurring in the environment and Its bioremediation can be performed by expressing the bacterial mer gene. Hg resistant bacteria possess mer operon in their genome. This operon contains functional genes, a promoter, a regulator, and an operator. Functional genes merA and merB encodes for Hg ion reductase. Reduction of As (V) to As (III) is

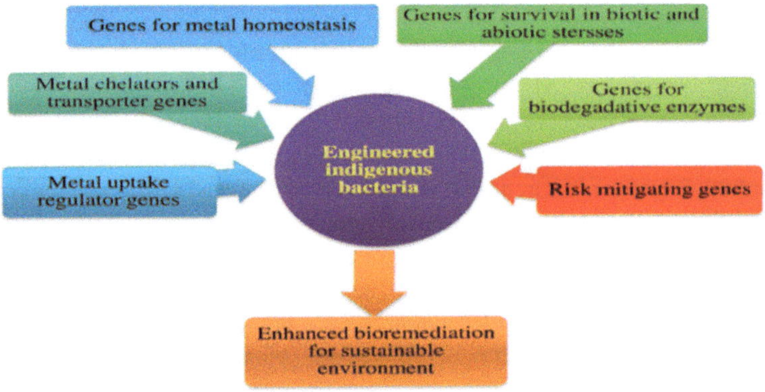

Fig. 2 Genetic engineering in indigenous bacteria to enhance the bioremediation capability of microbial strain (Singh et al. 2011)

not efficient as As (III) is more toxic than As (V). Therefore, the conversion of As into volatile As compounds like mono-, di-, and trimethylarsine is more convenient for bioremediation purposes. Transgenic bacteria are more capable of the conversion of As into volatilized form as compared to wild type. The *Thermus thermophilus* HB8 contains TTHB128 and TTHB127 which can produce arsenite oxidase. It can convert the toxic form of As to a nontoxic form (Upadhyay et al. 2018).

Several human activities like mining, smelting, combustion, disposal of batteries, and other materials generate Pb in the environment. Pb^{2+} is the most reactive form found in different forms like carbonate bound, Fe/Mn oxide bound, organic and residual phase. The presence of Pb in soluble form and exchangeable form is more dangerous to the environment than immobilized Pb.

Cd commonly occurs in ores with Zn, Pb, and Cu. It is released into the environment by various human activities like mining, electroplating, galvanizing, discharge from ceramic industrial waste, textile, and leather industries, and use of Cd batteries (Xu et al. 2020). Cd is highly toxic even at very low concentrations. According to various studies, Gram-negative bacteria are resistant to Cd ions. The *P. aeruginosa* PU21 biomass acts as a bioadsorbent and removes Cd, Cu, and Pb from contaminated water effectively, while dead cell biomass can adsorb Cd and Pb in an aqueous solution (Chellaiah 2018).

The *putida* can degrade camphor, octane, salicylate, and naphthalene due to the presence of XYL and NAH plasmid by conjugation by recombining parts of CAM and OCT. They can easily grow on crude oil as they can metabolize hydrocarbons efficiently. Genetic engineering or plasmid can modify microorganisms with the required catalytic potential that can remove environmental pollutants effectively.

11 Conclusion and Future Perspectives

Bioremediation is an eco-friendly and cost-effective technology to clean contaminated sites using microorganisms. The technology has several advantages over other physical and chemical methods as it doesn't require many resources and needs very less energy than conventional technology, and also does not form produce hazardous by-products.

Several bacteria capable of bioremediation have been identified, however, their degradation pathways and characteristics need to be assessed before application. Adverse effects of various environmental pollutions can be forecasted by using multidisciplinary technologies such as "Omics" (genomics, metabolomics, proteomics, and transcriptomics) that have contributed toward a better understanding of microbial identification, functions, metabolic, and catabolic pathways. Hence, the enzyme profile and the intermediate metabolites should be the subject of future studies based on genomics and proteomics that will contribute to a better understanding the microbial cellular functions and gene products. Tools like GeoClip, RNAseq, and mass spectrometry should be adopted to identify RNA and protein to get insights into cellular pathways and the identification of functional genes potentially involved in remediation pathways. Genetic engineering technology can be developed on larger scales for waste recycling, soil treatment, management of solid wastes, etc. Efforts should be made in biofilm-mediated bioremediation and microbial fuel cell in the bioremediation.

Future research on the application of environmental bacteria capable of degrading multiple pollutants should be focus for bioremediation, clean technologies genetic engineering, nanotechnology, and the use of meta-genomics and meta-proteomics analysis. The research and development in this direction will give future regulations, completing bioremediation targets, contaminant availability, and their adverse effect on the natural ecosystem and human health.

Although bioremediation offers an attractive and sustainable approach with high efficiency of detoxification along with low cost, the scientific community must provide innovative mechanisms based on optimizing enzyme production, metabolic pathways, and growth conditions to enhance bioremediation efficacy for better control and management of pollutants.

References

Abdel-Razek MA, Abozeid AM, Eltholth MM, Abouelenien FA, El-Midany SA, Moustafa NY, Mohamed RA (2019) Bioremediation of a pesticide and selected heavy metals in wastewater from various sources using a consortium of microalgae and cyanobacteria. Slov Vet Res 56:61–73. https://doi.org/10.26873/SVR-744-2019

Akintelu SA, Yao B, Folorunso AS (2021) Bioremediation and pharmacological applications of gold nanoparticles synthesized from plant materials. Heliyon 7:e06591. https://doi.org/10.1016/j.heliyon.2021.e06591

Amezcua-Allieri MA, Lead JR, Rodríguez-Vázquez R (2005) Impact of microbial activity on copper, lead and nickel mobilization during the bioremediation of soil PAHs. Chemosphere 61:484–491. https://doi.org/10.1016/j.chemosphere.2005.03.002

Arashiro SMT (2018) Lead absorption mechanisms in bacteria as strategies for lead bioremediation. Appl Microbiol Biotechnol 102:5437–5444

Atlas RM (2009) Bioremediation: applied microbial solutions for real-world environmental cleanup, 2005. Cent Eur J Public Health 6:145145

Azubuike CC, Chikere CB, Okpokwasili GC (2016a) Bioremediation techniques–classification based on site of application: principles, advantages, limitations, and prospects. World J Microbiol Biotechnol 32. https://doi.org/10.1007/s11274-016-2137-x

Azubuike CC, Chikere CB, Okpokwasili GC (2016b) Bioremediation techniques–classification based on site of application: principles, advantages, limitations, and prospects. World J Microbiol Biotechnol 32:1–18. https://doi.org/10.1007/s11274-016-2137-x

Bilal M, Iqbal HMN (2020) Microbial bioremediation is a robust process to mitigate pollutants of environmental concern. Case Stud Chem Environ Eng 2:1–4. https://doi.org/10.1016/j.cscee.2020.100011

Brim H, McFarlan SC, Fredrickson JK, Minton KW, Zhai M, Wackett LP, Daly MJ (2020) Engineering Deinococcus radiodurans for metal remediation in radioactive mixed wasteenvironments. Nature Biotechnol 18:85–90. https://doi.org/10.1038/71986

Chaudhary DS, Vigneswaran S, Ngo HH, Shim WG, Moon H (2003) Biofilter in water and wastewater treatment. Korean J Chem Eng 20:1054–1065. https://doi.org/10.1007/BF02706936

Chellaiah ER (2018) Cadmium (heavy metals) bioremediation by Pseudomonas aeruginosa: a minireview. Appl Water Sci 8:1–10. https://doi.org/10.1007/s13201-018-0796-5

Cheung KH, Gu JD (2007) Mechanism of hexavalent chromium detoxification by microorganisms and bioremediation application potential: a review. Int Biodeterior Biodegrad 59:8–15. https://doi.org/10.1016/j.ibiod.2006.05.002

Cho JY, Lee Park S, Lee H-J, Kim SH, Suh MJ, Ham S, Bhatia SK, Gurav R, Park S-H, Park K, Yoo D, Yang Y-H (2021) Polyhydroxyalkanoates (PHAs) degradation by the newly isolated marine Bacillus sp. JY14. Chemosphere 283:131172

Choi Y-K, Gurav R, Kim HJ, Yang YH, Bhatia SK (2020) Evaluation for simultaneous removal of anionic and cationic dyes onto maple leaf-derived biochar using response surface methodology. Appl Sci 10(9)

Coelho LM, Rezende HC, Coelho LM, de Sousa PAR, Melo DFO, Coelho NMM (2015) Bioremediation of polluted waters using microorganisms. Adv Bioremediation Wastewater Polluted Soil 1–22. https://doi.org/10.5772/60770

Comte S, Guibaud G, Baudu M (2008) Biosorption properties of extracellular polymeric substances (EPS) towards Cd, Cu, and Pb for different pH values. 151:185–193. https://doi.org/10.1016/j.jhazmat.2007.05.070

Concetta Tomei M, Daugulis AJ (2013) Ex-situ bioremediation of contaminated soils: an overview of conventional and innovative technologies. Crit Rev Environ Sci Technol 43:2107–2139. https://doi.org/10.1080/10643389.2012.672056

Dangi AK, Sharma B, Hill RT, Shukla P (2019) Bioremediation through microbes: systems biology and metabolic engineering approach. Crit Rev Biotechnol 39:79–98. https://doi.org/10.1080/07388551.2018.1500997

d'Errico G, Aloj V, Ventorino V, Bottiglieri A, Comite E, Ritieni A, Marra R, Censi SB, Flematti GR, Pepe O, Vinale F (2020) Methyl t-butyl ether-degrading bacteria for bioremediation and biocontrol purposes. PLoS ONE 15:1–16. https://doi.org/10.1371/journal.pone.0228936

Diaconu M, Pavel LV, Hlihor RM, Rosca M, Fertu DI, Lenz M, Corvini PX, Gavrilescu M (2020) Characterization of heavy metal toxicity in some plants and microorganisms—a preliminary approach for environmental bioremediation. N Biotechnol 56:130–139. https://doi.org/10.1016/j.nbt.2020.01.003

El-Naggar NEA, Hamouda RA, Mousa IE, Abdel-Hamid MS, Rabei NH (2018) Biosorption optimization, characterization, immobilization, and application of Gelidium amansii biomass for complete Pb2+ removal from aqueous solutions. Sci Rep 8:1–19. https://doi.org/10.1038/s41598-018-31660-7

Fernández PM, Viñarta SC, Bernal AR, Cruz EL, Figueroa LIC (2018) Bioremediation strategies for chromium removal: current research, scale-up approach and future perspectives. Chemosphere 208:139–148. https://doi.org/10.1016/j.chemosphere.2018.05.166

Frederick TM, Taylor EA, Willis JL, Shultz MS, Woodruff PJ (2013) Chromate reduction is expedited by bacteria engineered to produce the compatible solute trehalose. Biotechnol Lett 35:1291–1296. https://doi.org/10.1007/s10529-013-1200-z

Garbisu C, Garaiyurrebaso O, Epelde L, Grohmann E, Alkorta I (2017) Plasmid-mediated bioaugmentation for the bioremediation of contaminated soils. Front Microbiol 8:1–13. https://doi.org/10.3389/fmicb.2017.01966

Gidarakos E, Aivalioti M (2007) Large scale and long term application of bioslurping: the case of a Greek petroleum refinery site. J Hazard Mater 149:574–581. https://doi.org/10.1016/j.jhazmat.2007.06.110

Gonzalez-Muñoz MT, Martinez-Ruiz F, Morcillo F, Martin-Ramos JD, Paytan A (2012) Precipitation of barite by marine bacteria: a possible mechanism for marine barite formation. Geology 40:675–678. https://doi.org/10.1130/G33006.1

Gurav RG, Tang J, Jadhav JP (2016) Sulfitolytic and keratinolytic potential of Chryseobacterium sp. RBT revealed hydrolysis of melanin containing feathers. 3 Biotech 6(2):145

Gurav R, Lyu H, Ma J, Tang J, Liu Q, Zhang H (2017) Degradation of n-alkanes and PAHs from the heavy crude oil using salt-tolerant bacterial consortia and analysis of their catabolic genes. Environ Sci Pollut Res 24(12):11392–11403

Gurav R, Bhatia SK, Choi T-R, Park Y-L, Park JY, Han Y-H, Vyavahare G, Jadhav J, Song H-S, Yang P, Yoon J-J, Bhatnagar A, Choi Y-K, Yang Y-H (2019a) Treatment of furazolidone contaminated water using banana pseudostem biochar engineered with facile synthesized magnetic nanocomposites. Biores Technol 122472

Gurav R, Bhatia SK, Moon Y-M, Choi T-R, Jung H-R, Yang S-Y, Song H-S, Jeon J-M, Yoon J-J, Kim Y-G, Yang Y-H (2019b) One-pot exploitation of chitin biomass for simultaneous production of electricity, n-acetylglucosamine and polyhydroxyalkanoates in microbial fuel cell using novel marine bacterium Arenibacter palladensis YHY2. J Clean Prod 209:324–332

Gurav R, Bhatia SK, Choi T-R, Kim H-J, Lee H-J, Cho J, Ham S, Suh M-J, Kim S, Kim S-K, Yoo D, Yang Y-H (2021a). Seafood processing chitin waste for electricity generation in a microbial fuel cell using halotolerant catalyst Oceanisphaera arctica YHY1. Sustainability 13(15)

Gurav R, Bhatia SK, Choi T-R, Choi Y-K, Kim HJ, Song H-S, Lee SM, Lee Park S, Lee HS, Koh J, Jeon J-M, Yoon J-J, Yang Y-H (2021b) Application of macroalgal biomass derived biochar and bioelectrochemical system with Shewanella for the adsorptive removal and biodegradation of toxic azo dye. Chemosphere 264:128539

Gurav R, Bhatia SK, Choi T-R, Kim HJ, Choi Y-K, Lee H-J, Ham S, Cho JY, Kim SH, Lee SH, Yun J, Yang Y-H (2021c) Adsorptive removal of synthetic plastic components bisphenol-A and solvent black-3 dye from single and binary solutions using pristine pinecone biochar. Chemosphere 296:134034

Gurav R, Bhatia SK, Choi T-R, Kim HJ, Song H-S, Park S-L, Lee S-M, Lee H-S, Kim S-H, Yoon J-J, Yang Y-H (2020) Utilization of different lignocellulosic hydrolysates as carbon source for electricity generation using novel Shewanella marisflavi BBL25. J Clean Prod 124084

Hlihor RM, Gavrilescu M, Tavares T, Favier L, Olivieri G (2017) Bioremediation: an overview on current practices, advances, and new perspectives in environmental pollution treatment. Biomed Res Int 2017:3–5. https://doi.org/10.1155/2017/6327610

Hristozkova M, Geneva M, Stancheva I, Boychinova M, Djonova E (2016) Contribution of arbuscular mycorrhizal fungi in attenuation of heavy metal impact on Calendula officinalis development. Appl Soil Ecol 101:57–63. https://doi.org/10.1016/j.apsoil.2016.01.008

Janssen DB, Stucki G (2020) Perspectives of genetically engineered microbes for groundwater bioremediation. Environ Sci Process Impacts 22:487–499. https://doi.org/10.1039/c9em00601j

Janusz G, Pawlik A, Sulej J, Świderska-Burek U, Jarosz-Wilkolazka A, Paszczyński A (2017) Lignin degradation: microorganisms, enzymes involved, genomes analysis and evolution. FEMS Microbiol Rev 41:941–962. https://doi.org/10.1093/femsre/fux049

Kim DW, Suhaimi MA, Kim BM, Cho MH, Chen FF (2013) Rough cut machining for impellers with 3-axis and 5-axis NC machines. Lect Notes Mech Eng 7:609–616. https://doi.org/10.1007/978-3-319-00557-7_50

Kim JE, Bhatia SK, Song HJ, Yoo E, Jeon HJ, Yoon J-Y, Yang Y, Gurav R, Yang Y-H, Kim HJ, Choi Y-K (2020) Adsorptive removal of tetracycline from aqueous solution by maple leaf-derived biochar. Biores Technol 306:123092. https://doi.org/10.1016/j.biortech.2020.123092

Kumar V, Dwivedi SK (2021) Bioremediation mechanism and potential of copper by actively growing fungus Trichoderma lixii CR700 isolated from electroplating wastewater. J Environ Manage 277:111370. https://doi.org/10.1016/j.jenvman.2020.111370

Kumar SS, Kadier A, Malyan SK, Ahmad A, Bishnoi NR (2017) Phytoremediation and rhizoremediation: uptake, mobilization, and sequestration of heavy metals by plants. Plant-Microbe Interact Agro-Ecol Perspect 2:367–394. https://doi.org/10.1007/978-981-10-6593-4_15

Kumar S, Kaushik G, Dar MA, Nimesh S, López-chuken UJ, Villarreal-chiu JF (2018) Microbial degradation of organophosphate pesticides: a review. Pedosphere 28:190–208. https://doi.org/10.1016/S1002-0160(18)60017-7

Liu Q, Tang J, Bai Z, Hecker M, Giesy JP (2015) Distribution of petroleum degrading genes and factor analysis of petroleum contaminated soil from the Dagang Oilfield, China. Sci Rep 5:11068

Mahamuni PP, Patil PM, Dhanavade MJ, Badiger MV, Shadija PG, Lokhande AC, Bohara RA (2019) Synthesis and characterization of zinc oxide nanoparticles by using polyol chemistry for their antimicrobial and antibiofilm activity. Biochem Biophys Rep 17:71–80. https://doi.org/10.1016/j.bbrep.2018.11.007

Mahamuni-Badiger PP, Patil PM, Badiger MV, Patel PR, Thorat-Gadgil BS, Pandit A, Bohara RA (2020a) Biofilm formation to inhibition: role of zinc oxide-based nanoparticles. Mater Sci Eng C 108:1–74. https://doi.org/10.1016/j.msec.2019.110319

Mahamuni-Badiger PP, Patil PM, Patel PR, Dhanavade MJ, Badiger MV, Marathe YN, Bohara RA (2020b) Electrospun poly(3-hydroxybutyrate-co-3-hydroxyvalerate)/polyethylene oxide (PEO) microfibers reinforced with ZnO nanocrystals for antibacterial and antibiofilm wound dressing applications. New J Chem. https://doi.org/10.1039/d0nj01384f

Malla MA, Dubey A, Yadav S, Kumar A, Hashem A, Abd-Allah EF (2018) Understanding and designing the strategies for the microbe-mediated remediation of environmental contaminants using omics approaches. Front Microbiol 9. https://doi.org/10.3389/fmicb.2018.01132

Medfu Tarekegn M, Zewdu Salilih F, Ishetu AI (2020) Microbes are used as a tool for the bioremediation of heavy metal from the environment. Cogent Food Agric 6. https://doi.org/10.1080/23311932.2020.1783174

Mello IS, Targanski S, Pietro-Souza W, Frutuoso Stachack FF, Terezo AJ, Soares MA (2020) Endophytic bacteria stimulate mercury phytoremediation by modulating its bioaccumulation and volatilization. Ecotoxicol Environ Saf 202:110818. https://doi.org/10.1016/j.ecoenv.2020.110818

Nagda A, Meena M, Shah MP (2021) Bioremediation of industrial effluents: a synergistic approach. J Basic Microbiol. https://doi.org/10.1002/jobm.202100225

Ojuederie OB, Babalola OO (2017) Microbial and plant-assisted bioremediation of heavy metal polluted environments: a review. https://doi.org/10.3390/ijerph14121504

Pande V, Pandey SC, Sati D, Pande V, Samant M (2020) Bioremediation: an emerging effective approach towards environment restoration. Environ Sustain 3:91–103. https://doi.org/10.1007/s42398-020-00099-w

Raghu G, Balaji V, Venkateswaran G, Rodrigue A, Maruthi Mohan P (2008) Bioremediation of trace cobalt from simulated spent decontamination solutions of nuclear power reactors using E. coli expressing NiCoT genes. Appl Microbiol Biotechnol 81:571–578. https://doi.org/10.1007/s00253-008-1741-6

Rao MA, Scelza R, Scotti R, Gianfreda L (2010) Role of enzymes in the remediation of polluted environments. J Soil Sci Plant Nutr 10:333–353. https://doi.org/10.4067/S0718-95162010000100008

Rekik H, Zaraî Jaouadi N, Bouacem K, Zenati B, Kourdali S, Badis A, Annane R, Bouanane-Darenfed A, Bejar S, Jaouadi B (2019) Physical and enzymatic properties of a new manganese peroxidase from the white-rot fungus Trametes pubescens strain i8 for lignin biodegradation and textile-dyes biodecolorization. Int J Biol Macromol 125:514–525. https://doi.org/10.1016/j.ijbiomac.2018.12.053

Sarkar J, Kazy SK, Gupta A, Dutta A, Mohapatra B, Roy A, Bera P, Mitra A, Sar P (2016) Biostimulation of indigenous microbial community for bioremediation of petroleum refinery sludge. Front Microbiol 7:1–20. https://doi.org/10.3389/fmicb.2016.01407

Sharma B, Dangi AK, Shukla P (2018) Contemporary enzyme-based technologies for bioremediation: a review. J Environ Manage 210:10–22. https://doi.org/10.1016/j.jenvman.2017.12.075

Sharma R, Kumar R, Satapathy SC, Al-Ansari N, Singh KK, Mahapatra RP, Agarwal AK, Le HV, Pham BT (2020) Analysis of water pollution using different physicochemical parameters: a study of Yamuna River. Front Environ Sci 8:1–18. https://doi.org/10.3389/fenvs.2020.581591

Singh JS, Abhilash PC, Singh HB, Singh RP, Singh DP (2011) Genetically engineered bacteria: an emerging tool for environmental remediation and future research perspectives. Gene 480:1–9. https://doi.org/10.1016/j.gene.2011.03.001

Singh S, Kumar V, Upadhyay N, Singh J, Singla S, Datta S (2017) Efficient biodegradation of acephate by Pseudomonas pseudoalcaligenes PS-5 in the presence and absence of heavy metal ions [Cu(II) and Fe(III)], and humic acid. 3 Biotech 7. https://doi.org/10.1007/s13205-017-0900-9

Solano AMS, Garcia-Segura S, Martínez-Huitle CA, Brillas E (2015) Degradation of acidic aqueous solutions of the diazo dye Congo Red by photo-assisted electrochemical processes based on Fenton's reaction chemistry. Appl Catal B Environ 168–169:559–571. https://doi.org/10.1016/j.apcatb.2015.01.019

Srivastava J, Naraian R, Kalra SJS, Chandra H (2014) Advances in microbial bioremediation and the factors influencing the process. Int J Environ Sci Technol 11:1787–1800. https://doi.org/10.1007/s13762-013-0412-z

Stelting S, Burns RG, Sunna A, Visnovsky G, Bunt CR (2012) Immobilization of Pseudomonas sp. strain ADP: a stable inoculant for the bioremediation of atrazine. Appl Clay Sci 64:90–93. https://doi.org/10.1016/j.clay.2011.12.006

Suryawanshi SS, Kamble PP, Gurav R, Yang Y-H, Jadhav JP (2023) Statistical comparison of various agricultural and non-agricultural waste biomass-derived biochar for methylene blue dye sorption. Biomass Convers Biorefin 13:5353–5366

Tanvi DA, Pratam KM, Lohit RT, Vijayalakshmi BK, Devaraja TN, Vasudha M, Ramesh A, Chakra PS, Gayathri D (2020) Biosorption of heavy metal arsenic from Industrial Sewage of Davangere District, Karnataka, India, using indigenous fungal isolates. SN Appl Sci 2:1–7. https://doi.org/10.1007/s42452-020-03622-0

Tišma M, Šalić A, Planinić M, Zelić B, Potočnik M, Šelo G, Bucić-Kojić A (2020) Production, characterisation and immobilization of laccase for an efficient aniline-based dyedecolourization. J Water Process Eng 36:101327. https://doi.org/10.1016/j.jwpe.2020.101327

Upadhyay MK, Yadav P, Shukla A, Srivastava S (2018) Utilizing the potential of microorganisms for managing arsenic contamination: a feasible and sustainable approach. Front Environ Sci 6:1–11. https://doi.org/10.3389/fenvs.2018.00024

Urionabarrenetxea E, Garcia-Velasco N, Anza M, Artetxe U, Lacalle R, Garbisu C, Becerril T, Soto M (2021) Application of in situ bioremediation strategies in soils amended with sewage sludges. Sci Total Environ 766:144099. https://doi.org/10.1016/j.scitotenv.2020.144099

Vaccari M, Castro FD, Stolfini M (2020) Material flow analysis and heavy hydrocarbon removal in a full-scale biopile and soil washing plant in northern Italy. Waste Manag Res 38:966–977. https://doi.org/10.1177/0734242X20934176

Varjani SJ, Upasani VN (2017) Critical review on biosurfactant analysis, purification and characterization using rhamnolipid as a model biosurfactant. Bioresour Technol 232:389–397. https://doi.org/10.1016/j.biortech.2017.02.047

Vyavahare G, Jadhav P, Jadhav J, Patil R, Aware C, Patil D, Gophane A, Yang Y-H, Gurav R (2019) Strategies for crystal violet dye sorption on biochar derived from mango leaves and evaluation of residual dye toxicity. J Clean Prod 207:296–305

Wang T, Wang J, Jin Y (2007) Slurry reactors for gas-to-liquid processes: a review. Ind Eng Chem Res 46:5824–5847. https://doi.org/10.1021/ie070330t

Wang T, Yang Y, Zhao T (2019a). Genome Sequence of a 3, 2015. https://doi.org/10.1128/genomeA.00793-15.Copyright

Wang R, Lin JQ, Liu XM, Pang X, Zhang CJ, Yang CL, Gao XY, Lin CM, Li YQ, Li Y, Lin JQ, Chen LX (2019b) Sulfur oxidation in the acidophilic autotrophic Acidithiobacillus spp. Front Microbiol 10. https://doi.org/10.3389/fmicb.2018.03290

Wang X, Aulenta F, Puig S, Esteve-Núñez A, He Y, Mu Y, Rabaey K (2020) Microbial electrochemistry for bioremediation. Environ Sci Ecotechnol 1:100013. https://doi.org/10.1016/j.ese.2020.100013

Wu Y, Jing X, Gao C, Huang Q, Cai P (2018) Recent advances in microbial electrochemical system for soil bioremediation. Chemosphere 211:156–163. https://doi.org/10.1016/j.chemosphere.2018.07.089

Xu M, Liu Y, Deng Y, Zhang S, Hao X, Zhu P, Zhou J, Yin H, Liang Y, Liu H, Liu X, Bai L, Jiang L, Jiang H (2020) Bioremediation of cadmium-contaminated paddy soil using an autotrophic and heterotrophic mixture. RSC Adv 10:26090–26101. https://doi.org/10.1039/d0ra03935g

Yin X, Wang L, Zhang Z, Fan G, Liu J, Sun K, Sun GX (2017) Biomethylation and volatilization of arsenic by model protozoan Tetrahymena pyriformis under different phosphate regimes. Int J Environ Res Public Health 14. https://doi.org/10.3390/ijerph14020188
Yuan Y, Zhou L, Hou R, Wang Y, Zhou S (2021) Centimeter-long microbial electron transport for bioremediation applications. Trends Biotechnol 39:181–193. https://doi.org/10.1016/j.tibtech.2020.06.011
Zhan F, Li B, Jiang M, Yue X, He Y, Xia Y, Wang Y (2018) Arbuscular mycorrhizal fungi enhance antioxidant defense in the leaves and the retention of heavy metals in the roots of maize. Environ Sci Pollut Res 25:24338–24347. https://doi.org/10.1007/s11356-018-2487-z
Zhang T, Klapper I (2010) Mathematical model of biofilm induced calcite precipitation. Water Sci Technol 61:2957–2964. https://doi.org/10.2166/wst.2010.064

Phytochelatins: Heavy Metal Detoxifiers in Plants

Sonia Sethi

1 Introduction

Heavy metals are defined as elements with a relative density of more than 5 g cm^3. Arsenic (As), lead (Pb), chromium (Cr), cadmium (Cd), nickel (Ni), and mercury are heavy metals (HMs) that are widely connected with poisoning of people, plants, and other creatures (Hg). Heavy metal poisoning poses a huge hazard to all living forms on Earth, resulting in food chain pollution. Humans, animals, and plants are all poisoned by high levels of HMs in their tissues. The increased concentration of HMs in the soil is due to both natural and human-caused causes. Water, food, and air are three major routes for HMs to enter the human body. These hazardous metals bond to the cellular architecture of animals, causing biological activities to be disrupted. The toxicity of HMs varies based on a number of parameters, including exposure period, metal species reactivity, metal concentration, and the health state of those exposed.

Some heavy metals are nutritionally important (e.g., iron, cobalt, and zinc), whereas others are potentially toxic (ruthenium, silver, indium). In excessive levels or in particular chemical forms, metals in the second category might be regarded as potentially hazardous elements. In addition, certain heavy metals (mercury, cadmium, and lead) are extremely hazardous. Toilets, industrial wastes, agricultural runoff, occupational contacts, and dyes are all potential causes of heavy metal toxicity. Heavy metals are found in trace levels in the earth's crust, yet they are employed in a wide range of products, including cell phones, plastics, automobiles, and pesticides (Hubner et al. 2010).

Metals including cobalt, copper, chromium, iron, manganese, magnesium, molybdenum, nickel, selenium, and zinc have been shown to play key roles in several

S. Sethi (✉)
Dr. B. Lal Institute of Biotechnology, Malviya Industrial Area, Malviya Nagar, Jaipur, India
e-mail: soniakaura@drblal.com

M. P. Shah (ed.), *Advanced and Innovative Approaches of Environmental Biotechnology in Industrial Wastewater Treatment*,
https://doi.org/10.1007/978-981-99-2598-8_16

physiological and biological processes. As a result, a deficiency of these components might result in a variety of symptoms and disorders. Heavy metals have been defined as a category of metals with relatively large densities, atomic weights, or atomic numbers, despite the lack of a consistent definition. Heavy metal bioavailability is influenced by a variety of chemical and physical parameters, including the temperature of absorption and the physiological properties of the exposed species (Stern 2010).

Contamination of HMs is a major environmental hazard across the world. HM poisoning of ground water is also connected to the growth of cities, the rise of industry, and the extensive use of chemicals in agriculture. HMs are found in trace concentrations in the environment, i.e., 10 ppm, and are classified as trace elements. Because of their increased toxicity, As, Pb, Cd, Hg, and Cr are considered priority metals in terms of public health. The US Environmental Protection Agency has classed heavy metals as significant human carcinogens (USPEA). Although several key HMs are significant elements of a variety of important enzymes and play critical roles in many redox processes, their excessive exposure has been associated to a number of negative consequences and has been connected to a variety of disorders (WHO 1996).

2 Heavy Metals Toxicity

Heavy metals have been widely distributed in nature due to their various uses in industry, agriculture, medicine, and technology, prompting worries about their impact on human health and the environment. The degree of heavy metal toxicity is determined by a number of parameters, including dosage, contact pattern, chemical species, as well as the age, sex, genetics, and nutritional state of the individual who is exposed. Because of their high toxicity, arsenic, chromium, cadmium, lead, and mercury are significant metals in terms of public health. These metals are regarded to be systemic poisons that can cause organ failure even at low doses of exposure. As a result, worries regarding public health and environmental contamination from these heavy metals have grown in recent years.

Heavy metal exposure has risen in recent decades as a result of their extensive use in industry. Environmental pollution is also a major issue in mining, casting, and other related industries. Metallic ions can interact with biological components including DNA and nuclear proteins, causing apoptosis and carcinogenesis as a result of DNA damage and structural alterations (Flora et al. 2009). Reactive oxygen species and oxidative stress have a major role in the toxicity and carcinogenicity of heavy metals including arsenic, cadmium, chromium, lead, and mercury, according to laboratory study. Because of their systemic toxicity, which may harm multi-system organs even at low dosages, these metals have a substantial impact on overall health.

Arsenic, cadmium, chromium, lead, and mercury all contribute considerably to pollution in the environment. These elements enter the human body by ingestion, inhalation, and skin contact, causing a variety of health issues including cardiovascular disease, neurological and neurobehavioral abnormalities, diabetes, blood irregularities, and cancers of different forms. The health consequences of these metals vary depending on the metal and chemical composition. These effects are also dose-dependent over time. Heavy metal exposure has been linked to long-term health issues in humans, according to several studies. Some of these metals are hazardous in both acute and chronic forms. According to recent research, these hazardous ions may interact with physiologically vital metals like iron, calcium, and zinc, impairing their normal metabolic function (Flora et al. 2009).

Because of its capacity to inter-convert between Cu II to I oxidation states, copper (Cu) is a crucial cofactor for several enzymes involved in reactive oxygen species (ROS) management. Cu's toxicity is linked to the generation of reactive oxygen species (ROS), which causes oxidative stress. Many additional HMs, like Cu, are essential for the active functioning of biological circuits. Non-essential metals include As, aluminium (Al), antimony (Sb), barium (Ba), Ni, Cd, beryllium (Be), bismuth (Bi), Pb, Hg, indium (In), lithium (Li), vanadium (V), silver (Ag), tellurium (Te), platinum (Pt), tin (Sn), strontium (Sr), and uranium (U).

Millions of people in nations like India, Bangladesh, Mexico, Chile, Taiwan, and Uruguay are exposed to heavy metals on a daily basis. Heavy metal concentrations in soil rise as a result of pesticides, chemical fertilizers, and waste disposal. According to the USEPA (2020), the limit for HMs in soil and oral dosage is 0.77 mg kg^1 and 0.33 g kg^1 day^1, respectively, for As, 78 and 1 for Cd, 0.31 and 3 for Cr, 400 and N/A for Pb, 11 and N/A for mercury, and 1600 mg kg^1 and 20 g kg^1 day^1 for Ni (USEPA 2020). Heavy metal pollution is a severe concern in both the environmental and occupational environments.

Cd is typically found in soil at a value of 0.1 mg kg^1. Continuous use of Cd in industry has resulted in a significant increase in human exposure and contamination. In many wealthy nations, commercial usage of Cd has decreased in order to reduce pollution. For example, in the United States (US), daily Cd consumption is around 0.4 g kg^1 day^1, which is lower than the USEPA's recommended oral reference dosage (USEPA 2006). Cd has been classified as a human carcinogen by the US National Toxicology Program and the International Agency for Research on Cancer (IARC). Mercury is another common contaminant that can cause a variety of health problems. In the environment, there are inorganic, organic, and elemental forms of mercury, each with its own mode of toxicity. Methylmercury is the most prevalent organic form of mercury in the environment, and it is created as a result of methylation by microorganisms in soil and water.

Chromium is a naturally occurring element, and the health risks connected with it are dependent on the oxidation states in which it is present (high toxicity of the hexavalent form). In surface and drinking water, the recommended allowable limit of Cr (CrVI) is 50 g L^{-1} (WHO 1996). Cr concentrations in the atmosphere range between 1 and 100 ng cm^3, but can surpass this in locations where Cr is used in industry. In soil, chromium levels range from 1 to 3000 mg kg^{-1}, in sea water from 5 to 800 g L^{-1}, and in lakes and rivers from 26 to 5.2 mg L^{-1} (Yedjou et al. 2012). Cr has been shown to cause cancer in humans, although the mechanism has to be investigated further. The solubility of Cr's compound and its oxidation state are two important elements that influence its toxicity. As a result, the carcinogenicity of HMs is a major focus of study in terms of public health.

Arsenic is a naturally occurring element that may be found in low amounts in practically all situations. A considerable percentage of arsenic-containing chemicals are produced industrially for use as pesticides, herbicides, and fungicides in agriculture. These are also used in veterinary medicine to eradicate pumpkin worms from sheep and cow herds. In medical research, arsenic compounds have long been used to treat illnesses including syphilis, amoebic dysentery, and trypanosome.

Millions of people appear to be exposed to arsenic on a daily basis across the world, particularly in Bangladesh, India, Mexico, and Taiwan, where ground water is polluted with high levels of arsenic. Arsenic contamination is a major public health hazard. Several investigations have discovered substantial links between arsenic exposure and a variety of clinical illnesses, such as cardiovascular disease, peripheral vascular disease, diabetes, hearing loss, blood abnormalities (anaemia, leukopenia, and eosinophilia), and cancer (Centeno et al. 2006).

Lead is very hazardous to the liver, kidneys, central nervous system, and reproductive system, among other organ systems. Lead poisoning, on the other hand, appears differently in children and adults. Blood lead levels are linked to lower intellectual quotient, neurological impairments, and hearing loss in children. Acute lead poisoning causes brain damage, kidney damage, and digestive system illnesses in adults, while chronic lead poisoning affects blood cells, the central nervous system, the kidneys, and vitamin D metabolism. Lead also enters the bone structure when it combines with bone minerals. Lead poisoning has been proven to cause cellular deterioration by causing the creation of reactive oxygen species (Flora et al. 2007).

3 Heavy Metals (HMs) Detoxification

The initial stage in treating heavy metal poisoning is to identify and eradicate the source of exposure, as well as to keep the patient away from it. The first-line therapy involves symptomatic and supportive care, as well as ensuring that various organs of the body, including as the kidney, liver, respiratory, and cardiovascular systems, are monitored. Chelating agents are the most often used treatment (Trevor et al. 2010; Shah 2020).

Toxic metal ions cause oxidative stress in cells by releasing reactive oxygen species (ROS; Li et al. 2016). They cause DNA damage and/or inhibit DNA repair pathways, obstruct membrane function, disrupt nutritional homeostasis, and disrupt protein function and activity. Plant cells, on the other hand, have evolved a variety of adaptive systems to handle excess metal ions and use detoxifying processes to avoid involvement in harmful reactions.

Plants use techniques to avoid or minimize metal ion absorption in the apoplast by binding them to the cell wall or cellular exudates, or by impeding long-distance transport (Hasan et al. 2015). Cells, on the other hand, activate a complex network of storage and detoxification strategies when metal ions are present at high concentrations, such as chelation of metal ions with phytochelatins and metallothioneins in the cytosol, trafficking, and sequestration into the vacuole by vacuolar transporters.

4 Chelation Background

Chelation, which comes from the Greek word "chelos," or "claw," is the process of an organic molecule, the chelating agent, incorporating a mineral ion or cation into a complex ring structure. Sulphur, nitrogen, and/or oxygen are common electron-donor atoms on chelating molecules.

The strength of chemical bonds produced between chelators and metal ions in coordination complexes is determined by the elements involved and the stereochemistry. With a variety of metal ions that could bind competitively with the chelator (e.g., calcium, magnesium, zinc, copper, manganese, and other metals that typically exceed concentrations of toxic elements), the identity of the metal predominately bound by a chelating agent is determined by the chelator's accessibility to the tissues, the strength with which the metal is already bound in the tissues, the strength with which the metal binds to the chelator, and to some extent.

Chelators work by mobilizing metals from tissues while keeping the chelate moiety intact during circulation to the kidneys for urine excretion and the liver for bile excretion. Enterohepatic recirculation and reabsorption in the kidney are major sources of concern (Rooney 2007).

Another factor to consider is the chelate's solubility in water and lipids. Aqueous solubility improves blood transport and renal excretion, but a lipophilic chelator may have a stronger ability to chelate intracellular elements by penetrating cellular membranes (including those in the central nervous system). The bile may also be used to expel a lipophilic chelator in larger amounts. Active transport of intracellular metal complexes via "drug resistance proteins" may alter these generalizations (Thevenod 2010; Shah 2021).

5 Roles of Chelation in Natural Toxicokinetics

Metal-binding proteins, such as metallothioneins, are effective heavy metal chelators and play an important role in the body's natural reaction to these toxins (Klaassen et al. 2009). Glutathione is a biomarker for toxic metal overload and is involved in cellular response, transport, and excretion of metal cations (Geier et al. 2009).

Chelating chemicals are produced by both animals and plants, and the amount of metallothionein in diet can impact the bioavailability and metabolism of hazardous metals like cadmium (Pal and Rai 2010). Some foods have been recommended to help assist natural detoxification processes and prevent harmful metal absorption or reabsorption.

i. Food fibres, such as those found in cereals and fruits, can be utilized as a chelating therapy in conjunction with the primary treatment to prevent enterohepatic recirculation.
ii. Other natural polymers, such as algal polysaccharides alginate, chlorella, and citrus pectin, have been found to be potential heavy metal absorbents.
iii. Because hazardous metals have a strong affinity for sulfur-containing peptides, a diet rich in sulfur-containing foods such as alliums (a plant family that includes garlic and onion) and brassicas (like broccoli) is used to alleviate poisoning symptoms and accelerate heavy metal elimination.
iv. Cilantro (Coriandrum sativum) is a popular culinary and medicinal herb.

It inhibited the delta-aminolevulinic acid dehydratase (ALAD) enzyme and reduced lead absorption into bone in mice. In a recent experiment including 3- to 7-year-old children who had been exposed to lead, a cilantro extract was shown to be as efficient as placebo in boosting renal excretion (improvements across treatment and placebo groups were ascribed to improved diet during the intervention).

6 Heavy Metal Detoxification Through Phytoremediation

Human and animal exposure to HMs is mostly caused by rocks, soils, water, and the environment. Plants rely on these nutrients to survive, yet their sessile nature prevents them from avoiding pollutants. Plants have evolved their own detoxification and tolerance systems as a result. Plants absorb HMs and store them in various tissues in order to keep the concentration below dangerous levels. The buildup of hazardous contaminants occurs when people and animals consume these polluted plants. Plant roots allow HMs in the environment to enter the plant and translocate them to above-ground sections, disrupting the normal functioning and physiology of the plant and resulting in restricted development (Chauhan et al. 2020). The effect of HMs pollution in soil is a loss of soil fertility, a significant fall in agricultural productivity, and a decline in microbial diversity and activity of microorganisms (Kushwaha et al. 2015). To avoid food chain contamination, a thorough examination and knowledge of plant routes and processes is required.

Several strategies have been used to manage HMs contamination in soil at this moment, however they are insufficient. Thermal treatment, excavation and landfill, and acid leaching are not suited for commercial use and operations due to their prohibitive costs and limited efficiency. The use of plants, often known as phytoremediation, is an environmentally beneficial and cost-effective way for reducing hazardous HMs in soil. Plantation demonstrated to be an effective strategy for soil reclamation for HMs polluted locations. Phytoremediation aids in the restoration of natural habitats as well as the reduction of HMs stress and pollution in the environment. Plants have evolved methods to assist them resist HM stress, such as ROS homeostasis, chelation and sequestration of HMs, and exclusion of metal ions.

The microbial diversity of rhizospheric soil is well recognized to alter the transportation and availability of HMs to plants. Plant growth-promoting rhizospheric microorganisms supplemented in soil and seeds treated with them demonstrate many growth-promoting features and increase plant nutritional status. Microbes benefit the environment by producing siderophores, fixing nitrogen, releasing phytohormones, and increasing nutrient levels. There have been several studies on the use of rhizospheric microorganisms in conjunction with plants, a process known as rhizoremediation.

Phytoextraction (the use of metal accumulators to remove toxic HMs from soil), phytovolatilization (the production of volatile metal derivatives and evaporation through aerial parts), phytostabilization (plants reduce the bioavailability of toxic metals in soils), and rhizofiltration are all examples of phytoremediation (exclusion of toxicants from polluted water either through roots of plant or microorganisms associated with the rhizosphere). Plants engage their defence system in a variety of ways when exposed to HMs, including immobilization, compartmentalization of complexed metal ions, exclusion, and the production of stress-sensitive proteins and hormones (Chauhan et al. 2020).

Plant–microbe interactions improve the efficiency of phytoremediation. Microorganisms in the soil, particularly in the rhizosphere, play an important role in soil structure, nutrient loss prevention, toxicant detoxification, better plant growth and production, and insect control. As a result, the presence of rhizospheric bacteria boosts plants' ability to recover from HMs stress. Plants and rhizospheric microorganisms have a direct relationship in which plants provide carbon to microbes, which then aid in the removal of HMs from contaminated soil. Plants and rhizospheric bacteria also have an indirect connection in which plants promote microbial diversity while microorganisms use their metabolic activities to breakdown pollutants in the soil.

Plant growth-promoting microorganisms (PGPMs) equipped with HMs tolerance machinery, as well as nutrient supplementation, are proposed to be effective for HMs polluted soil restoration. Another key technique for removing HMs and protecting the food chain is to employ non-food crops that are not eaten by humans or other animals. In addition, using microorganisms in a consortium is a powerful strategy for reducing HMs stress in severely polluted locations. Plants grown in HMs-polluted soil are treated with PGPMs and nutrient additions, which have dual advantages in that they not only detoxify HMs-induced toxicity but also biofortify nutrients. The use of genetic engineering in the development of "microbial biosensors" is a new and exciting technology for HMs remediation and detection of contaminated sites (Joshi et al. 2011).

Stress caused by heavy metals in plants is mediated by a complicated signal transduction system, which is a two-step process that begins with the detection of the heavy metal (s). Plant vigour and capacity to deal with (metal) stress will be reduced if key nutrients are depleted. When a plant detects metals, it produces stress-related proteins and signalling molecules, which leads to the explicit activation of metal-responsive genes to alleviate the stress. As a result, the production of metal-specific legends (chelation) and subsequent compartmentalization of ligand–metal complexes in cells might be typical defensive mechanisms for heavy metal detoxification in plants and other animals.

Because these fungi have acquired resistance to HMs, mycorrhizal connections are known to boost the efficacy of phytoremediation in HMs-polluted soil. Fungal hyphae give plants with a huge absorption surface area, allowing them to absorb more water and nutrients. Mycorrhizae also serves as an exclusion barrier for HMs, limiting their absorption and accumulation in plants. Bioremediation is the most effective and environmentally acceptable method for removing harmful metals from the environment. It is critical to apply PGPRs in HM-polluted soil to prevent the use of excessive chemical fertilizers and to retain the soil's nutritional characteristics and structure. This is one of the most promising strategies for metalliferous environment cleanup, safe agricultural practises, and better metal tolerance mediated by bacteria (Mishra et al. 2017a).

Microbes aid in the decrease of metal bioavailability to plants through a variety of methods. Conversion of harmful metals into less toxic or non-toxic forms, enzymatic redox processes, metal chelation, bioaccumulation, and exclusion of HMs for improved survival in HM-polluted environments are examples of microbe resistance mechanisms. Microbial supplementation has a good influence on HMs-exposed plants because of their beneficial direct and indirect mechanisms of action, such as exopolysaccharide synthesis, biofilm formation, phytohormone production, and siderophores production. Nowadays, genetically modified microorganisms, also known as new phytomicrobial strategies, are widely used to improve HM amelioration and plant stress resistance. Due to the persistent and non-degradable character of HMs, the daily increase in concentration of HMs in soil and water has gotten a lot of attention across the world.

7 Enhanced Phytoremediation by Increasing Plant Capacities

Increase plant biomass production and, as a result, Cd phytoremediation rates, which may be done by soil amendments or plant alterations directly, is one of the easiest techniques for phytoremediation augmentation. Before planting, seed pre-treatments (priming) can improve seedling vigour and biomass output.

Seed priming is the controlled rehydration (imbibition) of seeds for metabolic activity induction without radicle emergence, followed by seed drying and re-imbibition before planting. It's commonly used to boost seed vigour, increase germination, and achieve germination uniformity, especially under stressful situations. The start of rehydration triggers "pre-germinative metabolism," which includes cellular processes including de novo nucleic acid and protein synthesis, phospholipid and sterol buildup, DNA repair, and antioxidant systems activation. Although the potential of plant priming in abiotic stress tolerance has been extensively investigated using various types of molecules added exogenously to plant organs (roots, leaves, etc.) with increased abiotic stress tolerance (Mishra et al. 2017a, b), there are only a few papers concerning how seed priming affects tolerance levels and what mechanism plants use to remember their "primed" state in seeds. Seed priming with various agents (water, proline, salicylic acid, silicic acid) enhanced biomass production by improving tolerance levels in salicylic acid under Cd stress (Antoniou et al. 2020).

Salicylic acid, a plant hormone, is one of the most researched chemicals in plant priming (SA). This hormone is important in a variety of metabolic activities, including antioxidant response to various abiotic stresses. Seed priming and plant priming with salicylic acid has been shown to impact the antioxidant state of plants, resulting in enhanced tolerance levels to heavy metal exposure (Choudhary et al. 2021). Exogenously administered salicylic acid can improve heavy metal tolerance and phytoremediation effectiveness in plants. Proline is another chemical that may be utilized as a priming agent. Proline is a stress signal in plants that have been exposed to abiotic stress.

Seed priming, which makes use of proline, can imprint seeds to protect them from abiotic stresses like heavy metals. Seed priming generates changes in metabolism that are remembered and transmitted to developing plants during heavy metal stress, although the exact mechanism is yet understood (Mladenov et al. 2021). The main benefit of this approach is how simple it is to do seed priming; it frequently involves basic processes like hydropriming (water pre-treatment of seeds), and it is regarded as an environmentally benign way of seed performance improvement. Because radicle emergence must be prevented, the timing of priming must be altered to ensure that metabolic activities are launched but no radicle emerges.

8 Uptake and Translocation of Heavy Metals Through Transporters

Metal ion deposition outside roots leads to metal ion entrance in roots and their translocation throughout the shoot via the xylem using the mass flow and diffusion technique. Metal transporters are involved in metal uptake and homeostasis in a variety of ways. Heavy Metal ATPases (HMAs), the ZIP (zinc/iron-regulated transporters) family, NRAMPs (natural resistance-associated macrophage protein), CDF (Cation Diffusion Facilitator), and Ca^{2+} are all members of this family. The CPx-type ATPases use energy in the form of ATPP1Btype to let hazardous metals including copper (Cu), lead (Pb), and cadmium (Cd) pass through the plasma membrane. The CDF transporters are involved in the cytoplasmic efflux of transition metal cations from the cytoplasm to the nucleus, such as Zn^{2+}, Cd^{2+}, Co^{2+}, Ni^{2+}, or Mn^{2+} (Koźmińska et al. 2018).

MHX, a member of the Ca^{2+}/ cation antiporter (CaCA) superfamily, is a vacuolar Mg^{2+} and Zn^{2+}/H^{+} exchanger that was found in larger quantities in the leaves of A. halleri than in A. thaliana, suggesting that it may play a role in Zn vacuolar storage. The NRAMP gene family in plants is very important since it is responsible for the absorption of the nutritionally important divalent cations Fe^{2+}, Mn^{2+}, Zn^{2+}, and Cd^{2+}, a toxic metal with no known significance in plant growth and development. ABC transporters are common transporters that play a role in a variety of physiological activities. With 29, 128, and 48 members in *S. cerevisiae*, Arabidopsis, and humans, respectively, this family is one of the biggest protein families. ATP-binding cassette (ABC) transporters have been found to play a function in Cd tolerance in plants (Takahashi et al. 2014).

Arsenic (As), mercury (Hg), and cadmium (Cd) resistance have been demonstrated using phytochelatins (PCs) transporters, ABC transporters (AtABCC1 and AtABCC2) [27]. ZIP transporters are also known as ZRT, IRT-like proteins, and they are involved in the translocation of various ions such as Cd, Zn, Fe, and Mn, depending on the substrate and their identity. They play a role in the transportation of Cd from the soil to the roots, as well as the subsequent transfer of Cd from the root to the shoot. Both HMA3 and HMA4, which are known to transport heavy metals (Cd and Zn) in Cd/Zn hyperaccumulators *A. halleri and N. caerulescens*, have their expression regulated (Park et al. 2012).

9 Phytochelatins: The Heavy Metal Chelator

Phytochelatins (PCs) are the best-studied heavy metal chelators in plants, notably in terms of Cadmium Cd tolerance. Plants, fungi, and all species of algae, including cyanobacteria, create phytochelatins, which are non-protein cysteine-rich oligopeptides synthesized by the enzyme phytochelatin synthase. Hayashi and his team originally found the peptide as Cd-binding complexes in fission yeast, S. pombe, subjected to Cd^{2+}, and dubbed them "cadystins." Hayashi and his colleagues identified Cadystins, A and B (c-Glu-Cys) n-Gly with $n = 2$ and 3, respectively, in 1984; however, Grill et al. (1985) (30) reported a widespread presence of the same peptides, as well as those with higher degrees of polymerization, and dubbed them "phytochelatins" ($n = 2$–11).

Phytochelatins are tiny cysteine-rich peptides that use thiolate coordination to bind heavy metal ions. With cadmium, they typically form a Mr 3600 complex. Metal ions such as cadmium, zinc, copper, lead, and mercury are thought to be accumulated, detoxified, and metabolized in plant cells by phytochelatins (Grill et al. 1985). [Glu(-Cys)]n-Gly ($n = 2$ to 8) is the general structure of this group of peptides. We found two of these peptides ($n = 2$ and 3) in the yeast *Schizosaccharomyces pombe* (Kondo et al. 1985) and expanded our results to the same class of chelating chemicals ($n =$ 2 to 8) seen in higher plants (Grill et al. 1986).

Phytochelatins bind to HMs such as As and Cd and transport them into the vacuole (Tripathi et al. 2013). Metallothioneins (MTs) are a cysteine-rich, low-molecular-weight metal binding peptide with a well-documented involvement in metal tolerance. By releasing acids, chelating agents, phosphate solubilization, and redox potential shifts, rhizospheric microorganisms modify the bioavailability of HMs (Mishra et al. 2017a, b). The activity of *Alcaligenes faeccalis and Pseudomonas fuorescens* is claimed to oxidise As to AsV and reduce CrVI–III (Kushwaha et al. 2015).

Cobbett and Goldsbrough (2002) have characterized the structure and function of both MTs and PCs in terrestrial plants, as well as processes for HM sequestration and compartmentalization. MTs are now divided into 15 evolutionary distinct groups (Capdevila and Atrian 2011), with plant MTs having the greatest information on phylogeny and structure of metal-binding domains (Leszczyszyn et al. 2013). PCs and their precursors, which can also be implicated in HM sequestration, have had their structure, biosynthesis routes, and activity control comprehensively examined (Filiz et al. 2019a).

There has been a lot of research done on this peptide, indicating the relevance of PC peptides in harmful ion sequestering of borderline class metals in plants, yeast, and microbes, and Cd has been identified as the primary activator of the enzyme PC synthase. There have been reports that the PC synthase enzyme has been activated for the production of the peptide from GSH. Chemical inhibitors were used to investigate heavy metal tolerance and its link to PCs. *S. pombe and A. thaliana* also produced GSH biosynthesis mutants, such as Buthionine sulfoximine (BSO) (Grill et al. 1987) and PC-deficient Cd-hypersensitive mutants.

Plants that have been exposed to non-tolerant and non-accumulator plants, as well as hypertolerant and hyperaccumulator plants, have shown a significant rise in GSH and PC synthesis, as well as increased PC synthesis (Gupta et al. 2008). PCs are thought to be involved in the chelation of vital metal ions when they are present in excess, according to reports. Zn is an essential metal that is found in the structure of around 10% of Zn-dependent proteins (Tennstedt et al. 2009). When the Zn threshold limit in plants is exceeded (between 100 and 300 lg g-1 dry weight, depending on plant species and physiological condition), unexpected binding of Zn ions to thiols or other functional groups causes several important proteins to be disrupted (Tennstedt et al. 2009).

Phytochelatins are synthesized enzymatically in the presence of heavy metals from glutathione via phytochelatin synthase activity (Filiz et al. 2019b). They are only found in plants and certain microorganisms, and their synthesis also is initiated/transformed in the entity of anionic (Ag, Au, Cd, Cu, Hg, Pb, and Zn) and cationic (As) metal(loid)s (Shukla et al. 2016). Among these metal(loid)s, particularly As and Cd are major inducers of phytochelatin gene activity. Glutathione is a kind of precursor and is considered to play role together with chelating agents in detoxification of free radicals existing in the cells because of heavy metal activities. Phytochelatin residues containing thiols or sulfonyl groups have a high potential for interactions with heavy metals, and phytochelatin–metal stability is dependent on metal deposition intracellularly (Filiz et al. 2019b).

Following the synthesis of the phytochelatin–metal complex, it is rapidly transported into the plant cellular vacuole through an ATP-binding cassette transporter family or an organic solute transporter family (Ghori et al. 2016). Heavy metal sequestration from the cytosol to the vacuole is also aided by certain transporter families (cation diffusion facilitator and natural resistance-associated macrophage protein). This isn't the only method plants deal with heavy metal detoxification. Under Cd stress, simultaneous increases in phytochelatin activity boost antioxidant synthesis in Brassica chinensis. Phytochelatins promote the synthesis of antioxidant enzymes (superoxide dismutase, ascorbate peroxidase, glutathione peroxidase, glutathione and glutathione reductase) and may help plants cope with reactive oxygen species generated by heavy metal stress (Filiz et al. 2019b).

Cd xylem loading is influenced by chelating agents (phytochelatins), vacuolar sequestration, and apoplectic barriers, as well as loading activity to the xylem, where the high cation exchange capacity of xylem cell walls governs metal ion transport. To reduce Cd toxicity in hyperaccumulating plants, plants use various detoxification mechanisms such as Cd vacuolar sequestration, Cd chelation (binding Cd to S-containing ligands—phytochelatins, glutathione, and metallothionines—cysteine-rich, metal-binding proteins) and Cd chelation (binding Cd to S-containing ligands—phytochelatins, glutathione, and Phytochelatins have a function in metal inactivation and accumulation processes, whereas metallothionines are restricted to the cytosol and perform a minor or non-existent role in Cd accumulation.

The metal must be ligated to nicotianamine, glutathione (GSH), or phytochelatins (PCs) if it is carried by phloem, and PCs have a high affinity for Cd binding. Cd is thought to be loaded into the phloem in the form of Cd-thiolate complexes, where the stability of the Cd-S bond reduces toxicity, and xylem-to-phloem transfer is important for Cd transport in plants. Metal movement from the root through the xylem is regulated by sulphur and acetate ligands, and the capacity to load Cd into xylem parenchyma cells is determined by transport protein activity. Heavy metal transporters located in the shoot cell membrane regulate Cd^{2+} absorption from the xylem to the shoot symplast (Ishimaru et al. 2021). Many elements of Cd transport and accumulation remain unknown, particularly in plants with varying levels of Cd tolerance, resistance, and accumulation capability; greater understanding of those mechanisms will lead to more effective use of hyperaccumulating plants in phytoremediation procedures.

Cd is sequestered in vacuoles in plant leaves to decrease its deleterious effects on photosynthesis and other functions, or it is detoxified by chelating chemicals (glutathione, phytochelatins, metallothioneins, and other cysteine-rich membrane proteins), according to (Karalija et al. 2021). Plants can also limit Cd absorption by excreting carboxylic acid (citric, malic) and histidine from their root exudates.

10 Phytochelatinsynthase

The PCS is made up of two domains, one with a conserved N-terminal domain that demonstrates -glutamylcysteine dipeptidyl transpeptidase activity and the other with a variable C-terminal domain that is involved in metal sensing (Vatamaniuk et al. 2004). The chemical activities of the N- and C-domains, as well as the catalytic mechanism of eukaryotic PCS, were discovered using AtPCS1 as a model. The N-terminal half of AtPCS1 is adequate for GSH deglycination and elongation of PC molecules, indicating that the N-terminal domain is responsible for the core catalysis. The shortened AtPCS1 lacking the C-terminal domain, on the other hand, is less thermostable and has lower PC synthetase activity than the full-length enzyme. Notably, in the presence of Zn^{2+}, the loss of the C-terminal region entirely inhibits the enzyme's PC synthesis activity and partially inactivates PC production. These findings indicate that the C-terminal domain is required for protein stability and serves as a metal sensor. The C-terminal end of AtPCS1 is essential for the enhancement of PC synthesis activity, according to further data. Multiple areas implicated in Zn-dependent and As-dependent activation of PC synthesis may be found in the residues from Asp373 to the C-terminal end of AtPCS1 (Uraguchi et al. 2018).

The ping-pong process is used to synthesize PCs, with one GSH serving as the low-affinity substrate for the first step and one metal-GSH conjugate serving as the high-affinity substrate for the second (Vatamaniuk et al. 2000). GSH occurs at a much greater level (millimolar) than heavy metal ions in conventional PC synthesis processes in vitro, which mimic the concentrations of GSH and metal ions in the cytosol (micromolar). More than 98 percent of total metal ions are presumably linked with GSH as bis(glutathionato) metal ions (metalGS2), and free Cd concentrations can be as low as 106 M in this scenario. GSH and CdGS2 are two distinct molecules for PC production in these conditions.

A Gly residue is eliminated immediately after GSH reaches the catalytic site of PCS to produce the Glu-Cys acyl-enzyme intermediate, and subsequently a metal-GS2 accepts the Glu-Cys unit to make PC2 ((Glu-Cys)2-Gly). PCs are elongated after PC2 is manufactured by employing previously synthesised PCs as acceptors to receive Glu-Cys. PCS catalyses the elongation of peptide chains from the C terminus to the N terminus.

10.1 Heavy Metal Accumulation by Engineering PC

Another technique for maintaining the balance of GSH metabolism in cells with constitutive PC synthesis is to use pathway engineering to co-express both GSH and PC synthesis pathways. To boost PC synthesis without diminishing the GSH pool, a kinetic model of GSH and phytochelatin synthesis in plants shows that at least two enzymes, -glutamylcysteine synthetase (-ECS) and PCS, should be enhanced

(Wawrzyński et al. 2006). In reality, the impacts of altered GSH/PC production routes in Escherichia coli and tobacco plant have been studied.

While single-gene expression in the PC production pathway showed little impact, E. coli cells co-overexpressing these enzymes collected much larger quantities of PCs and Cd^{2+}. These findings back up "gene stacking" as a strategy for improving heavy metal metabolism. Although co-overexpressing these three genes in tobacco boosted various non-protein thiol classes, the transgenic plants' Cd^{2+} accumulation remained unchanged compared to the natural type. These findings show that, in addition to the availability of precursors for PC synthesis, there are additional processes that restrict Cd accumulation in plants (Yan et al. 2020).

Overall, the genetic engineering techniques used to manipulate PC production have shown promise in terms of increasing plant performance in heavy metal phytoremediation. There are certain drawbacks, such as the intricacy of the stress response generated by heavy metals (Del Buono et al. 2020). While increased PC synthesis can help with heavy metal chelation, additional aspects such as subsequent vacuolar sequestration or the delicate balance of GSH metabolic pathways under heavy metal stress must be taken into account if heavy metal tolerance is to be achieved.

PCS has a wide substrate selectivity and may be used as a substrate with GS derivates. PCS isolated from plant species, for example, may absorb S-alkylated GSH like S-methyl-GS and S-hexyl-GS. xenobiotic GS-conjugates, and substantial side residues (abbreviated as GS-conjugates). Benzyl-, nitrophenyl-, phenylbenzyl-, uracil-, bimane-, and acetamido-fluorescein-groups are among the bulky S-residues of GSH that can be transformed to -Glu-Cys-conjugates by PCS. PCS, on the other hand, tends to convert the -Glu-Cys-conjugate intermediate to a hydrogen group when using GS-derivates with these bulky S-linked side residues. As a result, instead of polymerizing GS-conjugates, PCS processes their hydrolysis.

Because of its capacity to handle GS-conjugates, PCS also participates in the biodegradation of xenobiotic substances in addition to its important function in heavy metal detoxification. In plant cells, glutathione conjugation is a significant mechanism for inactivating xenobiotic substances. By transferring xenobiotics in the cytosol to glutathione (GSH), glutathione transferase (GST) detoxifies them (Printz et al. 2016). These GS-conjugates enter vacuoles quickly, where they are sequestered and degraded. AtABCC1/AtMRP1 and AtABCC2/AtMRP2, which also transport PC-metal complexes into vacuoles, help transport GS-conjugates for vacuolar sequestration in Arabidopsis. The following degradation of GS-conjugates is probably performed in the vacuoles due to the great effectiveness of this sequestration mechanism.

Glutamyl-transpeptidase (GGT) in the vacuole initiates the degradation of GS-conjugates by removing the -Glu residue to generate Cys-Gly-conjugates, and subsequently carboxypeptidase cleaves the Gly residue, resulting in the buildup of Cys-conjugates. When vacuolar sequestration is not an option, PCS might start the GS-conjugate degradation.

11 Genetic Engineering Approaches

Following the generation of three genetically engineered lines, a phytochelatin synthase gene from *Brassica napus* was cloned and transformed into *Arabidopsis thaliana* AtPCs1 mutant cad1–3 using the Agrobacterium-mediated transformation floral dip method. These transgene lines and the *A. thaliana* AtPCs1 mutant were morphologically and physiologically evaluated. The transgenic lines were shown to be quite effective in remediation in this study (Bai et al. 2019). Despite the fact that phytochelatin synthase genes are found in many plants, not all of them have been discovered and described at the functional and molecular levels.

Fan et al. (2018) isolated two phytochelatin synthase genes from *Morus alba* and used the Agrobacterium-mediated floral dip technique to transfer them into Arabidopsis and tobacco plants to test their heavy metal accumulation capacities. When transgenic *Arabidopsis* and tobacco plants reached the seedling stage, total root length, and fresh weight readings, as well as atomic absorption spectrophotometry data, were used to examine their increasing biomass and Zn/Cd accumulation capacity rates. Phytochelatin synthase genes were identified as prospective genetic resources for heavy metal phytoremediation, according to their findings. After converting the Vicia sativa phytochelatin synthase genes 1 homolog into A. thaliana, its functioning under Cd stress was studied. The pCAMBIA1304 vector was used to introduce the amplified PCs1 gene from total RNA under the control of the cauliflower mosaic virus 35 promoter. After insertion, the PCs1 gene was transformed using the Agrobacterium-mediated floral dip technique (Zhang et al. 2018).

New genes identified in many types of creatures (animals, plants, and microorganisms) are used in the possible manipulation of accumulating or degrading heavy metals during the production of new genetically engineered plants. Kühnlenz et al. (2015) conducted an intriguing transgenic study employing phytochelatin synthase genes originating from *Caenorhabditis elegans* and *Schizosaccharomyces pombe* for heavy metal detoxification in *Arabidopsis* as study material. In that study, CePCs genes were introduced into an AtPCs1 mutant line to investigate Cd accumulation, CePCs transcript, and phytochelatin content. Apart from members of the Brassicaceae family, poplar (*Populus tomentosa*) is a widely distributed model plant species that is actively employed in heavy metal phytoremediation.

After using the Agrobacterium-mediated leaf disc method to transfer the PtPCs gene to tobacco, wild and transgenic types were used to investigate Cd stress in terms of morphological and physiological indices (leaf relative electrolyte leakage, malondialdehyde content, total superoxide dismutase activity, chlorophyll content, and root activity), as well as Cd tolerance rate. PtPCs may have a role in Cd tolerance and accumulation, but not in Cd transport, according to the scientists (Chen et al. 2015b).

Glutathione-dependent phytochelatin synthesis may be produced in transgenics not only by transferring phytochelatin synthase genes, but also by inducing glutathione-dependent genes. XCD1, also known as MAN3, from *A. thaliana* Heynh columbia-0 was transferred to Arabidopsis mutant genotypes and examined for Cd accumulation and tolerance after the cloning of a kind of inducer gene. Transgenic and mutant Arabidopsis lines were studied for gene transcripts and protein expressions, Cd accumulation, glutathione, and phytochelatin content. By cross-talking with related genes, XCD1 regulates glutathione-dependent phytochelatin synthesis (Chen et al. 2015a).

Plant genetic engineering may be used to create hyperaccumulators that are more sophisticated, efficient, and durable. Plants that are good candidates for genetic engineering include those that produce a lot of biomass and have a lot of heavy metal storage capacity. Induction of gene overexpression, such as glutamylcysteine synthetase results in boosting heavy metal buildup, can be achieved by genetic engineering. The gshl gene from Escherichia coli is expressed in Brassica juncea transgenic plants, which produce larger levels of phytochelatins, glutathione, and nonprotein thiols, as well as improved heavy metal tolerance (Mladenov et al. 2021). Incorporating the HcNAS1 gene from Hordeum vulgare, which is responsible for the production of metal-chelating amino acids, into Arabidopsis can induce heavy metal accumulation (Koźmińska et al. 2018). The metallothionein gene IlMt2a from Iris lactea var. chinensis was also introduced into the Arabidopsis genome resulting in higher tolerance of Cd.

In addition to incorporating new genes into the plant genome, genetic engineering may be used to boost tolerance and accumulation by overexpressing distinct genes responsible for heightened heavy metal. Increased metal buildup in roots (phytostabilization) or shoots (phytoextraction) can be induced by overexpressing metal transport proteins in plants. Furthermore, genetic manipulation of phytochelatin synthetase (phytochelatin synthetase) and c-glutamyl cysteine synthetase (c-glutamyl cysteine synthetase) genes can result in increased heavy metal tolerance, as evidenced by higher Cd accumulation in transgenic tabacco (Nicotiana glauca and Nicotiana tabacum) (Banerjee and Roychoudhury 2021). Several research in the last several years have revealed that plants overexpressing metallothioneins transgenes increase heavy metal tolerance.

Gene silencing is a new strategy in genetic engineering for improving heavy metal tolerance. Small RNA molecules suppress gene expression and translation of target mRNA as part of this process. This strategy may be used in crops to ensure that no heavy metals accumulate in the plants, and Cd levels in grains were dramatically reduced by silencing the phytochelatin synthase gene. In contrast, silencing the gene producing the root-localized Cd-transporter OsNRAMP5 resulted in increased Cd translocation to the shoots (Saurabh et al. 2014). The most significant drawback of gene modification and genetic transformation is public acceptability; there is still apprehension about GMO plants, and the procedure of introducing such plants into open areas is lengthy and complicated.

References

Antoniou C, Rafaella X, Giannis C, Anastasis C, Khosrow K, Vasileios F (2020) Exploring the potential of nitric oxide and hydrogen sulfide (NOSH)-releasing synthetic compounds as novel priming agents against drought stress in *Medicago sativa* plants. Biomolecules 2020(10):120

Bai J, Wang X, Wang R, Wang J, Le S, Zhao Y (2019) Overexpression of three duplicated BnPCS genes enhanced Cd accumulation and translocation in Arabidopsis thaliana mutant cad1–3. Bull Environ Contam Toxicol 102:146–152

Banerjee A, Roychoudhury A (2021) Metallothionein-assisted phytoremediation of inorganic pollutants. Handbook of Bioremediation. Academic Press, Cambridge, MA, USA, pp 81–90

Capdevila M, Atrian S (2011) Metallothionein protein evolution: a miniassay. J Biol Inorg Chem 6:977–989

Centeno JA, Tchounwou PB, Patlolla AK, Mullick FG, Murakata L, Meza E et al (2006) Managing arsenic in the environment: from soil to human health. CSIRO Publishing, Australia, pp 311–318

Chauhan R, Awasthi S, Indoliya Y, Chauhan AS, Mishra S, Agrawal L, Srivastava S, Dwivedi S, Singh PC, Mallick S, Chauhan PS, Pande V, Chakrabarty D, Tripathi RD (2020) Transcriptome and proteome analyses reveal selenium mediated amelioration of arsenic toxicity in rice (Oryza sativa L.). J Hazard Mat 390:122122

Chen J et al (2015a) MAN3 gene regulates cadmium tolerance through the glutathione-dependent pathway in Arabidopsis thaliana. New Phytol 205:570–582

Chen YK, Liu YX, Ding YN, Wang XT, Xu JC (2015b) Overexpression of PtPCS enhances cadmium tolerance and cadmium accumulation in tobacco. Plant Cell Tissue Organ 121:389–396

Choudhary S, Zehra A, Mukarram M, Wani KI, Naeem M, Khan MMA, Aftab T (2021) Salicylic acid-mediated alleviation of soil boron toxicity in *Mentha arvensis* and *Cymbopogon flexuosus*: Growth, antioxidant responses, essential oil contents and components. Chemosphere 276:130153

Cobbett C, Goldsbrough P (2002) Phytochelatins and metallothioneins: roles in heavy metal detoxification and homeostasis. Annu Rev Plant Biol 53:159–182

Del Buono D, Terzano R, Panfili I, Bartucca ML (2020) Phytoremediation and detoxification of xenobiotics in plants: herbicide-safeners as a tool to improve plant efficiency in the remediation of polluted environments. A mini-review. Int J Phytoremediation 22(8):789–803

Fan W et al (2018) Two mulberry phytochelatin synthase genes confer zinc/cadmium tolerance and accumulation in transgenic Arabidopsis and tobacco. Gene 645:95–104

Filiz E, Saracoglu IA, Ozyigit II II, Yalcin B (2019a) Comparative analyses of phytochelatin synthase (PCS) genes in higher plants. Biotechnol Biotechnol Equip 33:178–194

Filiz E, Saracoglu IA, Ozyigit II, Yalcin B (2019b) Abiotic stress-induced regulation of antioxidant genes in diferent Arabidopsis ecotypes: microarray data evaluation. Biotechnol Biotechnol Equip 33:128–143

Flora SJ, Saxena G, Gautam P, Kaur P, Gill KD (2007) Response of lead-induced oxidative stress and alterations in biogenic amines in different rat brain regions to combined administration of DMSA and MiADMSA. Chem Biol Interact 170(3):209–220

Flora SJ, Mittal M, Mehta A (2009) Heavy metal induced oxidative stress & its possible reversal by chelation therapy. Altern Med Rev 14(1):87–88

Geier DA, Kern JK, Garver CR et al (2009) Biomarkers of environmental toxicity and susceptibility in autism. J Neurol Sci 280(1):101–108

Ghori Z, Iftikhar H, Bhatti MF, Nasar um M, Sharma I, Kazi AG, Ahmad P (2016) Chapter 15-Phytoextraction: the use of plants to remove heavy metals from soil. In: Ahmad P (ed) Plant metal interaction. Elsevier, Amsterdam, pp 385–409

Grill E, Winnacker E-L, Zenk MH (1985) Science 230:674–676

Grill E, Winnacker E L, Zenk MH (1986) FEBS Lett 197:115–120

Grill E, Winnacker E-L, Zenk MH (1987) Phytochelatins, a class of heavy-metal-binding peptides from plants, are functionally analogous to metallothioneins. Proc Natl Acad Sci USA 84:439–443

Gupta DK, Tripathi RD, Mishra S, Srivastava S, Dwivedi S, Rai UN, Yang XE, Huang H, Inouhe M (2008) Arsenic accumulation in roots and shoots vis-a-vis its effects on growth and level of phytochelatins in seedlings of Cicer arietinum L. J Environ Biol 29:281–286

Hasan M, Ahammed GJ, Yin L, Shi K, Xia X, Zhou Y et al (2015) Melatonin mitigates cadmium phytotoxicity through modulation of phytochelatins biosynthesis, vacuolar sequestration, and antioxidant potential in *Solanum lycopersicum* L. Front Plant Sci 6:601

Hubner R, Astin KB, Herbert RJ (2010) 'Heavy metal'—time to move on from semantics to pragmatics? J Environ Monit 12(8):1511–1514

Ishimaru Y, Takahashi R, Bashir K, Shimo H, Senoura T, Sugimoto K, Ono K, Yano M, Ishikawa S, Arao T (2021) Characterizing the role of rice NRAMP5 in manganese, iron and cadmium transport. Sci Rep 2:286

Joshi D, Mittal D, Shrivastav S, Shukla S, Srivastav AK (2011) Combined effect of N-acetyl cysteine, zinc, and selenium against chronic dimethylmercury-induced oxidative stress: a biochemical and histopathological approach. Arch Environ Contam Toxicol 61(4):558–567

Karalija E, Selović A, Dahija S, Demir A, Samardžić J, Vrobel O, Zeljković SĆ, Parić A (2021) Use of seed priming to improve Cd accumulation and tolerance in *Silene sendtneri*, novel Cd hyper-accumulator. Ecotoxicol Environ Saf 210:111882

Klaassen CD, Liu J, Diwan BA (2009) Metallothionein protection of cadmium toxicity. Toxicol Appl Pharmacol 238(3):215–220

Kondo N, Isobe M, Imai K, Goto T (1985) Agric Biol Chem 49:71–83

Koźmińska A, Wiszniewska A, Hanus-Fajerska E, Muszyńska E (2018) Recent strategies of increasing metal tolerance and phytoremediation potential using genetic transformation of plants. Plant Biotechnol Rep 12:1–14

Kühnlenz T, Westphal L, Schmidt H, Scheel D, Clemens S (2015) Expression of Caenorhabditis elegans PCS in the AtPCS1-defcient Arabidopsis thaliana cad1-3 mutant separates the metal tolerance and non-host resistance functions of phytochelatin synthases. Plant Cell Environ 38:2239–2247

Kushwaha A, Rani R, Kumar S, Gautam A (2015) Heavy metal detoxification and tolerance mechanisms in plants: implications for phytoremediation. Environ Rev 24(1):39–51

Leszczyszyn OI, Imam HT, Blindauer CA (2013) Diversity and distribution of plant metallothioneins: a review of structure, properties and functions. Metallomics 5:1146–1169

Li M, Ahammed GJ, Li C, Bao X, Yu J, Huang C et al (2016) Brassinosteroid ameliorates zinc oxide nanoparticles-induced oxidative stress by improving antioxidant potential and redox homeostasis in tomato seedling. Front Plant Sci 7:615

Mishra J, Singh R, Arora NK (2017a) Alleviation of heavy metal stress in plants and remediation of soil by rhizosphere microorganisms. Front Microbiol 8:1706

Mishra S, Mishra A, Kupper H (2017b) Protein Biochemistry and Expression Regulation of Cadmium/Zinc Pumping ATPases in the Hyperaccumulator Plants Arabidopsis halleri and Noccaeacaerulescens. Front Plant Sci 8:1–13

Mladenov V, Fotopoulos V, Kaiserli E, Karalija E, Maury S, Baranek M, Segal N, Testillano PS, Vassileva V, Pinto G et al (2021) Deciphering the epigenetic alphabet involved in transgenerational stress memory in crops. Int J Mol Sci 22:7118

Pal R, Rai JPN (2010) Phytochelatins: peptides involved in heavy metal detoxification. Appl Biochem Biotechnol 160(3):945–963

Park J, Song WY, Ko D, Eom Y, Hansen TH (2012) The phytochelatin transporters AtABCC1 and AtABCC2 mediate tolerance to cadmium and mercury. Plant J 69:278–288

Printz B, Lutts S, Hausman JF, Sergeant K (2016) Copper trafficking in plants and its implication on cell wall dynamics. Front Plant Sci 7:601

Romanyuk ND, Rigden DJ, Vatamaniuk OK, Lang A, Cahoon RE, Jez JM et al (2006) Mutagenic definition of a papain-like catalytic triad, sufficiency of the N-terminal domain for single-site core catalytic enzyme acylation, and C-terminal domain for augmentative metal activation of a eukaryotic phytochelatin synthase. Plant Physiol 141(3):858–869

Rooney JPK (2007) The role of thiols, dithiols, nutritional factors and interacting ligands in the toxicology of mercury. Toxicology 234(3):145–156
Saurabh S, Vidyarthi AS, Prasad D (2014) RNA interference: concept to reality in crop improvement. Planta 239:543–564
Shah MP (2020) Microbial bioremediation & biodegradation. Springer
Shah MP (2021) Removal of emerging contaminants through microbial processes. Springer
Shukla D, Trivedi PK, Nath P, Tuteja N (2016) Metallothioneins and phytochelatins: role and perspectives in heavy metal(loid) s stress tolerance in crop plants. In: Tuteja N, Gill SS (eds) Abiotic stress response in plants. Wiley, New York
Song Y, Jin L, Wang X (2017) Cadmium absorption and transportation pathways in plants. Int J Phytoremediat 19:133–141
Stern BR (2010) Essentiality and toxicity in copper health risk assessment: overview, update and regulatory considerations. J Toxicol Environ Health A
Takahashi R, Ishimaru Y, Shimo H, Bashir K, Senoura T, Sugimoto K, Ono K, Suzui N, Kawachi N, Ishii S et al (2014) From laboratory to field: *OsNRAMP5*-knockdown rice is a promising candidate for Cd phytoremediation in paddy fields. PLoS ONE 9:e98816
Tennstedt P, Peisker D, Bottcher C, Trampczynska A, Clemens S (2009) Phytochelatin synthesis is essential for the detoxification of excess zinc and contributes significantly to the accumulation of zinc. Plant Physiol 149:938–948
Thevenod F (2010) Catch me if you can! Novel aspects of cadmium transport in mammalian cells. Biometals 23:857–875
Trevor AJ, Katzung BG, Masters SB, Kruidering-Hall M (2010) Pharmacology examination & board review. McGraw-Hill Medical, New York, pp 469–483
Tripathi P, Tripathi RD, Singh RP, Dwivedi S, Chakrabarty D, Trivedi PK, Adhikari B (2013) Arsenite tolerance in rice (Oryza sativa L.) involves coordinated role of metabolic pathways of thiols and amino acids. Environ Sci Pollut Res Int 20(2):884–896
Uraguchi S, Sone Y, Ohta Y, Ohkama-Ohtsu N, Hofmann C, Hess N et al (2018) Identification of C-terminal regions in Arabidopsis thaliana Phytochelatin Synthase 1 specifically involved in activation by arsenite. Plant and Cell Physiol 59(3):500–509
U.S Environmental Protection Agency (EPA) (2006) Cadmium compounds
US Environmental Protection Agency (USEPA) (2020) Integrated risk information system of the US environmental protection agency. USEPA, Washington, DC, USA
Vatamaniuk OK, Mari S, Lu YP, Rea PA (2000) Mechanism of heavy metal ion activation of phytochelatin (PC) synthase. J Biol Chem 275:31451–31459
Vatamaniuk OK, Mari S, Lang A, Chalasani S, Demkiv LO, Rea PA (2004) Phytochelatin synthase, a dipeptidyltransferase that undergoes multisite acylation with γ-glutamylcysteine during catalysis. J Biol Chem 279(21):22449–22460
Wawrzyński A, Kopera E, Wawrzyńska A, Kamińska J, Bal W, Sirko A (2006) Effects of simultaneous expression of heterologous genes involved in phytochelatin biosynthesis on thiol content and cadmium accumulation in tobacco plants. J Exp Bot 57(10):2173–2182
WHO/FAO/IAEA (1996) Trace elements in human nutrition and health
Yan A, Wang Y, Tan SN, Mohd Yusof ML, Ghosh S, Chen Z (2020) Phytoremediation: a promising approach for revegetation of heavy metal-polluted land. Front Plant Sci 11:359
Yedjou CG, Patlolla AK, Sutton DJ (2012) Heavy metals toxicity and the environment Paul B Tchounwou. Published in final edited form as. EXS 101:133–164
Zhang X, Rui H, Zhang F, Hu Z, Xia Y, Shen Z (2018) Overexpression of a functional Vicia sativa PCS1 homolog increases cadmium tolerance and phytochelatins synthesis in Arabidopsis. Front Plant Sci

Applications of Bioremediation in Treatment of Environmental Pollution

Preeti Kumari, Sagnik Nag, Archna Dhasmana, Jutishna Bora, and Sumira Malik

1 Introduction

In recent years, several innovative approaches and technologies have been made to prevent environmental pollution and protect biodiversity, but the issues are still the same. Increased activities of humans on rapid industrialization, energy reserves, and dangerous farming methods have contributed to increased environmental pollution over the last few decades (Azubuike et al. 2016). The intensive usage of chemical xenobiotic substances and global industrialization such as heavy metals, solvents, petroleum products, and pesticides are the primary cause of pollution. Consequently, accumulation and release of these pollutants or toxins result in deterioration in natural texture, global climate, plant-soil productivity, and threat to life on this planet (Mishra et al. 2021). Recent advances in bioremediation techniques have played a significant role in effectively restoring contaminated sites by eco-friendly and cost-effective approaches (Azubuike et al. 2016). This paper contributes to different bioremediation techniques that can be utilized to treat different environmental pollutants.

Preeti Kumari and Sagnik Nag are equally contributed.

P. Kumari · J. Bora · S. Malik (✉)
Amity Institute of Biotechnology, Amity University Jharkhand, Ranchi, Jharkhand 834001, India
e-mail: smalik@rnc.amity.edu

S. Nag
Vellore Institute of Technology (VIT), Vellore, Tamil Nadu, India

A. Dhasmana
Himalayan School of Biosciences, Swami Rama Himalayan University, Dehradun, Uttarakhand, India

M. P. Shah (ed.), *Advanced and Innovative Approaches of Environmental Biotechnology in Industrial Wastewater Treatment*,
https://doi.org/10.1007/978-981-99-2598-8_17

2 Concept of Bioremediation

The international agency The United States Environmental Protection Agency (US EPA) defined the term bioremediation as the process of utilization of living microbes in the elimination of contaminants from water or soil using of nonharmful insects to eliminate agricultural pests or counteract illnesses of plants, trees, and soil (Wang et al. 2021). Bioremediation is carried out in two ways, either by plants or microbes (Fig. 1). In general, microbial remediation is the most focused approach that involves the metabolic machinery of the indigenous microbiome for the remediation of which Bioremediation is achieved by modifying the environment, such as through the application of fertilizer and aeration, to overcome the factors that limit hydrocarbon biodegradation rates by indigenous microbial communities. In the second technique, exogenous microbial populations are added in which the cultures of seeds are chosen for their hydrocarbon-degradation abilities. These are applicable for treating a variety of wastes, such as remediation of polluted groundwater and soil (Junior Letti et al. 2018).

Bacteria, protists, fungus, and other microbes have a significant role in constantly flouting down the biological matter in an unpolluted environment. On the other hand, Bioremediation helps by giving oxygen, fertilizers, and other conditions that help these pollution-eating organisms to develop quickly. As a result, these organisms would be able to break down organic pollution at a faster rate (Speight 2018). Microorganisms can be found naturally or introduced into a habitat or a specific

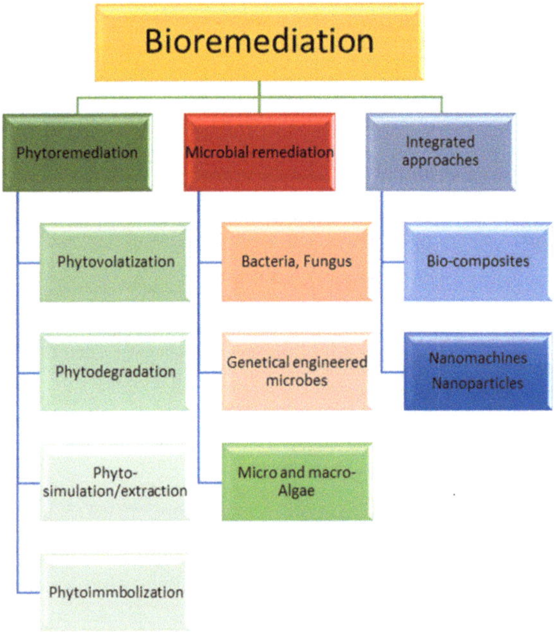

Fig. 1 Classification of bioremediation methods for waste treatment

remediation system to break down hazardous pollutants into less toxic biological and chemical forms. Bioremediation approaches mainly rely on using metabolic machinery to reduce the noxious properties of pollutants by broad-scale mineralization, pollutant immobilization, and (3) transformation into less toxic products (Pandey et al. 2009a, b). Bioremediation strategies can be divided into ex-situ and in-situ treatments. However, reliant on the type of pollutant and microorganisms used in these technologies, the microbes could be anaerobic, anaerobic, or both.

3 *In-situ* Bioremediation

Aerial contamination, inadvertent terrestrial soil or coastline chemical spills, or intentional contaminants deposition on the soil surface can cause topsoil contamination. Consequently, prolonged bleaching, oil spills, buried wastes, underground sewage tanks, and contamination are often not restricted to the superficial soil layer but spread into the subterranean layers of soil and groundwater system (Jørgensen 2007). Although a polluted site may appear stable, it may pose a future concern if not remedied. In-situ bioremediation involves a microbiological process to convert organic pollutants to inorganic substances such as water, methane, carbon dioxide, and salts in engineered otherwise natural environments. In theory, in-situ bioremediation aims to remove or attenuate pollutants in natural environments by utilizing the contagious metabolic capability to excavate the contaminated sample. Treatments through in-situ bioremediation do not require the removal of soil or samples, making them less expensive and, more crucially, resulting in less leakage of volatile chemical pollutants into non-polluted areas (Pandey et al. 2009a, b) (Table 1).

3.1 Intrinsic Bioremediation

The maximum communal environmental contaminants are petroleum hydrocarbons, and oil spills severely threaten marine and terrestrial ecosystems. Drilling oil, transport, storage, and processing may result in oil contamination that can occur accidentally or intentionally. Oil spills are a huge environmental threat since they severely harm the ecosystems around them (Hassanshahian et al. 2012). The most common reason for groundwater pollution is the leakage of natural fuel, i.e., gasoline and petroleum complexes, from underground storage tanks (USTs). Here, intrinsic Bioremediation is based on the ability of native microflora degradation the potential for toxins discharged into the subsurface and, simultaneously, reduce threats to the environment and public health (Borden et al. 1995). Intrinsic Bioremediation has recently been proposed as one of the promising cleanup strategies for sites that are contaminated by petroleum hydrocarbons. Pollutants present in marine sediment may show harmful effects on marine ecosystems and human health. Toxic chemicals can accumulate in the bodies of the benthic creature and thus can affect and poison predators

Table 1 Key limiting aspects to control the microbial efficiency of *in-situ* bioremediation

S. No	Factors		Reference
1	Low temperature	The soil's lower temperature is an efficient microbial breakdown of pollutants for a substantial part of the year	Romantschuk et al. (2000)
2	Anaerobic conditions	Anaerobic degradation is slow; certain chemicals are not destroyed anaerobically, while others are only partially decomposed and may produce hazardous molecules	Boopathy (2004)
3	Low levels of co-substrates and nutrients	The nutritional balance at a polluted site is frequently off. If the contaminant is a hydrocarbon, such as oil, there will almost certainly be a nitrogen shortage, but each site must be assessed individually, considering factors such as the contaminant's solubility to avoid overfertilization	Schaefer and Juliane (2007)
4	Bioavailability	Longitudinal dispersal of pollutants in degrading bacteria and contaminant solubility; this variable quantity is partially connected and is a crucial determinant determining degradation velocity independently and in combination	Pavlostathis (2003)
5	Lack of degradation activity	Synthetic, xenobiotic substances degraded by the biological pathway may not be preventing biodegradation, or the pollution may not activate metabolic enzymes encoding genes that are active on the compound	Romantschuk et al. (2000)

at the highest of the food chain. Because hydrocarbon components are neurotoxic organic pollutants and carcinogens, humans exposed to them are likely to suffer adverse effects (Catania et al. 2015).

Ordinary attenuation is essential as proven by several outlines of indication for a regulatory agency to allow the cleanup of a polluted site by natural attenuation. Antique groundwater and soil chemistry postulate the declining pollutant mass and concentration over time at pertinent sample facts is one such line of evidence. Besides, specific parameters, e.g., impurity sorption, solute dilution, volatilization, and abiotic and biotic degradation, may all contribute to this decline. The biological breakdown of hydrocarbons, also known as intrinsic Bioremediation, is particularly advantageous since it remediates hydrocarbons to free, non-toxic products of carbon dioxide, water, and methane relative to redistributing them (Dojka et al. 1998). Removing petroleum products from wetlands involves evaporation, photo-oxidation, suspension, microbiological degradation, and physio flushing.

Moreover, degradation and suspension are the principal removal methods once petroleum is absorbed into the sediment. Anoxia and nutrient availability can hinder petroleum biodegradation in wetland environments (Mills et al. 2003).

The principal types of machinery that regulate the remediation of organic hydrocarbon containment and lowering of pollutant concentrations are aerobic and anaerobic degradation processes. The electron acceptor utilized by subsurface bacteria in aerobic biodegradation is dissolved oxygen (D.O.). Anaerobic processes are biodegradation methods that utilize terminal electron acceptors such as carbon dioxide, nitrate, ferric iron, and sulfate. The synergistic effect of environment and microbes have the ultimate role in determining anaerobic biodegradation processes control (Kao 2000). For intrinsic Bioremediation, three approaches might be used. There has been a drop in pollutant absorptions laterally from the groundwater movement channel to the contaminated leakage site; there has been predictable damage of pollutants at the ground scale, and there has been microbial test centre information that supports the amount of Bioremediation. If there is a least or no possibility of the condition deteriorating and everyday actions to improve the condition in a more or less efficient manner, then perhaps all that is required is monitoring (Kao and Prosser 1999).

3.2 *Engineered Bioremediation*

Although certain pollutants are successfully destroyed by very varied and specialized microflora in the atmosphere, the majority of toxic contaminants are degraded gradually and hence predisposed to collect in the atmosphere. This growth can pose a severe threat in some circumstances. The usage of contemporary genetic engineering techniques and technologies has substantially advanced concepts of the genetics and biochemistry of genetically modified microorganisms (GEMs). It helps to create new strains with needed traits through genetic modification, expression, and metabolic regulation by enzyme specificity, binding, affinity, and intracellular localization, resulting in innovative approaches for finding microbes and their specific pollutants (Urgun-Demirtas et al. 2006).

The administration elaborates that acquiring approval for ecological disturbance, moderately than constraints to produce, is one barrier to designing genetically engineered microbes (GEMs) for various eco-friendly sustainable practices (Romantschuk et al. 2000). The University of Tennessee has achieved GEM's first large-scale waste treatment in collaboration with Oak Ridge National Laboratory. The GEM in question was a strain of *Pseudomonas fluorescens* known as HK44, which was released into a controlled soil environment. The native microbial strain of HK44 was generated from a polyaromatic hydrocarbon-contaminated gas plant site (PAHs). *P. fluorescens* HK44 was created by introducing the naphthalene catabolic plasmid pUTK21 into this strain. Over the last decade, there has been minimal progress in using GEMs in Bioremediation (Sayler and Ripp 2000) (Table 2).

Table 2 Genetically engineered microbes for Bioremediation and their targeted compounds

S. No	Microbial strain	Chimeric gene	Pollutants	References
1	*Pseudomonas putida PaW340 (pDH5)*	pDH5	4-chlorobenzoic acid	Massa et al. (2009)
2	*Escherichia coli* JM109 (pGEX-AZR)	Azoreductase	Decolorize azo dyes, C.I. Direct Blue 71	Jin et al. (2009)
3	*Escherichia coli* AtzA	Atrazine chlorohydrolase	Atrazine	Strong et al. (2000)
4	*Pseudomonas fluorescens* HK4	*lux CDABE*	Naphthalene	Sayler and Ripp (2000)
5	*Burkholderiacepacia L.S.2.4*	pTOD plasmid	Toluene	Barac et al. (2004)
6	*Comamonastestosteroni SB3*	pNB2::dsRed plasmid	3-chloroaniline	Bathe et al. (2009)
7	*Pseudomonas putida PaW85*	pWW0 plasmid	Petroleum	Jussila et al. (2007)
8	*Rhodococcus* sp. RHA1 (pRHD34::fcb)	fcbABC operon	2(4)-chlorobenzoate 2(4)-chlorobiphenyl	Rodrigues et al. (2006)
9	*Burkholderiacepacia VM1468*	pTOM-Bu61 plasmid	Toluene	Taghavi et al. (2005)
10	*Pseudomonas fluorescens* F113rifpcb*rrn*BP1::gfp-mut3	operon BPH, GFP	Chlorinated biphenyls	Boldt et al. (2004)
11	*Pseudomonas putida KT2442 (pNF142::TnMo-d-OTc)*	pNF142 gfp	Naphthalene	Filonov et al. (2005)

3.2.1 Biosparging

Although Raymond's use of in-well aerators or spargers to transport oxygen into groundwater was the first to suggest air sparging to help Bioremediation, the air did not likely enter the aquifer as a separate phase with his use (as is expected in the current practice of air sparging). Dieter Hiller first used Authentic air sparging in Germany in the 1780s. The early development that air sparging may provide is the substantial oxygen to facilitate biodegradation (Brown et al. 1996). Biosparging is inoculating pressurized air under the groundwater to raise oxygen levels in the water system and speed up the BOD level and aerobic digestion by the microorganisms. This pathway upsurges the interface between soil and groundwater by increasing partying in the saturated zone. Besides, the excellent efficiency and low cost of the miniatured air inoculation sites permit a high degree of flexibility in the structural design and creation (Vidali 2001).

Biosparging is a method for removing volatile organic chemicals (VOCs) such as benzene, toluene, and xylene from the environment (Kao et al. 2008). The main factors influencing the outcome are the contaminated site's intrinsic soil permeability, texture, and other physiochemical parameters, along with the pollutant physicochemical. Recent studies on this method used on-site contamination treatments, e.g., high to low-grade petroleum byproducts.

3.2.2 Bioventing

Another promising in-situ therapy remediation method is bioventing, which includes aeration and nutrient feedstock release to contaminated sites, thereby inducing the growth of the native microbes having degradation potential for xenobiotic toxic pollutants. It allows the passage of oxygen-rich air at low rates required for biodegradation, dropping volatilization, and toxic contaminant release into the atmosphere (Vidali 2001). The main design for bioventing systems is that the contaminants of concern are biodegradable under current site conditions, i.e., whether inhibition or toxicity is visible at the site, and second, whether the required terminal electron acceptor, i.e., oxygen, can be transported effectively within the soil to encourage aerobic contaminant biodegradation. The prime interrogation was measured by soil gas confirmation and in-situ inhalations, whereas another inquiry responded by measuring in-situ air permeability (Dupont 1993).

A blower with a succession of gaseous influent and effluent wells makes up an in-situ bioventing system, similar to a soil venting system. The well sites will be site-specific and gritty by whether the airborne will be pumped into or removed from the Earth. Bioventing can also be done in a soil pile design on excavated soils (c). the maximum level of the water table with the contamination below groundwater, bioventing is more advantageous because the lubricant captivated in the solid phase. Besides removing fuel vapours, bioventing may promote air diffusion and trigger oil-remediating microbial activity by compensating for oxygen deficit in bottomless subsurface layers (Gruiz and Kriston 1995). Bioventing for up to 2 months is sufficient and results in the highest degradation. Due to higher operating expenses and a lack of advances in biodegradation, increasing the bioremediation duration is not reasonable (Thomé et al. 2014).

3.2.3 Bioslurping

The leakage of petroleum hydrocarbons (P.H.) from subversive storage tanks is the primary source of soil and groundwater pollution (USTs). The requirement for redress technologies that are viable, fast, and adaptable in various physical situations has been highlighted by problems related to the cleanup of petroleum-contaminated areas. Bioslurping is the version and claim of vacuum-enhanced dewatering technologies to remediate hydrocarbon-contaminated sites among bioremediation approaches. It combines features of bioventing and permitted product retrieval to report two discrete contamination media, recovering permitted products while also bioremediating vadose zone soils; as a result, it can expand byproduct recovery without requiring large amounts of groundwater extraction. During free product removal and groundwater cleanup operations, slurping systems can be changed to regular bioventing activities using hydraulic control (Yen et al. 2003). The following are the key advantages of the bioslurping technology: (1) highly efficient free phase recovery, (2) minimal groundwater and free oil phase mixing, (3) minimal pumped liquid volume, (4) stimulation of the biodegradation process in the unsaturated zone of the subsurface, and (5) ability to be applied using a variety of methods (6) The ability to prevent soil vapour treatment when modest pumping rates are used, as well as (7) effective management of free phase mobility—expansion (Gidarakos and Aivalioti 2007).

3.2.4 Biostimulation

The entire quality of the atmosphere is intrinsically related to the worth of life on Earth, and emissions of determined, bioaccumulative, and hazardous substances negatively influence social health and the ecosystem. The migration of hazardous elements in the environment allows these contaminants to enter the tissues of plants, animals, and humans. The lack of awareness of environmental effects associated with the production, use, and disposal of hazardous polycyclic aromatic hydrocarbons (PAHs), originating from hazardous substances leads to degraded lands and pollution (Vidali 2001). They are organic molecules with many rings that are nonpolar, neutral, and hydrophobic. When growing on insoluble or immiscible chemicals, some microorganisms produce surface-active agents (biosurfactants), which are accessible as a substitute to chemical surfactants for increasing the bioavailability of hydrophobic pollutants (Straube et al. 2003).

Biostimulation recognizes and modifies physiochemical parameters (e.g., temperature, pH, moisture content, nutrient content) that slow down the biodegradation of pollutants by the afflicted site's indigenous microorganisms. The pollutant biodegradation may occur at substantial and acceptable rates once the variables have been adjusted suitably (Abdulsalam et al. 2010). Biostimulation's biological stabilization of biomass has recently expanded traction, as constant nutrient supply indicates increased activity of the native microbial consortia, resulting in improved contaminant degradation and biosolids stabilization. Biostimulation procedures are broadly used to improve biosolids' Bioremediation by allowing indigenous microorganisms to produce hydrolytic enzymes (Vaithyanathan et al. 2021).

3.2.5 Bioaugmentation

Bioaugmentation is green technology for removing organic pollutants from the environment. It is a practical approach to engineering bioremediation that involves inoculating particular habitats with microorganisms that have specific catalytic abilities. Bioaugmentation is mainly advised for areas with insufficient numbers of autochthonous microorganisms capable of degrading toxins and native populations lacking the catabolic trails required to metabolize pollutants. It is based on improving the catabolic capability of soil microbial populations for pollution breakdown (Cycoń et al. 2017). Bioremediation of polluted places depends on the microbial world's enormous metabolic capacities to change organic artificial toxins into fundamentally harmless or, at the very least, less hazardous substances. The most attractive aspect of this group of technologies is that it is compatible with the Earth's major natural recycling routes, known as biogeochemical cycles, making bioremediation a long-term and environmentally friendly cleanup method. Contempt, the presence of microorganisms capable of pollutant biodegradation in contaminated soils and aquifers, local environmental conditions in these places may be unfavourable (El Fantroussi and Agathos 2005). Adding microorganisms, native or exogenic, to dirty locations is a standard part of Bioremediation. Nonindigenous bacterial strains infrequently compete well enough with native inhabitants to develop and withstand healthy population levels, and most soils with long-term experience with biodegradable waste have indigenous microorganisms that effectively degrade if the land treatment unit is well managed, limiting the use of different microbial cultures in a terrestrial treatment unit (Vidali 2001).

3.2.6 Natural Attenuation (N.A.)

Natural reduction is an in-situ curative technology that uses the subsurface's inherent assimilating potential to clean up contaminated sites. As a result, it may be a cost-effective treatment. Natural attenuation is based on the idea that natural processes in the subsurface reduce key pollutants to the point where there is no significant risk to down gradient groundwater users or discharge (Christensen et al. 2000). Many research efforts are currently focused on using natural attenuation to restrict the migration of dissolved contaminants. N.A. utilization procedures to rehabilitate as-contaminated soils and groundwater by naturally occurring physicochemical and biological processes. Natural attenuation is assumed to trust exclusively on natural procedures that are not increased. However, promoting the processes involved in natural remediation, such as by adding organic matter (O.M.) and lime, can be an efficient cleanup strategy (Clemente et al. 2006). Immobilization of As by sorption to solid phases such as (hydro)oxides of iron (Fe), aluminium (Al), and manganese (Mn), organic matter, and clay minerals, intra-conversion of As (III) to As(V) induced by Fe and Mn (hydro)oxides and clay phases or natural organic matter (NOM), biotransformation, and hyperaccumulation of As in plants are the main processes involved in the N.A. of As.

4 *Ex-situ* Bioremediation

Recent advancements in bioremediation methods effectively restabilize spoiled areas in an environmentally suitable and profitable manner. Researchers have created and modelled diverse bioremediation methods; however, no solitary bioremediation technique serves as a "silver bullet" for restoring damaged habitats (Azubuike et al. 2016). However, ex-situ bioremediation techniques involve treating polluted soils away from the polluted area because the price determines the success of ex-situ bioremediation methods, pollutant complexity, topographical location, and the technique used.

Solid-phase Bioremediation by land farming, soil biopile, windrow, windrow-composting, and slurry-phase Bioremediation (bioreactors) are two ex-situ techniques. Ex-situ bioremediation, the primary remediation method for a broad spectrum of hydrocarbons, has recently gained much scientific attention. Bioaugmentation, cutting-edge petroleum-degrading microbes are added to the soil matrix, and biostimulation, in which essential nutrients or biosurfactants are introduced to stimulate microbial petroleum degradation, are the two main strategies used to degrade contaminated soils concluded that ex-situ bioremediation (Gomez and Sartaj 2013). Within the term ex-situ, a vast range of technologies exists, each with varying degrees of sophistication, all treatment excavated soil. However, the ex-situ technologies have several advantages, the most important of which is the ability to better control the remediation process due to the controlled reaction atmosphere is more manageable, and the treatment process is more expectable than an in-situ environment (Tomei and Daugulis 2012).

4.1 Slurry Phase Bioremediation

Lately, slurry phase bioremediation is an inexpensive ex-situ method that has attracted the interest of several researchers. In this Bioremediation, excavated contaminated soil is generally screened to remove debris and more extensive materials before being deposited in an on-site stirred tank reactor; the soil is blended with water to generate a slurry. In this system, a triphase blending condition is given by adding water to unearthed soil into the containment tank, which speeds up Bioremediation. The amount of solids in the slurry is determined by the kind of soil, the mixing and aeration equipment accessible, and the pollutant removal rates required. Soil moisture increases the amount of solubilized contaminant, resulting in increased bioavailability. Furthermore, soil slurry bioreactors may optimize and manage biodegradation's abiotic conditions (Partovinia et al. 2010).

Bio-slurry phase systems produced by bacteria or inoculated strains possess metabolic abilities to transform hazardous organic material and provide thoroughgoing influence by providing maximum filth rate while minimizing abiotic losses (Prasanna et al. 2008). Slurry-phase Bioremediation is much more beneficial than in-situ because mass transfer constraints between contaminates, nutrients, oxygen, and microbes get lowered.

4.1.1 Bioreactor

The Bioreactor is unprocessed container materials that are turned into the specified product(s) through a sequence of biological processes. Distinct functioning modes of the Bioreactor include batch, fed-batch, sequencing batch, continuous, and multistage bioreactors through the high-quality operating mode are influenced mainly by the low-cost investment and expenditure. Bioreactor parameters endorse normal cell processes by pretending and preserving their natural environment for optimal growth. Contaminated samples can be supplied into a bioreactor as a slurry; in any case, using a bioreactor to clean polluted soil provides several benefits over conventional ex-situ bioremediation procedures (Azubuike et al. 2016). Multiple bioreactor technologies have been developed to treat solid, liquid, and gaseous substrates polluted with a wide range of organic compounds. In reactor bioremediation, contaminated compact material or water is processed through a designed containment system. Usually, the biodegradation rate and amount in a bioreactor system are more significant than in-situ or solid-phase systems because the encircled atmosphere is more manageable and, therefore, more controlled and expectable (Vidali 2001).

Aerobic bioprocesses include aerophilic fermentation, organic wastewater, and hydrocarbon-impacted soil/sediments treatments. The stirred-tank bioreactors are mechanically circulated, with stirrers serving as the primary gas-dispersing equipment, and offer high mass transfer rates and adequate mixing. Bioreactors are pneumatically stimulated are either bubble columns or airlift bioreactors. The shear rate is the most favourable parameter for the influential culture of shear-sensitive and filamentous cells and bioreactor designs are versatile, allowing for the most significant biological degradation while reducing non-living.

4.2 Solid Phase Bioremediation

The solid phase, the Bioremediation treatment approach, uses an aboveground treatment area to remediate polluted soil. Solid-phase bioreactors are nutrient-added heaps of polluted soil. Solid-phase Bioremediation is based on the powered breakdown of polluted soil by abrasion and thorough fraternization of the components in an enclosed container. It demonstrates that nutrients, microbes, pollutants, oxygen, and water are constantly in contact (Kumar et al. 2018). Forced aeration requires very little energy, which makes solid-phase Bioremediation cheaper even when extensive treatment durations are necessary. However, the efficiency of solid-phase treatment is generally incomplete by non-uniform pollutant elimination and low rates and amount of degradation. For detoxification and disintegration of harmful toxic pollutants, solid-phase procedures include biopiling, land farming, composting, and biofiltration.

4.2.1 Biopilling

Biopiling is an ex-situ bioremediation procedure involving bio cells, biomounds, or bioheaps that involves burrowed soil, sludge, or sediments combined with soil amendments, deposited on a handling area, and remediated using forced aeration. Here entails piling polluted soil, sludge, or dry sediments into heaps and increasing aerophilic microbiome biodegradation by creating optimum multiplying conditions within the pile (Germaine et al. 2012). Usually, it refers to stacking the waste to be biotreated into heaps, up to 2–4 m, by adding nutrients and air. Biopiles can be stationary with a ventilation pipe attached, or they can be rotated or blended using specific machinery. A bulking agent, such as straw, sawdust, bark, wood chips, or some other organic material, can be added to biopiles. In biopiles, several organic pollutants have been effectively Bioremediation. This method has been shown to work in field experimental or full-scale applications, particularly for petroleum hydrocarbons (Jørgensen et al. 2000).

4.2.2 Land Farming

Land farming is an ex-situ approach in which land application/land treatment, tainted soil, sediment, or sludge is dug and dispersed on a prepared bed. Regularly turned over and ploughed for aeration until contaminants degrade via aerobic microbial operations in the soils due to aeration and adding moisture, minerals, and nutrients. This method treats the top 10–35 cm of soil (Kumar et al. 2018). This 'low-tech' biological treatment approach entails applying and propagating a more-or-less solid organic bioavailable waste on the soil surface and the waste's assimilation into the upper soil zone. Table 3 discusses the advantages and disadvantages of land farming (Maila and Cloete 2004).

4.2.3 Composting

Composting is one of the cheapest solutions for soil bioremediation since it is a primary process for stabilizing solid farming waste and municipal solid waste (MSW) through the breakdown of recyclable components by microbial populations. It is a biological breakup process in which microbes convert organic wastes into humus-like materials, a constant organic end product (compost). The polluted soil is pulled out and combined with bulking and organic ingredients in this procedure, e.g., biomass. The presence of organic elements promotes the growth of a diverse infectious population that, through enzymatic activity, converts organic matter to compost (Chen et al. 2015) because an improper ratio of polluted soil to biological additions might

Table 3 Aids and restraints of landfarming

Technology	Benefits	Constraints
Landfarming	• The short initial investment required • The low initial investment required • Easy to plan and apply technology • Huge soil quantities may be treated • Ex-situ application possible • Has a minimal influence on the environment • Low-energy consumption	• Only biodegradable pollutants may be removed • A considerable treatment area is required • Contaminant exposure is a possibility • Costs associated with excavation might be significant • Inadequate understanding of microbial processes or the unravelling of Bioremediation limiting factors

slow or stop microbial activity, it should be assessed. Organic additives, such as sewage sludge or compost, can help to speed up the breakdown of organic pollutants by augmenting nutrients and carbon sources in polluted soil (Namkoong et al. 2002). The microbial accessibility to the target pollutants in the Bioremediation of filthy soils via the composting process is determined by a range of physical, chemical, and biological parameters, with the amendment qualities playing a significant influence in influencing the process behaviour (Sayara et al. 2010).

4.2.4 Biofiltration

Biofiltration is a method that employs microorganisms that are immobilized and cultured on a biologically porous substrate. These medium serves as a physical provision for dynamic biomass and, in certain situations, as a basis of nutrients for growth (Oyarzún et al. 2003). Biological degradation, rather than physical treatment, eliminates biodegradable contaminants in a biofiltration system. Microbes (aerobic, anaerobic, facultative, bacteria, fungus, algae, and protozoa) eventually build on the superficial of the filter medium as the filtering process progresses, forming a biological film or thin layer known as a biofilm. Attachment of microorganisms, development of microorganisms, and decay and detachment of microorganisms are the three basic processes in a biofilter. Most biofilters work at a pH range of 6–8 with the fluctuation in pH during the operation of a biofilter.

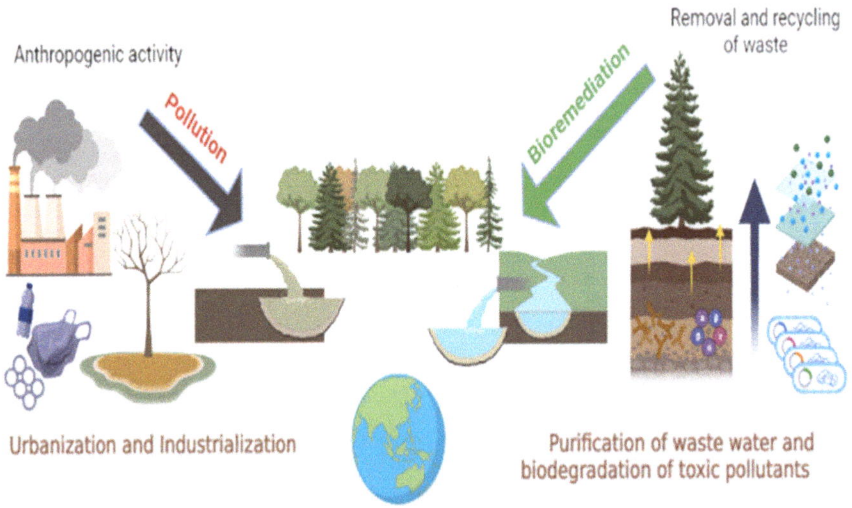

Fig. 2 Summary of the equilibrium process among deterioration and Bioremediation mediated by microbes

5 Conclusion

The different kinds of techniques, along with the microbial strains, are responsible for the process of Bioremediation. They maintain the equilibrium among environmental degradation caused by the pollutants developed through anthropogenic activities, urbanization, and industrialization, and environmental Bioremediation through removing toxic pollutants in wastewater with the removal and recycling of waste as described in Fig. 2.

References

Abdulsalam S, Bugaje IM, Adefila SS, Ibrahim S (2010) Comparison of biostimulation and bioaugmentation for remediation of soil contaminated with spent motor oil. Int J Environ Sci Technol 8(1):187–194. https://doi.org/10.1007/bf03326208

Azubuike CC, Chikere CB, Okpokwasili GC (2016) Bioremediation techniques–classification based on site of application: principles, advantages, limitations and prospects. *World J Microbiol Biotechnol*, 32(11). https://doi.org/10.1007/s11274-016-2137-x

Barac T, Taghavi S, Borremans B, Provoost A, Oeyen L, Colpaert JV, ... van der Lelie D (2004) Engineered endophytic bacteria improve phytoremediation of water-soluble, volatile, organic pollutants. Nat Biotechnol 22(5):583–588. https://doi.org/10.1038/nbt960

Bathe S, Schwarzenbeck N, Hausner M (2009) Bioaugmentation of activated sludge towards 3-chloroaniline removal with a mixed bacterial population carrying a degradative plasmid. Biores Technol 100(12):2902–2909. https://doi.org/10.1016/j.biortech.2009.01.06

Boldt TS, SÃ¸rensen J, Karlson U, Molin Sã¸, Ramos C (2004) Combined use of different GFP reporters for monitoring single-cell activities of a genetically modified PCB degrader in the rhizosphere of alfalfa. FEMS Microbiol Ecol, 48(2):139–148. https://doi.org/10.1016/j.femsec.2004.01.002

Boopathy R (2000) Factors limiting bioremediation technologies. Biores Technol 74(1):63–67. https://doi.org/10.1016/s0960-8524(99)00144-3

Boopathy R (2004) Anaerobic biodegradation of no. 2 diesel fuel in soil: a soil column study. Bioresour Technol, 94(2):143–151. https://doi.org/10.1016/j.biortech.2003.12.00

Borden RC, Gomez CA, Becker MT (1995) Geochemical indicators of intrinsic bioremediation. Ground Water 33(2):180–189. https://doi.org/10.1111/j.1745-6584.1995.tb00272.x

Boriŝev I, Boriŝev M, Jović D, Župunski M, Arsenov D, Pajević S, Djordjevic A (2020) Nanotechnology and remediation of agrochemicals. Agrochem Detect Treatment Remediat 487–533. https://doi.org/10.1016/b978-0-08-103017-2.00019-2

Brar SK, Verma M, Surampalli RY, Misra K, Tyagi RD, Meunier N, Blais JF (2006) Bioremediation of hazardous wastes—a review. Pract Period Hazard Toxic Radioact Waste Manag 10(2):59–72. https://doi.org/10.1061/(ASCE)1090-025x(2006)10:2(59)

Brown RA, Hinchee RE, Norris RD, Wilson JT (1996) Bioremediation of petroleum hydrocarbons: a flexible, variable speed technology. Remediat J 6(3):95–109. https://doi.org/10.1002/rem.3440060308

Catania V, Santisi S, Signa G, Vizzini S, Mazzola A, Cappello S, … Quatrini P (2015) Intrinsic bioremediation potential of a chronically polluted marine coastal area. Mar Pollut Bull 99(1–2):138–149. https://doi.org/10.1016/j.marpolbul.2015.07.0

Chen M, Xu P, Zeng G, Yang C, Huang D, Zhang J (2015) Bioremediation of soils contaminated with polycyclic aromatic hydrocarbons, petroleum, pesticides, chlorophenols and heavy metals by composting: Applications, microbes and future research needs. Biotechnol Adv 33(6):745–755. https://doi.org/10.1016/j.biotechadv.2015.05

Christensen TH, Bjerg PL, Kjeldsen P (2000) Natural attenuation: a feasible approach to remediation of ground water pollution at landfills? Groundwater Monitor Remediat 20(1):69–77. https://doi.org/10.1111/j.1745-6592.2000.tb00253.x

Clemente R, Almela C, Bernal MP (2006) A remediation strategy based on active phytoremediation followed by natural attenuation in a soil contaminated by pyrite waste. Environ Pollut 143(3):397–406. https://doi.org/10.1016/j.envpol.2005.12.011

Cycoń M, Mrozik A, Piotrowska-Seget Z (2017) Bioaugmentation as a strategy for the remediation of pesticide-polluted soil: a review. Chemosphere 172:52–71. https://doi.org/10.1016/j.chemosphere.2016.12

Dojka MA, Hugenholtz P, Haack SK, Pace NR (1998) Microbial diversity in a hydrocarbon- and chlorinated-solvent-contaminated aquifer undergoing intrinsic bioremediation. Appl Environ Microbiol 64(10):3869–3877. https://doi.org/10.1128/aem.64.10.3869-3877.1998

Dupont RR (1993) Fundamentals of bioventing applied to fuel contaminated sites. Environ Prog 12(1):45–53. https://doi.org/10.1002/ep.670120109

El Fantroussi S, Agathos SN (2005) Is Bioaugmentation a feasible strategy for pollutant removal and site remediation? Curr Opin Microbiol 8(3):268–275. https://doi.org/10.1016/j.mib.2005.04.011

Farhadian M, Vachelard C, Duchez D, Larroche C (2008) In situ bioremediation of monoaromatic pollutants in groundwater: a review. Biores Technol 99(13):5296–5308. https://doi.org/10.1016/j.biortech.2007.10.02

Filonov AE, Akhmetov LI, Puntus IF, Esikova TZ, Gafarov AB, Izmalkova TY, … Boronin AM (2005) The Construction and monitoring of genetically tagged, plasmid-containing Naphthalene-Degrading Strains in Soil. Microbiology 74(4):453–458. https://doi.org/10.1007/s11021-005-0088-6

Germaine KJ, McGuinness M, Dowling DN (2012) Improving phytoremediation through plant-associated bacteria. In: de Bruijn FJ (ed) Molecular ecology of the rhizosphere. Wiley-Blackwell, Hoboken

Gidarakos E, Aivalioti M (2007) Large scale and long term application of bioslurping: the case of a Greek petroleum refinery site. J Hazard Mater 149(3):574–581. https://doi.org/10.1016/j.jhazmat.2007.06.110
Gomez F, Sartaj M (2013) Field scale ex-situ bioremediation of petroleum contaminated soil under cold climate conditions. Int Biodeterior Biodegrad 85:375–382. https://doi.org/10.1016/j.ibiod.2013.08.003
Gruiz K, Kriston E (1995) In situ bioremediation of hydrocarbon in soil. J Soil Contam 4(2):163–173. https://doi.org/10.1080/15320389509383490
Hassanshahian M, Emtiazi G, Cappello S (2012) Isolation and characterization of crude-oil-degrading bacteria from the Persian Gulf and the Caspian Sea. Mar Pollut Bull 64(1):7–12. https://doi.org/10.1016/j.marpolbul.2011.11.0
Janssen DB, Stucki G (2020) Perspectives of genetically engineered microbes for groundwater bioremediation. Environ Sci Process Impacts. https://doi.org/10.1039/c9em00601j
Jin R, Yang H, Zhang A, Wang J, Liu G (2009) Bioaugmentation on decolorization of C.I. Direct Blue 71 by using genetically engineered strain *Escherichia coli JM109 (pGEX-AZR)*. J Hazard Mater, 163(2–3):1123–1128. https://doi.org/10.1016/j.jhazmat.2008.07.067
Jørgensen KS (2007) In situ bioremediation. Adv Appl Microbiol 61:285–305. https://doi.org/10.1016/s0065-2164(06)61008-3
Jørgensen KS, Puustinen J, Suortti A-M (2000) Bioremediation of petroleum hydrocarbon-contaminated soil by composting in biopiles. Environ Pollut 107(2):245–254. https://doi.org/10.1016/s0269-7491(99)00144-x
Junior Letti LA, DestéfanisVítola FM, de Melo V, Pereira G, Karp SG, Pedroni Medeiros AB, Ferreira da Costa ES, … Soccol CR (2018) Solid-state fermentation for the production of mushrooms. Curr Dev Biotechnol Bioeng 285–318. https://doi.org/10.1016/b978-0-444-63990-5.00014-1
Jussila MM, Zhao J, Suominen L, Lindström K (2007) TOL plasmid transfer during bacterial conjugation in vitro and rhizoremediation of oil compounds in vivo. Environ Pollut 146(2):510–524. https://doi.org/10.1016/j.envpol.2006.07.012
Kao C (2000) Control of BTEX migration by intrinsic Bioremediation at a gasoline spill site. Water Res 34(13):3413–3423. https://doi.org/10.1016/s0043-1354(00)00070-1
Kao CM, Prosser J (1999) Intrinsic bioremediation of trichloroethylene and chlorobenzene: field and laboratory studies. J Hazard Mater 69(1):67–79. https://doi.org/10.1016/s0304-3894(99)00060-6
Kao CM, Chen CY, Chen SC, Chien HY, Chen YL (2008) Application of in situ biosparging to remediate a petroleum-hydrocarbon spill site: field and microbial evaluation. Chemosphere 70(8):1492–1499. https://doi.org/10.1016/j.chemosphere.2007.08
Kumar V, Shahi SK, Singh S (2018) Bioremediation: an eco-sustainable approach for restoration of contaminated sites. Microbial Bioprospecting Sustain Dev 115–136. https://doi.org/10.1007/978-981-13-0053-0_6
Lee M, Swindoll CM (1993) Bioventing for in situ remediation. Hydrol Sci J 38(4):273–282. https://doi.org/10.1080/02626669309492674
Lin Z, Puls RW (2003) Potential indicators for the assessment of arsenic natural attenuation in the subsurface. Adv Environ Res 7(4):825–834. https://doi.org/10.1016/s1093-0191(02)00056-4
Maila MP, Cloete TE (2004) Bioremediation of petroleum hydrocarbons through landfarming: are simplicity and cost-effectiveness the only advantages? Rev Environ Sci Biotechnol 3(4):349–360. https://doi.org/10.1007/s11157-004-6653-z
Malhautier L, Khammar N, Bayle S, Fanlo J-L (2005) Biofiltration of volatile organic compounds. Appl Microbiol Biotechnol 68(1):16–22. https://doi.org/10.1007/s00253-005-1960-z
Massa V, Infantino A, Radice F, Orlandi V, Tavecchio F, Giudici R, … Barbieri P (2009) Efficiency of natural and engineered bacterial strains in soil slurry degradation of 4-chlorobenzoic acid. Int Biodeterior Biodegradat 63(1):112–115. https://doi.org/10.1016/j.ibiod.2008.07.006
Megharaj M, Venkateswarlu K, Naidu R (2014) Bioremediation. Encyclopedia Toxicol 485–489. https://doi.org/10.1016/b978-0-12-386454-3.01001-0

Mills MA, Bonner JS, McDonald TJ, Page CA, Autenrieth RL (2003) Intrinsic Bioremediation of a petroleum-impacted wetland. Mar Pollut Bull 46(7):887–899. https://doi.org/10.1016/s0025-326x(02)00367-3

Mishra M, Singh SK, Kumar A (2021) Role of omics approaches in microbial Bioremediation. Microbe Mediated Remediation Environ Contaminants 435–445. https://doi.org/10.1016/b978-0-12-821199-1.00036-5

Namkoong W, Hwang E-Y, Park J-S, Choi J-Y (2002) Bioremediation of diesel-contaminated soil with composting. Environ Pollut 119(1):23–31. https://doi.org/10.1016/s0269-7491(01)00328-1

Oyarzún P, Arancibia F, Canales C, Aroca GE (2003) Biofiltration of high concentration of hydrogen sulphide using *Thiobacillus thioparus*. Process Biochem 39(2):165–170. https://doi.org/10.1016/s0032-9592(03)00050-5

Pandey J, Chauhan A, Jain RK (2009a) Integrative approaches for assessing the ecological sustainability of in situ bioremediation. FEMS Microbiol Rev 33(2):324–375. https://doi.org/10.1111/j.1574-6976.2008.00133.x

Pandey J, Chauhan A, Jain RK (2009b) Integrative approaches for assessing the ecological sustainability ofinsitubioremediation. FEMS Microbiol Rev 33(2):324–375. https://doi.org/10.1111/j.1574-6976.2008.00133.x

Partovinia A, Naeimpoor F, Hejazi P (2010) Carbon content reduction in a model reluctant clayey soil: Slurry phase n-hexadecane bioremediation. J Hazard Mater 181(1–3):133–139. https://doi.org/10.1016/j.jhazmat.2010.04.106

Pavlostathis SG (2003) Water Air Soil Pollut Focus, 3(3):117–129. https://doi.org/10.1023/a:1023913330677

Prasanna D, Venkata Mohan S, Purushotham Reddy B, Sarma PN (2008) Bioremediation of anthracene contaminated soil in bio-slurry phase reactor operated in periodic discontinuous batch mode. J Hazard Mater 153(1–2):244–251. https://doi.org/10.1016/j.jhazmat.2007.08.063

Rodrigues JLM, Kachel CA, Aiello MR, Quensen JF, Maltseva OV, Tsoi TV, Tiedje JM (2006) Degradation of Aroclor 1242 dechlorination products in sediments by *Burkholderiaxenovorans LB400(ohb) and Rhodococcus* sp. Strain RHA1(FCB). Appl Environ Microbiol, 72(4):2476–2482. https://doi.org/10.1128/aem.72.4.2476-2482.2006

Romanchuk M, Strand I, Petänen T, Peltola R, Jonsson-Vihanne M, Koivula T, … Haahtela K (2000) Means to improve the effect of in situ bioremediation of contaminated soil: an overview of novel approaches. Environ Pollut 107(2):179–185. https://doi.org/10.1016/s0269-7491(99)00136-0

Ross D (1990) Slurry-phase bioremediation: case studies and cost comparisons. Remediat J 1(1):61–74. https://doi.org/10.1002/rem.3440010108

Sayara T, Sarrà M, Sánchez A (2010) Effects of compost stability and contaminant concentration on the bioremediation of PAHs-contaminated soil through composting. J Hazard Mater 179(1–3):999–1006. https://doi.org/10.1016/j.jhazmat.2010.03.104

Sayler GS, Ripp S (2000) Field applications of genetically engineered microorganisms for bioremediation processes. Curr Opin Biotechnol 11(3):286–289. https://doi.org/10.1016/s0958-1669(00)00097-5

Schaefer M, Juliane F (2007) The influence of earthworms and organic additives on the biodegradation of oil-contaminated soil. Appl Soil Ecol 36(1):53–62. https://doi.org/10.1016/j.apsoil.2006.11.002

Speight JG (2018) Mechanisms of transformation. React Mech Environ Eng 337–384. https://doi.org/10.1016/b978-0-12-804422-3.00010-9

Straube WL, Nestler CC, Hansen LD, Ringleberg D, Pritchard PH, Jones-Meehan J (2003) Remediation of polyaromatic hydrocarbons (PAHs) through landfarming with biostimulation and bioaugmentation. Acta Biotechnol 23(23):179–196. https://doi.org/10.1002/abio.200390025

Strong LC, McTavish H, Sadowsky MJ, Wackett LP (2000) Field-scale remediation of atrazine-contaminated soil using recombinant *Escherichia coli* expressing atrazine chlorohydrolase. Environ Microbiol 2(1):91–98. https://doi.org/10.1046/j.1462-2920.2000.00079.x

Taghavi S, Barac T, Greenberg B, Borremans B, Vangronsveld J, van der Lelie D (2005) Horizontal gene transfer to endogenous endophytic bacteria from poplar improves phytoremediation of toluene. Appl Environ Microbiol 71(12):8500–8505. https://doi.org/10.1128/aem.71.12.8500-8505.2005

Thomé A, Reginatto C, Cecchin I, Colla LM (2014) Bioventing in a residual clayey soil contaminated with a blend of biodiesel and diesel oil. J Environ Eng 140(11):06014005. https://doi.org/10.1061/(ASCE)ee.1943-7870.0000863

Tomei MC, Daugulis AJ (2012) Ex situ bioremediation of contaminated soils: an overview of conventional and innovative technologies. Crit Rev Environ Sci Technol 43(20):2107–2139. https://doi.org/10.1080/10643389.2012.672056

Urgun-Demirtas M, Stark B, Pagilla K (2006) Use of genetically engineered microorganisms (GEMs) for the bioremediation of contaminants. Crit Rev Biotechnol 26(3):145–164. https://doi.org/10.1080/07388550600842794

Vaish S, Pathak B (2020) Bionano technological approaches for degradation and decolourization of dye by mangrove plants. Biotechnol Utilization Mangrove Resour 399–412. https://doi.org/10.1016/b978-0-12-819532-1.00019-6

Vaithyanathan VK, Savary O, Cabana H (2021) Performance evaluation of biocatalytic and biostimulation approaches for the remediation of trace organic contaminants in municipal biosolids. Waste Manage 120:373–381. https://doi.org/10.1016/j.wasman.2020.11.046

Vidali M (2001) Bioremediation An Overview. Pure Appl Chem 73(7):1163–1172. https://doi.org/10.1351/pac200173071163

Wang S, Mulligan CN (2006) Natural attenuation processes for remediation of arsenic contaminated soils and groundwater. J Hazard Mater 138(3):459–470. https://doi.org/10.1016/j.jhazmat.2006.09.048

Wang M, Chen S, Jia X, Chen L (2021) Concept and types of bioremediation. Handbook Bioremediat 3–8. https://doi.org/10.1016/b978-0-12-819382-2.00001-6

Yen HK, Chang NB, Lin TF (2003) Bioslurping model for assessing light hydrocarbon recovery in contaminated unconfined aquifer. I: Simulation Analysis. Pract Period Hazard Toxic Radioact Waste Manag, 7(2):114–130. https://doi.org/10.1061/(ASCE)1090-025x(2003)7:2(114)

Combined Applications of Physico-Chemical Treatments in Treatment of Industrial Wastewater

Jutishna Bora, Ishani Saha, Vardan Vaibhav, Mayukh Singh, and Sumira Malik

1 Introduction

Water is a treasured resource, and there was a time when it could be procured without paying any cost. Nevertheless, in current times water is not accessible for either domestic use or commercial use. The cost of water for industrial use has increased so much that now it is considered equivalent to other raw materials being used in the industries. Water has several functions in every industry and aids in many processes. Nearly all the water used in industries produces industrial wastewater (Bhandari 2014). The remediation of wastewater is gaining more prominence as a consequence of declining sources of water, a rise in commercial, household, and agricultural water usage, and consequently, disposal of wastewater and the related costs accompanied by strict regulations for discharge which have decreased the permissible level of contaminants in waste streams (https://www.das-ee.com/en/wastewater-treatment/treatment-technologies/chemical-physical-processes/). Discharge of wastewater from industries in nature generates a significant impact and can lead to several other problems. This mainly is the case with chemical and allied process industries.

The chemical and related processes use water to a large extent, making them *water-intensive industries,* i.e., industries with high water consumption. In the industries producing fertilizers, the reactant used is water. Nitric acid production (used for nitrogen-based fertilizers) utilizing ammonia gas oxidation requires a reaction

J. Bora · S. Malik (✉)
Amity Institute of Biotechnology, Amity University Jharkhand, Ranchi, Jharkhand, India
e-mail: smalik@rnc.amity.edu

I. Saha · M. Singh
Amity Institute of Biotechnology, Amity University Kolkata, Kolkata, West Bengal, India

V. Vaibhav
Amity Institute of Biotechnology, Amity University, Noida, Uttar Pradesh, India

M. P. Shah (ed.), *Advanced and Innovative Approaches of Environmental Biotechnology in Industrial Wastewater Treatment*,
https://doi.org/10.1007/978-981-99-2598-8_18

between NO_2 and water. Likewise, the most frequently used wet process for manufacturing phosphoric acid (needed for phosphate fertilizers) also requires water as a reactant (Bhandari 2014). The amount of wastewater produced in several industries differs substantially depending on the process and is much higher in countries that still come under the developing category. This also depends on the contaminants' type and concentration inwastewater. For instance, the steel industry in India uses 25–60 m^3 of water per ton of steel, which is eight to ten times more than that in countries like the United States of America (i.e., developed countries) (Bhandari 2014). Waters used for cooling produced from the steel and coke industry, hence, comprise a considerable volume and can possess contaminants in the form of toxic constituents like cyanide, ammonia, phenols, and metals. In industries producing pharmaceutical products, wastewater is produced mainly from the equipment used in cleaning. Even though the volumes produced are not huge, the wastewater produced is polluted to a large extent because the wastewater produced has a high concentration of contaminants of organic nature found in clinical compounds, solvents, etc. (Bhandari 2014; https://www.das-ee.com/en/wastewater-treatment/treatment-technologies/chemical-physical-processes/).

The chemical and physical treatments usually occur in individual steps based on wastewater composition. Methods such as sedimentation, filtration, and centrifugal separation are successfully employed in treating wastewater with substances insoluble in water or colloids. Based on the constitution of wastewater, flotation is frequently used as a part of the physical treatment stage. In flotation, particles adhere to air bubbles smaller in size because of adhesion forces. A trustworthy, mechanical initial cleaning is necessary to treat sanitary wastewater to prevent damage in the following stages of treatment (https://www.das-ee.com/en/wastewater-treatment/treatment-technologies/chemical-physical-processes/). A previously soluble material is converted into an insoluble material that can be separated from the liquid via filtration when precipitation is occurring. Some other methods of removing contaminants are treating with U.V., ozone, ion exchange, and flocculation.

In states like Maharashtra, Rajasthan, etc., where water availability is very scarce, wastewater remediation becomes very important. There is a large amount of water usage in the municipal sector, and as a result, this sector generates a substantial amount of wastewater for discharge. The water and the wastes carried in water derived from houses, institutes, and commercial and industrial setups comprise municipal wastewater.

2 Physical Wastewater Treatment Processes

a. Coarse-and-Fine Materials Separation Utilizing Screens and Strainer

Wastewater screening, one of the earliest remediation methods, eliminates gross contaminants from the waste stream to protect downstream equipment from harm, mitigates interference with plant operations, and precludes unwanted floating material from entering primary settling tanks. The devices used for screening may be made up of parallel bars, rods, or wires, grating, wire mesh, or perforated plates, to block large suspended or floating materials. The openings are usually circular or rectangular, although they can be of any shape. Materials are screenings from bar racks and screens' manual or mechanical cleaning. These screenings are either disposed of by burial or by incineration. They can also be returned to the flowing waste after grinding them (Bhargava 2016). To eliminate solid pollutants from the wastewater, screens, and strainer can be employed. Solid pollutants such as nappies, hair, and wipes can be removed from the wastewater stream through these mechanical processes. The strainer separates textile fibres, plastic residues, and production residues such as potato peels and other wastes. Employment of coarse or fine screen is done, and is based on an area of application. They purify the wastewater through parallelly arranged rods. Screens, meshes, grits, perforations, etc., of different sizes, are present in the strainer. From the wastewater stream, solid materials as large as human excreta to as small as particles of sand are separated by using a coarse strainer of pore size greater than 20 mm to a micro strainer of pore size less than 0.05 mm (https://www.das-ee.com/en/wastewater-treatment/treatment-technologies/chemical-physical-processes/).

During the treatment of sanitary wastewater, the initial cleaning via mechanical methods holds utmost importance. The suspended fibres are causing a considerable problem in the wastewater, particularly the textile fibres of wet wipes and materials which are not woven (as they are tear-resistant). They seem to accumulate, possibly causing blockages and causing enormous damage to pumps and mixers.

b. Mechanical Separation of Solid Substances through Filtration

The separation of solid substances from fluids is achieved through filtration. Filters made up of textiles or metals are usually used in technical applications. Commonly used materials that function as filtration systems are sand filters, cloth fibres, and drum screens.

The suspended solids of organic and inorganic nature, sand particles, and dust particles from wastewater are eliminated by filtration systems. To deliver water for domestic and drinking purposes, filtration, usually in multistage processes, is used as well. A membrane plays the role of filter medium when it comes to membrane filtration, which is another mechanical separation process. This approach is generally employed to segregate extremely fine particles (https://www.das-ee.com/en/wastewater-treatment/treatment-technologies/chemical-physical-processes/ ; http://ngoenvironment.com/en/Wastewater-treatment-tec29-PHYSICAL-AND-CHEMICAL-WASTEWATER-TREATMENT-METHODS-d55.html).

c. Wastewater Treatment through Membrane Technology

The soluble and insoluble substances in wastewater are separated and concentrated by membrane filtration. This kind of separation is carried out under pressure. The membrane retains the particles and molecules of a particular dimension as the membrane has a particular pore size. Different membrane filtration techniques are applied to purify water, treating wastewater, processing water recycling, and collecting recyclable materials for the recovery of valuable substances.

1. **Microfiltration** is utilized to segregate particles and microorganisms such as bacteria and yeasts. It also finds application in cold sterilization and separation of emulsions composed of oil and water.
2. **Ultrafiltration** plays a significant role in the remediation of wastewater and drinkable water by separating particles, microbes, turbidity, and proteins from the water. In swimming pools, ultrafiltration is used to purify water, for instance. To substitute subsequent treatment steps or to enhance the capacity of treatment of biological wastewater plants, the step of ultrafiltration can be arranged directly inside or as a distinct stage after the activation tank while reassembling old wastewater remediation plants.
3. **Nanofiltration** includes the retention of viral particles, heavy metal ions, molecules, and ultra-fine particles.
4. **Reverse Osmosis** plays an essential role in remediating drinkable water in rural areas that are not linked to a network of pipes, concentrating wastewater in the landfill, removing salt from seawater, and decalcifying water present in boilers in power plants. In this method, the osmosis process is reversed via applying pressure through a semi-permeable membrane which further concentrates substances dissolved in liquids. Whenever the pressure being applied surpasses the respective osmotic pressure, it causes the solvent molecules to diffuse beyond the membrane where the dissolved material is already less concentrated. Ultrapure water is produced through reverse Osmosis as well (http://ngoenvironment.com/en/Wastewater-treatment-tec29-PHYSICAL-AND-CHEMICAL-WASTEWATER-TREATMENT-METHODS-d55.html).

d. Wastewater Treatment Through Flotation

Flotation eliminates particles suspended from the fluids carrying tiny bubbles of gas and transports the substances to the surface, removing the bubbles and the particles with the help of a clearing device. Floatation is used in separating oils, fats, and finely dispersed solids and particles in wastewater treatment. For a better accumulation of particles or droplets' function, the size of micro-bubbles should be small. Dissolved Air Flotation (DAF), an economically viable method, is frequently used in wastewater technology. The flotation process is supported by auxiliary agents like pushers, frothers, controllers, and collectors.

e. Solids Separation Through Sedimentation

The gravitational force is used to separate solid particles inside the sedimentation tanks. A sedimentation tank is a flattened, almost current-free tank especially articulated for the process of sedimentation. At the bottom of the tank, the solid particles settle down. A sedimentation tank or Wastewater clarifier has a critical role to play either after or before biological treatment processes to eliminate bulky sludge solids through separation from the liquid phase and settling (http://ngoenvironment.com/en/Wastewater-treatment-tec29-PHYSICAL-AND-CHEMICAL-WASTEWATER-TREATMENT-METHODS-d55.html). Whenever sedimentation is used before treatment by biological methods, the principal benefit is that it could help to substantially reduce BOD levels and, therefore, cause a decrease in load feed into the aeration pond.

The sedimentation process is used in different ways in wastewater treatment. The insoluble materials settle and form primary sludge, which is then concentrated in the digestion tower, where it is transformed without oxygen inside the preliminary cleaning tank. The transformation process generates digested sludge and fermentation gas, which, in its pure state like biogas, is transformed into electrical energy to meet energy demands. After the aerobically generated sludge has been separated from the wastewater by sedimentation inside the clarifier tank, it is added to the digestion tower. Particles that are bulkier than water are separated through sand traps and sludge collectors additionally (Bhandari 2014; http://ngoenvironment.com/en/Wastewater-treatment-tec29-PHYSICAL-AND-CHEMICAL-WASTEWATER-TREATMENT-METHODS-d55.html).

3 Wastewater Treatment Through Chemical Processes

1. **Neutralization**: To adjust the pH values, neutralization is employed in wastewater technology. Neutralizing commercial wastewater and after flocculation and precipitation, acids or alkalis are added per requirements.
2. **Oxidation/Reduction**: The reactions involving oxidation and reduction are often used in treating wastewater by chemical methods and the treatment of drinkable water. From the potable water, removal of pesticides and chlorinated hydrocarbons is achieved by oxidation processes involving ozone and hydrogen peroxide. For wastewater remediation, the complex biologically degradable substances are eliminated by employing the process of oxidation. Photochemical purification involves the generation of hydroxyl free radicals from H_2O_2 or O_3 and is exceptionally efficient. The materials such as cytostatic drugs, antibiotics, hormones, and other trace substances of anthropogenic origin are degraded by utilizing these Advances in Oxidation Processes (AOP) (https://www.das-ee.com/en/wastewater-treatment/treatment-technologies/chemical-physical-processes/).
3. **Adsorption and Chemisorption**: The phenomena of accumulation of substances on a solid's body surface is called adsorption. Adsorption is a physical process involving the adherence of molecules to boundary surface mediated by Van der Waals force. The phenomenon of binding substances to the surface of a solid through chemical bonding is called chemisorption. Unlike adsorption, chemisorption is irreversible. The soluble contents of wastewater cannot be adequately removed with cheap methods like flocculation, biological wastewater treatment, and precipitation; hence, such substances are removed by utilizing activated carbon. By adsorption on the surface of activated carbon, the colouring materials from textile-dying plants can be removed entirely. Trace elements of

human origin like leftovers of pharmaceutical industries and substances like adsorbable, organically bounds halogens (AOX) also bind to activated carbon.

To eliminate arsenic and heavy metals, activated doped carbon can be used. Iron hydroxide in granulated form is an excellent option to remove poisonous metalloid arsenic from drinkable water, polluted groundwater, and wastewater from industries. This method involves the reaction between iron hydroxide and arsenate ions to produce iron arsenate. This process is economical as well as effective (http://ngoenvironment.com/en/Wastewater-treatment-tec29-PHYSICAL-AND-CHEMICAL-WASTEWATER-TREATMENT-METHODS-d55.html).

4. **Precipitation**: Precipitation is the phenomenon of the separation of a previously soluble substance from a fluid. By adding appropriate agents, precipitation can be developed, which is a general method. The heavy metals can be transformed into metal hydroxides which are difficult to dissolve by precipitation.

The precipitation of anions can be in the form of calcium, iron, and aluminium. For example, the fluoride ion's separation is done by precipitation with calcium hydroxide suspension. The phosphate concentration is decreased by adding salts such as iron (II) sulfate, iron chloride, or aluminium chloride, while wastewater treatment in the treatment plant is carried out. The phosphate precipitation can be added either as a succeeding separate process step or as simultaneous precipitation into the biological treatment stage.

5. **Flocculation and Coagulation**: For elimination from water, the highly minute particles, which are in the form of a colloidal solution or suspended, are prepared through flocculation. Due to mutual electrical repulsion, more enormous-sized conglomerates are aggregated, provided that the surface charge of this very fine particulate is the same. Flocculation also finds uses in improving the properties of settling as well on drain sewage sludge. Employing iron and aluminium salts for flocculation allows the flocculating of phosphate at an equivalent time. ChemTreat flocculants include low, medium, and high relative molecular mass polymers.

Coagulation may be a process during which destabilization colloidal particles are destabilized within the fluid by adding salts, which reduce, neutralize, or invert the electrical repulsion between particles. Coagulants are often broadly classified as inorganic and organic. Coagulation is one of the most commonly employed methods in effluent treatment. However, using coagulants in wastewater treatment containing refractory pollutants is a complex problem, and no general solutions are available (Bhandari 2014).

6. **Ozonation Ozone (O_3)**: It is used to disinfect drinking water, remove effluents from wastewater treatment plants through a process known as ozonation, and degrade organic and inorganic contaminants in wastewater.
7. **Disinfection**: The objective of disinfection in wastewater treatment is to crucially lower the number of microorganisms in the water that will be discharged back into

the environment for later use as drinking water, bathing water, irrigation, and so on. The efficiency of disinfection is determined by the water quality (cloudiness, pH, etc.), the type of disinfectant utilized, disinfectant dosage (concentration and time), and other environmental factors (https://www.das-ee.com/en/wastewater-treatment/treatment-technologies/chemical-physical-processes/). The microorganism from the treated water should be removed, killed, or inactivated before it is discharged into any waterbody. There are various disinfection methods for wastewater treatment, but the two most popular are chlorine and U.V. light.

4 Treatment Technologies for Addressing the Removal of Industrial Effluents

India has established itself as a research and development centre, poised to lead the world into the fourth industrial revolution. The fourth level focuses on fusing technologies to advance future manufacturing procedures. Industrialization has aided India's economic development and growth compared to other emerging countries. With the expansion, industries and production have a hike in pollutants and chemicals discharged without treatment. India's various industries include textile, cement, mining, food processing, chemical, steel, and dairy. They create a combination of harmful and harmless contaminants that are either persistent or not. If not handled, these contaminants are highly carcinogenic or will become so in the future (Shah and Banerjee 2020). Chemical and synthetic dyes, heavy metals, and pesticides, all of which are persistent pollutants, are likely to be generated by these sectors. Treatment of wastewater is required to maintain BOD and COD levels while reducing the number of heavy metals in river water. The various treatment modalities are utilized alone or in combination (Shah and Banerjee 2020) (Table 1).

Unwanted components in the effluents are non-biodegradable and hazardous to the environment. The effluent composition varies depending on the industry and the pretreatment it has received before being discharged to a neighbouring land or aquatic body. The most typical metrics for evaluating wastewater include high COD, BOD, suspended particles, and heavy metals. The effluents' persistent nature causes them to accumulate in flora and wildlife. The sewage builds up over time, putting a strain on the ecosystem. Farmers require river water for irrigation. Therefore effluent streams released to surrounding rivers or lands impact nearby agriculture. As a result, hazardous components enter the food chain, directly hurting consumers. Once they have made their way into the food chain, they build up and can even kill live species. They decrease the water's purity, rendering it unfit for further usage (Shah and Banerjee 2020).

Table 1 The advantages and disadvantages of the physical and chemical methods for treating industrial wastewater.

Treatment procedure	Method name	Advantages	Disadvantages
Physical methods	Screening	Lesser equipment maintenance Ideal for smaller operations with limited screenings Less expensive Offer superior flow and screening capture	Clogging and excessive backwater levels High raking It enhanced labour costs It enhanced equipment maintenance costs
	Filtration	Elimination of pollutants such as ammonia, trace metals, volatile organic compounds, manganese, iron, and other toxins Minimize impurities	Unremoved pathogens and pollutants and contaminants Maintenance of the equipment is very critical
	Membrane technology	Allowance of large amounts of non-turbid water Inexpensive No chemicals are required The water sample can be cleaned of 90–100% of contaminants Less energy consumption This procedure can be used to sterilize heat-sensitive media	In membrane filtration, turbid water is not used A potential risk of bacterial overgrowth Glass filters are easily breakable Membrane filters are prone to crack Only liquid sterilization Nano-filters, if used in particular, are expensive to repair Clogging
	Floatation	Efficient for small particle removal and low-density particles Retention time is short	Initial capital expenses and energy costs are both high. The expenditures for maintenance and operation, as well as the chemicals required, are not insignificant. It is a pH-dependent process
	Sedimentation	Sedimentation is a method of reducing particle concentration in water. The use of sedimentation minimizes the need for coagulation and flocculation. Chemicals are typically necessary for coagulation and flocculation, although enhanced sedimentation decreases the need for additional chemicals	Only settable materials, such as larger microbes, sand, and silts, settle well; smaller microorganisms and clays do not; only moderate to low microbial reductions are seen. The pathogen reduction method is unreliable; solids are not efficiently removed by settling from waters; labour-intensive

(continued)

Table 1 (continued)

Treatment procedure	Method name	Advantages	Disadvantages
Chemical methods	Neutralization	Wastewater treatment by biological methods is improved due to its less effectiveness at extreme pH levels. R residual products are not released With the entire water filtration cost, such facilities' investment costs are relatively modest. Installation is relatively simple. Thus, it is a low-complexity project	When using specific salts to treat wastewater, hazardous gases may be produced when the neutralization product is introduced. This could endanger people's health and the environment. Oily and solid coatings must be removed frequently for cleaning. Reference side contamination is caused by interactions between silver ions in the electrodes and silver-interacting compounds in the solution. As a result, the electrode life is reduced. Another downside of acidification with sulphuric acid or hydrochloric acid is that the amount of sulfate or chloride in the effluent increases, causing discharge standards to be exceeded
	Oxidation/reduction	The ozone concentration left over after ozone production can be utilized to purify biological wastes. There are unaccessible limits for the treatment of input concentrations in the case of resistant COD and colour components. Up to 100%, both the parameters can be eliminated Chemical oxidation produces a good to exceptional yield. Screening should reveal whether this technology can be employed in a specific situation, maybe in conjunction with other treatments. It is also possible to achieve the desired yield by raising the oxidant dosage	The introduction or creation of oxidants requires a certain quantity of additional energy. Chemical oxidation's end products are oxidized contaminants on the one hand and residual oxidant concentrations or their breakdown products on the other These are released along with the wastewater that has been treated Before the wastewater may be discharged, it must be checked for excessive levels of oxidizing chemicals The requirement to accurately administer oxidants and catalysts contributes to the technique's complexity

(continued)

Table 1 (continued)

Treatment procedure	Method name	Advantages	Disadvantages
	1. (a) Adsorption	Effectiveness (at low contaminant concentrations), Selectivity (tailored adsorbents), adsorbent renewability Cost-effectiveness (low-cost adsorbents)	Due to its relatively high investment and material costs, it may be inefficient with an unsuitable combination of adsorbent and contaminant (valid for adsorbents such as commercial activated carbons and commercial activated alumina) Chemical adsorption is known to be completely reversible, and regeneration may require much energy
	(b) Chemisorption	The high affinity of sorption of the sorbent and process enhancement should be considered for the system's optimal condition, resulting in cost-effective operation and maintenance	
	Precipitation	Technology that is well-established, with equipment and chemicals readily available. Some treatment chemicals, mainly lime, are very low-cost. Fully enclosed systems are frequently self-contained and low-maintenance, requiring just the replacement of the chemicals utilized. A sophisticated operator is not always required	Calculating adequate chemical dosages is often impossible due to competing reactions, different levels of alkalinity, and other considerations. Overdosing can reduce the treatment's effectiveness. Working with caustic chemicals may require chemical precipitation, raising operator safety concerns The addition of treatment chemicals, mainly lime, can result in a 50 per cent increase in the volume of waste sludge. Transporting large amounts of chemicals to the treatment site may be necessary

(continued)

Table 1 (continued)

Treatment procedure	Method name	Advantages	Disadvantages
	Flocculation and coagulation	Coagulation or flocculation can remove certain pollutants from wastewater that would otherwise be impossible to eradicate without using chemicals. These tanks and dosage units do not demand much expenditure	This method's operational costs are a significant disadvantage. Large volumes of coagulant and flocculant are sometimes necessary to achieve the desired level of flocculation. A significant amount of physico-chemical sludge is also created, typically handled outside. These costs are addendum, especially when dealing with large amounts of wastewater
	2. Ionization	It is a highly effective and efficient water-softening technology There is no material penetration into the soft water Most heavy metals can be reusable Water treatment is done with the wastewater produced by ion exchange equipment	Because sodium ions are introduced into softened water, the amount of acidity in the water can rise. It may contaminate the water, making it unsafe Ion exchangers are the equipment that is utilized to soften the water. The fact is that they must be cleaned due to their high saturation level Ion exchangers have substantial operating expenses as well

(continued)

Table 1 (continued)

Treatment procedure	Method name	Advantages	Disadvantages
	3. Ozonation	In the inhibition of viruses and bacteria, ozone outperforms chlorine. The effluent must be exposed to ozone for a brief period (approximately 10 to 30 min). Because ozone decomposes quickly, it does not leave a harmful remnant in wastewater that needs to be eliminated after treatment. Unlike U.V. and chlorine disinfection, ozonation does not allow bacteria to grow again. Because ozone is manufactured on-site, there are fewer safety issues around handling and shipment. Dissolved oxygen (D.O.) concentration in released wastewater is increased through ozonation	Low doses may not inactivate some viruses, spores, and cysts. Other disinfection technologies are more complicated than ozonation. Because ozone is very reactive and corrosive, corrosion-resistant materials, such as stainless steel, are required. For low-quality wastewater, ozonation is not cost-effective. Off-gases from the workers must be eliminated to prevent worker exposure to ozone, which is exceedingly unpleasant and potentially hazardous. The cost of therapy is relatively high, as it requires capital and energy
	4. Disinfection	Low-cost and widely available chemicals	Chlorine is highly hazardous to animals, people, and aquatic life
	(a) Chlorine disinfection	Strong oxidizer, it can be pretty effective at rendering a large number of dangerous pathogens dormant	
	(b) U.V. disinfection	No hazardous chemicals exist to contend with because U.V. disinfection is a physical process. The treated water has no potentially hazardous byproducts. It is highly effective against most bacteria, viruses, spores, and cysts and is faster to treat than other tertiary wastewater treatment processes	Light disinfecting a solution can be hampered by high total suspended solids (TSS) levels. This is not a problem if the previous treatment process successfully eradicates TSS. Because low levels of U.V. light are ineffective against some spores, viruses, and cysts, longer contact durations or higher-intensity exposure are required. In the case of microorganisms, photoreactivation may occur, in which the organisms repair themselves after treatment if the U.V. radiation is not powerful enough

4.1 Types of Effluent Composition from Various Industries

4.1.1 Dairy Industry

The dairy industry is maximally responsible for effluent volume and organic loading among the food industries. The organic load is primarily composed of milk solids and protein that lead to the loss of the industry daily (Clarke and Steinle 1995). Because of the high suspended particles, BOD, COD, and grease contents, the organic load comprises mainly lipids, proteins, and carbohydrates, which require treatment and fluctuations in the flow rate due to the production cycle discontinuity of various products (Baskaran et al. 2003). The dairy sector revolves around the production of yogurt, pasteurized milk, cream, sour milk, cottage cheese, cream, butter, ice cream, cheese, milk, and whey powders, and a variety of other delicacies from raw milk (Britz et al. 2006). Because the sector is so diverse in handling, product processing, and packing, it generates wastes of varying quality and quantity that require proper disposal treatments (Rosenwinkel et al. 2005).

4.1.2 Pulp and Paper Industry

The paper business is one of the oldest, having existed since 105 A.D. Its users have increased since then. Papers have come a long way in terms of utility, and it has now used for more than just writing; they are also used in packaging, for hygiene, toiletries, and everyday tasks. This paper industry is a big consumer of resources like wood, energy, and water, and it is one of the industries that maximum effluent pollution to the environment (Thompson et al. 2001). The wastewater from pulping is strongly loaded with organic matter that consists of dissolved wood-derived components. Industrial wastes are divided into four categories: effluents, gases, particulates, and solid wastes. The effluent contains high COD, BOD, and chlorinated compounds like suspended solids, adsorbable organic halides (AOX), lignins, tannins, and sulphur compounds that cause significant damage to the water environment. The mill effluents are considered to be "the nest to waste chemicals" (Peck and Daley 1994) due to the presence of different pollutants like natural (wood extracts—tannins, lignins, resin acids) and xenobiotic (furans, dioxins, chlorinated lignins). Pollutants like dibenzofurans and polychlorinated dibenzodioxins are classified as toxins because they resist degradation and are persistent (Star 2006; Tekin et al. 2006). Liver function impairment can be seen as a result of these effluents in the aquatic environment (Oikari and Nakari 1982). DNA-damaging chemicals example, dioxins, and furans cause heritable genetic abnormalities and cancer (Brusick 1980); dioxins are classified as "human carcinogens" by the WHO (World Health Organization). The effluent gives the water bodies a black/brown appearance.

4.1.3 Leather Industry

Leather is a well-established industry that uses upholstery, shoe production, clothing, gloves, and leather goods. It is a significant primary sector of export revenues and growth, and its products are always in demand. Chemical and mechanical treatments are used to turn hides/skins into commercial products. Acids, alkalis, dangerous salts, and solvents are the chemicals used in the aqueous phase (Carvalho et al. 2013a).

4.1.4 Pesticides

Pesticides like organochlorine and organophosphorus are the most commonly utilized in agriculture. Organochlorine-based pesticides, for example, DDT, are very toxic and impact soil quality since they are non-biodegradable and accumulate in food chains, whereas organophosphorus pesticides are readily biodegradable. Most developed countries have limited non-degradable pesticides because they harm both nature and humans. Compared to developed countries, developing countries continue to employ inexpensive DDT-based insecticides that are readily available and accessible. Pesticide residues obtained from surface runoff, leaching, and empty container dumping reach aquatic bodies which has a deleterious impact on water quality, affecting humans and wildlife (Heidemann 1993; Horsfall Jr et al. 1999).

4.2 Integrated Approach for the Treatment of Industrial Effluent by Physico-Chemical Processes for a Sustainable Environment

4.2.1 Physico-Chemical Treatment of Dairy Waste

Dairy waste physico-chemical treatment aims to break down and minimize milk fat and protein colloids in effluents. Some of the most common physico-chemical methods include electrocoagulation (E.C.), coagulation, and membrane treatment, which are used extensively (Spliid and Køppen 1998). Coagulation/flocculation is employed in treating industrial effluents (Ahmad et al. 2019). The organic compounds in the effluents are eliminated, resulting in lower BOD and COD concentrations. The number of suspended and colloidal particles is also reduced, resulting in a significant reduction in turbidity. Coagulant addition further destabilizes particulate matter and causes floc development, which results in sedimentation or flotation (Carvalho et al. 2013b). Specific lactic acid bacteria (LAB) strains can be used to promote coagulation naturally in dairy effluent. Lactic acid generation from lactose fermentation causes milk protein denaturation by these microbes.

Furthermore, pretreatment of oxidation with ferrous sulphate ($FeSO_4$) or hydrogen peroxide (H_2O_2) results in an 80% fat removal from cheese industry effluent (Sarkar et al. 2006). Nowadays, a natural coagulant derived from plants called tannin is frequently used due to its superior efficiency compared to other inorganic coagulants and its use across a wide pH range. Another option for treating dairy wastewater is electrochemical treatment (E.C.). An external electric power source causes oxidation of the anodic material, as well as reduction and subsequent deposition of elemental metals from the cathodic end. In addition, a series of reactions produce metal hydroxide and polyhydroxides, which are responsible for pollutants removal from effluent because of an electrostatic attraction along with coagulation (Vlissides et al. 2012).

Microfiltration (M.F.), nanofiltration (N.F.), ultrafiltration (U.F.), dialysis, Reverse Osmosis (R.O.), and electrodialysis are membrane treatment techniques for dairy waste (Kushwaha et al. 2011). The process, as mentioned above, produces high-quality effluent that is appropriate for immediate reuse.

4.2.2 Physico-Chemical Treatment of Pharmaceutical Waste

Microfiltration, nanofiltration, ultrafiltration, Reverse Osmosis, membrane bioreactors, and electrodialysis reversal are the physico-chemical techniques used to remediate pharmaceutical waste (Bellona and Drewes 2007). The membrane's pore size causes problems with wastewater treatment since micropollutants can travel through it. As a result, reverse Osmosis has been successfully employed to treat the pollutants (Watkinson et al. 2007). Activated carbon (A.C.) is the most common technology for removing synthetic and natural pollutants, and it can be employed in powdered or granular form in the bed filter (Annesini et al. 1987).

4.2.3 Physico-Chemical Treatment of Textile Waste

With reactive and vat dyes, coagulation/flocculation is employed to remove dispersed dyes with low decolourization (Yeap et al. 2014) efficiently. Decolourization of different colours has been accomplished using adsorption and activated carbon. Adsorbents such as date stones, charcoal, and potato plant waste are also used to decolourize colours; however, the procedure's principal limits are the supply of these goods and their disposal. Under ideal conditions, the oxidation-based approach has been used to degrade dye effluent and pesticides partially or fully. Oxidation and Fenton chemistry are used in the advanced oxidation process. In addition, the microfiltration process has removed particles and colloidal colour. Unused auxiliary chemicals, organic pollutants, and other contaminants, on the other hand, escape through the membrane (Juang et al. 2013; Kołtuniewicz and Drioli 2008), limiting their use in dye treatment. In addition, nanofiltration, ultrafiltration, E.C., and reverse Osmosis have also been used for wastewater treatment.

4.3 Byproducts of Wastewater Treatment

Some of the significant byproducts obtained after the treatment of industrial wastewater are:

1. **Sewage Sludge**: A product of wastewater treatment in which the left wastewater enters the sewage system and flow into its treatment facilities. The solid and liquid wastes are separated through settling, where they are processed and decomposed by bacteria (https://www.centerforfoodsafety.org/issues/1050/sewage-sludge/harms-of-sewage-sludge-application). These processed solids namely sewage sludge, contain several known and unknown hazardous substances. These include household, medical, chemical, industrial wastes and metals percolated from the sewer pipes. These novel materials are created in the wastewater treatment plant as a result of the combination of chemicals and organic compounds (https://www.centerforfoodsafety.org/issues/1050/sewage-sludge/harms-of-sewage-sludge-application).
2. **Biosolids**: These are residual, semi-solid substances produced as a by-product during the sewage treatment of industrial or municipal wastewater (Kumar and Chopra 2016). The physico-chemical characteristics of the biosolids include 20% of fat, 50% of carbohydrate (in the form of starch, sugar, and fibre), 30–40% of organic solid matter, 3% of nitrogen, 1.5% of phosphorus, 0.7% of potassium, 10% to 20% of C/N in ratio and heavy metals such as Cu, Zn in increased levels (Xu 2014). Properly treated and processed sewage sludge that has been produced from wastewater treatment facilities turns into biosolids and nutrient-rich organic materials (Kumar and Chopra 2013; Kumar et al. 2016). Biosolids can be recycled and further utilized as fertilizer to improve and maintain the productivity of soils and stimulate plant growth. These management practices of sewage sludge increase the usability of the sewage sludge or biosolids and in a way reduce the harmful substances from the sewage sludge to prevent their discharge into aquatic resources like rivers, lakes, streams, etc. (Spinosa 2008; Rogers 2012). The biosolids generally vary in characteristics and comprises of organic and inorganic chemicals, toxic metals and pathogens.
3. **Biogas**: The anaerobic treatment of liquid wastes or wastewater provides the scope to rapidly reduce the organic content of the waste while minimizing the treatment process, energy consumption, and production of microbial biomass or sludge. Utilizing anaerobic digestion in the treatment of wastewater sludge, methane gas is produced and is known as biogas. It must be considered as a renewable source of energy and even as one of the promising solutions to the extensive environmental problem concerning waste handling, water pollution, CO_2 emission, etc. (https://biogas.ifas.ufl.edu/wastewater.asp). Reduction in the processing of sludge and energy consumption are the two attributes that perform the direct anaerobic pre-treatment of wastewater, making them economically feasible for municipal and industrial waste streams. For relatively warm wastewaters containing significant degradable organic compounds, immediate anaerobic treatment may also provide excess energy. However, even with low-strength

wastewater, significant energy savings can be achieved by avoiding the major cost of aeration. Anaerobic treatment effluents are often unsuitable for direct discharge into receiving waters without further treatment and require aerobic polishing. Reduction in aeration demand and sludge production in aerobic treatment followed by anaerobic pre-treatment may justify this treatment scheme (BioTEnMaRe 2014).

4. **Hydrogen Sulfide (H_2S)**: This stinky gas has been culminated as a result of septic conditions during the collection and treatment of wastewater and has also been recognized as a significant problem for municipal and industrial wastewater systems. It forms almost at every point in a system, be it interceptors, force mains, and lift stations, to holding tanks, mechanical dewatering equipment, and drying beds (Berktay and Nas 2007). Apart from its uncomfortable odour, H_2S also poses a severe problem to the structural integrity of the collection system. Heavy damage to capital has been caused due to corrosion by sulfuric acid (H_2SO_4) formed as a result of the interaction of H_2S with moisture (H_2O). The most significant concern is related to the safety hazards associated with H_2S, which is acutely toxic and causes death among workers in sanitary sewer systems. Although disagreeably pungent at first, it instantly damages the sense of smell, and a worker may not be aware of its presence. Even at low concentrations in the air, exposure to hydrogen sulfide has been linked to fatigue, headaches, eye irritation, sore throats, and other health-related problems (Berktay and Nas 2007).

4.4 Applications of Some Byproducts Obtained from Wastewater Treatment

1. **Sewage Sludge**: After being processed, the sewage sludge is either dried and disposed of in a landfill, applied as fertilizer to crops, or bagged and sold as "biosolid compost" to be used in landscaping and agriculture.

This chemical mixture, often comprised of hazardous substances, hormones, and pathogens, are applied to our food. Chemicals like PCBs, heavy metals, flame retardants, and endocrine disrupters—some carcinogenic—are not screened out. Instead, it is accumulated in the soil and is absorbed by crops, compromising human health.

2. **Biosolids**: Organic and Inorganic Compounds, Harmful Metals, and Microorganisms Are Present in Biosolids.

Because of its ubiquitous use in soil amendment, energy generation, and nutrient supply, it is frequently referred to as a resource. When sewage sludge enters a sewage treatment plant, wastewater is digested anaerobically, resulting in wastewater removal from the sludge.

Biosolids are the source of energy, material, and nutrients that can be used as raw materials for energy production, industrial processing, and soil amendment. Mono-incineration and incineration with energy recovery, anaerobic digestion with biogas production and aerobic composting, gasification, pyrolysis, and wet oxidation processes are among the methods that can turn sewage sludge into valuable output (Poulsen and Hansen 2003; BioTEnMaRe 2014).

3. **Biogas**: Approximately the same amount of carbon dioxide (CO_2) is produced throughout the biogas production process that crops could absorb during their development cycle or would be discharged naturally during trash decomposing. As a result, biogas plants provide exceptionally environmentally friendly electricity and heat. Biogas production has many advantages, such as being a climate-friendly and renewable heat source. Converting biogas into heat and electricity in combined heat and power (CHP) facilities is the most popular technique of using it for energy production. All new plants should include CHP systems for the most efficient usage of biogas. Electricity generation, cooking fuel as a sustainable energy source, waste management in agriculture, injection into a natural gas pipeline, and clean, renewable fuel for transport vehicles in biogas fuel cells are among the most critical uses of biogas (https://www.carusllc.com/water/wastewater/hydrogen-sulfide).

5 Conclusion

For nearly a century, physicochemical methods have been implemented. However, due to the high expense of creating vast amounts of sludge, biological processes superseded these procedures. They have recently been reintroduced for a variety of purposes, including the removal of phosphorus from effluent that is discharged to the sea, obtaining average effluent quality at a lower cost compared to traditional treatments, and for the use of water for agricultural irrigation, potabilization, and sludge (primary and secondary) conditioning in industrial water. The resurgence of these techniques is also due to a growing knowledge that treatment costs should be proportional to the desired efficiency, as advances in the synthesis of high-efficiency flocculation polymers have been made at a reduced cost.

Depending on the dose and kind of coagulant employed, it is possible to remove total suspended solids (TSS), COD, BOD, and nutrients using this method. Heavy metals can also be removed using these techniques. However, removal efficacy varies depending on the various metals and their concentration. These methods have recently been utilized to eliminate pathogens like helminth eggs. Furthermore, they are particularly effective at removing germs, protozoa, and viruses.

Biosolids and sewage sludge are vital in energy production, soil amendment, and agricultural development. However, there are certain disadvantages to their utilization, as applications in sewage sludge or biosolids use in underdeveloped nations are likely to differ significantly from those in developed countries. As a result, sludge or biosolids should be examined and tested for contaminants (pathogens, heavy metals, and other contaminants) before being used in agricultural, soil amendment, or other applications. Furthermore, depending on the production and quality of the sludge, the use of biosolids or sludge should depend on regular monitoring, sampling, and analysis. The most effective sludge management strategy should be developed with a primary focus on economic, technological, and societal constraints; however, researchers, scientists, and policymakers are increasingly considering the overall sustainability of sludge management for future sustainable development.

References

Ahmad T, Aadil RM, Ahmed H, ur Rahman U, Soares BC, Souza SL, Pimentel TC, Scudino H, Guimarães JT, Esmerino EA, Freitas MQ (2019) Treatment and utilization of dairy industrial waste: a review. Trends Food Sci Technol. https://doi.org/10.1016/j.tifs.2019.04.003

Annesini MC, Gironi F, Ruzzi M, Tomei C (1987) Adsorption of organic compounds onto activated carbon. Water Res. https://doi.org/10.1016/0043-1354(87)90065-0

Baskaran K et al (2003) Wastewater reuse and treatment options for the dairy industry. Water Sci Technol Water Supply 3(3):85–91

Bellona C, Drewes JE (2007) Viability of a low-pressure nanofilter in treating recycled water for water reuse applications: a pilot-scale study. Water Res. https://doi.org/10.1016/j.watres.2007.05.0

Berktay SA, Nas B (2017) Biogas production and utilization potential of wastewater treatment. 179–188. https://doi.org/10.1080/00908310600712489

Bhandari VM (2014) Industrial wastewater treatment, recycling and reuse. Adv Physico-Chem Methods Treatment Ind Wastewaters 81–140. https://doi.org/10.1016/B978-0-08-099968-5.00002-7

Bhargava A (2016) Physico-chemical waste water treatment technologies: an overview. 4(05):5308–5319. http://ijsae.in. https://doi.org/10.18535/ijsre/v4i05.05

BioTEnMaRe (2014) BioTEnMaRe. http://ww.biotenmare.com/.

Britz TJ et al (2006) Treatment of dairy processing wastewaters. Waste treatment in the food processing industry. Taylor & Francis, Boca Raton, pp 1–28

Brusick D (1980) Principles of genetic toxicology. Springer, New York

Carvalho F et al (2013a) Cheese whey wastewater: characterization and treatment. Sci Total Environ 445:385–396

Carvalho F, Prazeres AR, Rivas J (2013b) Cheese whey wastewater: characterization and treatment. Sci Total Environ. https://doi.org/10.1016/j.scitotenv.2012.12.038

Clarke E, Steinle D (1995) Health and environmental safety aspects of organic colorants. Rev Prog Color Relat Top 25(1):1–5

Heidemann E (1993) Fundamentals of leather manufacture. Roether, New York

Horsfall M Jr et al (1999) Speciation of heavy metals in inter-tidal sediments of the Okrika river system, Rivers state Nigeria. Bull Chem Soc Ethiop 13(1):1–10.

Juang Y, Nurhayati E, Huang C, Pan JR, Huang S (2013) A hybrid electrochemical advanced oxidation/microfiltration system using BDD/Ti anode for acid yellow 36 dye wastewater treatment. Sep Purif Technol. https://doi.org/10.1016/j.seppur.2013.09.042

Kołtuniewicz A, Drioli E (2008) Membranes in clean technologies. Wiley-VCH

Kumar V, Chopra AK (2013) Accumulation and translocation of metals in soil and different parts of French bean (*Phaseolus vulgaris* L.) amended with sewage sludge. Bull Environ Contamination Toxicol 92(1):103–108. https://doi.org/10.1007/s00128-013-1142-0

Kumar V, Chopra AK (2016) Effects of sugarcane press mud on agronomical characteristics of hybrid cultivar of eggplant (Solanum melongena L.) under field conditions. Int J Recycling Organic Waste Agric 5:149–162. https://doi.org/10.1007/s40093-016-0125-7

Kumar V, Chopra AK, Srivastava S (2016) Assessment of heavy metals in spinach (*Spinacia oleracea* L.) grown in sewage sludge amended soil. Commun Soil Sci Plant Anal 47(2):221–236. https://doi.org/10.1080/00103624.2015.1122799

Kushwaha JP, Srivastava VC, Mall ID (2011) An overview of various technologies for the treatment of dairy wastewaters. Crit Rev Food Sci Nutr. https://doi.org/10.1080/10408

Oikari A, Nakari T (1982) Kraft pulp mill effluent components cause liver dysfunction in trout. Bull Environ Contam Toxicol 28(3):266–270

Peck V, Daley R (1994) Toward a "greener" pulp and paper industry. Environ Sci Technol 28(12):524A-527A

Poulsen TG, Hansen JA (2003) Strategic environmental assessment of alternative sewage sludge management scenarios. Aalborg University, Department of Environmental Engineering

Rogers H (2012) Perspectives for biogas in Europe. Oxford Institute for Energy Studies, U.K.

Rosenwinkel K-H, Austermann-Haun U, Meyer H (2005) Industrial wastewater sources and treatment strategies. Environmental biotechnology: concepts and applications

Sarkar B, Chakrabarti PP, Vijaykumar A, Kale V (2006) Wastewater treatment in dairy industries—possibility of reuse. Desalination. https://doi.org/10.1016/j.desal.2005.11.015

Spinosa L (2008) Status and perspectives of sludge management. CNR, U.K.

Spliid NH, Køppen B (1998) Occurrence of pesticides in Danish shallow groundwater. Chemosphere 37(7):1307–1316

Shah M, Banerjee A (eds.) (2020) Combined application of physico-chemical & microbiological processes for industrial effluent treatment plant. Springer Nature Singapore Pte Ltd. https://doi.org/10.1007/978-981-15-0497-6_3

Star E (2006) U.S. Environmental Protection Agency and U.S. Department of Energy

Tekin H et al (2006) Use of Fenton oxidation to improve the biodegradability of a pharmaceutical wastewater. J Hazard Mater 136(2):258–265

Thompson G et al (2001) The treatment of pulp and paper mill effluent: a review. Bioresour Technol 77(3):275–286

Vlissides AG, Tsimas ES, Barampouti EM, Mai ST (2012) Anaerobic digestion of cheese dairy wastewater following chemical oxidation. Biosyst Eng. https://doi.org/10.1016/j.biosystemseng.2012.09.001

Watkinson AJ, Murby EJ, Costanzo SD (2007) Removal of antibiotics in conventional and advanced wastewater treatment: implications for environmental discharge and wastewater recycling. Water Res. https://doi.org/10.1016/j.watres.2007.04.005

Xu A (2014) Biogas from sewage sludge-Safe disposal of sewage sludge in the People's Republic of China. University of Rostock, Rostock

Yeap KL, Teng TT, Poh BT, Morad N, Lee KE (2014) Preparation and characterization of coagulation/flocculation behavior of a novel inorganic-organic hybrid polymer for reactive and disperse dyes removal. Chem Eng J. https://doi.org/10.1016/j.cej.2014.01.004

Traditional Treatment Methods for Industrial Waste

Jutishna Bora, Richismita Hazra, Sagnik Nag, and Sumira Malik

1 Introduction

The Industrial Revolution has revolutionized humankind in a way that humans have been able to advance further into the twenty-first century. Unfortunately, rapid industrialization, urbanization, and the rising living standards of people have led to massive environmental pollution, proliferating rapidly. Industrialization is inevitable and crucial for uplifting a nation's economy (Ashraf and Hanafiah 2019). However, with the increase in demand for raw materials, non-renewable resources are shrinking daily. Industrial waste is an all-encompassing term that refers to all types of waste stemming from industrial activities. Industrial waste describes the materials which can no longer be used after a manufacturing process is completed. It omits residues that are directly recycled at the place of production and contaminants that are immediately passed into water or air.

Waste is created by taking out raw ingredients, manufacturing materials and facilities, and handling waste and emissions. The industrial sectors that release waste are several kinds of factories such as mining, industrial chemicals, food production, consumer goods, cloth mills, printing, packaging, etc. (Gaur et al. 2020; Amasuomo and Baird 2016). Industrial waste can be deadly, combustible, eroding, and reactive. The most apparent visible ecological impact caused by waste is making the streets ugly by degrading the atmosphere and splendour of the city. However, more

J. Bora · S. Malik (✉)
Amity Institute of Biotechnology, Amity University Jharkhand, Ranchi, Jharkhand, India
e-mail: smalik@rnc.amity.edu

R. Hazra
Amity Institute of Biotechnology, Amity University Kolkata, Kolkata, India

S. Nag
Department of Biotechnology, School of Biosciences & Technology, Vellore Institute of Technology (VIT), Vellore, Tamil Nadu, India

M. P. Shah (ed.), *Advanced and Innovative Approaches of Environmental Biotechnology in Industrial Wastewater Treatment*,
https://doi.org/10.1007/978-981-99-2598-8_19

thoughtful consequences, which are frequently unrecognized, include the transmission of trash to water, earth water, land, and air (Khedkar and Singh 2018). They can be in solid, liquid, or gaseous form and may be hazardous or non-hazardous. It may or may not be biodegradable and comprise compounds such as organic synthetic substances or heavy metals resistant to treatment. Solid industrial wastes are those produced by industries in solid form. Coal ash from thermal power plants, blast furnace slag and steel melting slag from iron and steel mills, red mud and tailings from nonferrous industries, press mud from sugar factories, lime, and fertilizer from paper and pulp industries, and gypsum from allied industries are the prime generators of industrial solid wastes. In a liquid state, they are mainly in the form of industrial wastewater, potentially harmful to our environment. Harmful waste abandoned by industries puts human health at risk (Mathur et al. 2012). Spill waste can also block land vehicles and cause traffic accidents. There is again a definite threat of accumulation of heavy metals in the food hierarchy. High levels of toxic contaminants are often obtained in animals, birds, and humans, predominantly in those who repeatedly come in contact with such discards (Jayashree Dipak 2016). These wastes pose a severe threat to the ecosystem and human health if not managed safely or adequately since the synthetic compounds, when discharged into the environment, affect our ecosystem adversely. Wastes have appeared to be a by-product of growth that cannot be lost as sheer waste, and converting these wastes into utilizable raw materials could help control pollution from their disposal. Industrial waste can be classified as hazardous, non-hazardous, and wastewater (Lakkaboyana et al. 2019) (Table 1).

Based on their physical, chemical, and biological properties, hazardous wastes refer to those that possess the characteristics of toxicity, reactivity, ignitability, corrosiveness, infectiousness, or radioactivity. Toxic wastes are harmful even in trace amounts and may lead to acute or chronic health effects. Reactive wastes are chemically unstable that react with air or water violently. Ignitable wastes are characterized by their burning ability at relatively low temperatures leading to fire hazards, whereas radioactive wastes persist in the environment for many years and emit ionizing energy harmful to living organisms. Further, hazardous wastes are transported via transport vehicles to approved treatment and storage sites where they are treated physically,

Table 1 Environmental and health issues associated with wastes

Environmental and health issues	Examples
Environmental contamination	Water and air quality, used land, noise
Transmittable disease	Gastrointestinal infections, diarrhoea/loss of motion, respiratory infections, hepatitis, skin conditions, trachoma, jaundice, eosinophilia, etc.
Non-transmittable disease	Poisoning, pneumonia, tetanus, hearing defects/loss, dust-related respiratory diseases, etc.
Injury	Work-related injury by sharps like wood, needles, glasses, metals, etc.
Aesthetics	Bad smell, lack of cleanliness, dust particles, etc.

chemically, thermally, and biologically. Physical treatment methods for hazardous wastes include evaporation, sedimentation, flotation, filtration, and solidification. Chemical methods include ion exchange, precipitation, oxidation and reduction, and neutralization. The thermal treatment method includes high-temperature incineration. Land farming and other techniques of bioremediation involving the role of microbes are employed for the biological treatment of this category of wastes (Lakkaboyana et al. 2021).

Non-hazardous industrial wastes—Although non-hazardous wastes or ordinary industrial wastes do not pose a threat, they cause tremendous environmental damage. This category of waste classification finds similarities with household waste in its nature and composition. Hence, it presents no hazard and requires no particular treatment methodologies. They can either be recycled and reused or treated and disposed of, shielding the environment (Fig. 1).

According to a survey conducted recently, a large quantity of industrial wastewater is mixed into streams, beaches, and lakes, producing contamination that negatively impacts human life. Industrial wastewater consists of wastewater from distinct industrial sectors producing a blend of impurities. Hence, this processing should be explicitly arranged for a distinct sort of fluid.

Effluents discharged from some of the major industries are mentioned below:

- Metal industries discharge heavy metals and their compounds
- The electroplating industry produces toxic wastes, predominantly heavy metals, and their compounds

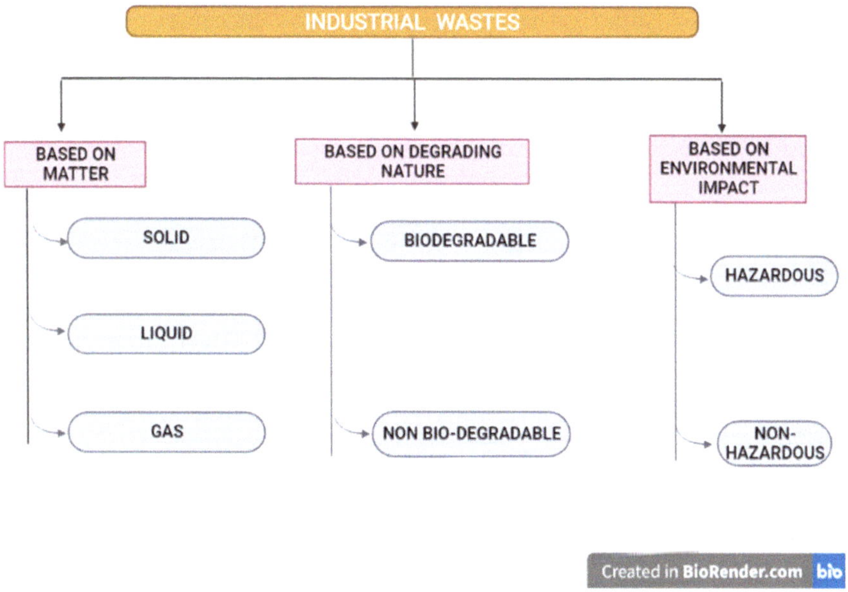

Fig. 1 Classification of industrial wastes

- Photography shops produce Ag compounds in large quantities
- Printing plants discharge inks and dyes
- Mash and paper industries discharge chloride compounds and dioxins
- Petrochemical industries release significant quantities of phenols and oils
- Food industries release organic compounds

2 Treatment of Industrial Wastes

2.1 Treatment of Industrial Solid Wastes

Total solids include both total suspended solids and total dissolved solids, which refer to the suspended and dissolved matter present in wastewater streams, respectively. The suspended solids can be potentially obtained after drying and weighing the residue collected after filtration since a filter can trap them. Volatile solids include such materials, which, when ignited, become free of volatile compounds. Volatile solids predominantly comprise organic compounds, mainly proteins, carbonates, and fats. Total dissolved solids are solids that can pass through a filter with a pore size of 0.45 μm. Settleable solids settle at the bottom of the tank and can be removed by sedimentation (Salgot and Folch 2018). Improper handling of solid wastes imposes risks to the health of humans, plants, animals, and birds. The solid waste decomposes into integral components, which is a frequent cause of local environmental pollution. Poisonous methane gas is released by decaying waste as a by-product of bacterial respiration. These gases contribute to greenhouse gases and climate change. Water penetrating through this solid waste (leachate) in the landfills is a menace to water present on the surface and under the ground (Alam and Ahmad 2013). Wastes from salt and chemical industries increase the salinity of soil and surface, and groundwater. Several industries release waste which causes the acidification of soil and water.

Direct health risks concern the workers in this field coming in direct contact with the waste. Indirect health risks from rats and insects are expected for the general people (Royal Commission on Environmental Pollution 1984). Other types of problems include inhalation of poisonous chemicals, weight during birth is reduced, Cancer, hereditary abnormalities, nervous system issues, and nausea and vomiting.

Solid waste is created from industrial sectors like mining, metallurgical, chemical, textile, pharma, food preservation and packaging, construction, and electronic industries (Soliman and Moustafa 2020). Solid wastes comprise dense parts of the waste material such as glass, synthetic materials, plastic, metals, and radioactive wastes. Solid waste treatment involves the following thing:

- Dumping is the precise and ultimate dumping of solid trash at landfill sites.
- Sanitary landfill—It is placing the waste without causing irritation to people's well-being. Here, the waste is spread to the smallest practical volume and covered with soil and plastics (Rinkesh 2018).

- Incineration—It is a process of turning waste into ash by burning at high temperatures. This method decreases the compact waste by 20–30% of the total amount.
- Pyrolysis—The burning waste in an oxygen or air-deficient environment. It cuts down the bulk waste and yields safe final products. It occurs under pressure and at a very high temperature of about 430 °C.

Electronic waste is any product containing electronic components that have come to the expiration of its usage. E-waste is generated from computers, computer devices like monitors, speakers, keyboards, printers, mobile phones, fax, telephone machines, televisions, etc. Processing of e-waste first comprises the disassembling of the apparatus into its constituent parts, often by hand or automated dismantling equipment. E-waste is treated to recover precious metals by incineration, cyanide leaching, and smelters. Magnets, Trommel screens, and eddy currents separate plastic, ferrous and nonferrous metals, and glass at a smelter. (Awasthi et al. 2018b; Ilankoon et al. 2018; Kishore and Monika 2010).

Electronics contain lethal substances; therefore, they must be picked up correctly when no longer desired. Technological advancements in electronics have led to the generation of e-waste not only from electronic industries but also from many other industries (though in small quantities) using electronic technology for their hassle-free running. The fine particles from e-waste can move long distances, creating adverse health hazards (chronic disease and cancer) for humans and animals. E-waste contains toxic components such as lead, cadmium, polybrominated flame retardants, and lithium which affect significant organs like the brain, heart, liver, kidney, and skeletal, nervous, and reproductive systems, causing diseases of these organs and congenital disabilities. (Awasthi et al. 2018a, b; Orisakwe et al. 2019). E-waste dumped in regular landfills causes heavy metals (mercury, lithium, lead, cadmium, barium, etc.) to seep through the earth, causing groundwater pollution and adulteration of crops (Rautela et al. 2021). Heavy metals from groundwater eventually reach water bodies, causing acidification and toxification of water. Dust particles or toxins (dioxins) are liberated into the air affecting air pollution. Over time, e-waste can create irreversible damage to the ecosystems.

2.2 *Treatment of Industrial Wastewater*

Industrial wastewater is broadly categorized into (1) organic industrial wastewater and (2) inorganic industrial wastewater. Organic industrial wastewater includes a broad range of chemical industries and large-scale chemical works. Although a considerable amount of water is utilized in almost all industries, a significantly less percentage is incorporated in their products. This causes the rest of it to find its way into the streams, natural water bodies, or municipal sewers as 'wastewater', leading to massive water pollution. This aqueous discard originates during an industrial manufacturing process or the cleaning activities associated with the same and is rendered useless. Its quality and volume vary significantly depending on the type of industry producing it. Equalization of the wastewater is required before its treatment since the discharge rate of wastewater is non-uniform. The water is held for a predetermined time in a continuously mixed basin for equalization. This creates an effluent possessing uniform characteristics.

Industrial wastewater released into the water bodies results in reduced dissolved oxygen, changes in water quality, bio-amplification of water organisms, and augmented nutrient elements (Environmental Canada 1997). The bacteria use dissolved oxygen (D.O.) to break down wastewater's organic and oxidation chemicals (Samer 2015) This water is used for watering crops leading to the buildup of chemicals in the soil. Wastewater effluent leads to changes in temperature, decreased sunlight infiltration (and thus reduced photosynthesis), physical harm to fish (Sperling 2007), and eutrophication of water sources, leading to the following penalties such as Clumps of algae, foul smell, and change in the colour of the water. The extraordinary growth of rooted aquatic vegetation interferes with navigation. Oxygen exhaustion results in the death of water organisms. Nitrate present in the wastewater causes depletion in the amphibian population.

Industrial wastewater is generated from various industries such as the food industry, battery production, power station, iron and steel industry, nuclear power plant, natural gas extraction, organic chemicals engineering, mines and quarries, petrochemicals industry, pulpwood industry, textile mills, etc. (Pavithra et al. 2019).

The following industrial sectors produce most organic industrial wastewater:

- Brewery and fermentation factories
- Textile plants
- Paper and cellulose manufacturing plants
- Tanneries and leather factories
- Soaps, synthetic detergents, dyes, glue production sectors
- Pharmaceutical and cosmetic industries

Inorganic industrial wastewater mainly originated from the following industrial sectors:

- Coal and steel industry
- Non-metallic minerals industry
- Iron picking works
- Electroplating plants
- Other commercial adventures

Wastewater generated from various industries may contain metallic chemicals such as cadmium, chromium, copper, lead, nickel, silver, and zinc; organic chemicals and cyanide; and oil and grease. Other metallic compounds may include mercury, arsenic, and selenium. Inorganic constituents might include heavy metals, nitrogen, phosphorus, chlorides, and sulfur, among other pollutants. Gases might include carbon dioxide, nitrogen, oxygen, hydrogen sulfide, and methane. Most industrial wastewaters contain chlorides. This may be due to the leaching of marine sediments and deposits or seawater pollution. Fluorides, when present in wastewater in excess amounts, can cause fluorosis or mottling of teeth. Organic compounds include fats, oil, grease, surfactants, proteins, pesticides, volatile organic compounds, and toxic chemicals. Volatile organic compounds include benzene, toluene, xylenes, trichloroethane, dichloromethane, and trichloroethylene. Nitrogen compounds, including nitrates and nitrites, are also present (Yang et al. 2019).

Treatment of industrial waste is essential as it reduces the amount of waste released into the environment. Loss of water caused by water pollution is reduced. Potentially unsafe chemicals are not released into the environment. Simplified waste disposal considerably reduces costs. (Soltani et al. 2017) Treatment of waste by standard procedures such as precipitation by chemicals, adsorption using carbon, settling chambers, filters, ion exchange, evaporation, and membrane treatment is operational (Wang and Chen 2009). The chemical treatment completely breaks unsafe waste into perishable gases or transforms the waste's chemical nature, such as neutralizing acidity or alkalinity. After neutralization, the waste can be mixed with solvents to be processed using ion exchange. The ion exchange section comprises strong acid cation and HCl for regeneration, weak base anion, and Caustic soda for regeneration. The solution got free of ions and fed into the buffer storage tank. This solution can be treated further in evaporation chambers maintained at high temperatures in a vacuum. Other than that, chemical precipitation converts the solution into a solid by making the solution supersaturated. Chemical precipitation can remove contaminants from municipal and industrial wastewater (Rajasulochana and Preethy 2016). Calcium hydroxide precipitation is a widely used method. Other than that, adsorption of gas, vapour, and liquid waste is performed on a solid surface (e.g. bauxite, silica gel, aluminium, carbon) using surface chemical force. The treatment process also depends on the type of waste used. Gaseous waste like dust particles is often collected in settling chambers and treated using fabric filters. Usually, 99% of 1 mm dry and free-flowing particles are filtered using fabrics. On the contrary, the electrostatic method uses electrostatic forces to collect waste gas on high-voltage discharge electrodes, which can be cleaned later. This technique is the most efficient for gas particles of any size. E-wastes are usually incarnated in the open air to recover precious metals. Nowadays, biological management of harmful substances has become widespread because of better performance and availability of raw materials at a low cost (Bhatia et al. 2017; Santagata et al. 2021). Microorganisms like bacteria (Kwok 2019), algae (Abdel-Raouf et al. 2012), fungi, and yeast (Jain et al. 2017) can effectively treat heavy metal-containing pollutants.

The factors affecting the architecture and planning for industrial wastewater treatment are:

- Non-uniform waste discharge, which becomes seasonal sometimes
- Presence of wastes in high concentration
- The non-biodegradable and toxic nature of wastes
- Large constituents of organic compounds
- High pH level
- Harmful substantial metals
- High salt level

- High turbidity due to the presence of impurities of inorganic compounds

Effective and efficient wastewater treatment requires a combination of distinct unit operations arranged sequentially to facilitate the further downstream unit processes. Wastewater from all sectors, including industrial wastewaters, is expected to meet a specified quality upon reaching the end of this cascade of unit operations. Although newer approaches for wastewater treatments have come up with enhanced accuracy, the traditional treatment processes include the following unit operations:

- Preliminary treatment or pretreatment
- Primary treatment
- Secondary treatment
- Tertiary treatment
- Sludge treatment

2.2.1 Pretreatment

Pretreatment of industrial wastewater refers to eliminating regular, ordinary, and toxic contaminants from industrial wastewater streams before their discharge into sewer systems. It is a significant consideration since the pollutants will likely seep into the soil and cause serious environmental and health hazards. The operations involved in this process are collection, segregation, adjustment, and decontamination. The primary objectives of industrial wastewater pretreatment include:

- To avoid any hindrance in the unit operations in the successive levels
- To block the influx of pollutants or contaminants that causes health impairments and environmental hazards.
- For reusing and recycling these waters.
- For improving safety and lowering costs of subsequent operations.

Evaluation of wastewater and assessment of its discharge circumstances are the key factors behind the pretreatment of wastewater. This stage includes flow measurement and helps remove materials that hinder the proper functioning of different downstream mechanical, chemical, or biological processes. The quality of the incoming sewage remains unchanged. Removal of relatively large-sized suspended materials such as rags and plastic bags is done via screens and gets collected in baskets that are either cleared manually or automatically. Regular cleaning of these baskets is essential since they would emit an undesirable odour when too much material gets collected on them.

Moreover, the presence of grit within the sewage will likely wear the mechanical equipment in contact with the sewage. Grit includes inorganic materials such as sand, gravel, cinder, eggshells, bone chips, seeds, and other solid materials with higher specific gravity than biodegradable solids in wastewater. So, grit removal is a crucial part of the pretreatment of industrial sewage. Apart from grit, oil and grease might be present in these wastewater streams, which, when combined with particulate matter, might hinder the downstream screening processes and require removal via oil and grease traps. If not removed regularly, oil and grease might also emit undesirable odours. Besides mechanical methods, various chemical methods, including pH adjusters, coagulants, and flocculants, might also be adopted for wastewater pretreatment (Quach-Cu et al. 2018; Gogate et al. 2020).

2.2.2 Primary Treatment

The preliminary treatment process is followed by the primary treatment process, which removes either floatable or settleable suspended solids. It is a slow process that usually relies on breaking larger particles into smaller ones that are removed easily via sedimentation and filtration. Most debris can be removed without gravity settling, while the rest of the colloidal and dissolved particles could be removed with the settleable suspended solids. Gravity clarifiers provide suitable conditions for the settleable particles to settle to the bottom of the chamber, forming a sludge layer (Gantz et al. 2020). A scrapper scraps off this sludge at the base of the gravity clarifier chamber, which is moved into a hopper from where it gets pumped to the sludge treatment process. The settled sewage overflows the outlet. Apart from gravity clarifiers, rotating and static fine screens with pore sizes around 0.8–2.3 mm might also be used in the primary wastewater treatment process. This fine screen is placed after the oil and grease trap if the water contains oil and grease. However, fine screens are not as efficient as primary clarifiers.

2.2.3 Secondary Treatment

The primary wastewater treatment process is followed by the secondary treatment process that removes colloidal and dissolved contaminants persisting after the preliminary and primary treatment stages. This stage mainly comprises biological treatment methods wherein the microbial population is employed for treating the wastewater. The biological system is often an anaerobic suspended growth process housed in a bioreactor. The bioreactor is designed to be complete-mix, plug-flow, or arbitrary flow. Apart from aerobic growth systems, attached growth systems, including trickling filters and rotating biological contactors, are also efficient in the secondary wastewater treatment process. In these systems, microorganisms form a biofilm on a support medium. It is porous with a large surface area to volume ratio. Such biofilms do not get submerged in sewage in the trickling filters or get intermittently submerged in a rotating biological contactor. Oxygen is collected from the atmosphere and is transferred to the liquid film that forms on the biofilm (Raju et al. 2020).

2.2.4 Tertiary Treatment

The secondary treatment process is followed by the tertiary treatment process, which is an advanced treatment process that aids in the removal of stubborn contaminants that pass through the previous treatment stages. Tertiary treatment processes involve physical or chemical separation techniques, including carbon adsorption, flocculation, advanced filtration, ion exchange, dechlorination, and reverse osmosis. The main goal of this stage of treatment is to provide a final polished and advanced treatment stage before its discharge or reuse (Cai et al. 2020).

2.2.5 Sludge Treatment

Sludge treatment is also a crucial part of the wastewater treatment procedure. The waste sludge can potentially be thickened in gravity thickeners followed by its aerobic or anaerobic digestion for the reduction of solid content. The sludge is rendered safer in terms of pathogenicity. The digested sludge is dewatered for reduction of moisture content which would also reduce the volume. Standard methodologies adopted include drying beds, filter presses, and centrifuges. Consolidation of sludge refers to the reduction in sludge volume by eliminating water and dissolved solids. Destruction of sludge refers to the oxidation of organic carbon content to carbon dioxide or the reduction of the same to methane (Gurjar 2021).

Industrial wastewater treatment describes the process and mechanisms involved in treating water or effluents contaminated by anthropogenic industrial or commercial activities before their discharge into the environment. This wastewater is treated to remove unwanted substances and contaminants to ensure that the industries comply entirely with regional industrial wastewater treatment standards.

2.2.6 Pretreatment methods

Flow-Equalization

Flow rate equalization is a pretreatment technique for buffering or equalizing the characteristics of industrial wastewater before they are made to enter the wastewater treatment process. This technique controls the hydraulic velocity or flow rate since waste streams vary significantly in the level of contaminants and flow rates. Equalization basins ensure a consistent wastewater flow, with uniform characteristics to further downstream processes by retaining flow fluctuations (Manderson 2018). This mixing basin has the following major functions:

- Prevention of sedimentation
- Prevention of stratification
- To avoid concentration variation
- Provide aeration
- Prevent odour
- Hinder the flow of highly toxic materials to the treatment plant
- Avoid shock loading of biological treatment plants

Neutralization

Wastewater neutralization is a crucial step in the industrial wastewater treatment process. Effluents discharged from industries often contain acidic or alkaline components that require neutralization before any further process or discharge. Neutralization refers to the treatment of effluents so that it is neither too acidic nor too alkaline. A pH of 6–9 should be maintained. This step in wastewater treatment makes industrial effluents compatible with municipal sewage, facilitating the process during joint treatment. A higher or lower pH (relative to the specified value) could cause harm or inactivate the microorganisms employed for the biological oxidation of organic matter. It would also hinder the corrosion of pipelines and equipment used (Forsido et al. 2019). The commonly adopted approaches for neutralization can be summarized as follows:

- Wastes are mixed to maintain a near-neutral pH
- Acid wastes are passed through limestone beds
- Acidic wastes are mixed with lime slurries
- Concentrated solutions of NaOH (caustic soda) or Na2CO3 (soda ash) are added to acidic effluents in proper proportions
- Blowing waste boiler flue to alkaline wastewater
- CO_2 is produced in alkaline water
- Sulphuric acid (H_2SO_4) is added to alkaline wastewater

2.2.7 Oil and Grease Removal

Industrial wastewater often contains oil and grease, which significantly cause ecological damage to aquatic organisms, plants, and animals. They are carcinogenic and mutagenic for humans as well. These oils form a layer on the water's surface, thereby decreasing the dissolved oxygen level in the water (Sanghamitra et al. 2021). Oil is found to be present in any of the three distinct forms:

- Free oil or floating oil—removed by skimming technique or gravity separation
- Emulsified oil—removed by the addition of chemicals to lower the pH level, followed by the addition of oxygen or nitrogen (dissolved) for removing the emulsified oils
- Dissolved oil—removed using biological methods

Significant types of oil/water separators include:

- Grease trap
- Skimmer
- Hydro-cyclone
- Centrifuge
- Dissolved air flotation
- Filtration

2.2.8 Primary treatment methods

Screening and grit removal

Screening is the first primary physical unit operation performed at the industrial wastewater treatment plant. Objects like rags, paper, plastics, and metals are removed by screening. Two types of screens are involved, namely coarse screen and fine screen. Coarse screens have 6 mm or more prominent openings for removing large solids, and fine screens have openings of 1.5 to 6 mm for removing materials that might restrict the proper operation and maintenance in downstream processes. Grit is solid particles that might cause undue mechanical damage to the process equipment (Sorber 2018).

Sedimentation

Sedimentation is the next step in conventional industrial wastewater treatment plants. It is a physical method that involves a mechanism allowing suspended material settles by gravity. Sedimentation enhances filtration by removing particulates, coarse dispersed phase, and coagulated and flocculated impurities. Sedimentation basins are constructed in various shapes and sizes, such as circular and rectangular tanks (Rebosura Jr et al. 2021).

The sedimentation basin primarily comprises four distinct zones:

- Inlet zone—incoming wastewater is distributed uniformly over the cross-sectional area of the tank
- Settling zone—the particles are left to settle down at the bottom of the tank
- Sludge zone—solids get collected at the bottom of the tank
- Outlet zone—clarified liquid is collected uniformly

Several factors influence the process of sedimentation. They are:

- The adequate depth of the tank
- Size, shape, and weight of particles
- Viscosity of water
- Temperature of water
- Area of tank
- The surface overflow rate
- Velocity of flow

2.2.9 Secondary treatment methods

Activated Sludge

One of the effective methods for the biological treatment of industrial wastewater is activated sludge, a flocculent culture of microorganisms developed in aeration tanks under controlled conditions (Xu et al. 2020). The microbes are of two types:

- Floc-forming species—they get quickly settled in the clarifiers
- Filament-forming species—they float on the tank's surface and are carried out along with the final effluent

This process involves converting suspended and particulate organic matter in the wastewater into end products that are harmless to the environment and new cell tissue or cell matter. The term 'floc' refers to the coagulated form of these microbes with organic matter in the wastewater. Some of the settled active organism that was made to flow out of the aeration tank with treated water is returned to the tank for mixing with the incoming wastewater. The microorganisms use oxygen which provides energy for cell growth and produces carbon dioxide and water; they function by adsorbing and absorbing the food matter.

Essential components of activated sludge:

- Primary clarifier for the separation of solids that are carried along with the wastewater
- The reactor containing microorganisms in suspension and aerated
- liquid–solid separation
- A sludge recycling system

Trickling Filter

A trickling filter is an aerobic biological wastewater treatment system based on the principle of biological oxidation of contaminants contained within industrial wastewater. This technique employs a population of microorganisms that includes aerobic, anaerobic, and facultative bacteria; algae; fungi; and protozoa that absorb the organic material present in the wastewater. When the contaminated effluent flows over the medium, the microbes present in the water get attached to the rock, slag, or any other surface, forming a film. This is followed by the degradation of the organic material by the microorganism in the outer portion of the slime layer. Thickening of the layer due to microbial growth blocks oxygen penetration, leading to the development of an anaerobic organism. As the biological film grows, microorganisms in the vicinity of the surface lose their ability to cling to the medium. A part of the slime layer falls off the filter, termed 'sloughing'. An under-drain system picks these up and transports them to a clarifier for elimination from the wastewater (Dang et al. 2022). Industrial trickling filters are of two types:

- Large tanks or enclosures filled with plastic packing or other similar media

- Vertical towers filled with plastic packing or other similar media
  - Sludge Incineration

Incineration is one of the renowned and widely implemented sludge thermo-chemical treatment methodologies used in industrial sectors. Predominantly configured as fluidized beds, sludge incinerators are employed for the oxidation of organic contents of the wastewater, which would also lead to the generation of thermal energy. Three main types of incinerators are used:

- Multiple hearths
- Fluidized bed
- Electric infrared

Key features of the process include (Hao et al. 2020):

- Carried out in a furnace using oxygen
- Temperature over 850 degrees Celsius is required
- Pathogens are destroyed
- Recyclable mineral by-products are obtained
- Negligible odour

Sewage sludge incinerators emit the following pollutants:

- Particulate matter
- Metals
- Carbon monoxide
- Nitrogen oxides
- Sulfur dioxide
- Unburned hydrocarbons

Chemical Precipitation

It is an efficient conventional removal system for heavy metal-containing waste effluent treatment. This simple system includes the addition of coagulating agents such as iron salts, lime, alum, and other organic polymers. The main controlling factor in the precipitation process is temperature and pH. Significant uses include (Peng and Guo 2020):

- Softening of wastewater
- Removal of phosphates
- Elimination of heavy metals

The most commonly used precipitation in industries includes:

- Sulfide precipitation
- Carbonate precipitation

- Sodium hydroxide precipitation

However, a significant drawback of this system is the production of a considerable volume of sludge containing toxic compounds.

2.2.10 Tertiary treatment methods

Adsorption

Adsorption is an effective method for the industrial wastewater treatment process based on the principle of interfacial attraction between an adsorbent and an adsorbate. This method can be divided into batch, semi-batch, and continuous modes. This consequence of surface energy has specific parameters or factors affecting it. They are:

- Temperature
- Surface area
- Agitation
- Pressure
- pH
- Characteristics of adsorbent
- Activation of solid adsorbent

Activated carbon is considered the utmost removal agent for the removal of heavy metal-containing industrial waste effluent, and adsorption can be broad of two types:

- Physical adsorption—it involves the intermolecular attraction caused by Vander Walls force between an adsorbent molecule and an adsorbate
- Chemical adsorption—it includes strong chemical reactions between an adsorbent and an adsorbate that result in forming a new electronic bond; this process is known as activated adsorption

Key benefits of the process include (Burakov et al. 2018):

- The adjustable design of the system machinery
- Easy operation
- High removal percentage
- Regeneration of adsorbent
- Reuse of adsorbent

Ion exchange

Ion exchange is a chemical process that helps in the removal of dissolved ionic contaminants from wastewater. This is an effective industrial wastewater treatment

approach where the undesirable ionic components of water are exchanged with non-objectionable ionic substances, provided both have a similar type of electrical charge (negative/positive). The process involves ion exchange resins which are synthetic copolymers of styrene and divinylbenzene/acrylic acid with functional groups (acids/alkalis/salts) and are small in size with micro-porous beads (Zhao et al. 2018). The ion exchange process is reversible since the resins are reusable.
Utilization:

- Softening
- Dealkalization
- De-solicitation
- Demineralization

Categorization of ion exchange resins:

- Strong acid cation
- Weak acid cation
- Strong base anion
- Weak base anion

3 Prospects and Challenges

3.1 Treatment of Gaseous Industrial Waste

Industrial emissions produce gaseous wastes. They include CO2, CH4, chlorofluorocarbons (CFCs), oxides of nitrogen (NOx), carbon monoxide (C.O.), oxides of sulfur (SOx), etc. Exhaust gases from industries contain large amounts of noxious gases such as NOx, CO, CO2, SO2, H2S, CS2, and sulfur-containing organic compounds. Several industries produce gaseous waste such as chemical, mining, oil and gas, electric power plants, nuclear plants, and other industries with distillation units. (Popov et al. 2019).

3.2 Effects of Industrial Gaseous Waste on the Environment

- Greenhouse gases like carbon dioxide are produced in vast amounts by burning fossil fuels in power systems and heating appliances (Gaffney and Marley 2009).
- Mists of sulphuric acid affect many things such as carbonate materials (marble and mortar); steel, copper, and aluminium metals are corroded; and cloth fabrics and plant tissue are damaged.
- Particulate matter from gaseous pollutants settles on surfaces, giving a dirty grey look to those surfaces. These particles also cause corrosion, acting as centres

for the spread of corrosion. Furthermore, particles reflect sunlight away from the earth, causing a small yet noticeable cooling of the Northern Hemisphere (Najjar 2011).

- Ozone compounds destroy the leaves of many vegetables. Ozone also affects citrus plants to drop their fruit early.

These gases cause severe environmental and health dangers. Therefore, correct steps to control gaseous wastes are needed. (Bianchini et al. 2018; Sparks 2012; "US EPA" n.d.) Some essential control measures are:

(i) Wet Scrubbers—They absorb gaseous pollutants like SO2, H2S, HC1, C12, NH3, etc.
(ii) Settling Chambers—Industries use settling chambers to collect dust particles of size more significant than a loom.
(iii) Cloth filters—About 99% of particulate matter can be filtered out using cloth filters of various sizes. Dry and free-flowing waste gas is passed through a filter bag, and the particles are collected inside.
(vi) Electrostatic precipitators—Electrostatic forces move the gas particles to the collection surface of charged electrodes. The waste gas is passed between high-voltage discharge electrodes. Most gas particles of almost all sizes become charged and collected on electrodes.
(v) Absorption—Gaseous pollutants like carbon dioxide are transferred into the suitable liquid and absorbed.
(vi) Adsorption—Carbon-based adsorption of gases and vapours is commonly used. The internal surface area of a solid governs the amount of adsorbed substances.

4 Challenges in Traditional Industrial Waste Treatment Methods

- The increasing amount of waste and pressure on the earth

In urban areas, the extent of the generated trash is growing because of the rise in the population and industrial and economic growth. This leads to pressure on the prevailing administration method, and an additional terrestrial area is required for dumping practices. Eg. In India, the amount of waste produced per capita is projected to rise at a rate of 1–1.33% yearly (Thakur et al. 2020).

- Inadequate resources

Usually, a negligible portion of the funds is allocated for the assembly and management of waste. Also, the individuals are not well trained in many cases leading to improper waste management practices. For a city with a million population, the cost of a modern sanitary landfill is assessed to be between 3 and 10 million US$ which is usually not allocated in most cases (Mingaleva et al. 2019).

- Inappropriate technology

The tools and machines currently used are usually established for general use or have been improvised from other industrial sectors. This leads to the underapplication of prevailing possessions and a fall in efficiency.

- Limited utilization of the private sector in recycling activities

Industrial Waste can be efficiently managed if the civic authority, the private sector industry, and the citizens can actively participate. However, recycling is considered a task or duty of the industry involved in producing waste. Thus, uncollected waste can be seen in many areas, which might be helpful for the private sector. (Rybnytska et al. 2018).

- Inefficient management system

The waste management programs are unplanned most of the time and run absurdly. The work rules are not identified, and the management staff is not appropriately guided (Khajuria et al. 2010).

5 Future Prospects of Industrial Waste Treatment

Waste-to-energy processes avoid waste in landfills and generate relatively green power sources. One of the most promising of these techniques is plasma gasification. Plasma technology is used in surface treatment and coating, reforming carbon dioxide and methane, removing volatile compounds, odour, and disinfection. In this technique, plasma heats waste to high temperatures and converts it into valuable gases like hydrogen, creating a sustainable fuel source (Dobslaw and Glocker 2020).

- Biocompatible Reagents: These include biopolymers such as Starch, Maltose, Sucrose, D-Glucose, Agar, Vitamins, Amino acid, Polysaccharides, Cellulose, Clay, and many others. They are known to improve stability and dispersion in Green Nanoparticles.
- Synthesis by Microorganisms: Microorganisms like Fungi, Bacteria, and Algae are used to synthesize green nanoparticles. Silver nanoparticles manufactured using microalgae are good antimicrobial agents and, therefore, widely used to disinfect water from microbes.

- Plant-mediated Synthesis: Various plants and their parts are used to synthesize nanoparticles and nanocomposites. Silver nanoparticles synthesized using P. thonningii leaf extract were used to remove heavy metals (magnesium, copper, lead, iron) from wastewater.

The nanomaterials of nanocomposites can be used to eliminate different environmental pollutants—such as heavy metals, dyes, chlorine, and phosphorus compounds, volatile compounds, etc. (Guerra et al. 2018). Nanomaterials are used for creating nanocomposite membranes by their embedment in the membrane matrix or by deposition on the surface. Nanocomposites improve the adsorption performance and interact more specifically with the targeted contaminants. Nanocomposites have been used for the removal of toxic pollutants for environmental remediation, such as heavy metal ions from drinking water and industrial wastes, organic contaminants and various dyes from wastewater, carbon nanocomposites, and nanofibers for air purification, etc. (Ahsan et al. 2018; Suhaimi et al. 2020). Nanomaterials can be synthesized by green technology and synthetic technology. Green technology uses raw materials from plants, microorganisms, or biomolecules for their synthesis. The green synthesis depends on (i) the choice of solvents, (ii) Reducing agents (iii) capping agents. Numerous chemicals have been used to produce colloidal metal nanoparticles from different precursors using chemical reducing and capping agents in both aqueous and nonaqueous solvents. The chemical routes include the polyol method, microemulsion, thermal decomposition, electrochemical method, and physical methods such as plasma method, vapour deposition, microwave irradiation, pulsed laser method, sonochemical reduction, radiolytic, and photochemical method (Cele 2020). Chemical synthetic routes produce chemical wastes; therefore, green technology is preferred chiefly for nanoparticle synthesis.

The application of nanotechnology has proved to be an adequate substitute compared to conventional liquid waste handling technologies because of their increased reactive surface and selection of compounds. In addition, nanomaterials can undertake chemical reactions that are impossible by conventional materials. Nanomaterial treatments are simple, decreasing the utility of traditional treatments and therefore reducing the energy, cost, and time (Dermatas et al. 2018; Kaza et al. 2018).

6 Conclusion

The significant industrial growth has led to rapid urbanization, a critical indicator of regional economic progress; however, many regions, particularly developing nations, have been plagued by its negative consequences—the deterioration of environmental degradation due to increased industrial pollution emissions. The continued urbanization and population explosion complement each other. Since industrialization, the invasion of scientific technologies and manufacturing of goods advanced tremendously, further stimulating industrial productivity and raising industrial pollutant emissions. The generation of industrial waste seems to be the by-product of the growing industrial activities.

Manufacturing wastes from various processes, such as sludges, residues, kiln, dust, slags, and ashes, are termed industrial waste. Food, metallurgy, and non-metallurgical industries generate most of these industrial wastes. The nature of the wastes varies based on the utilization of raw materials, production methods, and outlets of the products. Industrial wastes are classified into solids, liquids, and gaseous forms of waste. These wastes consist of inorganic and organic fractions, biodegradable, and non-biodegradable substances, recyclable or non-recyclable, etc. Industries discharge these toxic pollutants into the rivers and drains without efficient treatment. Therefore, major rising concerns are for industrial liquid wastes. The persistent deposition and leaching of these industrial pollutants such as heavy metals (lead, mercury, cadmium, zinc, copper, etc.), radionuclides (uranium, selenium, thorium, etc.), hazardous chemical compounds, ferroalloys, metallurgical, and blast furnace slags, toxic gases these prompt a global concern. Different traditional strategies of remediation are followed for the treatment of industrial wastewater. Most of the solids can be eliminated using simple sedimentation techniques. For very fine solids and solids closer to the density of water, methods such as ultrafiltration, flocculation using alum salts, or the addition of polyelectrolytes can be used. Removal of biodegradable organic material such as plant or animal origin is carried out using extended conventional wastewater treatments such as trickling filters or activated sludge. For heavy metal remediation, chemical precipitation techniques are used in combination with the other methods (e. g., sulfide precipitation with nanofiltration). The other expensive conventional techniques, such as oxidation–reduction and ion—exchanges, result in harmful sludge as a secondary contaminant. These physicochemical remediation methodologies are detrimental to the environment and require higher capital investment for these industrial wastewater treatments. The concept of greener technologies and methodologies such as bioremediation renders a natural solution that offers an alternative to destroying various contaminants using natural biological activity. It includes bio coagulation, bioleaching, biosorbents, bioaccumulation, and biotransformation phytoaccumulation are some of the novel sustainable approaches for treating wastewater. The help of microbes, fungi, plants, and algal bioremediation techniques over previously employed conventional ones will undoubtedly be the way ahead in developing a green, sustainable, cost-effective approach for the treatment of industrial wastes and bringing an ecological balance.

References

Abdel-Raouf N, Al-Homaidan AA, Ibraheem IBM (2012) Microalgae and wastewater treatment. Saudi J Biol Sci 19:257–275. https://doi.org/10.1016/j.sjbs.2012.04.005

Ahsan MdA, Jabbari V, Islam MdT, Kim H, Hernandez-Viezcas JA, Lin Y, Díaz-Moreno CA, Lopez J, Gardea-Torresdey J, Noveron JC (2018) Green synthesis of a highly efficient biosorbent for organic, pharmaceutical, and heavy metal pollutants removal: engineering surface chemistry of

polymeric biomass of spent coffee waste. J Water Process Eng 25:309–319. https://doi.org/10.1016/j.jwpe.2018.08.005
Alam P, Ahmad K (2013) Impact of solid waste on health and the environment. Int J Sustain Dev 2:165–168.
Amasuomo E, Baird J (2016) The concept of waste and waste management. J Manag Sustain 6:88. https://doi.org/10.5539/jms.v6n4p88
Article JF (2019) Industrial waste and municipal solid waste treatment and disposal. Environ Pollution Sci 377–391. https://doi.org/10.1016/B978-0-12-814719-1.00021-5
Ashraf MA, Hanafiah MM (2019) Sustaining life on earth system through clean air, pure water, and fertile soil. Environ Sci Pollut Res 26(14):13679–13680
Awasthi AK, Wang M, Awasthi MK, Wang Z, Li J (2018a) Environmental pollution and human body burden from improper recycling of e-waste in China: a short-review. Environ Pollut 243:1310–1316. https://doi.org/10.1016/j.envpol.2018.08.037
Awasthi AK, Wang M, Wang Z, Awasthi MK, Li J (2018b) E-waste management in India: a mini-review. Waste Manag. Res J Sustain Circ Econ 36:408–414. https://doi.org/10.1177/0734242X18767038
Bhatia D, Sharma NR, Singh J, Kanwar RS (2017) Biological methods for textile dye removal from wastewater: a review. Crit Rev Environ Sci Technol 47:1836–1876. https://doi.org/10.1080/10643389.2017.1393263
Bianchini A, Pellegrini M, Rossi J, Saccani C (2018) Theoretical model and preliminary design of an innovative wet scrubber for the separation of fine particulate matter produced by biomass combustion in small size boilers. Biomass Bioenergy 116:60–71. https://doi.org/10.1016/j.biombioe.2018.05.011
Burakov AE, Galunin EV, Burakova IV, Kucherova AE, Agarwal S, Tkachev AG, Gupta VK (2018) Adsorption of heavy metals on conventional and nanostructured materials for wastewater treatment purposes: a review. Ecotoxicol Environ Saf 148:702–712
Cai QQ, Wu MY, Li R, Deng SH, Lee BCY, Ong SL, Hu JY (2020) Potential of combined advanced oxidation—biological process for cost-effective organic matters removal in reverse osmosis concentrate produced from industrial wastewater reclamation: screening of AOP pretreatment technologies. Chem Eng J 389:123419
Cele T (2020) Preparation of nanoparticles. Eng Nanomater Health Saf 1–14. https://doi.org/10.5772/intechopen.90771
Dang BT, Nguyen TT, Bui XT, Ngo HH, Lin KYA, Tomoaki I, Saunders T, Huynh TN, Cao TND, Visvanathan C, Varjani S (2022) Non-submerged attached growth process for domestic wastewater treatment: influence of media types and internal recirculation ratios. Biores Technol 343:126125
Dermatas D, Mpouras T, Panagiotakis I (2018) Application of nanotechnology for waste management: challenges and limitations. Waste Manag Res 36:197–199. https://doi.org/10.1177/0734242X18758820
Dobslaw C, Glocker B (2020) Plasma technology and its relevance in waste air and waste gas treatment. Sustain Switz 12:1–39. https://doi.org/10.3390/su12218981
Environmental Canada (1997) Review of the impacts of municipal wastewater effluents on Canadian waters and human health. Ecosyst Sci
Forsido T, McCrindle R, Maree J, Mpenyana-Monyatsi L (2019) Neutralization of acid effluent from the steel manufacturing industry and removal of metals using an integrated electric arc furnace dust slag/lime process. S.N. Appl Sci. 1(12):1–6
Gaffney JS, Marley NA (2009) The impacts of combustion emissions on air quality and climate – from coal to biofuels and beyond. Atmos Environ 43:23–36
Gantz RG, Lipták J, Liu DH (2020) Equalization and primary treatment. In: Wastewater treatment (pp. 123–165). CRC Press
Gaur VK, Sharma P, Sirohi R, Awasthi MK, Dussap C-G, Pandey A (2020) Assessing the impact of industrial waste on environment and mitigation strategies: a comprehensive review. J Hazard Mater 398:123019. https://doi.org/10.1016/j.jhazmat.2020.123019

Gogate PR, Thanekar PD, Oke AP (2020) Strategies to improve biological oxidation of real wastewater using cavitation-based pretreatment approaches. Ultrason Sonochem 64:105016

Guerra FD, Attia MF, Whitehead DC, Alexis F (2018) Nanotechnology for environmental remediation: materials and applications. Molecules 23:1–23. https://doi.org/10.3390/molecules23071760

Gurjar BR (2021) Sludge treatment and disposal. CRC Press

Hanafiah MM, Mohamad NHSM, Abd Aziz NIH (2018) Salvinia molesta and Pistia stratiotes as phytoremediation agents in sewage wastewater treatment. Sains Malaysiana 47(8):1625–1634

Hao X, Chen Q, van Loosdrecht MC, Li J, Jiang H (2020) Sustainable disposal of excess sludge: incineration without anaerobic digestion. Water Res 170:115298

Ho C-C, Chen M-S (2018) Risk assessment and quality improvement of liquid waste management in Taiwan University chemical laboratories. Waste Manag 71:578–588. https://doi.org/10.1016/j.wasman.2017.09.029

Ilankoon IMSK, Ghorbani Y, Chong MN, Herath G, Moyo T, Petersen J (2018) E-waste in the international context – a review of trade flows, regulations, hazards, waste management strategies and technologies for value recovery. Waste Manag 82:258–275. https://doi.org/10.1016/j.wasman.2018.10.018

Jain A, Yadav S, Nigam VK, Sharma SR (2017) Fungal-mediated solid waste management: a review. In Prasad R (ed.), Mycoremediation and environmental sustainability, fungal biology. Springer International Publishing, Cham, pp. 153–170. https://doi.org/10.1007/978-3-319-68957-9_9

Jayashree Dipak D (2016) Review on the diseases caused due to improper handling of Biomedical Waste. Sci Eng 4:98–100

Karachaliou T, Protonotarios V, Kaliampakos D, Menegaki M (2016) Using risk assessment and management approaches to develop cost-effective and sustainable mine waste management strategies. Recycling 1:328–342. https://doi.org/10.3390/recycling1030328

Kaza S, Yao L, Bhada-Tata P, Van Woerden F (2018) What a waste 2.0: a global snapshot of solid waste management to 2050. World Bank Publications

Khajuria A, Matsui T, Machimura T, Morioka T (2010) Assessment of the challenge of sustainable recycling of municipal solid waste management in India. Int J Environ Technol Manag 13:171–187. https://doi.org/10.1504/IJETM.2010.034300

Khedkar, R., Singh, K., 2018. Food industry waste: a panacea or pollution hazard? In: Jindal T (ed.) Paradigms in pollution prevention, Springer briefs in environmental science. Springer International Publishing, Cham, pp. 35–47. https://doi.org/10.1007/978-3-319-58415-7_3

Kishore J, Monika, (2010) E-waste management: As a challenge to public health in India. Indian J Commun Med 35:382. https://doi.org/10.4103/0970-0218.69251

Krishna RS, Mishra J, Meher S, Das SK, Mustakim SM, Singh SK (2020) Industrial solid waste management through sustainable green technology: case study insights from steel and mining industry in Keonjhar, India. Mater Today Proc 33:5243–5249. https://doi.org/10.1016/j.matpr.2020.02.949

Kwok R (2019) Inner workings: how bacteria could help recycle electronic waste. Proc Natl Acad Sci 116:711–713. https://doi.org/10.1073/pnas.1820329116

Lakkaboyana SK, Khantong S, Asmel NK, Yuzir A, Yaacob WZW (2019) Synthesis of copper oxide nanowires-activated carbon (AC@ CuO-NWs) and applied for removal methylene blue from aqueous solution: kinetics, isotherms, and thermodynamics. J Inorg Organomet Polym Mater 29(5):1658–1668

Lakkaboyana SK, Soontarapa K, Marella RK, Kannan K (2021) Preparation of novel chitosan polymeric nanocomposite as an efficient material for the removal of Acid Blue 25 from aqueous environment. Int J Biol Macromol 168:760–768

Manderson TM (2018) Determination of the volume of flow equalization basin in wastewater treatment system. Civil and Environmental Research 10(4):34–41

Mathur P, Pathan P, Shobhawat S (2012) Need of biomedical waste management system in hospitals: an emerging review. Curr World Environ 117–124.

Mingaleva Z, Vukovic N, Volkova I, Salimova T (2019) Waste management in green and smart cities: a case study of Russia. Sustainability 12:94. https://doi.org/10.3390/su12010094
Najjar YSH (2011) Gaseous pollutants formation and their harmful effects on health and environment. Innov Energy Policies 1:1–9. https://doi.org/10.4303/iep/e101203
Orisakwe OE, Frazzoli C, Ilo CE, Oritsemuelebi B (2019) Public health burden of E-waste in Africa. J. Health Pollut. 9:190610. https://doi.org/10.5696/2156-9614-9.22.190610
Pavithra KG, Senthil Kumar P, Jaikumar V, Sundar Rajan P (2019) Removal of colorants from wastewater: a review on sources and treatment strategies. J Ind Eng Chem 75:1–19. https://doi.org/10.1016/j.jiec.2019.02.011
Peng H, Guo J (2020) Removal of chromium from wastewater by membrane filtration, chemical precipitation, ion exchange, adsorption electrocoagulation, electrochemical reduction, electrodialysis, electrodeionization, photocatalysis and nanotechnology: a review. Environ Chem Lett 1–14
Pintor AMA, Vilar VJP, Botelho CMS, Boaventura RAR (2016) Oil and grease removal from wastewaters: sorption treatment as an alternative to state-of-the-art technologies. A critical review. Chem Eng J 297:229–255. https://doi.org/10.1016/j.cej.2016.03.121
Popov SK, Putin SN, Burmakina AV, Stogov PA (2019) Integrated power and chemical accumulation of high-temperature gaseous waste from industrial installations. Therm Eng 66:116–126. https://doi.org/10.1134/S0040601519020034
Quach-Cu J, Herrera-Lynch B, Marciniak C, Adams S, Simmerman A, Reinke RA (2018) The effect of primary, secondary, and tertiary wastewater treatment processes on antibiotic resistance gene (ARG) concentrations in solid and dissolved wastewater fractions. Water 10(1):37
Rajasulochana P, Preethy V (2016) Comparison on efficiency of various techniques in treatment of waste and sewage water – A comprehensive review. Resour Effic Technol 2:175–184. https://doi.org/10.1016/j.reffit.2016.09.004
Raju S, Carbery M, Kuttykattil A, Senthirajah K, Lundmark A, Rogers Z, Suresh SCB, Evans G, Palanisami T (2020) Improved methodology to determine the fate and transport of microplastics in a secondary wastewater treatment plant. Water Res 173:115549
Rautela R, Arya S, Vishwakarma S, Lee J, Kim KH, Kumar S (2021) E-waste management and its effects on the environment and human health. Sci Total Environ 773:145623. https://doi.org/10.1016/j.scitotenv.2021.145623
Rebosura M Jr, Salehin S, Pikaar I, Keller J, Sharma K, Yuan Z (2021) The impact of primary sedimentation on the use of iron-rich drinking water sludge on the urban wastewater system. J Hazard Mater 402:124051
Redouane F, Mourad L (2016) Pollution characterization of liquid waste of the factory complex Fertial (Arzew, Algeria). J Air Waste Manag Assoc 66:260–266. https://doi.org/10.1080/10962247.2015.1123782
Rinkesh W (2018) What is solid waste management? Sources and methods of solid waste management – conserve energy future.
Royal Commission on Environmental Pollution (1984) 10th Report tackling pollution—experience and prospects. London.
Rybnytska O, Burstein F, Rybin AV, Zaslavsky A (2018) Decision support for optimizing waste management. J Decis Syst 27:68–78. https://doi.org/10.1080/12460125.2018.1464312
Salgot M, Folch M (2018) Wastewater treatment and water reuse. Curr Opin Environ Sci Health 2:64–74
Samer M (2015) Biological and chemical wastewater treatment processes. In Samer M (ed.), Wastewater treatment engineering. InTech. https://doi.org/10.5772/61250
Sanghamitra P, Mazumder D, Mukherjee S (2021) Treatment of wastewater containing oil and grease by biological method—a review. J Environ Sci Health A 56(4):394–412
Santagata R, Ripa M, Genovese A, Ulgiati S (2021) Food waste recovery pathways: challenges and opportunities for an emerging bio-based circular economy. A systematic review and an assessment. J Clean Prod 286:125490. https://doi.org/10.1016/j.jclepro.2020.125490

Soliman NK, Moustafa AF (2020) Industrial solid waste for heavy metals adsorption features and challenges; a review. J Mater Res Technol 9:10235–10253. https://doi.org/10.1016/j.jmrt.2020.07.045

Soltani A, Sadiq R, Hewage K (2017) The impacts of decision uncertainty on municipal solid waste management. J Environ Manage 197:305–315. https://doi.org/10.1016/j.jenvman.2017.03.079

Sorber CA (2018) Removal of viruses from wastewater and effluent by treatment processes. In: Viral pollution of the environment (pp. 39–52). CRC Press.

Sparks T (2012) Solid-liquid filtration: understanding filter presses and belt filters. Filter Sep 49:20–24. https://doi.org/10.1016/S0015-1882(12)70193-3

von Sperling M (2007) Wastewater characteristics, treatment and disposal, biological wastewater treatment series. IWA Publ. [u.a.], London.

Suhaimi S, Rosli AN, Ariffin AH, Wahab MHA, Idrus SZS (2020) Towards strategic e-waste management using green technology approach: a brief review on green product development process. J Phys Conf Ser 1529:022109. https://doi.org/10.1088/1742-6596/1529/2/022109

Thakur D, Ganguly R, Gupta AK, Ghali V (2020) Evaluation of existing solid waste management system in Una Town, India. In Ghosh SK (ed.), Sustainable waste management: policies and case studies. Springer Singapore, Singapore, pp. 367–381. https://doi.org/10.1007/978-981-13-7071-7_33

US EPA (n.d.) https://www.epa.gov/environmental-topics/land-waste-and-cleanup-topics (accessed 12.1.22)

Wang J, Chen C (2009) Biosorbents for heavy metals removal and their future. Biotechnol Adv 195–226.

Watanabe S, Ogi H, Atsuhiro Shibata NK (2019) Waste management in a hot laboratory of Japan Atomic Energy Agency – 2: condensation and solidification experiments on liquid waste. https://doi.org/10.5281/ZENODO.2643902

Xu Y, Lu Y, Zheng L, Wang Z, Dai X (2020) Perspective on enhancing the anaerobic digestion of waste activated sludge. J Hazard Mater 389:121847

Yang ZZ, Guo YN, Liu AL, Zhang J, Wang MY, Liu Y, Cao XL, Wu ZG, Wu LY, Wang CH, Cao JG (2019) Pollutant characteristics from wastewater of poplar pre-conditioning refiner chemical alkaline peroxide mechanical pulping pretreated with phanerochaete chrysosporium. BioResources 14(2):4792–4805

Zhao WY, Zhou M, Yan B, Sun X, Liu Y, Wang Y, Xu T, Zhang Y (2018) Waste conversion and resource recovery from wastewater by ion exchange membranes: state-of-the-art and perspective. Ind Eng Chem Res 57(18):6025–6039

Anthracene Removal from Wastewater Using Biotechnological Interventions

Moirangthem Singh Goutam and Madhava Anil Kumar

1 Introduction

A steep increase has been observed in the generation of wastewater, recently, due to an increase in industrialization and various other anthropogenic activities. As a result, not just water bodies but even fertile lands have started getting contaminated and have disappeared Ferronato and Torretta (2019). The major contaminants include dye effluents, heavy metals, plastic, polycyclic aromatic compounds (PAHs), etc., they not only cause harmful coloration of natural reservoirs but might result in eutrophication, which ultimately results in the loss of aesthetic beauty of nature Liu et al. (2009). Out of all the contaminants, PAHs are considered to be one of the major causes of pollution and is certainly a big threat to the organisms surviving in the water reservoirs Barman et al. (2017).

PAHs are compounds that have more than two aromatic rings mostly benzene ring and are emanated from the incomplete combustion of fossil fuel. PAHs are of two types based on their molecular weight: low molecular weight (LMW) PAH and high molecular weight (HMW) PAHs. The PAHs with two to three fused aromatic rings are considered as LMW-PAHs and those with four and more fused rings are HMW-PAHs Vasconcelos et al. (2011). Environmental Protection Agency (EPA) has included naphthalene, acenaphthylene, acenaphthene, fluorene, phenanthrene, anthracene, fluoranthene, which are LMW PAH, and pyrene, benzo[a]anthracene, chrysene, benzo[b]fluoranthene, benzo[k]fluoranthene, benzo[a]pyrene, benzo[g,h,i]perylene,

M. S. Goutam · M. A. Kumar (✉)
Analytical and Environmental Science Division & Centralized Instrument Facility, CSIR-Central Salt & Marine Chemicals Research Institute, Bhavnagar, India
e-mail: anilkumarm@csmcri.res.in

M. A. Kumar
Academy of Scientific and Innovative Research, Ghaziabad, India

M. P. Shah (ed.), *Advanced and Innovative Approaches of Environmental Biotechnology in Industrial Wastewater Treatment*,
https://doi.org/10.1007/978-981-99-2598-8_20

indeno[1,2,3-c,d]pyrene and dibenz[a,h]anthracene which are high priority pollutants. These PAHs have been found to be very persistent in the environment and therefore, are very much associated with health hazards and are considered as potentially toxic to humans Hussar et al. (2012). The main concern with these compounds is that they possess bio-recalcitrance properties, which are least degraded, and the exposure may consequently lead to the chromosomal aberrations Ahmed et al. (2012).

Heating and cooking are dominant domestic sources of PAHs contributing to the total emissions of PAHs in the environment. PAH from domestic sources, particularly in indoor ambience poses a major health concern Ravindra et al. (2006), Zhu et al. (2009). Total emission factors (EFs) for the PAHs depend on the drying status and compositions of raw coals used for cooking and heating Chen et al. (2004). Studies also reveal that the emissions of genotoxic PAHs from wood burning were two-fold high when compared with the emissions from charcoal incineration. PAH, emissions from industries are predominantly from the processing of raw materials and petrochemical handling activities. Vehicular exhaust is the major cause of PAH emissions and these emissions are attributed by the synthesis of aromatic molecules in fuel, engine deposits, and lubricants. Vehicular exhaust, especially diesel vehicles hold a significant contribution to PAH emission Ravindra et al. (2006). The biogenic sources of PAHs are from the burning of agro-based residues, where the burning of organic materials under uncontrolled combustion results in an escalated increase in the PAHs concentration in the ambient environment. Forest fires by lightning strikes, volcanic eruptions, and decaying organic matter are natural routes of PAH emissions. The concentration of PAH production depends on meteorological conditions, such as wind, temperature, relative humidity, and source characteristics.

2 Anthracene-Toxicity and Occurrence

Anthracene is a tricyclic aromatic hydrocarbon, usually a colourless compound with a mild aromatic odour and their oxidation forms anthraquinone, an intermediate in the production of dyes and pigments. Anthracene is a persistent and toxic soil contaminant found mainly in the waste-contaminated sites as well as threats to human and environmental health. Being hydrophobic in nature, anthracene exerts marine toxicity and tends to bioaccumulate in the food chain Lily et al. (2013). Anthracene has a common feature in common with the other carcinogenic compounds, thus they are considered as a model compound to understand the fate, bioavailability, biodegradability. Anthracene contamination is also attributed by the accidental spillage of petroleum products during transport and storage Owabor and Saniyo (2006).

Anthracene may enter into human body through ingestion and/or inhalation and binds with blood albumin. Anthracene binds to receptors (aryl hydrocarbon or glycine N-methyl transferase protein) mediated through the inducible cytochrome P450 enzymes. Furthermore, anthracene forms epoxide intermediates that covalently bind to DNA thereby causing mutagenesis Padros and Pelletier (2000), Uno et al. (2008).

Acute exposure to anthracene can cause skin irritation while inhalation can irritate respiratory passage and cause coughs. Chronic exposure may cause skin allergy followed by skin thickening and loss of skin pigmentation.

3 Need for Remediation of Anthracene Contaminates

State of the art relies upon the present chapter is an attempt to interpret anthracene biotransformation using different bioremedial measures. The technological tools particularly microorganisms, enzymes, and other biogenic materials those facilitating the faster removal of anthracene are also specially emphasized. The primary, secondary (biological) and tertiary treatment techniques are integrated to abate the problem of anthracene from environmental matrices.

Adsorption is a widely used and accepted technique as they exert process advantages and are considerably less sensitive to the pollutant's concentration Moradeeya et al. (2017). Saad et al. (2014) demonstrated anthracene removal using activated carbon, while Gupta and Kumar (2016) evaluated the performance of activated carbon prepared from banana peels for anthracene removal. Complete removal of anthracene was achieved by adsorbing anthracene onto vehicular tyres Gupta (2018). Cyclodextrins are cyclic oligosaccharides that are able to form complexes with hydrophobic molecules. They are non-toxic, biodegradable in the environment, and solubilize PAHs when added to aqueous medium Stokes et al. (2005). Vegetable oils have been proposed as an economical and environmentally friendly solvent to dissolve PAHs and has been shown to be as effective as organic solvents like acetone and dichloromethane Gong et al. (2005). Photolysis is one of the techniques employed for the remediation of PAHs and photolytic breakdown helps in the formation of oxidation products. Hassan et al. (2015) demonstrated the use of zinc oxide (ZnO) nanoparticles, prepared from leaf extract of *Corriandrum sativum* which showed almost 96% decomposition of anthracene at 25 °C, pH 7.0 and ultraviolet radiation for 240 min. Another similar study showed that ZnO nanoparticles degraded 74% of anthracene into anthraquinone, in 40 min and it was also proved that n-ZnO/p-MnO nanocomposites have higher photo-degradation efficiency Martínez-Vargas et al. (2019). Ozone treatment is one of the most important methods for degradation of PAHs and transforms PAHs into polar, more soluble oxygenated biodegradable intermediates. Organic solvents that can dissolve large amounts of ozone and have minimal toxicity make ideal extractants because they can be employed as good carrier to introduce ozone into the system Kulik et al. (2006). According to Stehr et al. (2001), ozonation had a detrimental impact on the biodegradation of benzo[a]pyrene and phenanthrene. Ozone treatment has various negative points, including the creation of intermediates that may be more harmful than the original chemical, even though it has proven to be an effective way of treating PAHs. Additionally, it has the ability to eliminate any existing native microbial degraders at the site. Different concentrations of peroxide along with ferrous iron (Fe II) act as catalyst in Fenton's reagent for oxidization of organic chemicals Flotron et al. (2005), Watts et al. (2002).

Surfactants are amphiphilic molecules with a hydrophilic polar head and are used to enhance PAHs solubility Mulligan (2001). The hydrophobic parts of the surfactant tend to associate to form a micelle with hydrophobic core. Surfactants solubilize hydrophobic contaminants by partitioning them into the hydrophobic core of the micelle Gao et al. (2007), Mata-Sandoval et al. (2002). Anionic and non-ionic surfactants are more commonly used in remediation because they are less likely to sorb onto soil surfaces Zhao et al. (2005).

4 Biotechnological Interventions for Anthracene Removal

Bioremediation of anthracene can be done by various methods such as biological methods such as enzymatic degradation and with the help of microorganisms. Bioremediation is an economically and environmentally attractive solution when compared with conventional treatment approaches. The most important stage in microbial remediation is to understand the complete metabolic pathway so that potentially toxic metabolites do not accumulate in the soil Van Herwijnen et al. (2003). The ability to remove PAHs by various groups of organisms including bacteria, fungi, and algae are reported extensively Van Herwijnen et al. (2003). Many bacterial strains from genera *Pseudomonas, Sphingobium, Nocardia, Rhodococcus, and Mycobacterium* are known to remediate anthracene contaminates under neutral conditions Ahmed et al. (2012). Somtrakoon et al. (2014) proved that more than 87% of the spiked anthracene was removed from both the bulk and rhizosphere soils, using ridge gourd (*Luffa acutangula*), a plant from the Cucurbitaceae family.

5 Biodegradation

Biodegradation and biotransformation are better than these conventional methods with respect to safety, cost, and removal efficiency Juwarkar et al. (2010), Kikani et al. (2021). It is employed to treat sites contaminated with PAHs and resident microorganisms are isolated and/or augmented from a completely different site. These microorganisms multiply in the population by utilizing the contaminant as a source of nutrition and can either completely mineralize or can form secondary metabolites Azubuike et al. (2016), Koshlaf and Ball (2017).

5.1 Biodegradation of Anthracene Using Fungi

Zebulun et al. (2011) utilized *Pleurotus ostreatus* for the removal of anthracene from soil and achieved 90% reduction in the anthracene levels in the soil. Ye et al. (2011) demonstrated that *Aspergillus fumigatus* was able to degrade anthracene with

Table 1 Anthracene degradation by different fungal species and their probable degradation products

Fungal species	Removal (%)	Possible products of degradation	References
Aspergillus fumigatus	60	Phthalic acid	Ye et al. (2011)
Pleurotus ostreatus	90	Anthrone, anthraquinone, phthalic acid	Zebulun et al. (2011)
Phanerochaete chrysosporium,	40	Anthrone, anthraquinone, phthalic acid	Jové et al. (2016)
Irpex lacteus	40		
Pleurotus ostreatus	15		
Aspergillus niger	31		
Penicillium simplicissimum	86		
Mucor racemosus	24		
Cladosporium sp. CBMAI 1237	71	Anthrone, anthraquinone, phthalic acid	Birolli et al. (2018)

a degradation efficiency of about 60%. *Pycnoporus sanguineus* H1 strain was used by Zhang et al. (2015), which was able to achieve 67.5% anthracene degradation. Anthracene was metabolized by internal cytochrome P450, mycelium-associated and extracellular laccase, and these enzymes together caused anthracene to be broken down. *Phanerochaete chrysosporium, Pleurotus ostreatus, Irpex lacteus*, and other lignolytic fungal species were evaluated by Jové et al. (2016) for their capacity to degrade anthracene. *Penicillium oxalicum*, which was isolated from a pond that had effluents from the outflow of an oil-storage tank in it, was used successfully by Aranda et al. (2017). *Cladosporium* sp. CBMAI 1237, a marine-derived fungus, may be able to breakdown anthracene up to 71%, according to Birolli et al. (2018). The list given in Table 1 enumerates the different reports of anthracene degradation using fungal species.

5.2 Biodegradation of Anthracene Using Bacteria

Biodegradation using bacterial species is considered as an effective strategy for the treatment of anthracene-contaminated sites. Ahmed et al. (2012) reported the use *Bacillus badius* D which exerted complete degradation of anthracene, while Lily et al. (2013) used *Brachybacterium paraconglomeratum* BMIT637C (MTCC 9445), isolated from automobile-contaminated soil and obtained 70.32% removal of 50 mg/100 mL anthracene in 10 days. Cui et al. (2014) used halophilic *Martelella* sp. Strain AD-3 for anthracene degradation. Swaathy et al. (2014) showed that *Bacillus licheniformis* (MTCC 5514), marine alkaliphile, was able to degrade more than 95% of 30 mg/100 mL anthracene. Goswami et al. (2017) experimented degradation of anthracene using oleaginous bacteria, *Rhodococcus opacus,* in the presence of heavy metals. They stated that the biodegradation efficiency of anthracene decreased during the biosorption of metals. The total time required for anthracene biodegradation increased from 144 to 216 h in the presence of iron, zinc, copper, and lead and increased even higher to 240 h in the presence of cadmium or nickel. Kurniati et al. (2019) isolated *Ochrobactrum intermedium* AMA9; which was able to degrade anthracene to 81.72 and 88.73% of 50 and 100 mg/L of anthracene. Moirangthem et al. (2020) proved that *Enterobacter ludwigii* strain GRAS-01 and *Pantoea agglomerans* strain GRAS-02 showed 72 and 69% removal of 50 mg/L anthracene, respectively as shown in Table 2.

5.3 *Biodegradation of Anthracene Using Mixed Culture*

The degradation efficiency is known to increase when more than one culture is used. Microorganisms tend to work synergistically in order to achieve better metabolism, as the intermediates produced by one, can serve as a substrate for another type of culture. *In-situ* biodegradation of any pollutant cannot be practically achieved using a single strain Janbandhu and Fulekar (2011), Moscoso et al. (2012). Arulazhagan et al. (2010) isolated three halotolerant strains: *Ochrobactrum* sp., *Enterobacter cloacae,* and *Stenotrophomonas maltophili,* and used them as a consortium. This consortium could degrade 49 and 95% of 3 mg/L anthracene in the presence of 30 mg/L sodium chloride in 2 and 4 days, respectively. Pugazhendi et al. (2017) utilized a halothermophilic consortium containing *Ochrobactrum halosaudia* strain AJH1, *Ochrobactrum halosaudia* strain AJH2 and *Pseudomonas aeruginosa* strain AJH3 to degrade 50 mg/L anthracene, in 4 days as shown in Table 3. Komal et al. (2017) isolated mixed bacterial culture from the soil contaminated with petroleum spills, Chennai. They reported that the consortium was able to degrade 50 mg/L anthracene in 24 h. Moirangthem et al. (2020) observed that the co-culture of *Enterobacter ludwigii* strain GRAS-01 and *Pantoea agglomerans* strain GRAS-02 was able to give 75% degradation of 50 mg/L anthracene.

6 Recent Advents in Tailoring Anthracene Biodegradation

The statistical optimization under the design of experiments (DoE) is the most favoured option for tailoring the process parameters Pandya and Kumar (2021), Bowden (2019). DoE being a multivariate technique facilitate the evaluation of effects/interactions among the process variables in minimum experiments Weissman and Anderson (2015), Vargason et al. (2017). Few reports are available on the anthracene degradation by DoE tools such as factorial designs; Plackett–Burman design (PBD), response surface methodology (RSM), and artificial neural network (ANN). Response surface estimation was used to improve the degradation of

Table 2 Anthracene degradation using different bacterial species and the probable products of biotransformation

Bacterial species	Removal (%)	Time (days)	Possible products reported	References
Bacillus badius D1	100	2.5	1,2-dihydroxyanthracene, 6,7 benzocoumarin, 1-methoxy-2-hydroxy-anthracene, phthalic acid, 9,10-dihydroxyanthracene, 9, 10-anthraquinone	Ahmed et al. (2012)
Brachybacterium paraconglomeratum BMIT637C (MTCC 9445)	70.32	10	–	Lily et al. (2013)
Martelella sp. strain AD-3	100	6	9,10-anthraquinone, 3-hydroxy-2-naphthoic acid, 6,7-benzocoumarin, naphthoic acid, salicylic acid, and gentisic acid	Cui et al. (2014)
Bacillus licheniformis MTCC 5514	> 95%	22	Naphthalene, naphthalene 2-methyl, phthalic acid, benzene acetic acid	Swaathy et al. (2014)
Ochrobactrum intermedium AMA9	88.73	5	–	Kurniati et al. (2019)
Enterobacter ludwigii strain GRAS-01	72	3	–	Moirangthem et al. (2020)
Pantoea agglomerans strain GRAS-02	69			

Table 3 Anthracene degradation using different bacterial co-culture/consortium

Bacterial co-culture/consortium	Removal (%)	Time (days)	References
Ochrobactrum sp., *Enterobacter cloacae* and *Stenotrophomonas maltophili*	95	4	Arulazhagan et al. (2010)
Ochrobactrum halosaudia, Ochrobactrum halosaudia and *Pseudomonas aeruginosa*	100	4	Pugazhendi et al. (2017)
Acclimatized bacterial mixed culture	100	1	Komal et al. (2017)
Enterobacter ludwigii strain GRAS-01 and *Pantoea agglomerans* strain GRAS-02	75	3	Moirangthem et al. (2020)

anthracene by the strain ANT3D of *Sphingobium yanoikuyae.* The parameters included were calcium chloride, potassium di-hydrogen phosphate, magnesium sulphate, di-potassium hydrogen phosphate, ammonium nitrate, and ferric chloride. Calcium chloride, potassium di-hydrogen phosphate, and di-potassium hydrogen phosphate were significant with the elimination of 47.18%, according to experimental investigations of PBD Rajpara (2016).

The efficiency of *Stenotrophomonas maltophila* to biodegrade phenanthrene, anthracene, and fluoranthene. The medium components were screened using PBD and the precise optimization of process variables was accomplished using RSM and ANN approaches Gosai et al. (2018). A bacterial mixed culture comprising *Lelliottia amnigena* strain BS8, *Bacillus oceanisediminis* strain BS1 and *B. circulans* strain BS7 was utilized for the degradation of phenanthrene, anthracene, and fluoranthene. The experiments were optimized using central composite design (CCD) under RSM. The factors optimized were sodium chloride, boric acid, ammonium chloride, and disodium hydrogen phosphate. The removal prediction by RSM and ANN were compared Sachaniya et al. (2020).

7 Conclusions

The chapter aims to explicitly discuss the problem of anthracene and its potential impacts on the ecology and environment. The problem of bioaccumulation and the health hazards of anthracene are critically reviewed. The advent of conventional treatment options such as physico-chemical treatment and the associated demerits of those techniques are highlighted. The biotechnological interventions such as enzymes, plant-based, and microbial-assisted degradation are elaborately discussed. The bacterial-aided degradation of anthracene either in pure strains as well as in consortia is discussed with the probable products of degradation. The tailoring of process parameters using the design of experiments is also discussed.

Acknowledgements The authors are thankful to Director, CSIR-CSMCRI for the support and the manuscript has been assigned CSIR-CSMCRI: 81/2021 registration.

References

Ahmed AT, Othman MA, Sarwade VD, Gawai KR (2012) Degradation of anthracene by alkaliphilic bacteria *Bacillus badius*. Environment and Pollution 1:97. https://doi.org/10.5539/ep.v1n2p97

Aranda E, Godoy P, Reina R, Badia-Fabregat M, Rosell M, Marco-Urrea E, García-Romera I (2017) Isolation of Ascomycota fungi with capability to transform PAHs: insights into the biodegradation mechanisms of Penicillium oxalicum. Int Biodeter Biodegr 122:141–150. https://doi.org/10.1016/j.ibiod.2017.05.015

Arulazhagan P, Vasudevan N, Yeom IT (2010) Biodegradation of polycyclic aromatic hydrocarbon by a halotolerant bacterial consortium isolated from marine environment. Int J Environ Sci Technol 7:639–652. https://doi.org/10.1007/BF03326174

Azubuike CC, Chikere CB, Okpokwasili GC (2016) Bioremediation techniques–classification based on site of application: principles, advantages, limitations and prospects. World J Microbiol Biotechnol 32:1–18. https://doi.org/10.1007/s11274-016-2137-x

Barman SR, Banerjee P, Mukhopadhayay A, Das P (2017) Biodegradation of acenapthene and naphthalene by *Pseudomonas mendocina*: process optimization, and toxicity evaluation. J Environ Chem Eng 5:4803–4812. https://doi.org/10.1016/j.jece.2017.09.012

Birolli WG, Santos DDA, Alvarenga N, Garcia AC, Romão LP, Porto AL (2018) Biodegradation of anthracene and several PAHs by the marine-derived fungus Cladosporium sp. CBMAI 1237. Mar Pollut Bull 129:525–533. https://doi.org/10.1016/j.marpolbul.2017.10.023

Bowden GD, Pichler BJ, Maurer A (2019) A design of experiments (DoE) approach accelerates the optimization of copper-mediated 18F-fluorination reactions of arylstannanes. Sci Rep 9:1–10. https://doi.org/10.1038/s41598-019-47846-6

Chen Y, Bi X, Mai B, Sheng G, Fu J (2004) Emission characterization of particulate/gaseous phases and size association for polycyclic aromatic hydrocarbons from residential coal combustion. Fuel 83:781–790. https://doi.org/10.1016/j.fuel.2003.11.003

Cui C, Ma L, Shi J, Lin K, Luo Q, Liu Y (2014) Metabolic pathway for degradation of anthracene by halophilic Martelella sp. AD-3. Int Biodeter Biodegr 89:67–73. https://doi.org/10.1016/j.ibiod.2014.01.012

Ferronato N, Torretta V (2019) Waste mismanagement in developing countries: A review of global issues. Int J Environ Res Public Health 16:1060. https://doi.org/10.3390/ijerph16061060

Flotron V, Delteil C, Padellec Y, Camel V (2005) Removal of sorbed polycyclic aromatic hydrocarbons from soil, sludge and sediment samples using the Fenton's reagent process. Chemosphere 59:1427–1437. https://doi.org/10.1016/j.chemosphere.2004.12.065

Gao Y, Ling W, Zhu L, Zhao B, Zheng QS (2007) Surfactant-enhanced phytoremediation of soils contaminated with hydrophobic organic contaminants: potential and assessment. Pedosphere 17:409–418. https://doi.org/10.1016/S1002-0160(07)60050-2

Gong Z, Alef K, Wilke BM, Li P (2005) Dissolution and removal of PAHs from a contaminated soil using sunflower oil. Chemosphere 58:291–298. https://doi.org/10.1016/j.chemosphere.2004.07.035

Gosai HB, Sachaniya BK, Dudhagara DR, Panseriya HZ, Dave B (2018) Bioengineering for multiple PAHs degradation using process centric and data centric approaches. Chemom Intell Lab Syst 179:99–108. https://doi.org/10.1016/j.chemolab.2018.04.019

Goswami L, Manikandan AN, Pakshirajan K, Pugazhenthi G (2017) Simultaneous heavy metal removal and anthracene biodegradation by the oleaginous bacteria Rhodococcus opacus 3 Biotech 7:1–9. https://doi.org/10.1007/s13205-016-0597-1

Gupta H and Kumar R (2016). Removal of PAH anthracene from aqueous media using banana peel activated carbon. Int. J. Sci. Res. Environ. Sci. 4, 0109–0114. https://doi.org/10.12983/ijsres-2016-p0109-0114

Gupta H (2018) Anthracene removal from water onto activated carbon derived from vehicular tyre. Sep Sci Technol 53:613–625. https://doi.org/10.1080/01496395.2017.1405038

Hassan SS, El Azab WI, Ali HR, Mansour MS (2015). Green synthesis and characterization of ZnO nanoparticles for photocatalytic degradation of anthracene. Adv Nat Sci: Nanosci Nanotechnol 6, 045012. https://doi.org/10.1088/2043-6262/6/4/045012

Hussar E, Richards S, Lin ZQ, Dixon RP, Johnson KA (2012) Human health risk assessment of 16 priority polycyclic aromatic hydrocarbons in soils of Chattanooga, Tennessee, USA. Water, Air, & Soil Pollut 223:5535–5548. https://doi.org/10.1007/s11270-012-1265-7

Janbandhu A, Fulekar MH (2011) Biodegradation of phenanthrene using adapted microbial consortium isolated from petrochemical contaminated environment. J Hazard Mater 187:333–340. https://doi.org/10.1016/j.jhazmat.2011.01.034

Jové P, Olivella MÀ, Camarero S, Caixach J, Planas C, Cano L, De Las Heras FX (2016) Fungal biodegradation of anthracene-polluted cork: a comparative study. J. Environ Sci. Health - Toxic/hazard. Subst. Environ. Eng. 51:70–77. https://doi.org/10.1080/10934529.2015.1079114

Juwarkar AA, Singh SK, Mudhoo A (2010) A comprehensive overview of elements in bioremediation. Rev Environ Sci Biotechnol 9:215–288. https://doi.org/10.1007/s11157-010-9215-6

Kikani M, Bhojani G, Amit C, Madhava AK (2021) Chemo-metrically formulated consortium with selectively screened bacterial strains for ameliorated biotransformation and detoxification of 1,4-dioxane. J Hazard Mater 413:125456. https://doi.org/10.1016/j.jhazmat.2021.125456

Komal J, Kumar MA, Thiruvengadaravi KV, Nilavunesan D, Premkumar MP, Kumar VV, Sivanesan S (2017) Indigenously acclimatized bacterial consortium for anthracene biotransformation. Energy Sources A: recovery Util. Environ. Eff. 39:528–537. https://doi.org/10.1080/15567036.2016.1235060

Koshlaf E, Ball AS (2017) Soil bioremediation approaches for petroleum hydrocarbon polluted environments. AIMS Microbiol. 3:25. https://doi.org/10.3934/microbiol.2017.1.25

Kurniati TH, Rusmana I, Suryani A, Mubarik NR (2019) Biosurfactant-producing and anthracene-degrading bacteria from oil contaminated soil. J Phys Conf Ser 1402:033049. https://doi.org/10.1088/1742-6596/1402/3/033049

Kulik N, Goi A, Trapido M, Tuhkanen T (2006) Degradation of polycyclic aromatic hydrocarbons by combined chemical pre-oxidation and bioremediation in creosote contaminated soil. J Environ Manage 78:382–391. https://doi.org/10.1016/j.jenvman.2005.05.005

Lily MK, Bahuguna A, Bhatt KK, Dangwal K (2013) Degradation of anthracene by a novel strain *Brachybacterium paraconglomeratum* BMIT637C (MTCC 9445). Int J Environ Sci 3:1242

Liu WX, Dou H, Wei ZC, Chang B, Qiu WX, Liu Y, Tao S (2009) Emission characteristics of polycyclic aromatic hydrocarbons from combustion of different residential coals in North China. Sci Total Environ 407:1436–1446. https://doi.org/10.1016/j.scitotenv.2008.10.055

Mata-Sandoval JC, Karns J, Torrents A (2002) Influence of rhamnolipids and Triton X-100 on the desorption of pesticides from soils. Environ Sci Technol 36:4669–4675. https://doi.org/10.1021/es011260z

Martínez-Vargas BL, Cruz-Ramírez M, Díaz-Real JA, Rodríguez-López JL, Bacame-Valenzuela FJ, Ortega-Borges R, Reyes-Vidal Y, Ortiz-Frade L (2019) Synthesis and characterization of n-ZnO/p-MnO nanocomposites for the photocatalytic degradation of anthracene. J Photochem Photobiol a: Chem 369:85–96. https://doi.org/10.1016/j.jphotochem.2018.10.010

Mulligan CN, Yong RN, Gibbs BF (2001) Surfactant-enhanced remediation of contaminated soil: A review. Eng Geol 60:371–380. https://doi.org/10.1016/S0013-7952(00)00117-4

Moirangthem GS, Dholakiya RN, Madhava AK (2020) Anthracene biotransformation using bacterial concoction of *Enterobacter ludwigii* strain GRAS-01 and *Pantoea agglomerans* strain GRAS-02. Int J Environ Anal Chem:1–13. https://doi.org/10.1080/03067319.2020.1803849

Moradeeya PG, Kumar MA, Thorat RB, Rathod M, Khambhaty Y, Basha S (2017) Nanocellulose for biosorption of chlorpyrifos from water: chemometric optimization, kinetics and equilibrium. Cellulose 24:1319–1332. https://doi.org/10.1007/s10570-017-1197-x

Moscoso F, Teijiz I, Deive FJ, Sanromán MA (2012) Efficient PAHs biodegradation by a bacterial consortium at flask and bioreactor scale. Bioresour Technol 119:270–276. https://doi.org/10.1016/j.biortech.2012.05.095

Myers RH, Montgomery DC, Anderson-Cook CM (2016) Wiley Series in Probability and Statistics, 4th edn. John Wiley & Sons, New York

Owabor CN, Saniyo E (2006) Degradation of anthracene: Influence of adsorbents from inorganic activation. Afr J Biotechnol 5:165–169. https://doi.org/10.5897/AJB2006.000-5026

Padros J, Pelletier E (2000) In vivo formation of (+)-anti-benzo [a] pyrene diol-epoxide–plasma albumin adducts in fish. Mar Environ Res 50:347–351. https://doi.org/10.1016/S0141-1136(00)00064-7

Pandya DK, Kumar MA (2021) Chemo-metric engineering designs for deciphering the biodegradation of polycyclic aromatic hydrocarbons. J Hazard Mater 411:125154. https://doi.org/10.1016/j.jhazmat.2021.125154

Pugazhendi A, Qari H, Basahi JMAB, Godon JJ, Dhavamani J (2017) Role of a halothermophilic bacterial consortium for the biodegradation of PAHs and the treatment of petroleum wastewater at extreme conditions. Int Biodeter Biodegr 121:44–54. https://doi.org/10.1016/j.ibiod.2017.03.015

Rajpara RK, Dudhagara DR, Bhatt JK, Gosai HB, Dave BP (2016) Plackett-Burman design for the screening of media component for anthracene degradation by *Sphingobium yanoikuyae* strain ANT3D. Int J Marine Biol Res 1:1–4. http://dx.doi.org/10.15226/24754706/1/2/00108

Ravindra K, Bencs L, Wauters E, De Hoog J, Deutsch F, Roekens E, Bleux N, Bergmans P, Van Grieken R (2006) Seasonal and site-specific variation in vapour and aerosol phase PAHs over Flanders (Belgium) and their relation with anthropogenic activities. Atmos Environ 40:771–785. https://doi.org/10.1016/j.atmosenv.2005.10.011

Saad MEK, Khiari R, Elaloui E, Moussaoui Y (2014) Adsorption of anthracene using activated carbon and *Posidonia oceanica*. Arab J Chem 7:109–113. https://doi.org/10.1016/j.arabjc.2013.11.002

Sachaniya BK, Gosai HB, Panseriya HZ, Dave BP (2020) Bioengineering for multiple PAHs degradation for contaminated sediments: Response surface methodology (RSM) and artificial neural network (ANN). Chemom Intell Lab Syst 202:104033. https://doi.org/10.1016/j.chemolab.2020.104033

Stehr J, Müller T, Svensson K, Kamnerdpetch C, Scheper T (2001) Basic examinations on chemical pre-oxidation by ozone for enhancing bioremediation of phenanthrene contaminated soils. Appl Microbiol Biotechnol 57:803–809. https://doi.org/10.1007/s13205-015-0293-6

Stokes JD, Paton GI, Semple KT (2005) Behaviour and assessment of bioavailability of organic contaminants in soil: relevance for risk assessment and remediation. Soil Use Manag 21:475–486. https://doi.org/10.1079/SUM2005347

Somtrakoon K, Chouychai W, Lee H (2014) Comparing anthracene and fluorene degradation in anthracene and fluorene-contaminated soil by single and mixed plant cultivation. Int J Phytoremediation 16:415–428. https://doi.org/10.1080/15226514.2013.803024

Swaathy S, Kavitha V, Pravin AS, Mandal AB, Gnanamani A (2014) Microbial surfactant mediated degradation of anthracene in aqueous phase by marine Bacillus licheniformis MTCC 5514. Biotechnol Rep 4:161–170. https://doi.org/10.1016/j.btre.2014.10.004

Uno S, Dragin N, Miller ML, Dalton TP, Gonzalez FJ, Nebert DW (2008) Basal and inducible CYP1 mRNA quantitation and protein localization throughout the mouse gastrointestinal tract. Free Radic Biol Med 44:570–583. https://doi.org/10.1016/j.freeradbiomed.2007.10.044

Van Herwijnen R, Springael D, Slot P, Govers HA, Parsons JR (2003) Degradation of anthracene by Mycobacterium sp. strain LB501T proceeds via a novel pathway, through o-phthalic acid. Appl Environ Microbiol 69:186–190. https://doi.org/10.1128/AEM.69.1.186-190.2003

Vargason T, Howsmon DP, McGuinness DL, Hahn J (2017) On the use of multivariate methods for analysis of data from biological networks. Processes 5:36. https://doi.org/10.3390/pr5030036
Vasconcelos U, França FPD, Oliveira FJ (2011) Removal of high-molecular weight polycyclic aromatic hydrocarbons. Quim Nova 34:218–221. https://doi.org/10.1590/S0100-40422011000200009
Watts RJ, Stanton PC, Howsawkeng J, Teel AL (2002) Mineralization of a sorbed polycyclic aromatic hydrocarbon in two soils using catalyzed hydrogen peroxide. Water Res 36:4283–4292. https://doi.org/10.1016/S0043-1354(02)00142-2
Weissman SA, Anderson NG (2015) Design of experiments (DoE) and process optimization: a review of recent publications. Org Process Res Dev 19:1605–1633. https://doi.org/10.1021/op500169m
Ye JS, Yin H, Qiang J, Peng H, Qin HM, Zhang N, He BY (2011) Biodegradation of anthracene by Aspergillus fumigatus. J Hazard Mater 185:174–181. https://doi.org/10.1016/j.jhazmat.2010.09.015
Zhang S, Ning Y, Zhang X, Zhao Y, Yang X, Wu K, Yang S, La G, Sun X, Li X (2015) Contrasting characteristics of anthracene and pyrene degradation by wood rot fungus Pycnoporus sanguineus H1. Int Biodeter Biodegr 105:228–232. https://doi.org/10.1016/j.ibiod.2015.09.012
Zebulun HO, Isikhuemhen OS, Inyang H (2011) Decontamination of anthracene-polluted soil through white rot fungus-induced biodegradation. Environmentalist 31:11–19. https://doi.org/10.1007/s10669-010-9284-7
Zhao B, Zhu L, Gao Y (2005) A novel solubilization of phenanthrene using Winsor I microemulsion-based sodium castor oil sulfate. J Hazard Mater 119:205–211. https://doi.org/10.1016/j.jhazmat.2004.12.009
Zhu L, Lu H, Chen S, Amagai T (2009) Pollution level, phase distribution and source analysis of polycyclic aromatic hydrocarbons in residential air in Hangzhou. China. J. Hazard. Mater. 162:1165–1170. https://doi.org/10.1016/j.jhazmat.2008.05.150

Correction to: Advanced and Innovative Approaches of Environmental Biotechnology in Industrial Wastewater Treatment

Maulin P. Shah

Correction to:
M. P. Shah (ed.), *Advanced and Innovative Approaches of Environmental Biotechnology in Industrial Wastewater Treatment*, https://doi.org/10.1007/978-981-99-2598-8

The original version of the book was inadvertently published without incorporating the author's proof corrections in Chapters 10, 11, and 15 which have now been corrected. The book has been updated with the changes.

The updated original versions of these chapters can be found at
https://doi.org/10.1007/978-981-99-2598-8_10
https://doi.org/10.1007/978-981-99-2598-8_11
https://doi.org/10.1007/978-981-99-2598-8_15

M. P. Shah (ed.), *Advanced and Innovative Approaches of Environmental Biotechnology in Industrial Wastewater Treatment*,
https://doi.org/10.1007/978-981-99-2598-8_21

MIX
Papier aus verantwortungsvollen Quellen
Paper from responsible sources
FSC® C105338

If you have any concerns about our products,
you can contact us on
ProductSafety@springernature.com

In case Publisher is established outside the EU,
the EU authorized representative is:
Springer Nature Customer Service Center GmbH
Europaplatz 3, 69115 Heidelberg, Germany

Printed by Libri Plureos GmbH
in Hamburg, Germany